Allerton   May '92

*Tides, Surges
and Mean Sea-Level*

Tides, Surges
and Mean Sea-Level

David T. Pugh

Natural Environment Research Council
Swindon, UK

JOHN WILEY & SONS

Chichester · New York · Brisbane · Toronto · Singapore

Library of Congress Cataloging-in-Publication Data:

Pugh, D. T.
  Tides, surges, and mean sea level.

  Bibliography: p.
  Includes index.
  1. Tides. 2. Storm surges. 3. Sea level.
  I. Title.
  GC301.2.P84 1987   551.47'08   87–2111
ISBN 0 471 91505 X

British Library Cataloguing in Publication Data:

Pugh, David T.
  Tides, surges and mean sea level.
  1. Tides
  I. Title
  525'.6   GC301.2

  ISBN 0 471 91505 X

Typeset by Bath Typesetting Limited,
Printed and bound in Great Britain by Bath Press, Avon

Contents

* Indicates that section may be omitted by non-mathematical readers.

v

Preface

Moving water has a special fascination, and the regular tidal movements of coastal seas must have challenged human imagination from earliest times. Indeed, the ancients who were able to link the regular movements of the sea to the movements of the sun and moon regarded tides as a tangible terrestrial manifestation of the powers of the celestial gods. For them the tides had religious significance; for us there are obviously many practical and scientific reasons for needing to know about and understand the dynamics of the oceans and coastal seas.

Modern practical studies are concerned with problems of marine transport, coastal erosion and the design of coastal defences against flooding. Interest in mean sea-level changes has recently focused attention on the possibility of significant increases over the coming century as a result of global warming. Scientifically, in addition to their position as a branch, perhaps the oldest branch, of physical oceanography, tides have a controlling influence on many marine biological and geological processes. One of the fascinations of studying tides is the diversity of the applications.

During the past two decades, mainly as a result of developments in instrument design and in the computer sciences, our understanding of tides and their related phenomena has made considerable advances. The aim of this book is to present modern tidal ideas to those who are not tidal specialists, but for whom some tidal knowledge is involved in their own professional or scientific field. These include hydrographers, marine and coastal engineers, geologists who specialize in beach or marine sedimentation processes, and biologists concerned with the ways in which living organisms adapt to the rhythms of the sea.

Some of the material presented here was originally prepared for a course of lectures in Liverpool University to third-year Combined Honours students, whose academic backgrounds varied from pure mathematics to field geology. This range of previous scientific experience inevitably causes difficulties in determining the level of mathematical treatment which should be adopted. In this book I have tried to keep the mathematics as simple as is consistent with a proper physical explanation, while developing the non-mathematical discussions in an essentially independent yet parallel way. Sections which may be omitted by the non-mathematical reader are marked with an asterisk.

Discussions of the engineering applications of tides are based on several years of experience giving advice to consultants and to Government Departments on behalf of the Proudman Oceanographic Laboratory at Bidston Observatory. Many of my colleagues on the staff of the Laboratory have given me advice and constructive criticism during the preparation. To them and to many other friends and associates I extend my grateful thanks. In particular, it is a pleasure to thank Graham Alcock, M. Amin, Trevor Baker, David Blackman, Peter Claridge, Hilary Faull, Roger Flather, Tony Heathershaw, John Howarth, Ian James, Kathy Jones, Trevor Norton, Lesley Rickards, Ralph Rayner, Alun Thomas, Phillip Williamson, and Philip Woodworth for helpful comments on earlier versions of the text. It is also a pleasure to acknowledge specific or general guidance during the development of my ideas from Duncan Carr Agnew, David Cartwright, Mike Collins, Keith Dyer, Norman Heaps, Chester Jelesnianski, Geoff Lennon, Nick McCave, Robin Pingree, John Simpson, Ian Vassie and Klaus Wyrtki. Finally, I am grateful to my family for tolerating and even encouraging this personal indulgence through the evenings and weekends of four long winters.

Cirencester, June, 1987   DAVID PUGH

List of Symbols

a	earth radius
A_1, A_s	right ascensions of the moon and sun
c	wave speed ... $(gD)^{\frac{1}{2}}$ in shallow water
C_1, C_s	hour angles of the moon and sun
d_1, d_s	declinations of the moon and sun
D	water depth
e_1, e_s	eccentricity of lunar and solar orbits
f	Coriolis parameter ... $2\,\omega_s \cos \varphi$
f_x	nodal amplitude factor for constituent x
F_s, F_b	surface and bottom stresses in the X direction
g	gravitational acceleration
g_x	phase lag of harmonic constituent x on Equilibrium Tide at Greenwich (usually expressed in degrees)
g_C, g_{AC}	phases of clockwise and anticlockwise components of current
g_{ux}, g_{vx}	phases of Cartesian current components
G_s, G_b	surface and bottom stresses in the Y direction
h	geocentric mean ecliptic longitude of the sun
H_x	amplitude of harmonic constituent x of tidal levels
i, j	integers
l	length variable
L	length of ocean basin
m_e, m_1, m_s	mass of earth, moon, sun
m, n	particular values in a series
M, N	total numbers in series
N	ascending nodal lunar longitude
p	longitude of lunar perigee
p'	longitude of solar perigee
P_A	atmospheric pressure at the sea surface
P_z	pressure at depth z
q	current speed
Q_C, Q_{AC}	amplitudes of clockwise and anticlockwise components of currents
r	distance, variously defined
R_1, R_s	lunar and solar distances from the earth
s	geocentric mean ecliptic longitude of the moon
s	standard deviation of a variable
$\mathbf{S}(t)$	meteorological surge component of sea-level or current variations
t	time

$\mathbf{T}(t)$ tidal component of sea-level or current variations

u_x nodal phase factor for harmonic constituent x

u, v current components in the X and Y directions

V_x astronomical phase angle of harmonic constituent x in the Equilibrium Tide, relative to the Greenwich meridian

W wind speed

x, y, z coordinates of a point

X, Y, Z Cartesian coordinate system . . . Z is positive vertically upwards

$Z(t)$ mean sea-level

α dimensionless ratio variously defined

β a general angular measure

$\varepsilon_1, \varepsilon_s$ ecliptic latitudes of the moon and sun

ζ displacement of water level from the mean

θ direction to which current and wind flows (either clockwise from north or anticlockwise from east)

λ_1, λ_s true ecliptic longitudes of the moon and sun

ρ_A, ρ air and sea water densities

σ_x angular speed of constituent x (usually in degrees per mean solar hour)

τ_s, τ_b unresolved surface and bottom stresses

Υ First Point of Aries

φ latitude of a point on the earth

ω_x angular speed of constituent x (usually in radians per mean solar hour)

$\omega_1, \omega_2, \omega_3$ } angular speeds of astronomical variables . . . see Table 3 : 2
$\omega_4, \omega_5, \omega_6$ } usually in radians

ω_s angular speed of the earth's rotation on its axis . . .

$$\omega_1 + \omega_3 = \omega_0 + \omega_2 = \omega_s.$$

Ω Equilibrium Tidal Potential

Harmonic constituents are shown in heavy type thus $\mathbf{X}_2$ to denote their vector property (H_x, g_x).

Overbars denote time-averaged values.

CHAPTER 1

Introduction

I must go down to the seas again, for the call of the running tide
Is a wild call and a clear call that may not be denied.

John Masefield, 'Sea Fever'*

The old proverb 'Time and tide wait for no man' emphasizes the importance of tides from early times. The word *tide* derives from the Saxon *tid* or time, as used in seasonal titles such as Eastertide or Yuletide; the modern German, *Gezeiten* for tides and *Zeiten* for times, maintains the connection even more strongly. Where necessary, people adapted their lives to cope with the tidal cycles instinctively and without any basic understanding. Gradually, however, they must have become aware of certain regular patterns in the motions of the sea which could be related to the movements of the moon and sun. The power which this knowledge gave them to make crude predictions, and so to plan their journeys several days ahead, would have been a spur to further analysis and understanding. However, without the discipline of hypotheses which could be proved or disproved by critical observations, that is, without the scientific method, such a deeper understanding was not possible.

Books dealing with the science of tidal phenomena are comparatively rare. However, unified treatments of general interest are found in Darwin (1911), Marmer (1926), Doodson and Warburg (1941), Dronkers (1964), Lisitzin (1974) and Forrester (1983). Accounts are also found in several books on physical oceanography, including those by Defant (1961), Von Arx (1962), and Pond and Pickard (1978). More popular accounts are given by Defant (1958), Macmillan (1966) and Redfield (1980).

In this introductory chapter we consider the development of some early tidal theories, showing that different places can have very different tidal patterns. We relate the patterns at five selected sites to the motions of the moon and sun, and discuss their modification by the weather. In addition we develop some basic definitions, and finally we discuss some simple statistical ways of summarizing tidal behaviour at any particular site.

*Reprinted by permission of the Society of Authors as the literary representative of the Estate of John Masefield, and with permission of Macmillan Publishing Company from POEMS by John Masefield (New York: Macmillan, 1953).

1:1   Early ideas and observations

Recent excavations in the Indian district of Ahmedabad have revealed a tidal dockyard which dates back to 2450 BC. There is no evidence that the Harappans, who had enough tidal experience to design and build the dock, related the local tidal movements to the moon or sun. The earliest reference which makes the connection occurs in the Samaveda of the Indian Vedic period, which extended from 2000 to 1400 BC (Panikkar and Srinivasan, 1971; Rizvi, personal communication). The large twice-daily tides of this region of the Indian Ocean amazed the army of Alexander the Great as it travelled in 325 BC southward along the river Indus towards the sea, because they were only familiar with the small tides of the Mediterranean Sea.

The recorded histories of those civilizations which bordered the Mediterranean understandably make few references to tides. No doubt the absence of large changes in coastal sea-levels would have made navigation easier, and helped the development of communication and trade between states in the very small ships of the times. Even in the Mediterranean, however, strong tidal currents run through certain narrow straits and passages, including the Strait of Messina where current speeds in excess of $2.00 \, \mathrm{ms^{-1}}$ are found. Between the mainland of Greece and the island of Euboea, the periodic reversals of flow through the Euripus are said to have greatly perplexed Aristotle during the years before his death in 322 BC. Earlier, in a summary of ancient lore, he had written that the 'ebbings and risings of the sea always come around with the moon and upon certain fixed times'. About this time Pytheas travelled through the Strait of Gibraltar to the British Isles, where he observed large tides at twice-daily periods. He is said to have been the first to report the half-monthly variations in the range of the Atlantic Ocean tides, and to note that the greatest ranges, which we call spring tides, occurred near to new and full moon. He also recorded the strong tidal streams of the Pentland Firth between Scotland and Orkney.

When the Roman writer Pliny the Elder (AD 23–79) compiled his *Natural History*, many other aspects of the relationships between tides and the moon had been noted. He described how, for twice-daily tides, the maximum tidal ranges occur a few days after the new or full moon, how the tides of the equinoxes in March and September have a larger range than those at the summer solstice in June and the winter solstice in December, and how there was a fixed interval between lunar transit and the next high tide at a particular location. Earlier, in his *Geography*, Strabo (*c*54 BC to *c* AD 24) tells of tides in the Persian Gulf which have their maximum range when the moon is furthest from the plane of the equator, and are small and irregular when the moon passes through the equatorial plane. We now know (see Chapter 3) that this behaviour is characteristic of tides which are dominated by once-daily cycles.

Not all of the facts reported in these early writings are confirmed by modern measurements. Pliny the Elder includes an account of how the moon's influence is most strongly felt by those animals which are without blood, and how the blood of man is increased or diminished in proportion to the 'quantity of her light'; leaves and vegetables were also believed to feel the moon's influence. All seas were said to be purified at the full moon, and Aristotle is credited with the law that no animal dies except when the tide is ebbing. This particular legend persisted in popular culture, and even as recently as 1595, Parish Registers in the Hartlepool area of the North of England recorded the phase of the tide along with the date and time of each death.

Even 2000 years ago the historical records show an impressive collection of observed tidal patterns (Harris, 1897–1907). However, the ideas advanced by the philosophers of that time, and for the following 1600 years, to explain the connection between the moon and the tides were less valid. Chinese ideas supposed water to be the blood of the earth, with tides as the beating of the earth's pulse; alternatively tides were caused by the earth breathing. Arabic explanations supposed the moon's rays to be reflected off rocks at the bottom of sea, thus heating and expanding the water, which then rolled in waves towards the shore. One poetic explanation invoked an angel who was set over the seas: when he placed his foot in the sea the flow of the tide began, but when he raised it, the tidal ebb followed. During this long period there was a decline in critical thought, so that the clear statements by the classical writers were gradually replaced by a confusion of supposed facts and ideas. One notable exception was the Venerable Bede, a Northumbrian monk, who described around 730 how the rise of the water along one coast of the British Isles coincided with a fall elsewhere. Bede also knew of the progression in the time of high tide from north to south along the Northumbrian coast.

By the mid-seventeenth century three different theories were being seriously considered. Galileo (1564–1642) proposed that the rotations of the earth, annually around the sun and daily about its own axis, induced motions of the sea which were modified by the shape of the sea-bed to give the tides. The French philosopher Descartes (1596–1650) thought that space was full of invisible matter or ether. As the moon travelled round the earth it compressed this ether in a way which transmitted pressure to the sea, hence forming the tides. Kepler (1571–1630) was one of the originators of the idea that the moon exerted a gravitational attraction on the water of the ocean, drawing it towards the place where it was overhead. This attraction was balanced by the earth's attraction on the waters for 'If the earth should cease to attract its waters, all marine waters would be elevated and would flow into the body of the moon'. Arguments over the relative merits of these three theories continued for several years, subject to the important scientific criterion that a valid theory must account for the observed tidal phenomena. Gradually as the ideas of a heliocentric system of planets, each rotating on its own axis, became established, and as the laws of the planetary motion and gravitational attraction were developed, Kepler's original ideas of the moon's gravity causing tides became the most plausible. Simple ideas of gravitational attraction were, however, unable to explain why the main oceans of the world experienced not one, but two tides for each transit of the moon.

A major advance in the scientific understanding of the generation of tides was made by Isaac Newton (1642–1727). He was able to apply his formulation of the law of gravitational attraction: that two bodies attract each other with a force which is proportional to the product of their masses and inversely proportional to the square of the distance between them, to show why there were two tides for each lunar transit. He also showed why the half-monthly spring to neap cycle occurred, why once-daily tides were a maximum when the moon was furthest from the plane of the equator, and why equinoctial tides were usually larger than those at the solstices. This impressive catalogue of observed tidal features which could be accounted for by a single physical law firmly established gravitational theory as the basis for all future tidal science. We will consider its detailed applications in subsequent chapters, but it should be noted that despite these considerable achievements, the theory accounted for only the broad features of observed tides. The details of the tides at any particular place are governed by the complicated responses of the ocean to these gravitational forces.

Further historical developments are summarized in subsequent chapters, but the interested reader is referred to Deacon (1971) for a coherent historical account.

1:2   Tidal patterns

Before the development of appropriate instrumentation, sea-level observations were confined to the coast and were not very accurate. Modern instruments, many of which will be described in the next chapter, have enabled a systematic collection of tidal data which shows that regular water movements are a feature on all the shores of the oceans and their adjacent seas. These regular water movements are seen as both the vertical rise and fall of sea-level, and the to and fro movements of the water currents. Levels at more than 3000 sites have been analysed and their tidal characteristics are published by the International Hydrographic Organization in Monaco. Less elaborate analyses for around 1000 further sites are available in published Tide Tables.

The two main tidal features of any sea-level record are the range, measured as the height between successive high and low levels, and the period, the time between one high (or low) level and the next high (or low) level. The tidal responses of the ocean to the forcing of the moon and sun are very complicated and both of these tidal features vary greatly from one site to another. Figure 1:1(a), which shows the tides for March 1981 at five sites, clearly illustrates this variability. The details of the relationships between the tides and the movements of the moon and sun are developed in Chapter 3, but Figure 1:1(b) shows the lunar variables for the same month. In this section we describe the observed sea-level variations at these five sites and relate them to the astronomy in a more general way.

In most of the world's oceans the dominant tidal pattern is similar to that shown for Bermuda in the North Atlantic, and for Mombasa on the African shore of the Indian Ocean. Each tidal cycle takes an average of 12 hours 25 minutes, so that two tidal cycles occur for each transit of the moon (every 24 hours 50 minutes). Because each tidal cycle occupies roughly half of a day, this type of tide is called semidiurnal. Semidiurnal tides have a range which typically increases and decreases cyclically over a fourteen-day period. The maximum ranges, called spring tides, occur a few days after both new and full moons (syzygy, when the moon, earth and sun are in line), whereas the minimum ranges, called neap tides, occur shortly after the times of the first and last quarters (lunar quadrature). The relationship between tidal ranges and the phase of the moon is due to the additional tide-raising attraction of the sun, which reinforces the moon's tides at syzygy, but reduces them at quadrature. The astronomical cycles are discussed in detail in Chapter 3, but Figure 1:1(b) shows that when the moon is at its maximum distance from the earth, known as lunar apogee, semidiurnal tidal ranges are less than when the moon is at its nearest approach, known as lunar perigee. This cycle in the moon's motion is repeated every 27.55 solar days. Maximum semidiurnal ranges occur when spring tides (syzygy) coincide with lunar perigee (Wood, 1986), whereas minimum semidiurnal ranges occur when neap tides (quadrature) coincide with lunar apogee. Semidiurnal tidal ranges increase and decrease at roughly the same time everywhere, but there are significant local differences. The maximum semidiurnal tidal ranges are in semi-enclosed seas. In the Minas Basin in the Bay of Fundy (Canada), the semidiurnal North Atlantic tides at Burncoat Head have a mean spring range of 12.9 m. The mean spring ranges at Avonmouth in the Bristol Channel (United Kingdom) and at Granville in the Gulf of St Malo (France) are 12.3 m and 11.4 m, respectively. Elsewhere, in Argentina the Puerto Gallegos mean spring tidal range is 10.4 m; at the Indian port of Bhavnagar in the Gulf of Cambay it is 8.8 m and the Korean port of Inchon has a mean spring range of 8.4 m. More generally, however, in the main oceans the semidiurnal mean spring tidal range is usually less than 2 m.

Close examination of the tidal patterns at Bermuda and Mombasa in Figure 1:1(a) shows that at certain times in the lunar month the high-water levels are alternately higher and lower than the average. This behaviour is also observed for the low-water levels, the differences being most pronounced when the moon's declination north and south of the equator is greatest. The differences can be accounted for by a small additional tide with a period close to one day, which adds to one high water level but subtracts from the next one. In Chapters 3 and 4 we will develop the idea of a superposition of several partial tides to produce the observed sea-level variations at any particular location.

In the case of the tide at Musay'id in the Persian Gulf, the tides with a one-day period, which are called diurnal tides, are similar in magnitude to the semidiurnal tides. This composite type of tidal regime is called a mixed tide, the relative importance of the semidiurnal and the diurnal components changing throughout the month as plotted in Figure 1:1(a). The diurnal tides are most important when the moon's declination is greatest but reduce to zero when the

6

moon is passing through the equatorial plane, where it has zero declination. The semidiurnal tides are most important after new and full moon; but unlike the diurnal tides, they do not reduce to zero range, being only partly reduced during the period of neap tides.

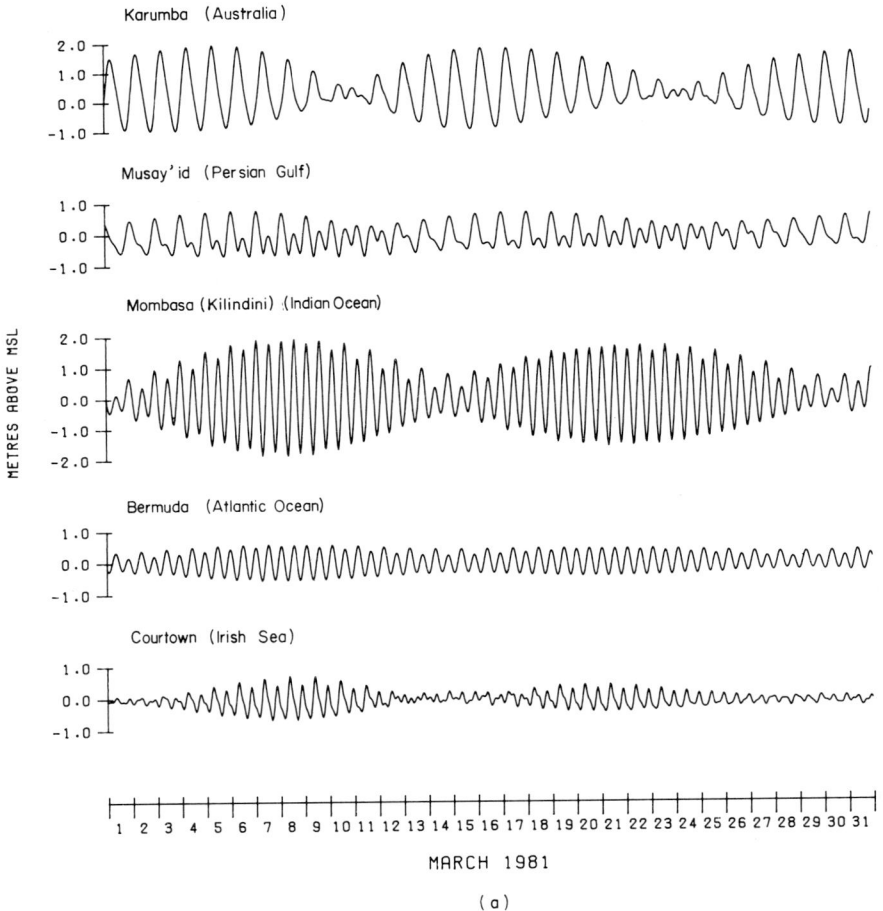

Figure 1:1 (a) Tidal predictions for March 1981 at five sites which have very different tidal regimes. At Karumba the tides are diurnal, at Musay'id they are mixed, whereas at both Kilindini and Bermuda semidiurnal tides are dominant. The tides at Courtown are strongly distorted by the influence of the shallow waters of the Irish Sea. (b) The lunar characteristics responsible for these tidal patterns. Solar and lunar tide producing forces combine at new and full moon to give large spring tidal ranges. Lunar distance varies through perigee and apogee over a 27.55-day period. Lunar declination north and south of the equator varies over a 27.21-day period. Solar declination is zero on 21 March.

LUNAR CHANGES

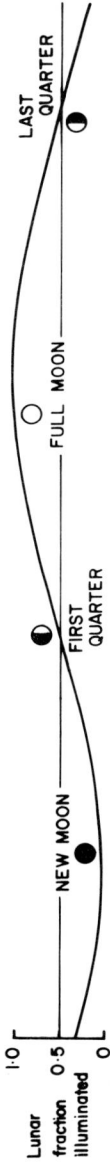

Lunar fraction illuminated

NEW MOON — FULL MOON — FIRST QUARTER — LAST QUARTER

Synodic or Lunar month = 29.5306 days

Lunar distance x10⁵ km

PERIGEE — APOGEE — Mean

Anomalistic month = 27.5546 days

Lunar declination (°)

Nodical month = 27.2122 days

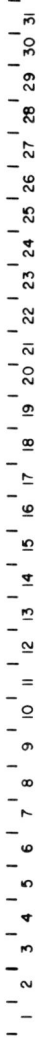

MARCH 1981
(b)

In a few places the diurnal tides are much larger than the semidiurnal tides. Karumba in the Australian Gulf of Carpentaria is the example shown in Figure 1:1(a). Here the tides reduce to zero when the moon's declination is zero, increasing to their largest values when the moon is at its greatest declination, either north or south of the equator (Figure 1:1(b)). Diurnal tides are also found in part of the Persian Gulf, the Gulf of Mexico and part of the South China Seas. The Chinese port of Pei-Hai has the world's largest diurnal tidal range, with a difference of 6.3 m between the highest and lowest predicted tidal levels.

The tidal ranges on the relatively shallow continental shelves are usually larger than those of the oceans. However, very small tidal ranges are observed in some shallow areas, often accompanied by curious distortions of the normal tidal patterns. Figure 1:1(a) shows the curves for Courtown on the Irish coast of the Irish Sea, where the range varies from more than a metre at spring tides to only a few centimetres during neap tides. At Courtown when the range is very small, careful examination shows that four tides a day occur. These effects are due to the distorted tidal propagation in very shallow water. Shallow-water distortions, which are discussed in detail in Chapter 7, are also responsible for the double high water feature of Southampton tides and for the double low waters seen at Portland, both in the English Channel, where semidiurnal tides prevail. Double low waters also occur along the Dutch coast of the North Sea from Haringvlietsluizen to Scheveningen, where they are particularly well developed at the Hook of Holland. Double high waters are also found at Den Helde in the North Sea and at Le Havre in the English Channel.

Tidal currents, often called tidal streams, have similar variations. Semidiurnal, mixed and diurnal currents occur, usually having the same characteristics as the local changes in tidal levels, but this is not always so. For example the currents in the Singapore Strait are often diurnal in character but the elevations are semidiurnal. The strongest currents are found in shallow water or through narrow channels which connect two seas, such as the currents through the Straits of Messina and those through the Euripus which perplexed Aristotle.

Currents in channels are constrained to flow either up or down the channel axis, but in more open waters all directions of flow are possible. During each tidal period the direction usually rotates through a complete circle while the speeds have two approximately equal maximum and two approximately equal minimum values. Figure 1:2 shows the distribution of tidal currents over one semidiurnal cycle at the Inner Dowsing light tower in the North Sea. Each line represents the speed and direction at a particular hour. Because each measurement of current is a vector described by two parameters, their variations are more complicated to analyse than changes of sea level.

Although this book is concerned with movements of the seas and oceans, two other geophysical phenomena which have tidal characteristics are of interest. In tropical regions there is a small but persistent 12-hour oscillation of the atmospheric pressure (Section 5:5) with a typical amplitude of 1.2 millibars, which reaches its maximum values near 1000 hours and 2200 hours local time; this produces a small tidal variation of sea levels. Also, accurate measure-

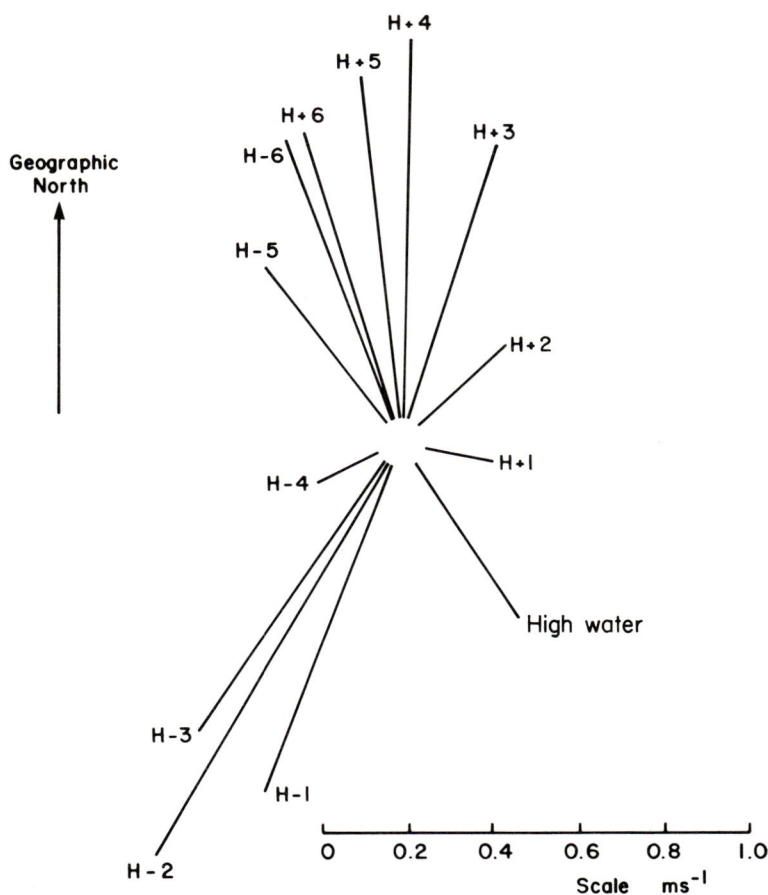

Figure 1:2 Spring tidal current curve, Inner Dowsing light tower, North Sea. Each line represents the velocity of the current at each hour from six hours before to six hours after local high water. Semidiurnal tides are strongly dominant. The distance from the centre represents the speed, and the bearing gives the direction of water flow.

ments of the earth's gravity field and of the tilt of the earth's crust show vertical movements of the land surface relative to the centre of the earth, which are tidally induced. These earth tides may have extreme amplitudes of tens of centimetres. It should be remembered that measured coastal sea levels are actually the difference between the vertical land movements and the vertical movements of the sea surface. For scientific computations of tides on a global scale the earth tides must be included because of their effects on ocean dynamics. However, for practical purposes only the observed movements relative to the land need be considered.

1:3 Meteorological and other non-tidal disturbances

The regular and predictable pattern of the tides is modified to a greater or lesser extent, by irregular factors, the principal ones being the atmospheric pressure and the winds acting on the sea surface. Figure 1:3(a) shows how the regular semidiurnal pattern of sea levels at Cromer in the North Sea was modified by the weather over a few days in January 1982. These irregular slow changes, known as surges, are plotted in the same diagram on an enlarged scale. On 29 January the coincidence of a spring tide and a large increase in the levels due to the weather has caused an exceptionally high total level. Figure 1:3(b) shows the tide gauge record for flooding at Gulfport, Mississippi due to hurricane Elena in 1985.

Historically there have been many disastrous coastal floodings caused by the coincidence of large meteorologically induced surges and large or even moderately high tides (see Chapter 6). Thus in November 1885 New York was inundated by high sea levels generated by a severe storm which also caused flooding at Boston. More than 6000 persons were drowned in September 1900 when the port of Galveston in Texas was overwhelmed by waters which rose more than 4.5 m above the mean high-water level, as a result of hurricane winds blowing at more than 50 m s^{-1} for several hours. Even these disasters were surpassed by the Bangladesh tragedy of 12 November 1970 when winds of 60 m s^{-1} raised sea levels by an estimated 9 m. Flooding extended over several low-lying islands, drowning hundreds of thousands of people. Severe storms, floods, and loss of life occurred in the same region in both 1876 and 1897, and more recently in 1960, 1961 and 1985.

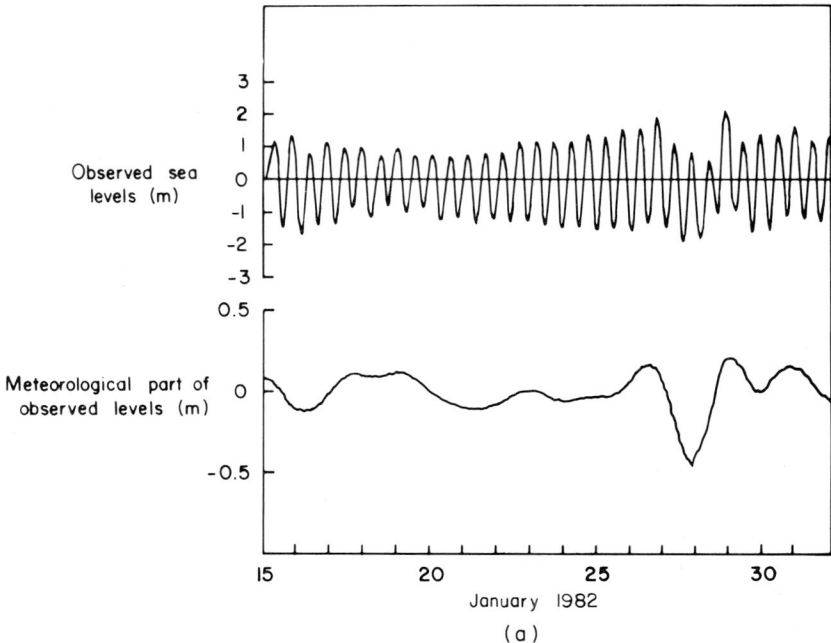

(a)

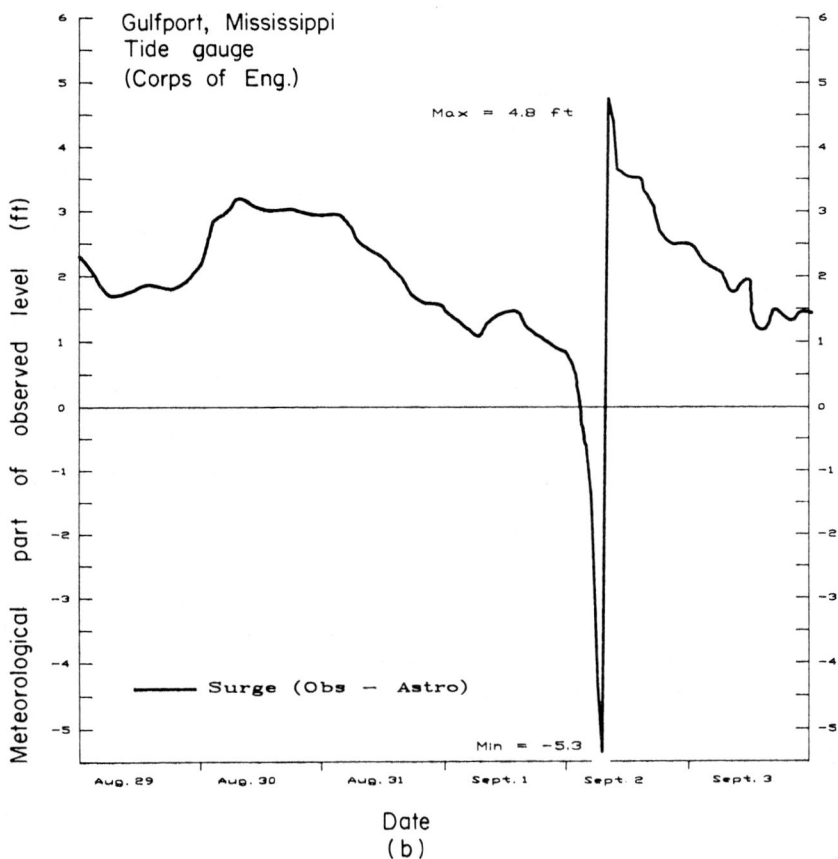

Figure 1:3 (a) Observed and meteorological variations of sea-level, Cromer, North Sea, January 1982. The vertical scale for the surges is a factor of five greater than the scale for the observed levels. (b) Tide gauge record for Gulfport, Mississippi during hurricane Elena, 1985 (Supplied by C. P. Jelesnianski).

The largest surges occur when hurricane winds blow for a long time over large expanses of shallow water. Whilst tropical storms cause the most extreme local flooding, storms at higher latitudes can also produce very large surges. In January 1953 catastrophic flooding on both the English and Dutch coasts of the North Sea occurred as a result of a depression tracking into the North Sea across the north of Scotland. This surge, which exceeded 2.6 m at Southend, coincided with slightly less than average spring tides, otherwise even more damage would have occurred. The coastal regions of the North Sea are prone to this type of flooding and there were previous events in 1929, 1938, 1943, 1949, and a particularly severe case in Hamburg in 1962. Similarly, along the Atlantic coast of the United States the Ash Wednesday storm of 7 March 1962 flooded many low-lying barrier islands, causing millions of dollars of damage.

These dramatic extremes and the resulting coastal flooding are rare events, but there is always a continuous background of sea-level changes due to the weather, which raise or lower the observed levels compared with the levels predicted. At higher latitudes these effects are greater during the stormy winter months. Knowledge of the probability of occurrence of these extreme events is an essential input to the safe design of coastal defences and other marine structures, as discussed in Chapter 8.

The tsunamis generated by submarine earthquakes are another cause of rare but sometimes catastrophic flooding, particularly for coasts around the Pacific Ocean. Tsunamis are sometimes popularly called 'tidal waves' but this is misleading because they are not generated by tidal forces nor do they have the periodic character of tidal movements (see Section 6:8). The naturalist Charles Darwin in *The Voyage of the Beagle* describes how, shortly after an earthquake on 20 February 1835, a great wave was seen approaching the Chilean town of Concepcion. When it reached the coast it broke along the shore in 'a fearful line of white breakers', tearing up cottages and trees. The water rose to 7 m above the normal maximum tidal level. Near the earthquake source the tsunami amplitudes are much smaller; the amplification occurs in the shallow coastal waters and is enhanced by the funnelling of the waves in narrowing bays. The term tsunami is of Japanese origin which is appropriate as their coast is particularly vulnerable to this type of flooding. For example, in 1933 the Japanese Bay of Sasu was struck by a tsunami which flooded to more than 13 m. Tsunamis can travel long distances across the oceans: waves from the 1883 Krakatoa island explosion in the Indonesian islands between Java and Sumatra were observed on the South Atlantic island of South Georgia by a German expedition 14 hours later. In the Atlantic Ocean tsunamis are comparatively rare, but the flooding after the 1755 Lisbon earthquake is well documented: at Newlyn in south-west England levels rose by 3 m in a few minutes, several hours after the seismic tremors.

1:4   Some definitions of common terms

It is now appropriate to define more exactly some of the common tidal and non-tidal terms as they will be used throughout this book; the Glossary on page 458 contains precise summaries. The first important distinction to make is between the popular use of the word *tide* to signify any change of sea-level, and the more specific use of the word to mean only the regular periodic variations.

Although any definition of tides will be somewhat arbitrary, it must emphasize this periodic and regular nature of the motion, whether that motion be of the sea surface level, currents, atmospheric pressure or earth movements. We define *tides* as periodic movements which are directly related in amplitude and phase to some periodic geophysical force. The dominant geophysical forcing function is the variation of the gravitational field on the surface of the earth, caused by the regular movements of the moon–earth and earth–sun systems.

Movements due to these forces are termed *gravitational tides*. This is to distinguish them from the smaller movements due to regular meteorological forces which are called either *meteorological* or more usually *radiational tides* because they occur at periods directly linked to the solar day. It can be argued that seasonal changes in the levels and in the circulation of sea water due to the variations of climate over an annual period are also regular and hence tidal.

Any sequence of measurements of sea-level or currents will have a tidal component and a non-tidal component. The *non-tidal component* which remains after analysis has removed the regular tides is called the *residual*, or *meteorological residual*. Sometimes the term *surge residual* is used, but more commonly the term *surge* or *storm surge* is used for a particular event during which a very large non-tidal component is generated, rather than to describe the whole continuum of non-tidal variability.

Periodic oscillations are described mathematically in terms of an amplitude and a period or frequency:

$$\mathbf{X}(t) = H_x \cos (\omega_x t - g_x) \qquad (1:1)$$

where $\mathbf{X}$ is the value of the variable quantity at time t, H_x is the amplitude of the oscillation, ω_x is the angular speed which is related to the period T_x by: $T_x = 2\pi/\omega_x$ (ω_x is measured in radians per unit time) and g_x is a phase lag relative to some defined time zero.

For tidal studies H_x may have units of metres (rarely feet) for levels, or metres per second (m s^{-1}) for currents. In its simple form equation (1:1) can only represent the to-and-fro currents along the axis of a channel. If the direction is variable as in Figure 1:2, then equation (1:1) may define the flow along a defined axis. The speed and direction of the total flow is completely specified if the currents along a second axis at right angles to the first are also defined.

Tidal *high water* is the maximum tidal level reached during a cycle. The observed high water level may be greater or less than the predicted tidal level because of meteorological effects. Similarly *low water* is the lowest level reached during a cycle. The difference between a high level and the next low water level is called the *range*, which for the simple oscillation defined by equation (1:1) is twice the amplitude.

Ocean tides and most shelf sea tides are dominated by semidiurnal oscillations, for which several descriptive terms have been developed through both popular and scientific usage (see Figure 1:4). *Spring tides* are semidiurnal tides of increased range which occur approximately twice a month at the time when the moon is either new or full. The *age* of the tide is an old term for the lag between new or full moon and the maximum spring tidal ranges. The average spring high water level taken over a long period is called *Mean High Water Springs* (MHWS) and the correspondingly averaged low water level is called *Mean Low Water Springs* (MLWS). Formulae are available for estimating these useful parameters directly from the results of tidal analyses (Section 4:2: 6), to avoid laborious searches through series of tidal predictions. *Neap tides* are the semidiurnal tides of small range which occur near the time of the first and last lunar quarters, between spring tides. *Mean High Water Neaps* and *Mean Low Water Neaps* are the average high and low waters at neap tides. These too may be estimated directly from tidal analyses.

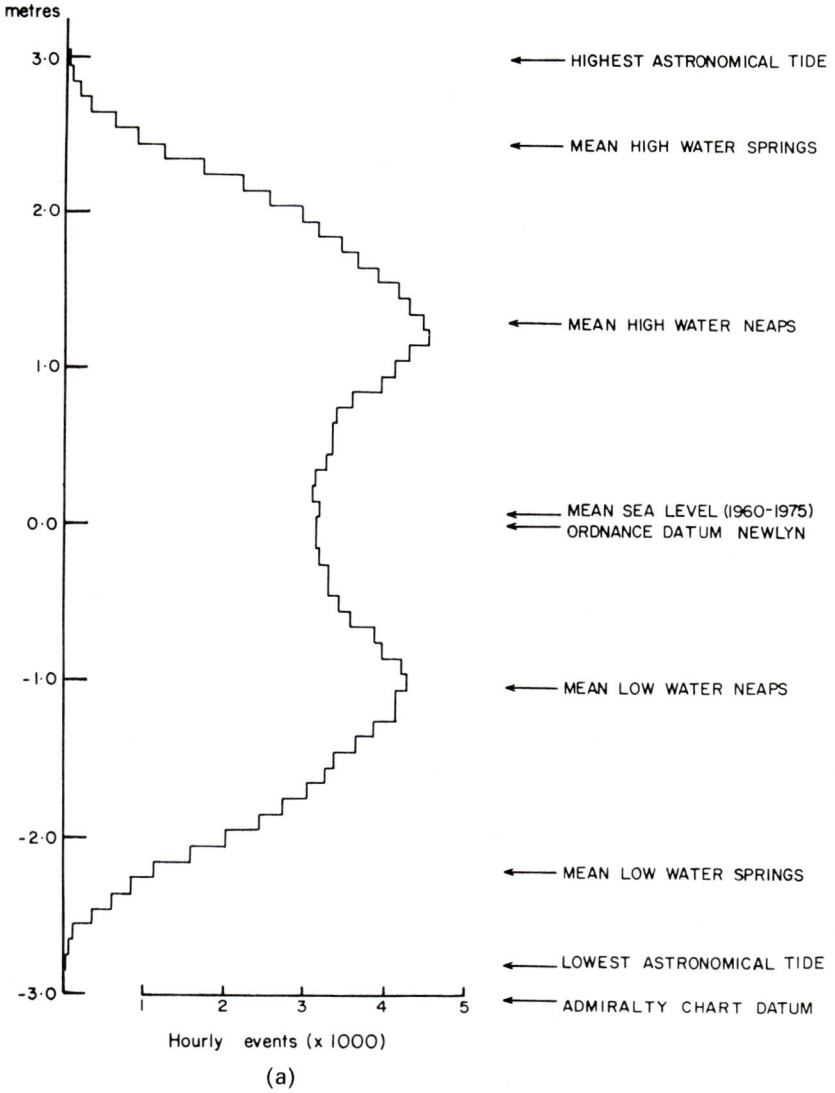

(a)

metres

5·0

← HIGHEST ASTRONOMICAL TIDE

4·0

3·0 ← MEAN HIGHER HIGH WATER

2·0

← MEAN SEA LEVEL

1·0 ← MEAN LOWER LOW WATER

0·0 ← CHART DATUM
← LOWEST ASTRONOMICAL TIDE

| 1 | 2 | 3 | 4 | 5 |

Hourly events (×100)

(b)

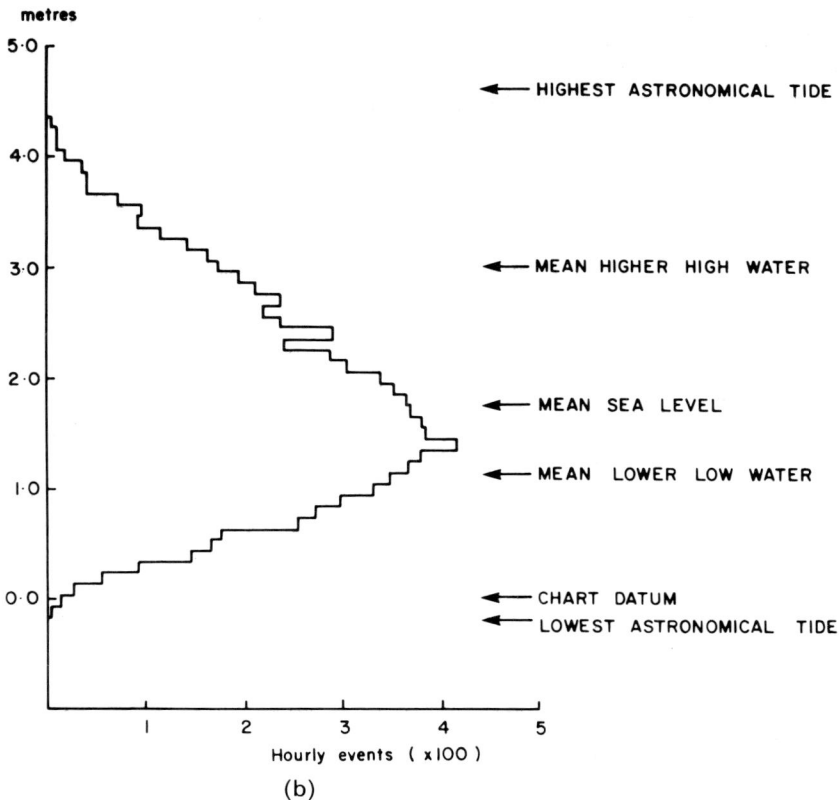

Figure 1:4   Frequency distribution of hourly tidal levels (a) at Newlyn (1951–69) relative to Chart Datum. For the semidiurnal Newlyn tides there is a distinct double peak in the distribution. For (b) the diurnal Karumba tides (1982) there is a single peak. HAT and LAT were not reached at Karumba in 1982.

Where the tidal regime is mixed the use of Mean High Water Springs and other semidiurnal terms becomes less appropriate. Mean High Water and Mean Low Water are frequently used. *Mean Higher High Water* (MHHW) is the average level over a long period of the higher high water level which occurs in each pair of high waters in a tidal day. Where only one high water occurs in a tidal day, this is taken as the higher high water. *Mean Lower Low Water* (MLLW) is the average of the lower level in each pair of low water levels in a tidal day. Mean Lower High Water and Mean Higher Low Water are corresponding terms which are seldom used.

These terms become increasingly difficult to determine exactly where the tide is dominated by diurnal oscillations (Figure 1:4(b)). Mean High Water (MHW) and Mean Low Water (MLW) averaged for all high and low levels respectively may be computed as datums for survey work, but difficulties arise when considering how to treat the multiple small high and low levels which occur during the time when the diurnal range falls to near zero, as for example at Karumba in Figure 1:1(a). Diurnal tides show a semi-monthly variation in their range from a maximum to zero and back to a maximum range again because of the monthly cycle in the lunar declination. Some authorities have suggested that these diurnal modulations should be called Diurnal Spring and Diurnal Neap Tides (Doodson and Warburg, 1941), but this usage is uncommon, being usually reserved for the semidiurnal range changes related to the phase of the moon.

Levels defined by analysis of long periods of sea-level variations are used to define a reference level known as a tidal datum. These are often used for map or chart making or for referring subsequent sea-level measurements. For geodetic surveys the *Mean Sea Level* is frequently adopted, being the average value of levels observed each hour over a period of at least a year, and preferably over about 19 years, to average out cycles of 18.6 years in the tidal amplitudes and phases (see Chapter 3), and to average out effects on the sea levels due to weather. In Britain the Ordnance Survey refer all levels to Ordnance Datum Newlyn (ODN) which is the mean sea level determined from six years of continuous records at Newlyn, Cornwall, extending from May 1915 to April 1921. Present United Kingdom mean sea levels are approximately 0.1 m higher than the 1915–1921 mean level. In the United States the National Geodetic Vertical Datum was established by assuming that local sea levels corresponded to the same geodetic zero at 26 selected tide gauge stations on the Pacific and Atlantic coasts of the United States and Canada. This assumption is not valid for the accuracy of modern survey techniques, as we shall discuss in Section 9:7.

The *Mean Tide Level* (MTL), which is the average value of all the heights of high and low water, is sometimes computed for convenience where a full set of hourly levels is not available. In general this will differ slightly from the mean sea-level because of tidal distortions in shallow water.

Hydrographers prefer to refer levels on their charts to an extreme low water Chart Datum, below which the sea-level seldom falls, in order to give navigators the assurance that the depth of water shown on charts is the minimum available whatever the state of the tide. *Lowest Astronomical Tide*, the lowest level that can be predicted to occur under any combination of astronomical conditions, is adopted for charts prepared by the British Hydrographic Department. The Canadian Hydrographic Department has adopted the level of lowest normal tides. The United States National Ocean Service (NOS) uses a Chart Datum which is defined in some areas of mixed tides as the mean of the lower of the two low waters in each day (MLLW), in some semidiurnal tidal areas as mean low water springs (MLWS), and elsewhere as the lowest possible low water. As the charts are revised, MLLW is being adopted by NOS as the standard datum for all locations (see Appendix 5). Throughout the world, published predictions of tidal levels are given relative to the local Chart Datum so that by adding them to the water depth shown on the chart, the navigator knows the total depth available for his vessel. Because the Chart Datum is defined in terms of the local tidal characteristics, it is not a horizontal plane. Geodetic datums, however, are theoretically horizontal planes to within the accuracy of the survey which established them. As a result, the amount by which Chart Datum falls below Survey Datum is not constant, being greatest where the tidal range is greatest.

Although for practical navigation requirements tidal predictions are normally published as levels above Chart Datum, for scientific and engineering purposes it is normally more convenient to consider levels relative to a Mean Sea Level datum. The tidal levels plotted in Figure 1:1(a) are relative to local Mean Sea Level.

If a long series of hourly tidal predictions are examined for the frequency with which each level occurs, certain values are found to be more probable than others. Highest and Lowest Astronomical Tides occur very seldom whereas, in the case of semidiurnal tides as illustrated by a histogram of 19 years of Newlyn predictions in Figure 1:4(a), the most probable levels are near Mean High Water Neaps and Mean Low Water Neaps. Water level is changing relatively quickly when passing through the mean level and so this appears as a minimum in the frequency histogram, whereas it pauses at the high-water and low-water levels. As a general rule one of the hump maxima is noticeably greater than the other especially in shallow water regions. Figure 1:4(b) shows the corresponding distribution for a year of predicted hourly tides at Karumba where diurnal tides prevail. Note that the double-humped distribution characteristic of semidiurnal tides (with a single dominant constituent) has been replaced by a continuous and slightly skew distribution. During the single year analysed at Karumba the diurnal tides failed to reach either the Highest or Lowest Astronomical Tide levels. This was because the range of lunar declinations, which has an 18.6-year cycle, was relatively small in 1982. Several of the other tidal terms defined here are also shown relative to the distributions of Figure 1: 4. In statistical terms these distributions of levels are called probability density functions.

*1:5 Basic statistics of tides as time series

The full process of tidal analysis will be considered in Chapter 4, but there are certain basic statistical ideas which may be used to describe tidal patterns without applying elaborate analysis procedures. The probability density functions of tidal levels shown in Figure 1:4 are one useful way of representing some aspects of tides. In this section we will consider ideas of the mean and the standard deviation, the variance and spectral analysis of a series of sea-level measurements made over a period of time.

The general representation of the observed level $\mathbf{X}(t)$ which varies with time may be written:

$$\mathbf{X}(t) = Z_0(t) + \mathbf{T}(t) + \mathbf{S}(t) \qquad (1:2)$$

where $Z_0(t)$ is the mean sea-level which changes slowly with time, $\mathbf{T}(t)$ is the

*Indicates that this section may be omitted by non-mathematical readers.

tidal part of the variation and $S(t)$ is the meteorological surge component.

A major part of this book will be concerned with separate discussions of the different terms of this equation.

Measurements of sea levels and currents are conveniently tabulated as a series of hourly values, which for satisfactory analysis should extend over a lunar month of 709 hours, or better over a year of 8766 hours. Suppose that we have M observations of the variable $X(t)$ represented by $x_1, x_2, \ldots, x_M$, then the mean is given by the formula:

$$\bar{x} = \tfrac{1}{M}(x_1 + x_2 + x_3 + \cdots + x_M) \tag{1:3}$$

or in conventional notation:

$$\bar{x} = \tfrac{1}{M}\sum x$$

Every value of the variable differs from the mean of the series of observations by an amount which is called its deviation. The deviation of the mth observation is given by

$$e_m = x_m - \bar{x}$$

Note that the average value of e must be zero because of the way in which the mean value $\bar{x}$ is calculated. A better way of describing the extent by which the values of x vary about the mean value $\bar{x}$ is to compute the *variance* s^2, the mean of the squares of the individual deviations:

$$s^2 = \tfrac{1}{M}\sum (x_m - \bar{x})^2 \tag{1:4}$$

which must always have a positive value. The square root of the variance, s, is called the standard deviation of the distribution of x about $\bar{x}$.

A further extremely useful technique called Fourier analysis represents a time series in terms of the distribution of its variance at different frequencies. The basic idea of Fourier analysis is that any function which satisfies certain theoretical conditions, may be represented as the sum of a series of sines and cosines of frequencies which are multiples of the fundamental frequency $\sigma = (2\pi/M\Delta t)$:

$$X(t) = Z_0 + \sum_{m=1}^{M/2} A_m \cos m\sigma t + \sum_{m=1}^{M/2} B_m \sin m\sigma t \tag{1:5}$$

where the coefficients A_m and B_m may be evaluated by analysis of M values of $X(t)$ sampled at constant intervals Δt. Z_0 is the average value of $Z_0(t)$ over the period of observations. The theoretical conditions required of the function $X(t)$ are satisfied by sea levels and currents and by any series of observations of natural phenomena. An alternative form of equation (1:5) is:

$$X(t) = Z_0 + \sum_{m=1}^{M/2} H_m \cos (m\sigma t - g_m) \tag{1:6}$$

where

$$H_m = (A_m^2 + B_m^2)^{\frac{1}{2}}$$

and

$$g_m = \arctan (B_m/A_m)$$

H_m and g_m are the amplitude and phase lag of the mth harmonic constituent of the function $\mathbf{X}(t)$. Note that the phase lags and angular speeds must be expressed consistently in terms of radian or degree angular measure; see Section 4:2 for a discussion of the usual tidal notations.

The variance of this function about the mean value Z_0 is given by squaring the terms within the summation symbol of equation (1:6), and averaging over the period of observation. If this multiplication is carried through it is found that all of the cross-product terms are zero, except for terms of the form:

$$H_m^2 \cos^2 (m\sigma t - g_m)$$

the average value of which, over an integral number of cycles is $\frac{1}{2} H_m^2$. The total variance in the series is therefore given by:

$$\frac{1}{2} \sum_{m=1}^{M/2} H_m^2$$

This powerful statistical result which states that the total variance of the series $\mathbf{X}(t)$ is the sum of the variance at each harmonic frequency is very important for tidal analyses.

The methods of tidal analysis to be described in Chapter 4, which enable separation of the tidal and surge components of the series $\mathbf{X}(t)$, have a condition that the two components are statistically independent. This means that the sum of their individual variances gives the variance in the total observed series:

$$\underbrace{\sum_{k=1}^{M} (\mathbf{X}(k\Delta t) - Z_0)^2}_{\text{Total variance}} = \underbrace{\sum_{k=1}^{M} (\mathbf{T}^2(k\Delta t)}_{\text{Tidal variance}} + \underbrace{\sum_{k=1}^{M} \mathbf{S}^2(k\Delta t)}_{\text{Surge variance}}$$

Furthermore, the variance of the tides may be computed as the sum of the variance in each frequency element. The important conclusion is that the total tidal variance in a series of observations is the diurnal variance plus the semidiurnal variance plus the variance at other higher and lower frequencies. In shallow water there may be tidal energy at higher frequencies because some semidiurnal tidal energy is shifted, mainly to quarter-diurnal and sixth-diurnal tidal periods (see Chapter 7). These concentrations of tidal energy in groups of similar period or frequency, which we discuss in detail in Chapter 4, are called tidal *species*.

We now consider how the tidal regime at a particular location may be characterized by the way in which the variance is distributed among the different species. Table 1:1 summarizes the distributions of variance for four sites. In Honolulu the total variance is small, and contains nearly equal contributions from the diurnal and semidiurnal species. Tides from higher species due to shallow water distortions are negligible because Honolulu is near the deep ocean. At Mombasa the diurnal tides are larger than at Honolulu, but the tidal curves are dominated by the large semidiurnal tides. There is a narrow continental shelf region adjacent to the Mombasa coast which generates a small variance at higher frequencies. At Newlyn the diurnal tides are very small and the semidiurnal tides are very large. Shallow-water tides are a significant factor, due to the effects of the extensive continental shelf which separates Newlyn from the Atlantic Ocean. Courtown has an exceptional tidal regime; the annual and other long period changes of level are unusually large, but the semidiurnal tides are unusually small for a continental shelf site.

Table 1:1   Distributions of variance at representative sea-level stations. Long-period tidal variations include annual, semi-annual and monthly changes. Units are cm^2.

	Tidal				Non-tidal	Total
	Long-period	Diurnal	Semidiurnal	Shallow-water		
Honolulu 1938–1957	9	154	157	0	35	355
Mombasa 1975–1976	5	245	7 555	2	19	7 826
Newlyn 1938–1957	17	37	17 055	100	191	17 400
Courtown 1978–1979	116	55	284	55	222	732

Mombasa and Honolulu have the smallest non-tidal or surge effect in the observed levels. The largest surge residuals are found in the higher latitudes where storms are more severe, and in regions of extensive shallow water.

The variance distribution shown in Table 1:1 can be taken a stage further. Figure 1:5 shows the results of a frequency distribution of the variance in a year of hourly sea-level measurements at Mombasa. Theoretically, given a year of 8766 hourly observations, 4383 separate frequency components could be determined, but such a fine resolution gives a spectral plot which is noisy and irregular. A more satisfactory presentation is obtained by averaging the variance over several adjacent frequency components.

In the case illustrated, the values are averaged over 60 elements, and plotted against a logarithmic vertical scale to accommodate the great range of values obtained. The significance of the semidiurnal and diurnal tides is clear. Most of the non-tidal variance is contained in the low frequencies, equivalent to periods of 50 hours or longer. Although weak, the fourth-diurnal tides stand out clearly above the background noise and there is a hint of a spectral peak in the vicinity of the third-diurnal tides near 0.12 cycles per hour. The averaged values of variance plotted within the diurnal and semidiurnal species conceal a fine structure of variance concentration at a few particular frequencies, which will be considered in more detail in Chapter 3.

The measurement of variance is conveniently related to the energy contained in a physical system and descriptions of tidal dynamics (see Sections 5:4:1 and 7:9) are often expressed in terms of energy fluxes and energy budgets. The variance of a series of sea levels may be related to the average potential energy

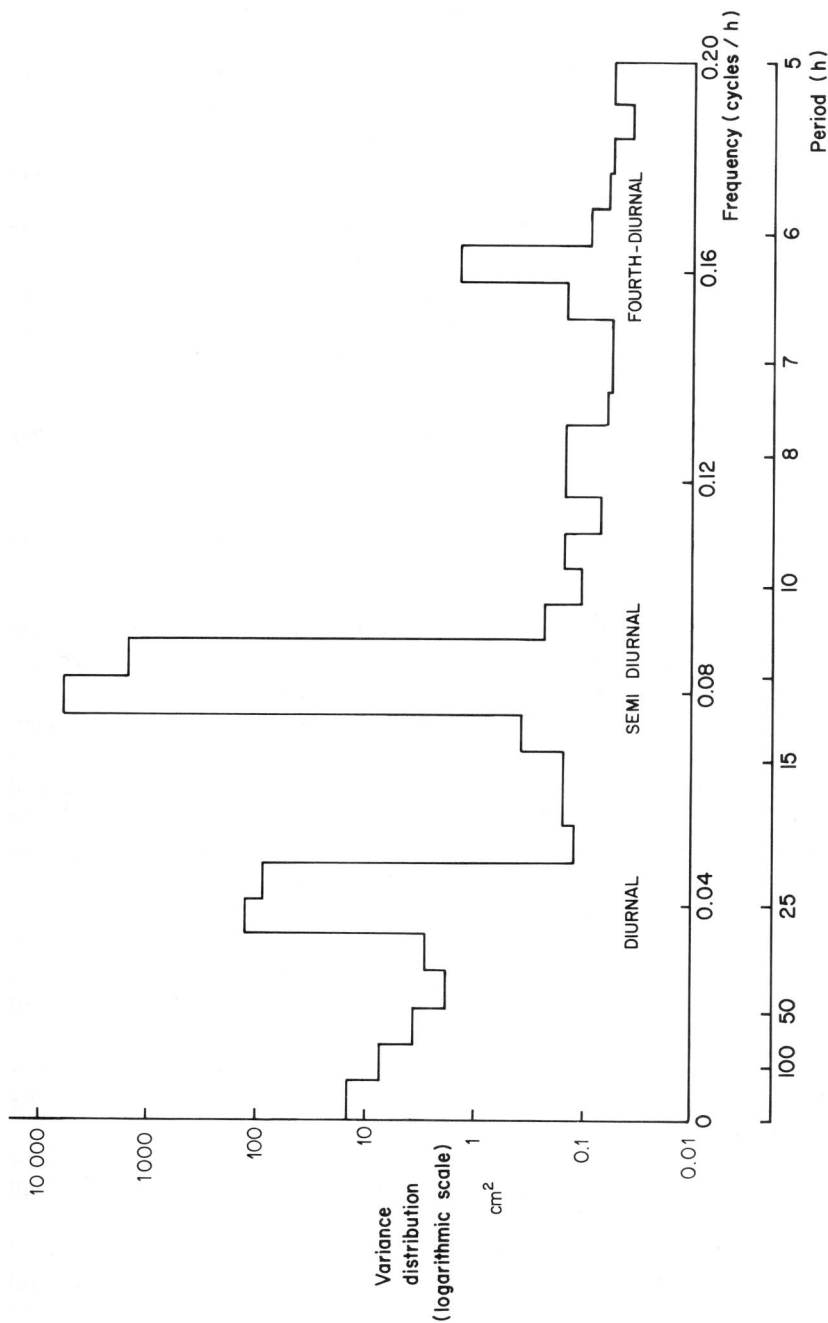

Figure 1:5 Distribution of variance in observed levels over one year (1975–6) at Mombasa. Semidiurnal tidal changes of level are the most significant, but diurnal tidal changes are also apparent. There is a small peak of variance in the fourth-diurnal tidal species due to distortions of the tide in shallow water.

of the water mass relative to a regional mean sea-level, and for current speeds the variance may be related to the average kinetic energy of the water movements.

The purpose of tidal analysis is to represent the variations in terms of a few significant parameters. In Chapters 3 and 4 we discuss the available techniques in more detail, but we first summarize, in the next chapter, different methods for measuring the variations of sea levels and currents in order to produce reliable data for subsequent analysis.

CHAPTER 2

Observations and Data Reduction

Indeed, I think that as we go on piling measurements upon measurements, and making one instrument after another more perfect to extend our knowledge of material things, the sea will always continue to escape us.

Belloc, *The Cruise of the Nona*

2:1 The science of measurement

The ocean is its own uncontrollable laboratory and the oceanographer who measures the properties of the sea is an observational rather than an experimental scientist. Technically the necessity of making measurements at sea presents many challenges in terms of the logistics of travel to the site, for deployment of the equipment, and for its safe and reliable operation in a frequently hostile environment.

This chapter summarizes methods of measuring changes of sea levels and currents over tidal and longer periods, with particular emphasis on new techniques such as satellite altimetry and high-frequency radar. One of the major instrumental difficulties in the measurement of long-period variations of both sea levels and currents is the ability to resolve the longer period variations from shorter period variations due to short-period wind-induced waves. These waves often have amplitudes which are much greater than the accuracy demanded for the long-period averaged values. It is not unusual to require measurements of tidally changing sea-level to an accuracy of 0.01 m in the presence of waves of 1.0 m amplitude, or measurements of mean currents of 0.03 m s^{-1} where speeds fluctuate by more than 1.0 m s^{-1} over a few seconds due to wave activity. The designer of instrument systems must overcome the problem of averaging these wave effects in a way which produces no residual distortion.

Coastal measurements of sea-level have a long history; many countries operate networks of tide gauges at selected sites to monitor mean sea-level changes and to assess the risks of flooding from the sea. The British network of some thirty gauges is an example of a network which transmits the data automatically to a central point for checking and eventual archiving in a special data bank. The National Ocean Service of the United States operates a similar system with more than one hundred and fifty gauges, some of which transmit the data via satellites. The Intergovernmental Oceanographic Commission has now established a global sea-level network, which consists of some 250 nationally operated gauges, to monitor long-period changes in global sea-level. Many series of sea-level observations extend over several decades, and although few of them extend over more than a hundred years, they represent one of the longest series of ocean measurements. Compared with many other types of ocean measurements, the equipment for measuring sea-level is cheap and easy to maintain.

Measurements of sea levels and currents away from the coast are more difficult because there are no obvious fixed reference points. They are also far more expensive because ships are needed to deploy and recover the equipment. Even a modest specially-equipped research vessel costs upwards of $10 000 per day to operate. As a result of these practical and financial difficulties, few measurements offshore extend over more than a year, and even measurements over a month have become practicable only since the development of automatic digital recording equipment.

When planning a measuring programme to monitor sea levels and currents in a particular area, several design factors must be considered. Where should measurements be made? For how long? To what accuracy? Is datum stability and connection to a national levelling datum necessary for the coastal levels? Ultimately, can these requirements be met within acceptable cost limits? The range of equipment now available extends from relatively cheap, imprecise short-term meters to accurate, long-term internally-recording, and inevitably expensive, equipment sold in a specialized international market. For a relatively crude local hydrographic survey the former may suffice, but for scientific studies the superior instruments and accuracies are essential. Target accuracies of 0.01 m in levels and 0.03 m s^{-1} in currents are possible with a little care. Unfortunately it is easy to make bad measurements which appear to have this accuracy, but which contain unidentified systematic or random error.

When we measure any quantity we are trying to describe it in terms of numbers; we may say that it is big, but by comparison with what other entity? We may say that it has a value of 43, but this is meaningless without properly specified units. We may say that it has a value of 43 metres, but it would be more precise to specify the value as 43.26 metres. It *may* also be more accurate to give this more precise value, but only if the instruments and procedures used were adequate. Oceanographers generally prefer to use SI units (UNESCO, 1985a); factors for converting to these units are given in Table 2:1.

It is important to realize that all measurements are comparative, and that no measurement is perfectly accurate. The comparative nature of measurements implies a system of units and standards established by international agreement and testing. Both length and time are now defined in terms of atomic radiation, the metre in terms of the Krypton-86 atom, and the second in terms of the Caesium-133 atom; secondary standards are adequate for marine physical measurements but it is important that measuring systems are regularly and carefully calibrated against these. An important distinction must be made between the absolute accuracy of a measurement, which is its accuracy compared with an absolute international standard or datum, and the relative accuracy or sensitivity of the measuring system, which is the accuracy with which the difference between individual measurements in a set has been determined. For example, a tide gauge may measure sea-level changes to 0.01 m, but because of inaccurate levelling or poor maintenance, its accuracy relative to a fixed datum may be in error by 0.05 m.

Table 2:1   Conversion factors to SI units.

Levels	Multiply by
Conversion to metres	
From feet	0.3 048
fathoms	1.8 288
Pressures	
Conversion to pascals (newtons per square metre)	
From dynes per square centimetre	0.1 000
bars	100 000
millibars	100
pounds per square inch	689 4.8
Speeds	
Conversion to metres per second	
From feet per second	0.3 048
knots	0.5 144
kilometres per day	0.0 116

Despite their apparent diversity all measuring systems have some common characteristics. Firstly, some physical parameter exists to be measured. For sea-level this is the contrast between air and sea water of their density, of the velocity with which sound travels within them, or of their electromagnetic properties. This physical parameter must be measured by some sensor, which is ideally very sensitive to the parameter itself, but quite insensitive to any other changes. The density contrast between air and water enables the interface to be located by a simple float. The contrast between the electromagnetic properties of air and water enables the level to be detected by eye. Other examples will be discussed later. The signal from the sensor must then be transmitted to a recording system, which will have its own characteristics and limitations. This telemetry may at one extreme be via satellite to a remote computer centre, or at the other extreme by human speech to someone who writes the values in a notebook. The choice of sensor, transmitter and recording medium must depend on the applications of the data and the accuracy and speed with which it is required. Flood warning systems need rapid data transmission, whereas for compilations of statistics of sea-level changes over several years, a delay may be of no consequence.

2:2   Sea levels, bottom pressures

Methods for measuring sea-level may be divided into two categories: those appropriate for coastal measurements, and those which can be used off-shore (Howarth and Pugh, 1983; Forrester, 1983; UNESCO, 1985b). Until quite recently all measurements of any duration were made at the coast which, although technically easier, can introduce local distortions. Coastal currents and waves interact with the coastal topography to produce local gradients on the sea surface. Mean sea levels at one site may be higher or lower by several centimetres, compared with the corresponding levels a few kilometres away along the coast, or outside rather than inside a harbour. Differences of 0.03 m would not be exceptional, and so the selection of a suitable site for coastal measurements is important.

The site selected should ideally be connected by relatively deep water to the area for which the measured levels are intended to be representative. It should experience the full tidal range, if possible, and not dry out at low waters. Sites affected by strong currents or directly exposed to waves should be avoided because of their local effects on sea-level. So should sites near headlands and in harbours with very restricted entrances. Practical requirements include ease of access, proximity to a permanent benchmark for good datum definition, and the availability of suitable accommodation for the recording instruments. Piers and lifeboat stations can be good sites but they are not always available.

2:2:1   Tide poles

Levels have been measured using vertically mounted poles for thousands of years. The ancient Egyptians linked their Nilometers to their temples, from where the priests gave warnings of imminent flooding. Tide poles or staffs are cheap, easy to install, and may be used almost anywhere. Many harbours have levels engraved in the walls. Along a broad drying beach a flight of tide poles, all connected to the same datum, may be used. For these, and for all measurements of level, there should be a careful connection of the gauge zero to a permanent shore benchmark, which is usually designated the *Tide Gauge Bench-mark* (TGBM). If possible at least three auxiliary secondary benchmarks should also be identified to guard against the TGBM being destroyed. In calm conditions it should be possible to read a pole to 0.02 m, but the accuracy deteriorates in the presence of waves. Experiments show that experienced observers can achieve a standard deviation for a set of readings of 0.05 m in the presence of waves of 1.5 m height. However, systematic errors may also be present due to operator bias and because of different daytime and night-time illumination. Waves may be averaged by taking the average between crest and trough levels; but the apparent trough level for an observer viewing at an angle over the sea surface may actually be the crest of an intermediate wave which is obscuring the true trough at the pole. In this case the averaged level will be too high.

Reading accuracy in the presence of waves may be increased by fitting a transparent tube (Perspex) alongside the pole, which connects to the sea through a narrower tube preventing immediate response to external level changes. For a transparent tube of internal diameter 0.025 m, a connecting tube 2.7 m long of internal diameter 0.004 m gives an averaging time of 30 seconds, sufficient to still most waves. The level in the transparent tube is easily read against the graduated pole, particularly if some dye is mixed into the tube water.

Although notable series of measurements have been made with tide poles, the tedium of reading over several months makes poles unsuitable for long-term measurements; however, they are often the best choice for short-term surveys of only limited accuracy. Commercially available gauges, consisting of a series of spaced electronic water-level sensors, which record both waves and tidal changes of level are available as an alternative to reading by eye.

2:2:2 Stilling-well gauges

The vast majority of permanent gauges installed since the mid-nineteenth century consist of an automatic chart recorder: the recording pen is driven by a float which moves vertically in a well, connected to the sea through a relatively small hole or narrow pipe (Great Britain, 1979; Forrester, 1983). The limited connection damps the external short-period wave oscillations. The basic idea was described by Moray in 1666. He proposed that a long narrow float should be mounted vertically in a well, and that the level to which the float top had risen be read at intervals. The first self-recording gauge, designed by Palmer, began operating at Sheerness in the Thames estuary (Palmer, 1831).

Figure 2:1 shows the essential components of such a well system, as used by the United States National Ocean Service. The well is attached to a vertical structure which extends well below the lowest level to be measured. At the bottom of the well is a cone which inserts into a support pipe; this pipe is fixed in the sea bottom. Six pairs of openings in the support pipe give little resistance to the inflow of water, but the narrow cone orifice restricts flow, preventing the rapid fluctuations due to waves from entering the well. However, the slower tidal changes which it is intended to measure do percolate. For most tide and wave regimes an orifice to well diameter ratio of 0.1 (an area ratio of 0.01) gives satisfactory results. Where waves are very active the area ratio may be further reduced to perhaps 0.003. Several alternative well arrangements are possible; these include a hole in the side of the well, or where necessary, a straight or syphon pipe connection (Lennon, 1971; Noye, 1974; Seelig, 1977). Copper is often used to prevent fouling of the narrow connections by marine growth. A layer of kerosene may be poured over the sea water in the well where icing is a winter problem.

The float should be as large as possible to give maximum force to overcome friction in the recording system, but should not touch the well sides. A 0.3 m

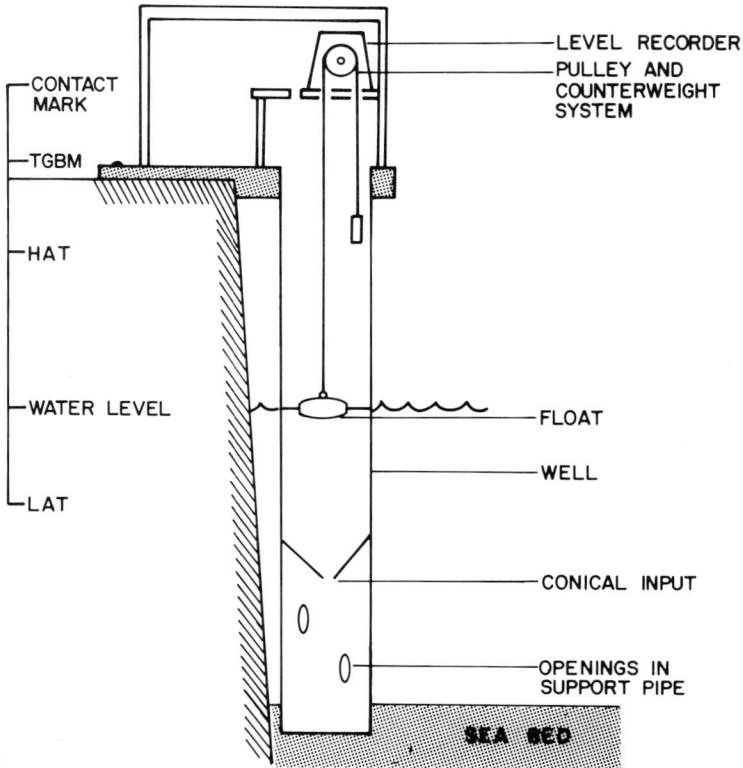

Figure 2:1 A basic stilling-well system used by the USA National Ocean Service. The cone orifice diameter is typically 10 per cent of the well diameter, except at very exposed sites. The well diameter is typically 0.3–0.5 m.

well with a 0.15 m float is a satisfactory combination. Vertical float motion is transmitted by a wire or tape to a pulley and counterweight system. Rotation of this pulley drives a series of gears which reduce the motion and drive a pen across the face of a chart mounted on a circular drum (Figure 2:2). Usually the drum rotates once in 24 hours, typically at a chart speed of 0.02 m per hour; the reduced level scale may be typically 0.03 m of chart for each metre of sea-level. Alternative forms of recording include both punched paper tape and magnetic tape.

Routine daily and weekly checks are an essential part of a measuring programme. In addition to timing checks, the recorder zero should be checked periodically by using a steel tape to measure the distance from the contact point level to the water surface. The instant the end touches the water may be detected by making the tape and the sea water part of an electrical circuit which includes a light or meter which operates on contact. For diagnosing errors in gauge operation, it is useful to plot the difference between the probed and the recorded levels against the true levels as measured by the probe. For a perfect

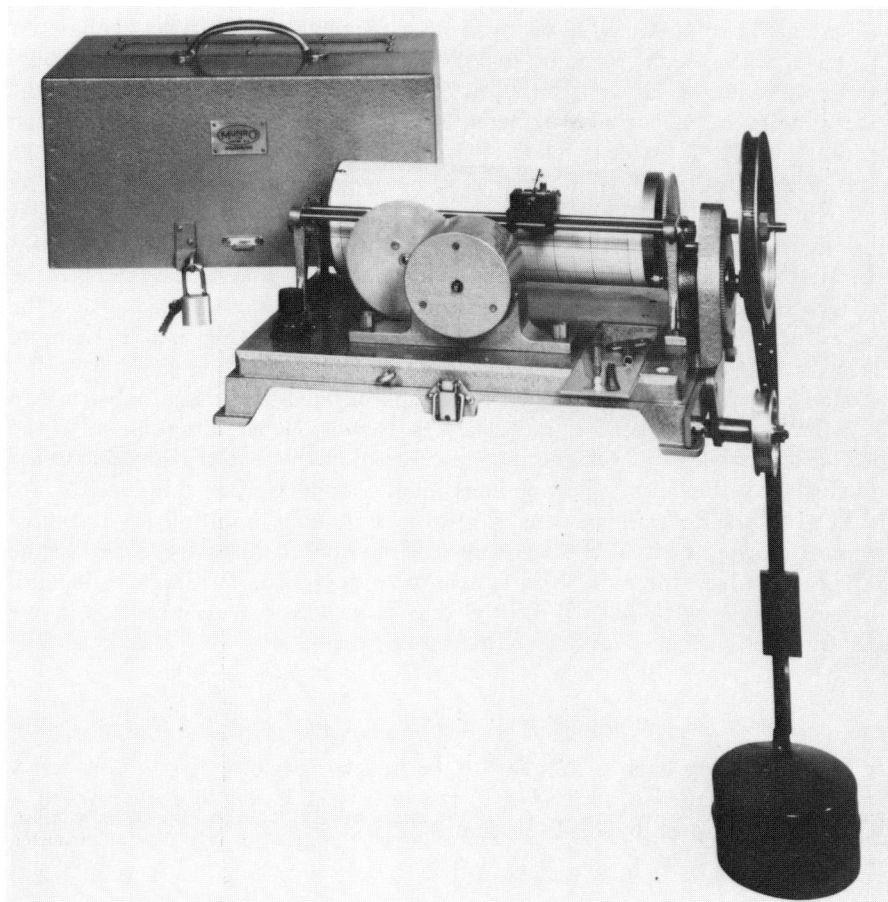

Figure 2:2   An example of the traditional stilling-well float-operated gauge. (*Supplied by R. W. Munro Ltd.*)

gauge the difference is always zero, but a wrongly calibrated gauge will show a gradual change in the difference. If the gauge mechanism is very stiff, the difference will not be the same for the rising and falling tide. In addition to regular checks of gauge zero against the contact point, the level of the contact point should be checked against a designated Tide Gauge Bench-mark (TGBM) and a series of auxiliary marks at least once a year.

Although stilling-well systems are robust and relatively simple to operate, they have a number of disadvantages: they are expensive and difficult to install, requiring a vertical structure for mounting over deep water. Accuracies are limited to about 0.02 m for levels and 2 minutes in time because of the width of the chart trace; charts can change their dimensions as the humidity changes. Reading of charts over long periods is a tedious procedure and prone to errors.

There are also fundamental problems associated with the physical behaviour of the wells. Two such errors are due to water level differences during the tidal cycle, and with draw-down due to flow of water past the well. In an estuary the water salinity and hence density increases with the rising tide. The water which enters the well at low tide is relatively fresh and light. At high tide the average water density in the well is an average of the open-sea density during the rising tide, while the well was filling. However, the water in the open sea at high tide is more dense and so the well level must be higher than the open-sea level for a pressure balance across the underwater connection. As an extreme example, the well level could be 0.12 m higher than the external level on a 10 m tide where the density increases from that of fresh water to that of normal sea water (Lennon, 1971).

Because the well, or the structure to which it is fixed, presents an obstruction to currents, it will distort the underwater pressures. Hence, the pressure at the connection between the well and the sea may not be the true pressure due to the undisturbed water level. The openings in the support pipe of Figure 2:1 are arranged to reduce this pressure distortion. A major limitation for the most accurate measurements is the behaviour of the flow through the orifice in the presence of waves; there can be a systematic difference in the levels inside and outside the well, particularly if the flow due to the waves is less restricted in one direction than another, as with a cone-shaped orifice.

2:2:3 Pressure measuring systems

An alternative approach is to measure the pressure at some fixed point below the sea surface and to convert this pressure to a level by using the basic hydrostatic relationship (see also equation (3:21)):

$$P = P_A + \rho g D \tag{2:1}$$

where P is the measured pressure, at the transducer depth, P_A is the atmospheric pressure acting on the water surface, ρ is the mean density of the overlying column of sea water, g is the gravitational acceleration, and D is the water level above the transducer.

The measured pressure increases as the water level increases. This relationship was first used to measure water depths by Robert Hooke and Edmund Halley in the seventeenth century. At the coast the underwater pressure may be transmitted to a shore-based recorder through a narrow tube. In simple systems the pressure is sensed by connecting the seaward end to a partially inflated bag or an air-filled drum open at the base, in which pressures adjust as water enters and leaves during the tidal cycle.

Better overall performance and datum stability are obtained with the pneumatic system shown in Figure 2:3 (Pugh, 1972); gauges based on this principle are called bubbler gauges or gas-purging pressure gauges. Compressed air or nitrogen gas from a cylinder is reduced in pressure through one or two valves so that there is a small steady flow down a connecting tube to escape through an orifice in an underwater canister, called a pressure-point or bubbler orifice chamber. At this underwater outlet, for low rates of gas escape, the gas pressure is equal to the water pressure P. This is also the pressure transmitted along the tube to the measuring and recording system, apart from a small correction for pressure gradients in the connecting tube.

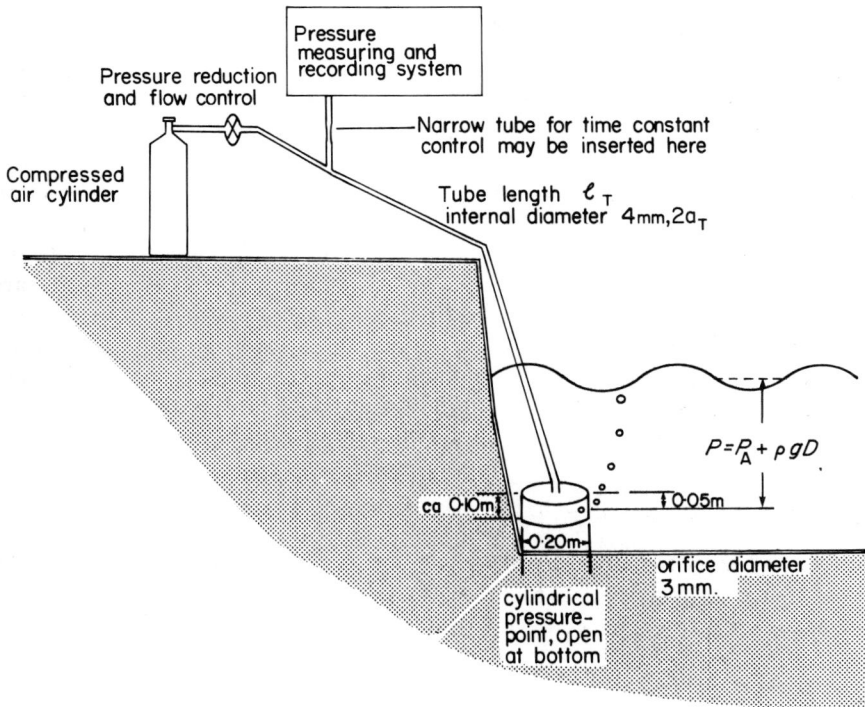

Figure 2:3   A basic pneumatic bubbling system for tube lengths less than 200 m.

Several forms of recorder have been used, including temperature-compensated pressure bellows which move a pen over a strip chart, a mercury manometer with a float arrangement as for a stilling-well, and a variety of electronic transducers which permit a digital signal to be recorded on magnetic tape. Figure 2:4 shows a gas control and pressure recording system. The normal procedure is to measure the pressure using a differential transducer which responds to the difference between the system pressure and the atmospheric pressure, so that only the water head pressure is recorded. If g and ρ are known, the water level relative to the pressure-point orifice datum may be calculated. For most sites a suitable constant value of ρ may be fixed by observation. However, in estuaries the density may change significantly during a tidal cycle. In this case a linear increase of density with level is usually an adequate adjustment.

Figure 2:4 The flow control and pressure recording equipment of a basic bubbling system. The gas cylinder may supply either nitrogen or compressed air. Ordinary SCUBA cylinders are very suitable.

The underwater pressure-point is designed to prevent waves forcing water into the connecting tube. If this happened large errors would result. The critical parameter is the ratio between the total volume of air in the pressure-point and connecting tube, and the area of the pressure-point cross-section: this ratio should not exceed 0.20 m; the actual shape of the pressure-point cross-section is not important. For tube lengths up to 200 m the pneumatic system shown in Figure 2:3 will be accurate to within 0.01 m water head equivalent. The rate at which air is forced through the tube should be sufficiently large to prevent water backing-up into the pressure-point and tube system during the most rapid rate of sea-level rise, but too rapid a rate will result in the air bottle having to be replaced more often than necessary. Another constraint is that the pressure drop in the connecting tube, which must be kept small, is proportional to the flow rate. Adequate flow of gas is maintained if the rate in millilitres per minute is set slightly in excess of:

$$\left(\frac{\text{system volume in ml}}{600}\right) \times (\text{maximum rate of water level increase in m h}^{-1})$$

For the system shown in Figure 2:3 and a tidal range of 5 m, a flow of 10 ml per minute will suffice. The error due to pressure drop in the tube depends on both the tube length and the rate of flow which in turn must increase as greater tube length increases the system volume. The error increases in proportion to $(l_T/a_T)^2$, where l_T is the tube length and a_T is the tube radius. For this reason special care is necessary when designing for connecting tube lengths in excess of 200 m.

The advantages of a pneumatic bubbler system include the stability of a clearly defined datum, the cheap expendable nature of the vulnerable under-water parts, and the possibility of having the underwater pressure point some hundreds of metres from the recorder. Systems with tube lengths of 400 m have been designed for use in special cases; the range enables measurements at virtually all coastal sites provided that the tubing is protected as it passes through the zone of breaking waves. On beaches protective outer tubing weighted with scrap chain is very effective for burying the air tube. The pressure point may be fixed to an auger rod twisted into the sand. Levelling to the pressure-point datum requires special equipment or the co-operation of a diver to hold the levelling stave. The major disadvantage is that faulty operation, for example due to a damaged or partly flooded tube, is difficult for an unskilled operator to detect.

Some gauges dispense with the pressure-point and connecting tube, by having the pressure transducer mounted in the water, with an electric cable connected to a recorder ashore. The disadvantages of this system are the difficulty of precise datum definition and the cost of replacing relatively expensive underwater components.

2:2:4 Reflection-time gauges

The time taken by a pulse of sound to travel from the source to a reflecting surface and back again is a measure of the distance from the source of the reflector. If a sound source is mounted above a stilling well the travel time may be used to replace the float and wire assembly. This travel time is given by:

$$t_p = \frac{2l_z}{C_a}$$

where l_z is the distance to be measured and C_a is the velocity of sound in air. For dry air (at 10 °C and one atmosphere pressure) $C_a = 337.5 \text{ m s}^{-1}$ so that a change in l_z of 0.01 m corresponds to a change in t_p of 0.000 059 s. Corrections must be made for the variations of C_a with air temperature, pressure and humidity. For example, if the temperature falls to 0 °C, C_a decreases to 331.5 m s^{-1} and it is necessary to reduce the calibration factor of the gauge by

1.8 per cent to compensate—one way of making these adjustments is to record continually the travel time of a sound pulse over a known standard distance. This method is difficult to apply without a stilling-well because reflections are directed away from the receiver if they are incident on an irregular or sloping sea surface. The United States National Ocean Service is gradually introducing a new sea-level monitoring system based on this type of acoustic gauge.

An alternative method is to place the transmitter and receiver on the sea-bed where they operate like an inverted ship's echo sounder. The velocity of sound in sea water (10 °C and 35 ppt) is 1490 m s^{-1}. To achieve an accuracy of 0.01 m the timing must be accurate to within 0.000 013 s. At 0 °C the calibration factor must be reduced by 2.8 per cent; in fresh water a reduction of 3.2 per cent is necessary.

The principle of timing a reflected signal may also be applied using a pulse of electromagnetic radiation. This travels at 299.792 $\times$ 10^6 m s^{-1} in air (at 10 °C and a pressure of one atmosphere) so that timing to an accuracy of 0.6 $\times$ 10^{-10} s is necessary to resolve a level difference of 0.01 m. Careful design is necessary to avoid reflections from surrounding structures and in one design a pair of copper tubes is used as a waveguide. Lasers have also been used in some combined wave and tide gauges. Sea water attenuates electromagnetic signals too rapidly for the inverted echosounder technique to be used. These systems all have the disadvantage of needing a vertical structure on which to mount the equipment.

2:2:5 Open-sea pressure gauges

Measurements of sea-level variations along the coast define conditions in only a limited area of sea. There are several applications for which the changes in levels at sites well away from the coast, even in the deep ocean, are required. These applications include hydrographic surveys where depths must be adjusted for the changing sea-level, navigation of ships or oil rigs through shallow channels, the design of oil rigs, the operation of flood warning systems and the scientific study of the hydrodynamic behaviour of oceans and shelf seas. Where an offshore structure is not available, the only stable reference point is the sea-bed. Instruments have been developed to descend to the bottom, where they can remain for up to a year, recording the pressure of the overlying water column. This pressure, which is related to water level by the hydrostatic equation (2:1), includes atmospheric pressure. Operation of these gauges requires a high level of technical skill for deployment and recovery of the equipment and to ensure reliability and precise calibration (Cartwright *et al.*, 1980).

Figure 2:5 shows a special gauge developed by the United Kingdom Institute of Oceanographic Sciences for making pressure measurements to depths greater than 4000 m. The whole assembly is contained within a protective framework below which a heavy tripod ballast frame is attached by a special release hook. Several glass spheres, each mounted within a pair of hemispheri-

Figure 2:5   A deep-sea pressure gauge developed by the UK Institute of Oceanographic Sciences ready for deployment. The instrument has a resolution of 0.01 m to depths in excess of 4000 m. Less elaborate systems are available commercially for general shallow-water applications.

cal caps, are attached to give buoyancy to the unit. The equipment is lowered into the sea by crane from a research vessel, and released so that it falls slowly to the sea-bed. Here it remains for the period of measurement, at the end of which the research ship returns to the site guided by accurate satellite navigation. The gauge is relocated by means of coded acoustic signals and instructed to fire a small pyrotechnic device which separates the main equipment from the ballast frame. The gauge rises slowly to the sea surface; in 4000 m of water this may take more than an hour. There is often keen competition for the first sighting! Additional methods of location include acoustic tracking and an automatic flashing light mounted on top of the main frame.

Less elaborate gauges are available commercially for measurements on the continental shelf. In shallow water it is often better to have the gauge attached to a ground line, which may be marked and recovered by a surface float, so that it can be recovered without acoustic commands. If the surface float is lost, the ground line can still be recovered by trawling a hook along the sea-bed. A further refinement for hydrographic survey work in shallow water is to transmit the bottom pressure to a surface buoy, either acoustically or by conducting cable; from this buoy the signal is radioed to the survey ship, so that corrections for sea surface level changes can be made as the survey proceeds.

The accuracy of these gauges is limited by the pressure sensing transducer itself. The ideal transducer would be sensitive only to pressure changes, but in practice temperature changes also affect their signal, and so separate corrections for this are also necessary. A second important requirement is to minimize the transducer zero drift over the time of deployment. On the continental shelf, in depths of less than 200 m this drift can be reduced to a few centimetres a month, which is comparable with the drift due to gradual settling of the ballast frame into the sediment. In depths of 4000 m the drifts are greater, and more care is necessary when making corrections for temperature effects. This is one problem which the oceans themselves help to minimize, because temperature changes at these depths are usually less than 0.1 °C over very long periods.

There are several different designs for pressure transducers: the most popular types have either a strain gauge sensor or a quartz crystal sensor. If the transducer's electronic circuit is designed to give an oscillating output whose frequency changes with pressure, then the average pressure over a sampling period may be determined by counting the oscillations. This is a very useful way of averaging out wave pressure variations in very shallow water; in deep water the sea does its own averaging as these waves are not felt at the sea-bed.

2:2:6 Satellite altimetry

Measurement of sea levels away from the coast using bottom pressure gauges is slow and expensive. At least two cruises are required, one for deployment and one for recovery. An alternative method is to measure the distance of the sea surface below an orbiting satellite, in the same way that a ship's echo sounder measures water depth, or the shore-based reflection-time gauges measure sea-level changes. A satellite orbits the earth many times in a single day, covering a large area very quickly, and generating enormous quantities of data. The complexities of interpreting data from these satellites requires oceanographers to pay particular attention to the accurate determination of satellite orbits and the exact shape of the mean sea-level surface (Robinson, 1985). These factors are also relevant to discussions of tidal forces and the geoid (Chapter 3) and global mean sea-level (Chapter 9), (Marsh *et al.*, 1986; Woodworth and Cartwright, 1986).

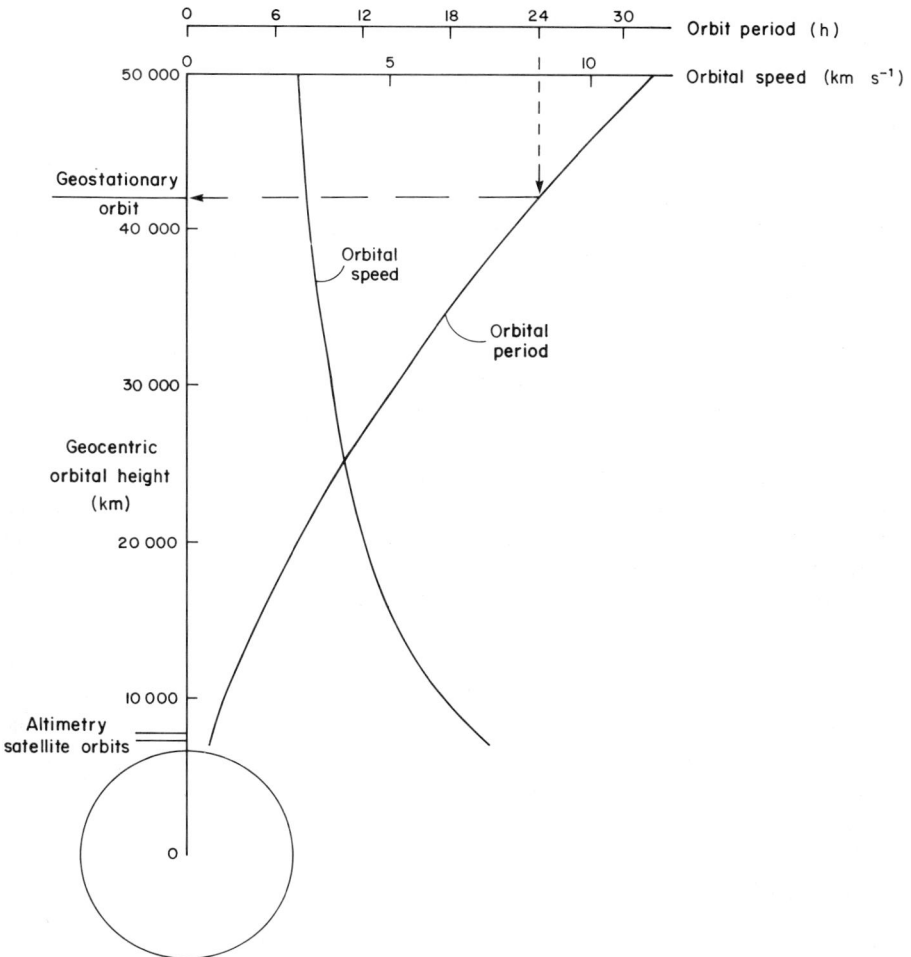

Figure 2:6 The orbital periods and orbital speeds for circular satellite orbits at different geocentric elevations, assuming a spherical earth.

If the earth were a perfect sphere with its mass uniformly distributed, then a satellite would orbit in an ellipse with the centre of the earth as one focus: for altimetry a nearly circular orbit is best. The acceleration necessary to maintain the satellite in its orbit is provided by the force of the earth's gravitational attraction. At lower orbits the greater gravitational attraction balances a larger orbital speed, whereas at higher orbits lower orbital speeds and longer periods of rotation maintain a dynamic equilibrium. For each orbital radius there is a definite speed and orbital period. Figure 2:6 shows corresponding speeds and orbital periods for circular orbits around a spherical earth of uniform mass distribution. The only orbit which would allow the satellite to measure altitudes above the same fixed point of the ocean is the orbit in the plane of the equator, at an elevation of 42 200 km from the centre of the earth. However, this elevation is too high and gives too limited a coverage. For altimetry, the best orbits are at elevations between 800 km and 1300 km above the earth's surface, at geocentric elevations from 7170 to 7670 km. At these altitudes, where each satellite orbit takes about 100 minutes, the atmospheric drag on the satellite is acceptably low, while the power required for transmission and reception of detectable pulses is not too high. Small changes in orbit height and inclination allow fine tuning of the ground coverage so that it repeats at regular intervals; any differences in the levels measured by repeated orbits must be due to sea surface variability. There are also more subtle constraints imposed by the need to avoid orbits which cross the same area of ocean at periods close to those found in the tides. For example, if this happened at an exact multiple of 24 hours the solar tide would appear frozen as a permanent distortion of the mean sea-level surface.

As with the ground-based electromagnetic pulsed systems, the return of a single pulse would have to be timed to better than 10^{-10} s for the satellite to sea surface distance to be accurate to 0.01 m. The first effective altimeter was flown on Skylab S-193, launched in 1973. This had a range resolution of better than 1 m. Improved techniques for comparing the pulses and increasing the power enabled the GEOS 3 satellite, launched in 1975, to achieve a range resolution of better than 0.50 m and a ground resolution of 3.6 km. SEASAT, which was launched in 1978, had a range resolution of better than 0.1 m and a ground resolution of 2.5 km over a calm sea. These range resolutions are based on the average of several separately timed pulses. Adjustments of the travel times are made for the atmospheric and ionospheric conditions through which the pulse travels. If the sea surface is smooth the return pulse has a sharply defined leading edge, but as the sea becomes more disturbed the leading edge is extended. The midpoint of this leading edge is taken as the return time. Corrections are made for the bias introduced because more energy is reflected from the wave troughs than from the crests. Additionally, wave heights can be estimated from the slope of the leading edge of the return echo. As a further refinement it is also possible to relate the reflected pulse strength to the surface wind speed. The continuous monitoring of the sea levels achieved by SEASAT over a 70-day period was terminated by an electrical failure, but GEOS-3 lasted for several years.

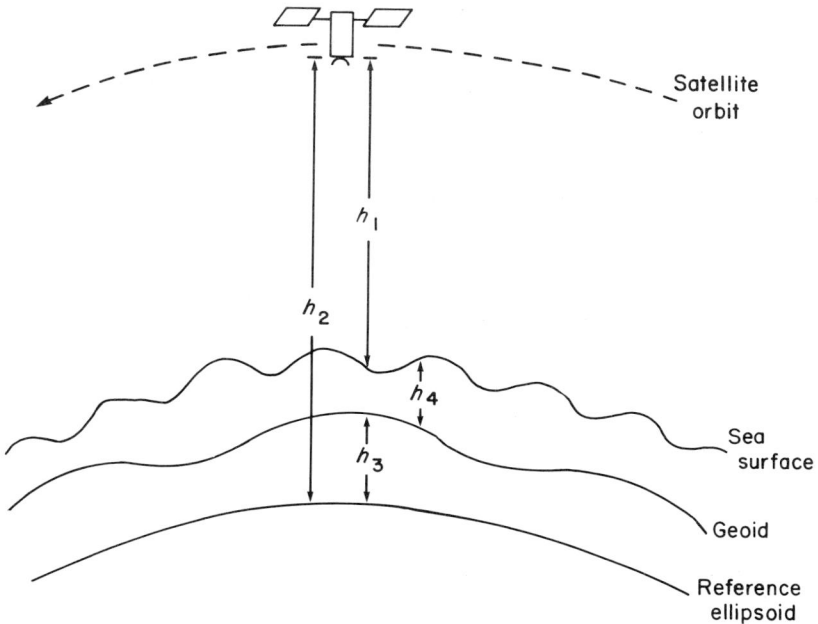

Figure 2:7 Schematic diagram of the parameters which must be known to determine the sea-level relative to the geoid by satellite altimetry.

Although the altimeter measures the distance from its orbit position to the sea surface, for oceanographic work the sea-level required is measured relative to the geoid. The geoid is the 'level' surface of equal gravitational potential which the sea surface would assume if there were no disturbing forces: in fact it has undulations of more than 100 m from a smooth ellipsoidal representation of the earth's surface (see Section 3:5) due to an uneven distribution of density in the upper mantle and crust of the earth. Figure 2:7 shows the parameters involved in interpreting a measured level h_1, of a satellite altimeter above the sea surface. The elevation of the satellite itself above the reference ellipsoid, h_2, must be known accurately, and geoid elevation h_3 must also be known in order to determine the required sea level h_4:

$$h_4 = h_2 - h_1 - h_3$$

Determination of both h_2 and h_3 present problems.

The simple circular orbits whose characteristics are shown in Figure 2:6 cannot be achieved in practice because several forces act to disturb the dynamics of the satellite. The earth is not a sphere of radial mass distribution; there is a flattening at the poles, the polar radius being 43 km less than the equatorial radius. For the relatively low orbits used for altimetry this has the effect of slowly rotating (precessing) the plane of the satellite orbit in absolute space. Satellites with heights around 1000 km in near polar orbits have orbital planes which precess at about $1°$ per day. At these levels the variations in gravity due to smaller scale near-surface density anomalies are also significant, but not sufficiently well known.

The drag of the atmosphere on the satellite and the pressure due to solar radiation are two other forces which cannot be precisely modelled. Atmospheric drag lowers the orbit radius by an amount estimated at tens of centimetres in each orbit, depending on the design of the satellite, with an uncertainty of perhaps 20 per cent. Other smaller orbit distortions are introduced by changes of satellite mass after manoeuvres, by the gravitational attractions of the moon and sun, and even by the changes of gravity due to the water mass movements of the ocean tides.

Because of these uncertainties the satellite positions and velocities must be tracked at intervals relative to fixed ground stations. Two systems which may be used for this tracking are ranging by laser, which requires clear skies and is accurate to within 0.05 m, and radio ranging systems which can measure through clouds and which may eventually be even more accurate. Errors in the geocentric coordinates of these ground stations must also be considered, and allowances made for solid-earth tidal movements. The position of the satellite between the known fixed positions must be calculated from estimates of the orbit dynamics. An ideal distribution of tracking stations would have an even global distribution, but in practice there is a concentration of stations in the northern hemisphere. Over short tracks the errors in satellite positions are small compared with oceanographic level changes but over thousands of kilometres relative errors of 1.0 m or more are present in the calculations of the satellite elevation h_2. However, future satellite altimeter systems which measure h_1 and h_2 to better than 0.05 m are technically feasible.

The problem of removing the elevation of the geoid undulations remains. These undulations have been defined by careful analysis of the way in which they cause the plane of satellite orbits to rotate in space over long periods. The accuracies of these determinations are typically 2 m for wavelengths greater than 2000 km. However, the variations of sea-level h_4 are also of this order. Because satellite orbits are only sensitive to the long wavelength changes, at shorter wavelengths other methods must be used to determine the geoid. Direct measurements of surface gravity can be used. The accuracy of ship-mounted gravimeters allows a definition of 0.1 m but the rate of data collection is too slow to be generally applicable. In the future it may be possible to determine the local geoid deformations by measuring the distances between a pair of low-orbit satellites. At present other ways of avoiding the need for an exact knowledge of h_3 must be considered.

On the time scale of dynamic ocean processes the geoid remains constant so that changes of the sea-level h_4 with time can be monitored by repeated orbits over the same track. SEASAT orbits were designed to repeat at 3-day intervals. Another way of improving the coverage is to look at points where north-going and south-going tracks cross. These techniques can only work if the only uncertainty is in the geoid elevation h_3; errors in the satellite elevations must be low compared with the sea-level changes h_4 which are to be monitored.

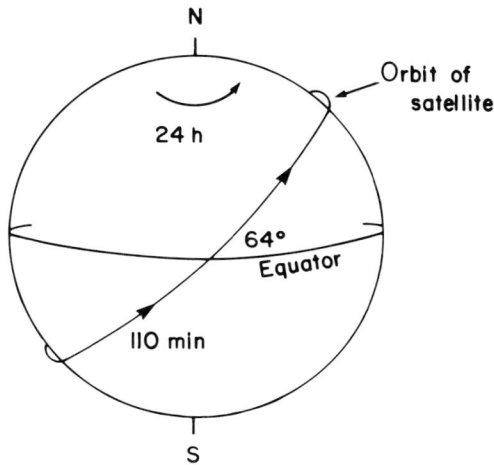

Figure 2:8   The relationship between the earth's rotation and the proposed TOPEX satellite orbit inclined at 64° to the equator.

A major proposal for the next generation of satellite altimeters has been prepared by the TOPEX group of the United States National Aeronautics and Space Administration (TOPEX, 1981); launch as the joint USA/French TOPEX/Poseidon mission is planned for 1991. It is intended that the global variability of sea-level will be studied from time scales of 20 days to 5 years on spatial scales from 30 km to the width of the ocean basins. NASA emphasizes the need for rapid processing of the vast quantities of data which would be generated. A network of global sea-level stations will enable ground-truth checks on these variations (Section 11:7) as well as serving many other purposes. The choice of the proposed orbit parameters is a compromise between frequently repeated orbits and detailed spatial resolution. Figure 2:8 shows how the 24-hour rotation of the earth under the 100-minute satellite orbit enables a coverage from 64° S to 64° N, with the pattern shown in Figure 2:9. The exact phase relationship between the 24-hour rotation and an integral number of satellite orbits is controlled by a fine adjustment of the orbit height and period to allow orbits to repeat over the same earth swath every 10 days. There will be no coverage at higher latitudes than 64° which means that the polar regions are not monitored. However, an orbit at a higher inclination would give track crossings at very acute angles, which would make comparisons less favourable. The TOPEX satellite has been designed to minimize atmospheric drag. The European Space Agency plans to launch a remote sensing satellite ERS-1, also early in the 1990s, which will carry a number of sensors, including an altimeter having a resolution of better than 0.10 m. Initially ERS-1 will have a 3-day repeat cycle of ground coverage.

Satellite altimetry is a new and very important development. For defining the sea surface off-shore it promises major progress over the next few decades.

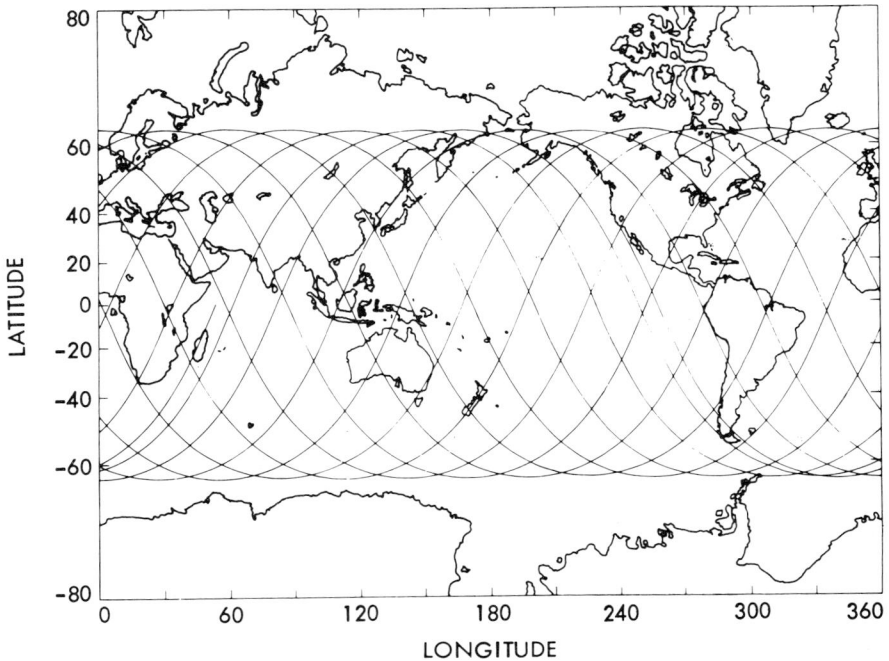

Figure 2:9   The ground track traced out by a satellite at an altitude of 1300 km with an orbital inclination of 64° during a one-day period of an exactly repeating ten-day cycle. (*Courtesy of the Jet Propulsion Laboratory, California Institute of Technology, Pasadena, California.*)

However, the more traditional techniques will still be necessary for ground-truth comparisons. More fundamentally, long-term trends in sea-level can only be determined by monitoring datum stability to better than 0.01 m over decades, a demand which present satellite technology cannot satisfy.

2:3   Current measurements

The flow of water may be represented mathematically as a velocity vector defining the direction and speed of movement. Strictly the movement in the vertical and in two horizontal directions should be monitored, but the long-period vertical movements which are normally two orders of magnitude less than the horizontal movements are not usually considered. In this brief review we describe only those techniques which are useful for measuring currents which vary over tidal and longer periods. These currents are due to the combined action of several forces, including tides and winds. Winds act only at the surface so their effects at depth are delayed, whereas tidal pressure forces act through the whole water depth. Near the bottom currents are slowed by the drag of the sea-bed. The result of these different forces is to produce a variation of current velocity with depth, called the *current profile*. For many purposes the average or depth mean current is adequate: the current measured by a single meter at mid-depth in the water column is often taken as the depth mean current. Experiments have shown that in shallower water the current at a level 40 per cent of the total depth above the sea-bed gives a better estimate of the depth mean current.

Current speeds were traditionally measured in nautical miles per hour (knots), but metres per second is now routine for science and engineering. Mariners usually prefer knots because they relate easily to ship speeds. For tidal currents an accuracy of 0.03 m s^{-1} is adequate but to monitor slow water drift a more accurate measurement than the 2.6 km per day which this represents is necessary. It is useful to remember that since 1 knot is 0.514 m s^{-1}, halving the speed in knots gives a good approximation to the speed in metres per second.

Directions can be given in terms of the points of the compass or more usually in degrees measured clockwise from geographic north. The convention is to define the direction towards which the current is flowing; this is the opposite to the convention which defines the wind direction as that from which it blows. The wind convention was established on land where the local weather is strongly influenced by that of the region from which the wind blows, whereas the currents convention was established by navigators who were concerned with where the drift was taking their ship.

There are two different ways to describe the water movements (Von Arx, 1962; Neumann, 1968; Dobson *et al.*, 1980). The *Eulerian* method measures the speed and direction at a geographically fixed point, and represents the whole flow pattern in a region by a grid of current vectors. Alternatively the movement of a particular parcel of water can be tracked as it moves from place to place, giving *Lagrangian* flow. The two types of flow can be related mathematically. Each has useful applications: for example, studies of sewage or radioactive waste dispersion would normally require the Lagrangian flow, whereas an engineer designing an offshore platform or pipeline would prefer Eulerian currents. Most scientific analyses are best applied to Eulerian flows, and these are also the easier to measure continuously over long periods.

One of the problems in making these measurements is the same as for measuring sea levels: how are the more rapid fluctuations to be averaged in an unbiased way so that the longer-period changes are monitored without distortion? Designing current meters for deployment in the sea, where absolute positions and directions are not easily fixed, and where instruments must resist buffeting and corrosion without themselves distorting the flow they try to measure, is more difficult than designing sea-level recorders. Although some long-term records of currents have been made from moored light vessels, few records which extend for more than a year have been collected because of these difficulties and because of the cost involved in the deployment and recovery of the equipment.

2:3:1 Drifters and dyes

Most of the current speeds and directions on nautical charts have been determined by observing the drift of a special log, weighted to float vertically, from an anchored ship. These measurements are representative of the mean motion to the depth of the log, which may be as much as 10 m, and as such give the likely drift of a ship which has an equivalent draught. Readings of the log movement, which should be taken at least once an hour, give the Eulerian flow at the fixed ship's position. Traditionally the log speed was determined by counting the rate at which knots in the paying-out line were counted; thus the name for the units of flow. If the log is allowed to float freely over several hours and its position tracked either from a following ship or from the shore, then the Lagrangian current is measured.

Currents at different depths may be measured by releasing a series of drogues which have a parachute or sail designed to remain in suspension at known distances below a surface marker buoy. The surface buoy should have a low profile to avoid wind drag, but should be easily visible, or capable of tracking on a radar screen or by satellite navigation systems. The main problems of using float tracers are the difficulty of constant tracking, keeping them in the area of interest which may require lifting and relocation, and reducing the buoy motion due to waves and due to wind drag. The latter may be 1 per cent of the wind drag even for a well-designed buoy. The dispersion or spread of the water, which is an important parameter for estimating the rate at which pollutants are diluted may be found by starting a group of drifters in a small source area and monitoring their subsequent scattering. An effective but very simple way of investigating Lagrangian near-surface movements is to release several floating cards in the same place at a known time. People who find these cards on the shore read a message which asks that they be returned to the originating laboratory in return for some small reward. Of course there is no way of knowing how long the card spent travelling to the shore and how long it spent there before being found, so that only a minimum current speed can be estimated. Movements of water and material near the bottom can be similarly tracked by releasing plastic Woodhead bottom drifters which are designed to float just above the sea-bed, with a weighted tail which drags along the bed to prevent them floating up to higher levels. These too may be tracked acoustically for short periods.

In the middle layers of the deep ocean acoustic neutrally-buoyant Swallow floats can give a very good idea of the water motions. Neutrally-buoyant floats located in the SOFAR acoustic transmitting layer of the ocean, where sound can travel without vertical spreading and be detected over thousands of kilometres, can transmit their movements to a network of acoustic receiving stations for a year or more, and over very long distances.

For studies of dispersion within more confined areas such as estuaries and near shore, various dyes or radioactive tracers may be released and their concentrations monitored. Rhodamine and fluorescein give good water mass identification without affecting water properties or the movements to be measured, because only small quantities are required. These dyes have been followed through dilution over many kilometres, but larger scale experiments are impractical because of the large volumes of tracer release necessary to allow detection. Certain perfluorinated chemicals, including perfluorodecalin and sulphur hexafluoride have been developed as ultra-sensitive marine tracers; in one experiment a release of 200 g of the former in the English Channel was traced over tens of kilometres (Watson *et al.*, 1987). Very sensitive techniques for detecting the radioactive isotopes released from nuclear fuel reprocessing plants have enabled seasonal and long-term drift patterns to be ascertained. For example, the radio-caesium isotopes released at Sellafield into the Irish Sea have been traced around the coast of Scotland and into the North Sea (McKinley *et al.*, 1981).

Many modern research ships have accurate equipment for measuring their movement through water. They can also determine their true position to within a few metres by satellite fixes. The difference between the absolute movement and the movement of the ship through the water is the water current. These modern techniques apply an identical principle to that used by the American pioneer Matthew Maury (1806–1863). Maury produced charts of ocean currents from the differences between the positions which navigators estimated from their ship's direction and speed, and the true ship position determined by astronomical fixes.

2:3:2 Rotor current meters

Many meters have been designed to measure current speed by counting the rate of rotation of a propeller or similar device suspended in the flow. The current direction relative to magnetic north is indicated by a magnet within the instrument: a large fin forces the meter to align with the direction of the flow. The simplest method of operating such instruments is to connect them by electrical cable to a moored ship and to display the readings in the laboratory. By raising or lowering the meter and taking readings at several depths it is possible to determine the variations of current with depth, the *current profile*. The tedium and expense of continuing these measurements limits the length of data which can be collected by a moored ship to only a few tidal cycles, insufficient for a good tidal current analysis. Moored rotor meters can record at pre-selected intervals for periods up to a year. Measurements taken over a lunar month enable reasonable tidal current analysis to be undertaken. Figure 2:10(a) shows an example of one of the many commercially available instruments, and Figure 2:11 shows a typical mooring rig configuration used for deployment.

The rotor may be mounted to spin about either a vertical or a horizontal axis. The meter illustrated has a Savonius rotor which is mounted to spin about a vertical axis. This has the advantage of avoiding a directional sensitivity in the rotor, but it means that the speeds of currents from all directions are added indiscriminately. Near the surface in the presence of rapidly oscillating currents due to waves, the fin is unable to reverse and so the forward and backward currents are both added into the rotor motion. If a propeller-type rotor with a horizontal axis is used, forward and backward flows cause the rotor to turn in different directions so that oscillating flows are correctly averaged. The meter must be correctly balanced in the water which, because of the different buoyancy, is not the same as being correctly balanced in the air: final balancing adjustments must be made in sea-water. Correct balancing also allows the magnetic compass to operate freely on its pivot. All meter parts including batteries must be non-magnetic to avoid small distortions of the earth's magnetic field. Similar distortions can also affect measurements of current directions from ships or near to steel offshore platforms. The difference between magnetic north and geographic north is allowed for in the data processing (see Section 2:4).

Meters must be calibrated for both speed and direction, if possible both before and after deployment. Speed calibrations are made in a towing tank or a flume. They are more stable than direction calibrations, which are made by rotating the meter at a site where the direction of magnetic north is known.

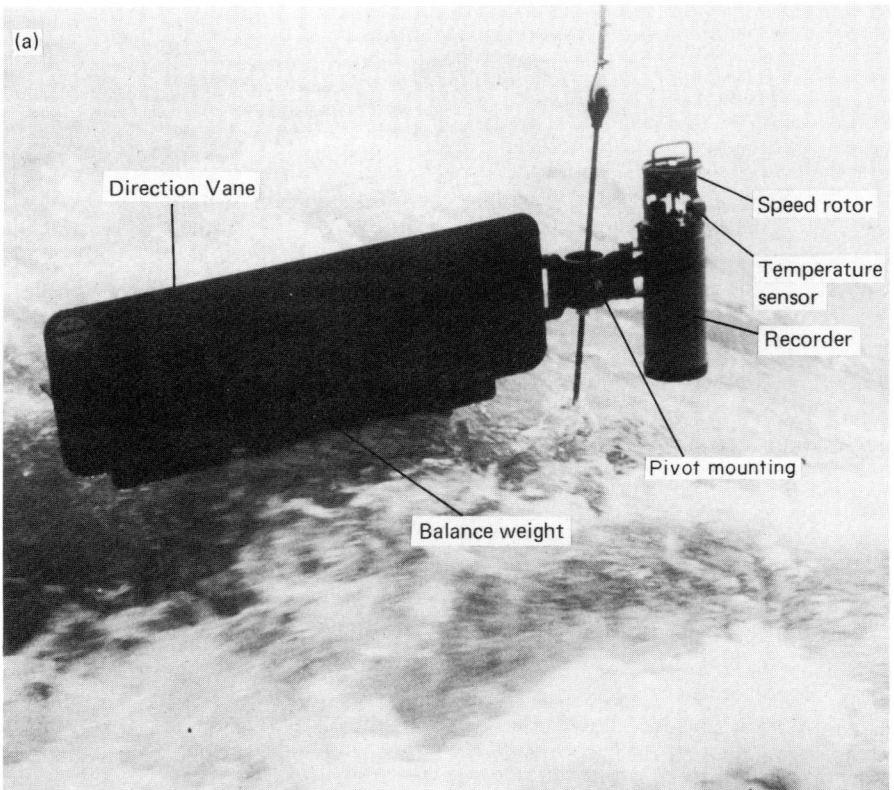

(a)

Direction Vane

Speed rotor

Temperature sensor

Recorder

Pivot mounting

Balance weight

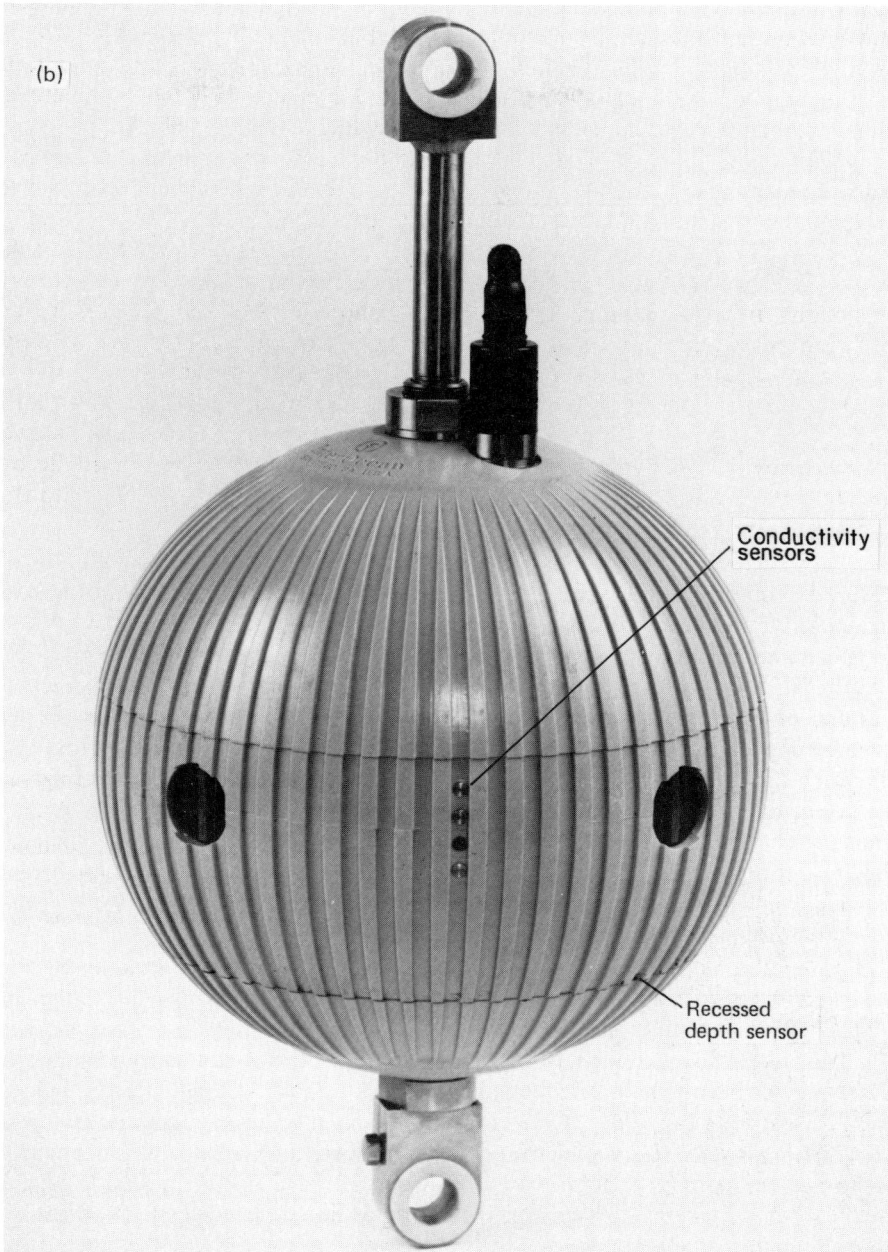

Figure 2:10   (a) A current metre fitted with a Savonius rotor. The position of the weights below the direction fin is adjusted so that the meter is balanced about its pivot point on the tensioned wire when submerged in water. (*Supplied by W. S. Ocean Systems.*); (b) an electromagnetic current meter: the S4 meter has titanium electrodes and a solid-state memory. (*Supplied by InterOcean Systems Inc.*).

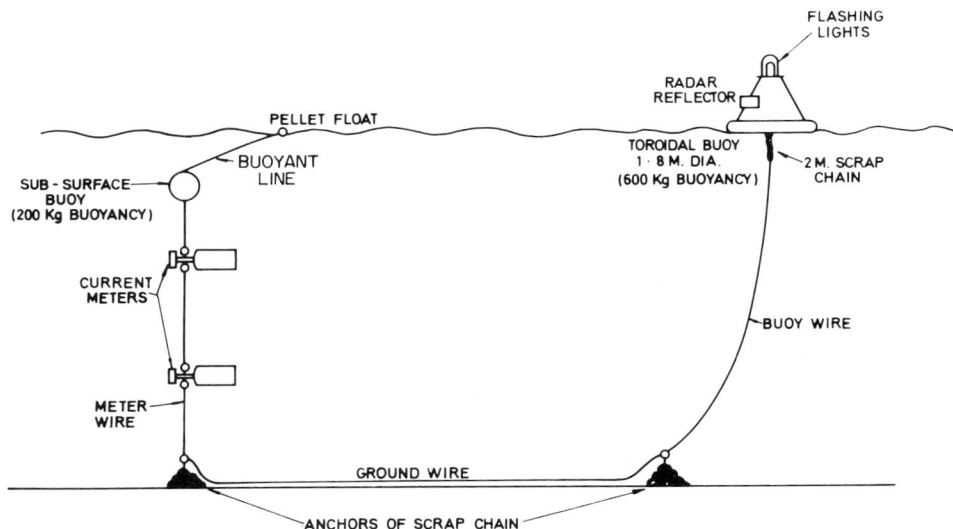

Figure 2:11   Diagram of a U-shaped current meter mooring system.

Overall calibration and system accuracy of 0.02 m in speed and 1° in direction are possible with careful use of standard instruments.

Meters must be properly suspended in the water column at the required depth. One method is shown in Figure 2:11. The U-shaped mooring with two clumps of chain as anchors is suitable for water depths less than 100 m. At greater depths a pop-up acoustic release system may be fitted above the first anchor. The sub-surface buoy below which the meters hang should be at sufficient depth for the wave motions to be negligible. In practice this means that meters cannot be deployed in this way within about 15 m of the surface. Because the meters and the wires are constantly in motion and liable to stress corrosion in the sea water, careful choice of materials is essential if the lifetime of the deployment is to extend for a month or even a year.

The length of time for the deployment and the recording capacity of the instrument control the choice of sampling interval. The instrument shown in Figure 2:10(a) is capable of measuring every ten minutes for 70 days; the total rotor counts over each 10-minute period, and the instantaneous magnetic direction are recorded on magnetic tape. For long-period deployments, values may be recorded each hour. Quartz crystal clocks give timing accurate to a few seconds a month. Many meters include additional sensors to measure water temperature, salinity and pressure. Pressure can be a useful indicator of the meter's behaviour on the mooring wire, a rapid increase in pressure indicating that the meter is being depressed to deeper levels by the drag of the current on the mooring.

Alternative meters have been developed which have two horizontal propeller speed-sensors set at right angles. The magnetic directions of the two axes are measured by a flux-gate compass, and the flow along these axes is resolved into north–south and east–west components. Because the meters are not sensitive to their exact alignment they are more reliable, especially in the near-surface zone where wave motions are present. They also avoid the problems introduced where speeds are averaged over several minutes but directions are taken only as single instantaneous values at the end of each sample (Weller and Davis, 1980).

2:3:3 Acoustic travel-time meters

With a rotor current meter there is a possibility of the rotor becoming fouled by stray material such as weed or fishing line. An alternative is to measure the travel time in both directions of an acoustic pulse between a pair of trans-ducers. The travel times will generally be unequal because of a component of current from one sensor towards the other. One neat application of this principle, the 'sing-around' current meter, converts the travel times for the pulses into frequencies by using the arrived pulse to trigger the next pulse in the train: the difference in the two frequencies is directly proportional to the current. If two pairs of sensors are mounted with their transmission paths at right angles, the two components of current measured define the complete horizontal current vector. A third pair of transducers enables the vertical currents to be included. The basic problem with these acoustic measurements is the very small time differences which must be resolved. For a transducer pair separation of 0.10 m and a speed resolution of 0.01 m s^{-1} the timing must be accurate to within 10^{-9} s. Very careful design is necessary for avoiding drift in the electronic sensors, and for exactly defining the instant of pulse arrival. Because these meters respond very rapidly to changes of current they may be used to measure rapid turbulent changes of flow.

2:3:4 Electromagnetic methods

Sea water is an efficient conductor of electricity because it contains dissolved salts. When an electrical conductor moves through a magnetic field, a voltage gradient is produced across the direction of the flow. This voltage difference, measured by a pair of electrodes, may be related to the rate of movement of the sea-water conductor. Electromagnetic current sensors fall into two categories, those which use a locally generated magnetic field and those which use the earth's magnetic field. The voltage differences to be sensed are small compared with the electrochemical voltages generated at the electrodes: to combat this, electrode noise may be reduced by having silver electrodes or by reversing the magnetic field and hence the electrical polarity. A commercial example of a current meter which uses this principle is shown in Figure 2:10(b). The voltage is sensed by two pairs of titanium electrodes located symmetrically on each side of the spherical housing: the sphere is both the instrument housing and the sensor. There is no mechanical motion to interfere with the flow, and the small and light construction creates less drag and requires less buoyancy in the mooring system. The instrument illustrated has a solid-state memory which enables data to be recorded and retrieved without opening the instrument. Voltages across straits have been monitored using telephone cables to give the average flow between the two shores. The conversion factor from voltage difference across the strait to sea current through the strait depends on the electrical resistance of the underlying rocks, so calibration against measurements by some direct means is necessary. Flows through the Dover Straits between France and England give calibration factors of 1.0 V $(m\ s^{-1})^{-1}$ for the residual flows but values 30 per cent lower for tidal flows.

2:3:5 Remote sensing of currents

The difference of sea-level across a strait may also be related to current speed through the strait. Steady flows on a rotating earth actually follow a curved path in absolute non-terrestrial coordinates; a pressure gradient perpendicular to the direction of flow is necessary to give the force which changes the momentum of the water. The dynamic relationship for steady flow in the absence of any other forces (see Section 3:8 and equation (3:23) for a fuller discussion) is:

$$\Delta\zeta = \Delta x \frac{2\omega_s \sin\varphi}{g} v \tag{2:2}$$

where $\Delta\zeta$ is the level difference between two stations, Δx is their horizontal separation, ω_s is the angular rate of rotation of the earth, φ is the latitude, g is gravitational acceleration and v is the horizontal current speed at right angles to the line joining the two stations.

When equation (2:2) is satisfied the gradients and currents are said to be in *geostrophic balance*. For flow through a channel 40 km wide, at $60°$ N, a change of 0.04 m s^{-1} in v results in an increase in level of 0.02 m to the right of the direction of flow. If the absolute difference in levels is not known, then only changes in the current v can be monitored. In ocean dynamics this method has been used for many years to compute geostrophic flows relative to an assumed level of no motion. Tidal flows are not in geostrophic balance because they are not steady for a period which is long compared with the period of rotation of the earth so level differences cannot be interpreted in terms of tidal currents.

Another remote sensing technique makes use of the Doppler reflection principle whereby a signal returning from a reflector which is moving relative to the observer is slightly shifted in frequency from that of the original signal. Small particles or bubbles in moving sea-water give a weak reflected signal. If the transmitted acoustic signal has a frequency ω, the change of frequency in the returning signal is:

$$\Delta\omega = \frac{2\omega v}{C_s} \tag{2:3}$$

where v is the speed component of the reflection towards the source and C_s is the speed of sound in the transmitting medium. Two orthogonal acoustic beams allow both speed and direction to be computed.

Acoustic Doppler meters may be mounted on a ship's hull or on the sea-bed. The advantage of transmitting at a high frequency to measure a larger value of $\Delta\omega$, is offset by the reduction in operating range because the high-frequency signals are rapidly attenuated. Typical ranges are 300 m at 75 kHz and 30 m at 1 MHz. By averaging over several pulses accuracies of 0.05 m s^{-1} are possible.

A different Doppler technique uses the returns of radar signals to a shore-based transmitter/receiver by reflection from the surface waves (Shearman, 1986). Even in very calm conditions there are always enough waves of the correct wavelength, half the wavelength of the radio transmission, travelling both towards and away from the shore station to give Bragg resonant back-scattering. The transmitted signal is returned with both positive and negative Doppler shifts due to the separate reflections from the waves travelling towards and away from the receiver. These two waves do not usually have equal speeds because the water on which they travel also has a component velocity towards or away from the transmitter which must be added to one speed and subtracted from the other. If the basic wave speed is C_ω, then the resultant speeds are $(v + C_\omega)$ and $(v - C_\omega)$. There are two Doppler-frequency shifts from (2:3)

$$\Delta\omega_1 = \frac{2\omega(v + C_\omega)}{C_e} \qquad \Delta\omega_2 = \frac{2\omega(v - C_\omega)}{C_e}$$

where C_e is now the speed of electromagnetic waves. In the spectrum of the returning signal, the mean frequency between these two Doppler shifted lines is displaced from the transmission frequency by:

$$\frac{2\omega v}{C_e} \tag{2:4}$$

This displacement is a direct measure of the water speed v, which may be determined, independently of the wave speed C_ω. The United States National Oceanic and Atmospheric Administration has developed a portable system, the Coastal Ocean Dynamics Application Radar (CODAR) (Barrick et al., 1977; Lipa and Barrick, 1986) which uses a pair of transmitter/receivers to give two components of current speed. Operating at 25.6 MHz, this equipment can map 800 vectors in 2.4 × 2.4 km squares. Such a system is well suited to producing synoptic maps of surface currents in estuaries or within 70 km of the shore. However, the resolution is not usually more than 0.10 m s^{-1}: there is a design trade-off between the resolution of current speed differences, the spatial resolution and the time over which the currents are averaged in these systems. The United Kingdom Rutherford–Appleton Laboratory has developed an Ocean Surface Current Radar (OSCR) based on the same Doppler principle (Prandle, 1987). The major design difference between the OSCR and CODAR systems is in the antenna. OSCR uses a 90 m beam-forming receive antenna of 16 equally spaced elements. Signals from 16 beams each 6° wide can be identified, and within each beam separate 'bin' lengths of 1.2 km can be resolved. Figure 2:12(a) shows the beam coverage for an experiment using a dual OSCR system in the Irish Sea. Measurements over 30 days were analysed to give current ellipses, as shown in Figure 2:12(b) (see Section 4:4:2).

52

The wide range of techniques available for measuring currents may confuse anyone planning an occasional experiment. Although each has advantages and disadvantages the most commonly used system consists of a moored recording meter or series of meters fitted with rotors. Savonius-type rotors should be avoided within the surface layer influenced by waves. High-quality direct measurements in the near-surface layer remain a challenge for the design engineer, but such measurements are important for pollution dispersion studies, for studies of momentum transfer through the sea surface from winds, and for evaluating the forces due to combined wave and current effects on offshore structures.

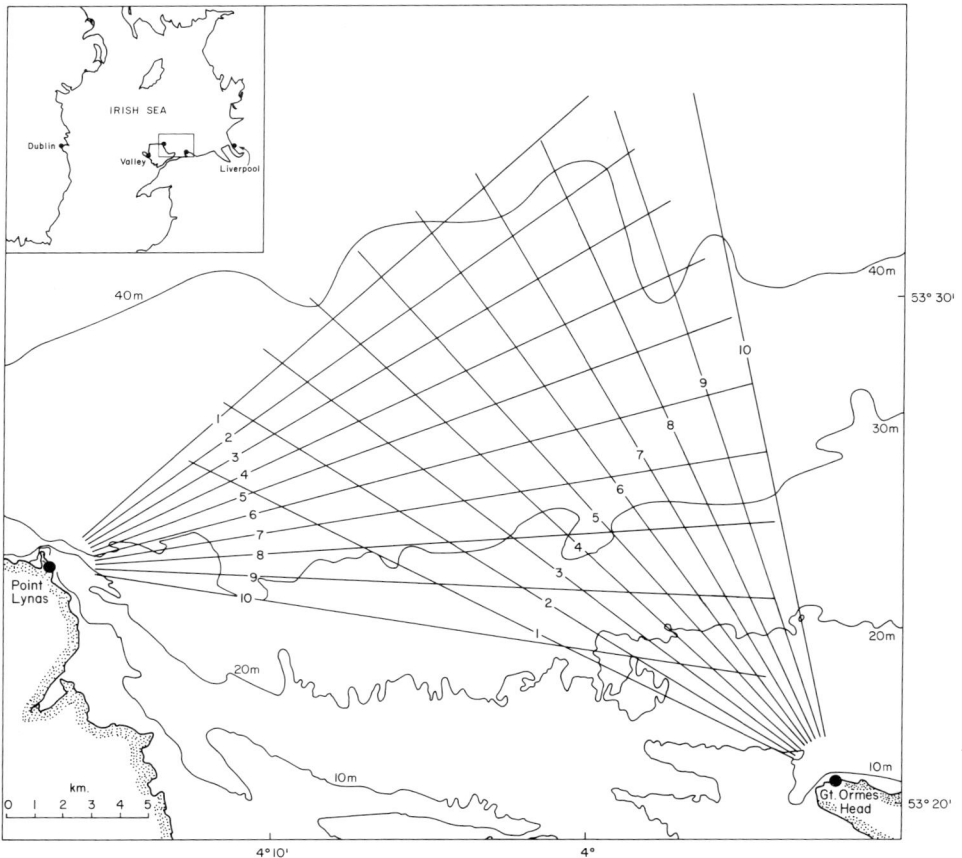

(a)

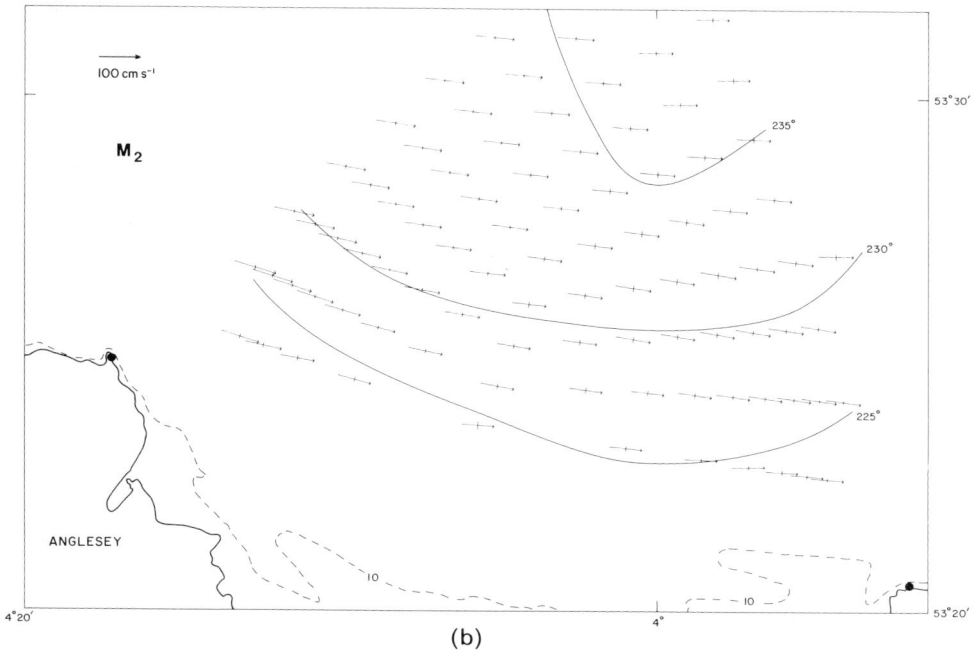

Figure 2:12 (a) The spatial coverage achieved by a deployment of the Ocean Surface Current Radar (OSCR) in the Irish Sea. Radars were located at Point Lynas and Great Ormes Head and (b) the **M₂** current ellipse parameters determined by analysis of 30 days of data. Arrows show direction of flood flow; solid contours show phase of maximum flood flow. (Prandle, 1987)

2:4 Data reduction

Before the raw data recorded by the instrumental system is ready for scientific analysis it must go through a process of checking and preparation known as data reduction (UNESCO, 1985b). The recorded data may be in the form of a chart recording such as that shown in Figure 2:13, a trace on photographic film, a punched paper tape, or a digital signal on magnetic tape or in a solid-state memory. This data must be read either by eye or by a mechanical translator, checked for recording errors, and adjusted for calibration factors and for control clocks running fast or slow. Gaps are interpolated if necessary before filtering to obtain values at standard times. Usually, but not invariably, oceanographic changes with a strong tidal component are presented for analysis in the form of hourly values, on the hour, either in local time or in Greenwich Mean Time. When the data has been processed, it is highly desirable that it is preserved with proper documentation in a permanent computer-based data bank.

54

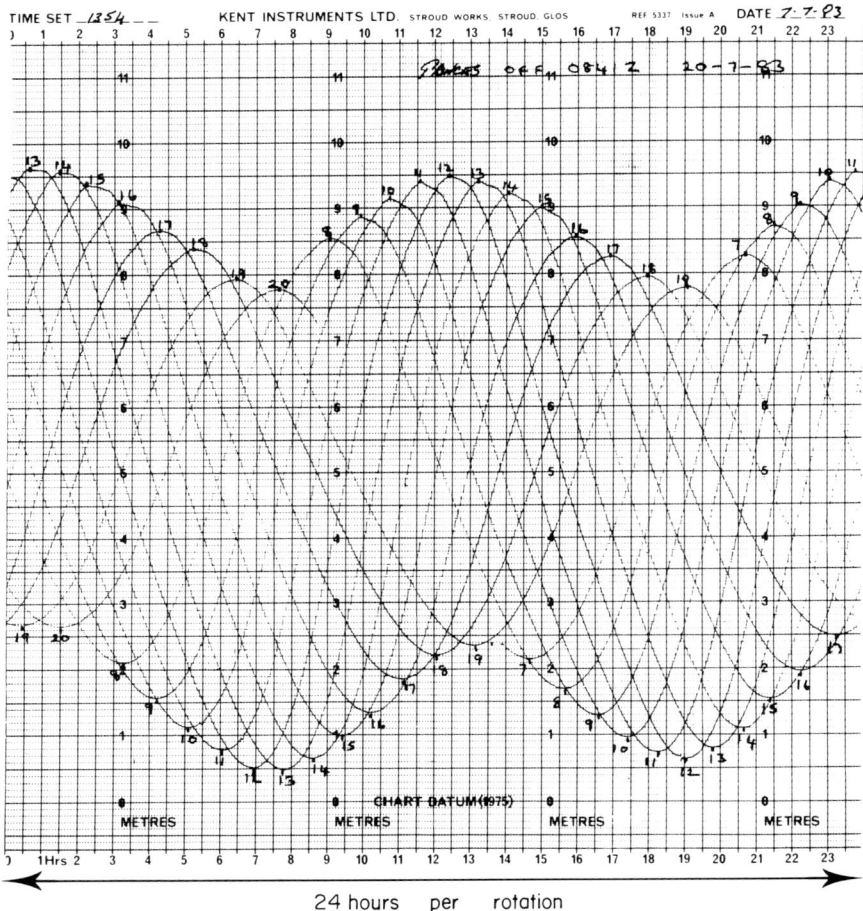

Figure 2:13 A typical 14-day chart record from a float gauge operating in the Mersey Estuary, Liverpool. The highest water levels always occur near noon and midnight at this site.

Most oceanographic equipment now uses quartz crystal timing control, accurate to a few seconds a month. Earlier equipment and some chart tide gauges of older design use a clockwork drive which is only accurate to a few minutes a month, and which may have a temperature-sensitive rate. If a start time and an end time are defined for a record, then the best approximation is to assume a steady time gain or loss through the period and to adjust intermediate times by linear interpolation. When reading values from a chart such as that shown in Figure 2:13, a timing error may be corrected at the reading stage by reading the trace to the right or left of the hour lines. If the gauge was running fast on removal the trace must be read to the right of the hour line; if running slow the reading is to the left. Charts can be digitized rapidly on a digitizing table. Where daily check sheets are completed as a monitor of the gauge performance, it is not usually advisable to use these for day by day adjustments of zero levels and times, unless the operators are known to be particularly reliable. Neither is it good practice to encourage operators to reset the gauge zero and timing on a daily basis. The best results are obtained where the instrument is checked daily for the absence of gross errors such as a dry pen or a stopped chart drive, but is otherwise untouched from the start to the end of a record.

Digitizing at hourly intervals is a long established compromise between the excessive labour of more frequent sampling and the poor curve representation which less frequent sampling would afford. However, if there are fluctuations due to local seiches (see Section 6:7) on the record with periods less than 2 hours, it is necessary either to sample more frequently or to smooth through these by eye during the reading process.

A fundamental rule of spectral analysis states that if the record contains oscillations of a particular period T then the sampling interval must be not greater than $T/2$ to resolve them in the spectrum. If the sampling is slower, the energy at the period is said to be aliased, producing distorted values of the spectral energy in the longer period lower frequency oscillations. On a chart, smoothing by eye through seiches is an acceptable alternative to very rapid sampling; digital instruments can be designed to average a pressure or rotor count over the time between samples, which removes the aliasing problem. In this case the correct time to be allocated to the sample is the central time of the averaging period.

Current meter data should be converted to speeds and directions, if possible after recalibrating the instruments to check the stability of the calibration constants. If a meter has counted rotor turns over a 10-minute period, the calculated speed applies to a time 5 minutes before the recording time. The instantaneous 10-minute readings of current direction at the start and end of each 10 minutes may be averaged to give a direction for this central time. The recorded values which will be relative to magnetic north must be adjusted to geographic north by adding the magnetic variation. Speeds and directions should be checked at this stage for unexpected features such as spikes, but for further processing and analysis it is usual to convert to current components in the north–south and east–west directions, with the flow to the north and to the east being defined as positive. A current speed in a direction measured clockwise from geographic north resolves as illustrated below.

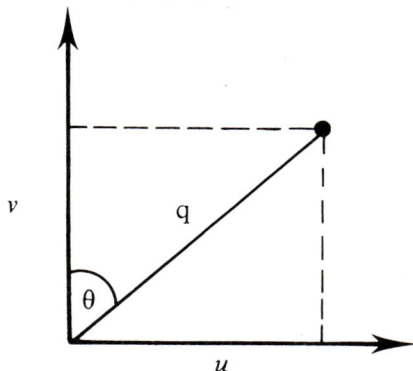

an eastward component $u = q \sin \theta$

a northward component $v = q \cos \theta$

The process of checking both levels and current components for recording errors depends on some preconceptions about the parameter being recorded. Clearly an isolated sea-level 10 m higher than anything previously observed is an error, but smaller less obvious errors should be considered with care and only rejected if there is independent evidence of malfunction. Otherwise there is a danger of disregarding real but unexpected events. If the original observations are not on charts, it is useful to plot them at this stage, ideally through an interactive computer terminal. The first check is to identify values which lie outside the expected range, or which appear as isolated spikes in the data. Computer files may be checked against values on either side by ensuring that the difference between a value at one hour does not differ by more than a specified tolerance τ from a function fitted through the surrounding values. For example, a 7-point Lagrangian fit requires each X_t value to satisfy the condition:

$$| - 0.0049\, X_{t-7} + 0.0410\, X_{t-5} - 0.1709\, X_{t-3} + 0.6836\, X_{t-1}$$
$$+ 0.5127\, X_{t+1} - 0.0684\, X_{t+3} + 0.0068\, X_{t+5} - X_t | < \tau \qquad (2:5)$$

Where tides are dominant a simpler method is to check against values at similar states of the tide on previous and subsequent days:

$$|\tfrac{1}{6}(- X_{t-50} + 4 X_{t-25} + 4 X_{t+25} - X_{t+50}) - X_t| < \tau \qquad (2:6)$$

Any value which fails this test should be marked for further scrutiny. However, because any isolated wrong value will also be used to fit smooth curves to check other values, it may also automatically cause these to be incorrectly flagged as errors.

A more sensitive check for records dominated by tides is to remove the tidal part by subtracting predicted tidal values (see Chapter 4) from observed values and to plot these residual values against time. Some common errors in tide gauge operation have well-defined characteristics in these plots. Figure 2:14 shows a residual plot with a timing error (1), a datum shift (2), two successive days' curves on the chart interchanged (3), and an isolated error of 1 m (4). When a chart is left on a 24-hour rotating drum as in Figure 2:13, for many days, interchange due to incorrect identification of the correct day is possible.

Gauges sampling at discrete intervals over a long period sometimes record a few samples more or less than the number expected. The discrepancy is much greater than errors due to gain or loss of the quartz timing. The whole record may be useless if the time of the missing or extra scans cannot be identified by closer inspection of the records. One way of recovering the data is to make a detailed day-by-day comparison of tidal predictions with the observations. The tidal predictions may be for the site concerned or for a nearby site. Diurnal and semidiurnal harmonics are fitted using the least-squares techniques described in Chapter 4, to 25 hourly values from midnight to midnight for each day (the D_1 and D_2 parameters discussed in Chapter 7). The time differences between the observed and predicted harmonics for each day are plotted to identify sudden jumps. In the case of a semidiurnal regime the variation of time differences for a

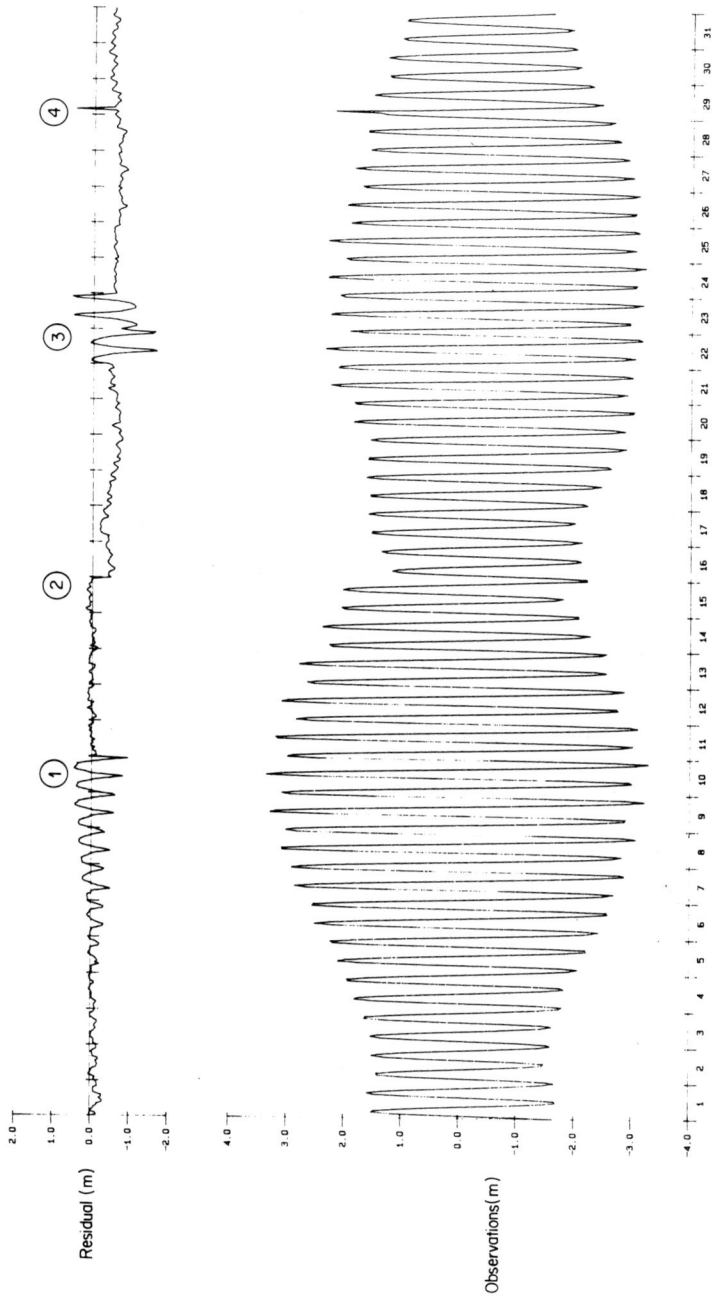

Figure 2:14 An example of non-tidal residuals plotted to show characteristic errors due to: (1) a steady timing gain of 20 minutes over a week, (2) a datum shift of 0.5 m, (3) wrong identification of daily traces on the chart, and (4) an isolated error of 1 m.

correct record would be less than 5 minutes. Sudden jumps at the sampling interval of 15 minutes are clearly identified and artificial scans may be added or extra scans may be removed from that part of the record. To overcome this difficulty some instruments have an independent time channel.

A separate problem relates to the filling of gaps in records. Gaps may extend from a few hours to several days, usually due to instrument failure. There will also be small gaps between successive records which need to be filled to obtain a complete series. Gaps of a few hours may be interpolated by eye. Larger gaps can be filled by fitting a function through the observed values on either side, if the gap is not too large. The 25-hour function of equation (2:6) can be used for gaps up to a day. Gaps should only be interpolated if there is a special need such as the application of a numerical filter with a broad window, when too much data would be lost by treating the original observations as two separate series. A more elaborate interpolation method is to add predicted tidal observations to an interpolation of the observed low-frequency meteorological changes.

To convert digital values at 5-minute, 10-minute, 15-minute and 30-minute intervals to hourly values, the formulae in Appendix 1 may be used. Timing adjustments to give values exactly on the hour can be made using a convenient curve fitting formula. This function is then recalculated at the corrected time to give the on-hour value. Other more or less elaborate functions are also available.

The final stage of data reduction is to deposit the data in an oceanographic data bank, together with details of the methods used for measurement, the steps taken in the processing, and clear indications of the data which has been interpolated or which is unreliable in some way. Although this may seem a tedious additional requirement, the high cost of making oceanographic measurements means that data should be available for use in as many investigations as possible.

CHAPTER 3

Forces

for to recognise causes, it seemed to him, is to think, and through thoughts alone
feelings become knowledge and are not lost, but become real and begin to mature.

Hermann Hesse, *Siddhartha*.

The essential elements of a physical understanding of tide and surge dynamics
are contained in Newton's laws of motion and in the principle of conservation
of mass. Newton's first law asserts that a body, which for our purposes can be
an element of sea water, continues at a uniform speed in a straight line unless
acted upon by a force. The second law relates the rate of change of motion or
momentum, to the magnitude of the imposed force:

$$\text{Acceleration} = \frac{\text{force}}{\text{mass}}$$

with the acceleration taking place in the direction of the force. For this law to
be valid, motion, for example in a straight line, must be observed in an external
system of space coordinates, not on a rotating earth. It therefore follows that
motion which appears to be in a straight line to an observer on a rotating earth
but which follows a curved path when observed from space can only be
produced by an additional terrestrial force, which may not be immediately
apparent.

In the first chapter we distinguished between two sets of forces, and the
different movements of the seas which they produce. The tidal forces, due to the
gravitational attraction of the moon and sun, are the most regular and most
precisely defined in the whole field of geophysics. Like the tidal motions which
they produce, they are coherent on a global scale. In contrast, the meteorological forces due to atmospheric pressures and to winds are no more predictable
than the weather itself and their influence during any particular storm is limited
to a more local area.

The first part of this chapter develops the basic equations for tidal forces.
The second part shows how all the forces acting on the sea can be incorporated
into hydrodynamic equations derived from Newton's second law; together with
the principle of conservation of mass, these provide a physical and mathematical basis for the analyses in the chapters which follow.

Readers who do not require a more elaborate mathematical development may omit Section 3:2 which deals with potential theory. They should read Section 3:1 which outlines the basic principles of tidal forces, Section 3:4 which discusses the astronomical reasons for observed tidal patterns, and perhaps scan Section 3:3 on the more detailed movements of the moon–earth–sun system. General descriptions are also given in Doodson and Warburg (1941) and Forrester (1983). Doodson and Warburg also give a more mathematical treatment, and other more advanced treatments are found in Doodson (1921), Lamb (1932), Schureman (1976), Godin (1972) and Franco (1981).

GRAVITATIONAL TIDAL FORCES

3:1 Gravitational attraction

Newton's law of gravitation states that any particle of mass m_1 in the universe attracts another particle of mass m_2 with a force which depends on the product of the two masses and the inverse of the square of their distance apart, r:

$$\text{Force} = G \frac{m_1 m_2}{r^2} \tag{3:1}$$

where G is the universal gravitational constant whose value depends only on the chosen units of mass, length and force. Although this law does not hold strictly for very accurate work where relativity effects become important, it is more than adequate for all tidal computations. G has units $M^{-1} L^3 T^{-2}$, and a numerical value of $6.67 \cdot 10^{-11} \, \text{N m}^2 \, \text{kg}^{-2}$ in MKS units. The total gravitational attraction between two large masses such as the earth and the moon is the vector sum of forces between innumerable pairs of particles which constitute the two bodies. Fortunately the total forces can be calculated by assuming that for each body the total mass is concentrated at a single point which for a sphere is at its centre. The net force of attraction between the earth and moon becomes simply:

$$\text{Force} = G \frac{m_e m_1}{R_1^2}$$

where R_1 is the distance between the two centres and m_e and m_1 are the total masses of the earth and moon.

Table 3:1 summarizes the several physical constants of the moon–earth–sun system. The scales and distances are hard to appreciate. In relative terms, if the moon is represented by a table-tennis ball, the earth may be represented by a sphere with the same radius as a table-tennis bat 4 m distant, and the sun may be represented by a sphere of 15 m diameter some 2 km away.

Consider for the moment only the moon–earth system. The two bodies will revolve about their common centre of mass, as shown for two point masses in

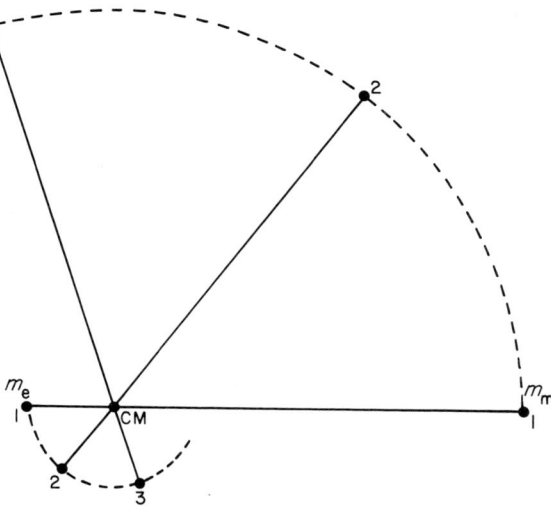

Figure 3:1 Part of the orbits of the earth and moon about their common centre of mass.

Figure 3:1, with a period which is called the *sidereal period*. The centre of mass of the moon–earth system actually lies within the earth, because the earth is 81 times more massive than the moon. However, to make things clearer, the relative scales in the diagrams which follow have been distorted. Both masses continue to move about their common centre of mass, the necessary acceleration of each body towards this centre of revolution being produced by their mutual attraction. The sidereal period for the moon–earth system is 27.32 days, defined as one sidereal month.

As the two bodies rotate in the system, each individual element of each body moves in a circle which has the same radius as the circle described by its centre of mass. The motion is illustrated in Figure 3:2, where the different parts of the letter Y travel around the centre of mass in circles which have the same radius. It is necessary to remember that in this argument we are not considering at the moment the rotations of the bodies about their own axes; the important influence of this rotation on the generation of tides will be introduced later. The fact that all particles describe circles of the same radius is not trivial, nor is it immediately obvious. An alternative illustration, recommended by Pond and Pickard (1978), is given by spreading one's hand flat on a table, and moving it so that the end of the thumb rotates in a horizontal circle, while the arm points in the same direction. All the other parts of the hand will also move around in a circle of the same radius.

As already indicated, for the earth, Figure 3:2 is slightly misleading because the revolution is shown about a point outside the earth whereas the actual motion is about an axis which passes through the sphere at a distance of 4671 km from the centre; however, the principle of equal circles of revolution for all the particles still applies.

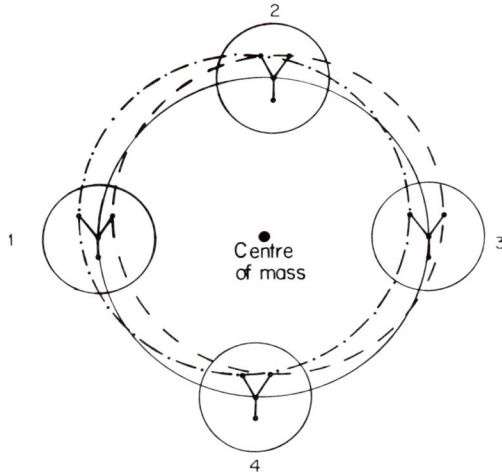

Figure 3:2 The earth circles about the centre of mass of the earth–moon system, without any rotation in absolute space. Each element of the earth travels around in circles which have the same radius.

The force necessary to give each particle of the earth the acceleration to perform this revolution is the same as for the particle at the centre of the earth: for such a particle the gravitational force provides this necessary acceleration. For those particles nearer the moon than the centre particle, the gravitational attraction is greater than that necessary to maintain the orbit. For those further away the forces are weaker. The differences between the forces necessary for the orbit, and the forces actually experienced generate the tides on the surface of the earth.

Consider a particle of mass m located at P_1 in Figure 3:3, on the earth's surface. From equation (3:1) the force towards the moon is:

$$G \frac{mm_1}{(R_1 - a)^2}$$

whereas the force necessary for its rotation is the same as for a particle at O:

$$G \frac{mm_1}{R_1^2}$$

The difference between these is the *tide producing force* at P_1:

$$Gmm_1 \left[\frac{1}{(R_1 - a)^2} - \frac{1}{R_1^2} \right]$$

$$= \frac{Gmm_1}{R_1^2} \left[\frac{1}{(1 - \frac{a}{R_1})^2} - 1 \right]$$

The term within the brackets can be expanded, making use of the approximations:

$$\left(\frac{a}{R_1}\right)^2 \ll 1 \qquad \text{since} \qquad a/R_1 \approx 1/60$$

and expanding $[1/(1-\alpha)^2] \approx 1 + 2\alpha$ for small α

to give a net force towards the moon of:

$$\text{Tidal force at } \mathbf{P}_1 = \frac{2Gmm_1a}{R_1^3} \tag{3:2}$$

Similar calculations for a particle at $\mathbf{P}_2$ show that gravitational attraction there
is too weak to supply the necessary acceleration. There is a net force away from
the moon:

$$-\frac{2Gmm_1a}{R_1^3}$$

The net force at $\mathbf{P}_3$ is directed towards the earth centre. The strength of this
force is found by making use of the approximation for $\sin(\hat{\text{OM}}\text{P}_3) = a/R_1$. The
force along $\mathbf{P}_3\text{M}$ is:

$$\frac{Gmm_1}{R_1^2}$$

and the component of this force towards O is:

$$\frac{Gmm_1a}{R_1^3} \tag{3:3}$$

The net effect is for particles at both $\mathbf{P}_1$ and $\mathbf{P}_2$ to be displaced away from the
centre of the earth, whereas particles at $\mathbf{P}_3$ are displaced towards the centre.
This results in an equilibrium shape (assuming static conditions) for a fluid
earth which is slightly elongated along the axis between the centres of the moon
and the earth.

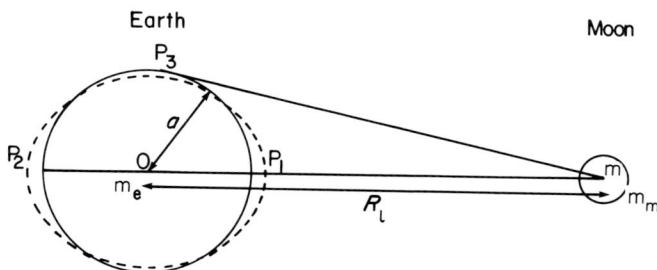

Figure 3:3 Diagram to show positions in the earth–moon system which are used to derive the
tidal forces. The separation is distorted but the relative diameters of the earth and moon are to
scale.

If we consider Figure 3:3 to show an equatorial section of the earth, and now introduce the rotation of the earth on its own axis (rotation about 0), each point on the circumference will pass through two maximum and two minimum levels for each daily rotation. This results in two tides a day, the semidiurnal tides.

The diurnal tides are generated because, except for the special case of the moon in the equatorial plane, the maxima and minima in each daily rotation are unequal in amplitude.

These simple arguments have shown that the tide producing forces depend on the finite radius of the earth a, the mass of the moon m_1, and on the inverse cube of the distance R_1.

Table 3:1 Astronomical constants.

		Symbol
The moon		
Mass	$7.35 \cdot 10^{22}$ kg	m_1
Mean radius	1738 km	
Mean distance from earth	384 400 km	
	= 60.3 earth radii	$\overline{R_2}$
The earth		
Mass	$5.97 \cdot 10^{24}$ kg	m_e
	= 81.3 lunar masses	
Equatorial radius	6378 km	a
Mean distance from sun	149 600 000 km	
	= 23 460 earth radii	$\overline{R_s}$
Mean distance from centre of earth to earth–moon mass centre	= 4671 km	
The sun		
Mass	$1.99 \cdot 10^{30}$ kg	m_s
	= 332 946 earth masses	
Radius	696 000 km	

We can simplify equation (3:2) by substituting for G, the universal gravitational constant. The gravitational force on a particle of mass m on the earth's surface is given by equation (3:1):

$$mg = \frac{Gmm_e}{a^2} \tag{3:4}$$

so the tidal force at P_1 may be written:

$$2mg\left(\frac{m_1}{m_e}\right)\left(\frac{a}{R_1}\right)^3 \tag{3:5}$$

From the values in Table 3:1, the acceleration is approximately:

$$2g\left(\frac{1}{81.3}\right)\left(\frac{1}{60.3}\right)^3 = 11.2 \times 10^{-8}\,g$$

so that the value of g is very slightly reduced at P_1 and P_2. A man weighing 100 kg would weigh 11.2 mg less as he passed through these positions!

In the same way that we have calculated the tidal forces due to the moon, the tidal forces due to the sun are calculated by replacing m_1 and R_1 by m_s and R_s in equation (3:5). The acceleration is:

$$2g\,(332\,946)\left(\frac{1}{23\,460}\right)^3 = 5.2 \times 10^{-8}\,g$$

The solar tidal forces are a factor of 0.46 weaker than the lunar tidal forces. The much greater solar mass is more than offset by its greater distance from the earth.

The other planets in the solar system produce negligible tidal forces. Venus, whose mass is $0.82\,m_e$, and whose nearest approach to the earth is approximately 6500 earth radii, will give a maximum tidal acceleration of:

$$2g\left(0.82\right)\left(\frac{1}{6500}\right)^3 = 6.0 \times 10^{-12}\,g$$

which is only 0.000 054 of the moon's tidal acceleration. Jupiter, whose mass is $318\,m_e$ has maximum accelerations which are an order of magnitude less than those of Venus because of Jupiter's greater distance from the earth. Of course tides are not only a terrestrial phenomenon, but are experienced on other celestial bodies. The role of tides in the development of the dynamics of the moon–earth system is discussed in Section 10:4.

* 3:2   The tidal forces—a fuller development

3:2:1   Potential fields

A more general development of the tidal forces than that outlined in Section 3:1 makes use of the concept of the *gravitational potential* of a body; gravitational potential is the work which must be done against the force of attraction to remove a particle of unit mass to an infinite distance from the body. The potential at P on the earth's surface in Figure 3:4 due to the moon is:

$$\Omega_p = -\frac{Gm_1}{MP} \tag{3:6}$$

Our definition of gravitational potential, involving a negative sign, is the one normally adopted in physics, but there is an alternative convention often used in geodesy, which treats the potential in equation (3:6) as positive. The advantage of the geodetic convention is that an increase in potential on the surface of the earth will result in an increase of the level of a free water surface (Garland, 1965). Potential has units of $L^2\,T^{-2}$.

Earth   Moon

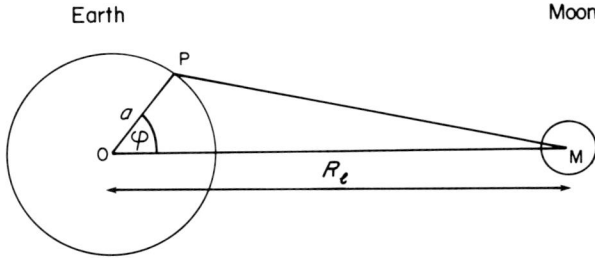

Figure 3:4   Diagram to show the location of the general point P on the earth's surface.

The advantage of working with gravitational potential is that it is a scalar property which allows simpler mathematical manipulation. In particular, the gravitational force on a particle of unit mass is given by $-$ grad Ω_p

Applying the cosine law to $\triangle$ OPM in Figure 3:4:

$$MP^2 = a^2 + R_1^2 - 2aR_1 \cos \varphi$$

we have:

$$\Omega_p = -\frac{Gm_1}{R_1} \left\{ 1 - 2\frac{a}{R_1} \cos \varphi + \frac{a^2}{R_1^2} \right\}^{-\frac{1}{2}}$$

which may be expanded as a series of Legendre polynomials:

$$\Omega_p = -\frac{Gm_1}{R_1} \left\{ 1 + \frac{a}{R_1} \mathscr{P}_1 (\cos \varphi) + \frac{a^2}{R_1^2} \mathscr{P}_2 (\cos \varphi) + \frac{a^3}{R_1^3} \mathscr{P}_3 (\cos \varphi) + \cdots \right\}$$

The terms in $\mathscr{P}_n (\cos\varphi)$ are Legendre polynomials: (3:7)

$$\mathscr{P}_1 = \cos \varphi$$
$$\mathscr{P}_2 = \tfrac{1}{2}(3 \cos^2 \varphi - 1)$$
$$\mathscr{P}_3 = \tfrac{1}{2}(5 \cos^3 \varphi - 3 \cos \varphi)$$

The tidal forces represented by the terms in this potential are calculated from their spatial gradients $-$ grad $(\mathscr{P}_n)$. The first term in equation (3:7) is constant (except for variations in R_1) and so produces no force. The second term produces a uniform force parallel to OM because differentiating with respect to $(a \cos \varphi)$ yields a gradient of potential:

$$-\frac{\partial \Omega_p}{\partial (a \cos \varphi)} = -\frac{Gm_1}{R_1^2}$$

This is the force necessary to produce the acceleration in the earth's orbit towards the centre of mass of the moon–earth system. The third term is the major tide producing term. For most purposes the fourth term may be neglected because $(a/R_1) \approx 1/60$, as may all the higher terms.

The effective tide generating potential is therefore written as:

$$\Omega_p = -\tfrac{1}{2} Gm_1 \frac{a^2}{R_1^3} (3 \cos^2 \varphi - 1)$$

(3:8)

The force on the unit mass at P corresponding to the potential may be resolved into two components:

$$\text{vertically upwards:} \quad -\frac{\partial \Omega_p}{\partial a} = 2g\Lambda_1 (\cos^2 \varphi - \tfrac{1}{3})$$

(3:9)

$$\text{horizontally in direction of increasing } \varphi: \quad -\frac{\partial \Omega_p}{a\delta\varphi} = -g\Lambda_1 \sin 2\varphi$$

where

$$\Lambda_1 = \tfrac{3}{2} \frac{m_1}{m_e} \left(\frac{a}{R_1} \right)^3$$

and equation (3:4) is used to substitute for G.

Λ_1 is very small (8.4×10^{-8}) so that compared with the forces due to the earth's gravity, these tidal forces are also very small. They vary slowly as R_1 varies. The vertical forces produce small changes in the weight of a body, as previously discussed, but it is the small horizontal forces which produce the tidal accelerations necessary to produce the water movements. Figure 3:5 shows the distribution of these horizontal *tractive* forces on the surface of the earth.

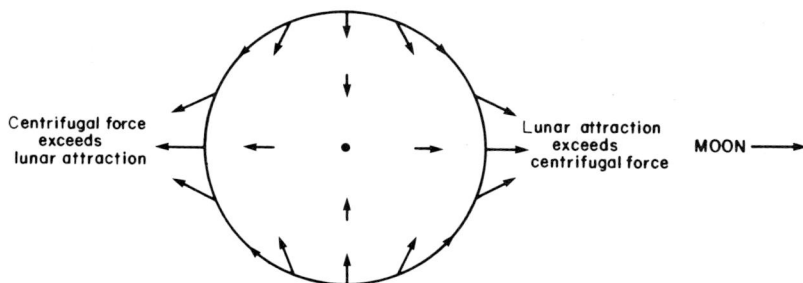

Figure 3:5 The distribution of the horizontal tidal tractive forces on the earth's surface.

The tractive forces over the earth's surface vary as the earth rotates and as the moon moves in its orbit. It is possible to regard these variations as due to small periodic rotary tilting of the horizontal plane at each point on the surface.

The lunar angle φ must be expressed in suitable astronomical variables. These are chosen to be the declination of the moon north of the equator d_1, the north latitude of P, φ_p, and the hour angle of the moon, which is the difference in longitude between the meridian of P and the meridian of the sub-lunar point V, as shown in Figure 3:6 (see also Section 3:3:4).

The angle φ is related to the other angles by a standard formula in spherical trigonometry which we will not prove here (see for example, Smart 1940):

$$\cos \varphi = \sin \varphi_p \sin d_1 + \cos \varphi_p \cos d_1 \cos C_p$$

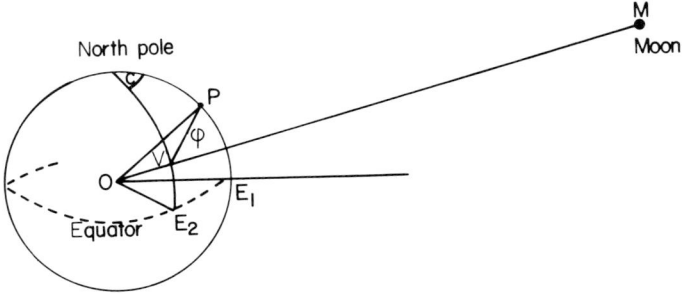

Figure 3:6 Location of P relative to the sub-lunar point V. The angle P_0V is φ.

Substituting for $\cos\varphi$ in equation (3:8), and with some further rearrangement eventually yields:

$$-\Omega_p = \tfrac{3}{2}\, ag\, \frac{m_1}{m_e}\left(\frac{a}{R_1}\right)^3 \left[\tfrac{3}{2}\,(\sin^2 d_1 - \tfrac{1}{3})(\sin^2 \varphi_p - \tfrac{1}{3})\right.$$
$$+\, \tfrac{1}{2}\sin 2\,d_1 \sin 2\,\varphi_p \cos C_p \tag{3:10}$$
$$\left. +\, \tfrac{1}{2}\cos^2 d_1 \cos^2 \varphi_p \cos 2\,C_p\right]$$

where d_1 is the lunar declination, φ is the latitude and C_p is the hour angle of P.

It is usual to compute from this formula an expression for the *Equilibrium Tide*. The Equilibrium Tide is defined as the elevation of the sea surface that would be in equilibrium with the tidal forces if the earth were covered with water to such a depth that the response is instantaneous. The Equilibrium Tide bears no spatial resemblance to the real observed ocean tide, but its development is an essential part of these discussions because it serves as an important reference system for tidal analysis.

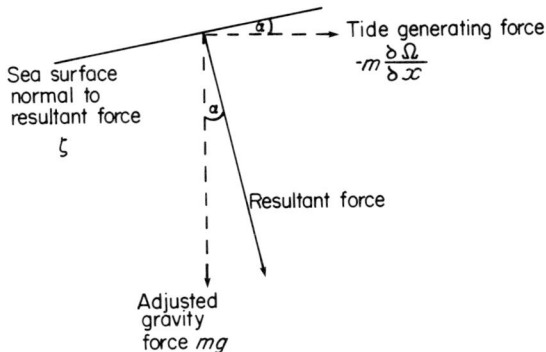

Figure 3:7 The relationship between the Equilibrium water surface, the tide generating force, and the normal earth's gravity force.

3:2:2 The Equilibrium Tide

In the equilibrium theory of tides the free surface is assumed to be a level surface under the combined action of the earth's gravity and the tidal disturbing force. Strictly, we should adjust the earth's gravity for the centrifugal effects of its rotation, and consider apparent gravity, but the difference is not important here. The tractive tide generating forces may be regarded as causing a deflection of the vertical (Proudman, 1953; Franco, 1981) as shown in Figure 3:7. The force is $-(\delta\Omega/\delta x)$ where x is a direction at right angles to the direction of undisturbed gravity.

Considering the magnitude of the forces:

$$\tan \alpha = -\left(\frac{\partial \Omega_p}{\partial x}\right)/g$$

and also:

$$\tan \alpha = \left(\frac{\partial \bar{\zeta}}{\partial x}\right)$$

so that

$$g \frac{\partial \bar{\zeta}}{\partial x} + \frac{\delta \Omega_p}{\delta x} = 0$$

or:

$$\frac{\partial}{\partial x} (g\bar{\zeta} + \Omega_p) = 0$$

and similarly:

$$\frac{\partial}{\partial y} (g\bar{\zeta} + \Omega_p) = 0$$

Integrating over a finite area gives:

$$g\bar{\zeta} + \Omega_p = \text{constant.} \qquad (3:11)$$

but if the integral is taken over the whole area of the ocean surface so that the total volume of water is conserved, then the constant is zero. Applying this condition to equation (3:10) the Equilibrium Tide becomes:

$$\bar{\zeta} = a \left(\frac{m_1}{m_e}\right) [C_0(t) (\tfrac{3}{2} \sin^2 \varphi_p - \tfrac{1}{2}) + C_1(t) \sin 2 \varphi_p + C_2(t) \cos^2 \varphi_p]$$

where the time dependent coefficients are:

$$C_0(t) = \left(\frac{a}{R_1}\right)^3 (\tfrac{3}{2} \sin^2 d_1 - \tfrac{1}{2})$$

$$C_1(t) = \left(\frac{a}{R_1}\right)^3 (\tfrac{3}{4} \sin^2 2 d_1 \cos C_p)$$

$$C_2(t) = \left(\frac{a}{R_1}\right)^3 (\tfrac{3}{4} \cos^2 d_1 \cos 2 C_p) \qquad (3:12)$$

The three coefficients characterize the three main species of tides: the long-period species, the diurnal species at a frequency of one cycle per day (cos C_p) and the semidiurnal species at two cycles per day (cos 2 C_p). The magnitudes of all three species are modulated by a common term which varies inversely as the cube of the lunar distance R_1.

The long-period tidal species is also produced because of the monthly variations in lunar declination d_1. It has maximum amplitude at the poles and zero amplitude at latitudes 35° 16', north and south of the equator.

The diurnal species is modulated at twice the frequency with which the lunar declination varies and it has a maximum amplitude when the declination is a maximum. Spatially it has maximum amplitude at 45° latitude and zero amplitude at the equator and the poles. The variations north and south of the equator are in opposite phase.

The semidiurnal species is also modulated at twice the frequency of the lunar declination, but is a maximum when the declination is zero. It has maximum amplitude at the equator and zero amplitude at the poles.

The amplitudes of the Equilibrium Tides are small. For the semidiurnal tide at the equator when the declination is zero the amplitude, calculated using the values in Table 3:1, is 0.27 m. However, the observed ocean tides are normally much larger than the Equilibrium Tide because of the dynamic response of the ocean to the tidal forces, as discussed in Chapter 5. The observed tides have their energy at the same frequencies as the Equilibrium Tide. The importance of the Equilibrium Tide lies in its use as a reference to which the observed phases and amplitudes of harmonic tidal constituents can be related, when preparing models for tidal prediction as described in Chapter 4. It also gives an indication of the important harmonic constituents to be included in a correct tidal analysis model.

The Equilibrium Tide due to the sun is expressed in a form analogous to equation (3:12) with m_1, R_1 and d_1 replaced by m_s, R_s and d_s. The resulting amplitudes are smaller by a factor of 0.46 than those of the lunar tides, but the essential details are the same.

3:2:3   The yielding earth

Although the ocean tides are the obvious terrestrial manifestation of the effects of tidal forces, sensitive instruments can also observe the effects as movements of the solid earth. Earth tides are important because they affect the ocean tides, because precise geodetic measurements require tidal corrections, and because they can be used to investigate the elastic properties of the earth.

If the earth were totally fluid the free surface would respond by adapting to the shape of the Equilibrium Tide. Conversely if the earth were totally rigid, there would be no surface deformation. In fact the earth responds elastically to the imposed forces. Moreover, because the natural modes of oscillation of the earth have periods of about 50 minutes or less, the response of the solid earth to the tidal forces can, to a good approximation, be regarded as in static equilibrium. The natural periods of the oceans are much longer than this and similar to the period of the tidal forcing, which is one of the reasons why the ocean response is much more complicated.

The Equilibrium Tide amplitude Ω_p/g (from equation (3:11)) is a second order spherical harmonic, and it has been shown that for this case the elastic response of the earth is a surface distortion of amplitude $h\Omega_p/g$, where h is a known elastic constant. Further, the redistribution of mass increases the gravitational potential by an amount which enhances the Equilibrium level by $k\Omega_p/g$, where k is another known elastic constant. The combined effect is a change in the height of the Equilibrium level above the solid earth of:

$$(1 + k - h)\,\frac{\Omega_p}{g}$$

A tide gauge mounted on the sea-bed would therefore sense a static Equilibrium ocean response which is lower than the true Equilibrium Tide by a factor $(1 + k - h) = 0.69$. Of course, this is a hypothetical case because the oceans do not respond in this static way. The factors k and h are called Love numbers after the mathematician who introduced them.

Both k and h are measures of the total elastic behaviour of the solid earth, and have been estimated from values based on seismic studies. They may also be deduced by measuring the tidal variations in the tilt of the earth's surface relative to the direction of gravity, and by measuring the variation in gravity itself. Both these variations are exceedingly small, but they can be measured accurately with very sensitive instruments. The tilt of the land surface is attenuated from the Equilibrium tilt by a factor $(1 + k - h)$ and the variations in measured gravity at a fixed point are increased by a factor $(1 + h - \frac{3}{2}k)$. Separate determinations of these two factors should allow the two unknowns k and h to be calculated. However, despite very careful experimental procedures, with tilt meters accurate to 5×10^{-8} degrees (one hundredth of the tidal signal), and gravity meters accurate to $10^{-9}\,g$, the results obtained from different sites show considerable scatter. The reasons for this are threefold. Firstly there are local anomalies in the response to the tidal forcing due to near-surface geology. Secondly, although traditionally the stable temperatures experienced in mines made them attractive for installing the very precise but temperature sensitive instruments, the tilt responses of a tunnel within the earth are not exactly the same as those of the upper surface. The third reason for variable local tidal responses is the ocean tide itself, which by its loading and unloading of the earth's crust, particularly in the vicinity of the observing earth–tide station, cause non-Equilibrium vertical displacements and tilts. The Cornwall peninsula in south-west England, for example, has a semidiurnal vertical movement of 0.10 m range due to the loading of the tides in the adjacent Celtic Sea. Calculations have shown that the effects on the local gravitational potential of tides in more distant oceans can also be significant.

For semidiurnal frequencies the published results for a wide variety of models of the earth's interior structure are within the ranges:

$$(1 + k - h) = 0.680 \rightarrow 0.695$$
$$(1 + h - \tfrac{3}{2}k) = 1.155 \rightarrow 1.165$$

and

$$h = 0.604 \rightarrow 0.630$$
$$k = 0.299 \rightarrow 0.310$$

Very precise geodetic work which depends on the direction of the apparent vertical must allow for the tidal tilts, especially when working near a tidal coastline.

The calculated Love numbers are at semidiurnal frequencies, but there are good physical reasons for expecting some variations in their amplitudes for different frequencies of tidal forcing. At periods of several years there is a transition between an elastic earth response and a viscous earth response because the earth cannot resist the slow imposed stresses. For such a viscous earth the response to very long period tides, such as the 18.6-year nodal tide discussed in Chapter 9, should give valuable geophysical information. Unfortunately, for the lengths of sea-level records available, the observations are too noisy for the very long period tidal responses to be determined with sufficient accuracy to make this a useful geophysical method of determining the low-frequency non-elastic properties of the earth. Further details are beyond the scope of this account but the interested reader is referred to the excellent review given by Baker (1984).

* 3:3   The moon–earth–sun system

In order to calculate the total Equilibrium Tide on the earth as a reference for tidal analysis it is necessary to define the change with time of the coordinates of the moon and sun, which are used as a reference for tidal analyses. The full relative movements of the three bodies in three dimensions are very complicated and their elaboration is beyond the scope of this account. Fuller details are given in Schureman (1976), Smart (1940), Doodson and Warburg (1941), Kaula (1968) and Roy (1978). However, there are several obvious features which we may consider. We begin by recognizing that there are two alternative reference systems which may be used to define the astronomical coordinates.

The most natural reference system for a terrestrial observer is the equatorial system, in which *declinations* are measured north and south of a plane which cuts the earth's equator. Angular distances around the plane are measured relative to a point on this celestial equator which is fixed with respect to the stellar background. The point chosen for this system is the *vernal equinox*, also called the 'First Point of Aries', which is represented by the symbol Υ. The angle, measured eastwards, between Υ and the equatorial intersection of the meridian through a celestial object is called the *right ascension* of the object. The declination and the right ascension together define the position of the object on the celestial background. In fact the vernal equinox is not absolutely fixed with respect to the stars, but moves slowly against the stellar background with a period of 26 000 years, a movement called the *precession of the equinoxes*.

The second system uses the plane of the earth's revolution around the sun as a reference. The celestial extension of this plane, which is traced by the sun's annual apparent movement, is called the *Ecliptic*. Conveniently, the point on this plane which is chosen for a zero reference is also the vernal equinox ϒ, at which the sun crosses the equatorial plane from south to north near 21 March each year. Celestial objects are located by their ecliptic latitude and ecliptic longitude. The angle between the two planes, of 23° 27′, is called the obliquity of the ecliptic, and is usually represented as ε. Figure 3:8 shows the relationship between these two coordinate systems.

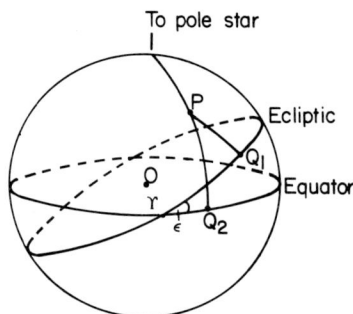

Figure 3:8 The relationship between equatorial and ecliptic coordinates. ϒ is the Vernal Equinox, the 'First Point of Aries'. ε is the obliquity of the ecliptic.

3:3:1 Elliptical motion, satellites

In Section 3:1, for simplicity, we considered bodies moving in circular orbits, but more generally Newton showed from his law of gravitation that two masses orbiting in space under the sole influence of their mutual attraction travel in ellipses. Each mass moves in its own ellipse, and each ellipse has the centre of mass of the two bodies as a focus. This is Kepler's first law of planetary motion, which he derived from observations.

In Figure 3:9 the mass at C is travelling in orbit around a mass at A, in an ellipse which has foci at A and B. The geometry of an ellipse is defined by AC + CB = constant. The distance of the mass from the focus at A is a maximum when the mass is at C_A and a minimum when the mass is at C_P. These positions are called the apogee and perigee of the orbit. The ratio OA/OC_P is called the *eccentricity*, e, of the ellipse. For a circular orbit A and B coincide at O, and the eccentricity is zero. Orbits which are very narrow and elongated

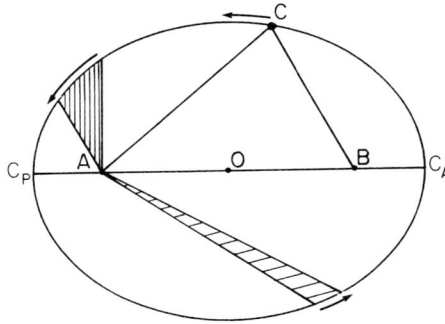

Figure 3:9   The properties of ellipses and elliptical orbits.

have an eccentricity close to unity. The ratio between the separation of the masses at the apogee and perigee is:

$$\frac{AC_A}{AC_P} = \frac{OC_A(1+e)}{OC_P(1-e)} = \frac{(1+e)}{(1-e)} \tag{3:13}$$

The distance $OC_A = OC_P$ is called the semi-major axis.

Kepler's second law states that the radius vector AC sweeps out equal areas in equal times. This is shown in Figure 3:9 where the two shaded segments have equal areas. This means that for an observer at A the planet will have its maximum angular speed at the point of nearest approach C_P and its minimum angular speed at apogee, C_A.

Kepler's third law relates the period of a complete orbit to the semi-major axis of the ellipse. From the law of gravitation this relationship may be derived theoretically:

$$(\text{period})^2 = 4\pi^2 \frac{r^3}{G(m_1 + m_2)} \tag{3:14}$$

where r is the semi-major axis and G, m_1, and m_2 are the same as in equation (3:1). The periods for earth satellite orbits discussed in Section 2:6 may be checked against this formula.

In astronomy the orbit described by a body and the position of the body in the orbit can be completely described by six parameters or orbital elements. The size and shape of the orbit are described by the semi-major axis and the eccentricity. Three parameters are necessary to define the orientation of the orbit with respect to a coordinate system. For example the plane of the ecliptic in Figure 3:8 is defined relative to the equatorial plane by the orientation of their line of intersection OΥ and by the obliquity ϵ. For an ellipse it is also necessary to define the direction of perigee in the plane of the orbit. The sixth element of the motion in an ellipse is the time of passage of the orbiting body through perigee.

3:3:2 The earth–sun system

The Equilibrium Tide in equation (3:12) is expressed in terms of the distance, declination and hour angle of the tide producing body. We need values for R_s, d_s, and C_s. These three parameters are easier to define for the earth–sun system than for the moon–earth system because the solar motions are always in the plane of the ecliptic, which means that the declination in ecliptic coordinates is always zero. The eccentricity of the earth's orbit about the sun is 0.0168.

From orbital theory the solar distance R_s may be shown to be given approximately by:

$$\frac{\overline{R}_s}{R_s} = (1 + e \cos (h - p'))$$

(3:15)

where $\overline{R}_s$ is the mean solar distance, h is the sun's geocentric *mean* ecliptic longitude (which increases by $\sigma_3 = 0.0411°$ per mean solar hour) and p' is the longitude of solar perigee, called perihelion, which completes a full cycle in 21 000 years.

The *true* ecliptic longitude of the sun increases at a slightly irregular rate through the orbit in accord with Kepler's second law. The true longitude λ_s is given to a first approximation by:

$$\lambda_s = h + 2e \sin (h - p')$$

(3:16)

where the value of λ_s is expressed in radians to facilitate the development of the harmonic expansions discussed in Chapter 4.

The right ascension is calculated from the ecliptic longitude and ecliptic latitude. Both the ecliptic longitude and the right ascension are zero when the sun is in the First Point of Aries (♈) at the vernal equinox, and also at the autumnal equinox. They both have values of $\pi/2$ when the sun has its maximum declination north of the equator in June, and both have values of $3\pi/2$ when the sun has its maximum declination south of the equator in December. Between these times there are small regular differences due to the obliquity of the orbit. The effect can be shown (Smart, 1940) to be represented to a first approximation by:

$$A_s = \lambda_s - \tan^2(\varepsilon_s/2) \sin 2 \lambda_s$$

(3:17)

where ε_s is the solar declination ecliptic latitude. The difference between the right ascension of the mean sun and the right ascension of the true sun (sundial time) at any time is called the *equation of time*. It may be calculated from equations (3:16) and (3:17) to give the annual variation plotted in Figure 3:10. During the annual cycle, differences of more than 15 minutes occur between clock-time and solar-time. In tidal analysis these differences are accounted for by a series of harmonic terms, as discussed in Chapter 4.

The solar declination in equatorial coordinates is given in terms of the ecliptic longitude of the sun:

$$\sin d_s = \sin \lambda_s \sin \varepsilon_s$$

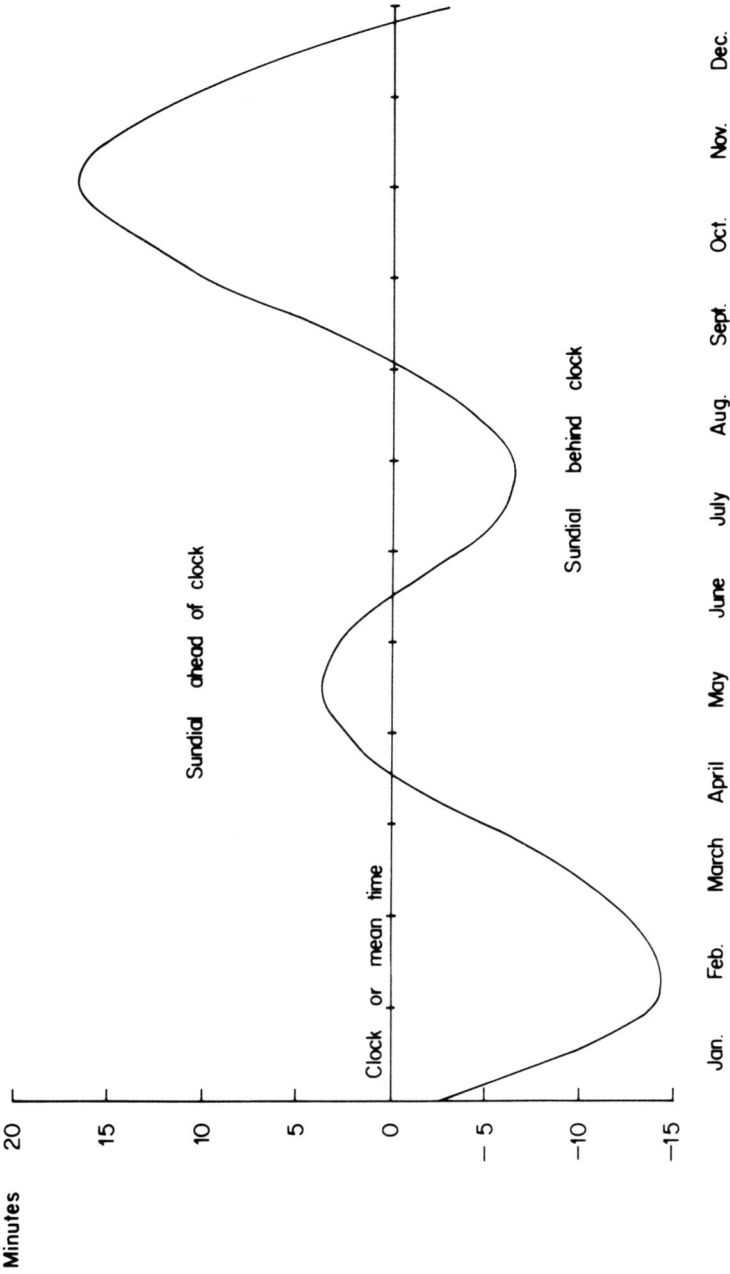

Figure 3:10 The annual equation of time, showing the difference between clock time and solar time, due to the solar Right Ascension being affected by the ellipticity and obliquity of the earth's orbit.

This is greatest when $\lambda_s = \pi/2$ and the maximum declination is $\varepsilon_s = 23°\,27'$, the angle which defines the Tropic of Cancer. When $\lambda_s = -(\pi/2)$ the sun is overhead on the Tropic of Capricorn at the time of its maximum declination $23°\,27'$ south of the equator.

3:3:3 The moon–earth system

The motions of the moon are more difficult to define in ecliptic or equatorial coordinates than those of the sun, because the plane of the motion is inclined at a mean angle of $5°\,09'$ to the plane of the ecliptic. Moreover, this plane rotates slowly over a period of 18.61 years. The ascending node, at which the moon crosses the ecliptic from south to north moves backwards along the ecliptic at a nearly uniform rate of $0.053°$ per mean solar day. This regression completes a revolution in 18.61 years. Viewed in equatorial coordinates when the *ascending* node corresponds to the First Point of Aries (♈), the moon's maximum declination north and south of the equator during the next month will be $(23°\,27' + 5°\,09') = 28°\,36'$; 9.3 years later, when the *descending* node is at the vernal equinox the maximum lunar declination is only $(23°\,27' - 5°\,09') = 18°\,18'$. Clearly the terms in the Equilibrium Tide which depend on the lunar declination will have a pronounced 18.61-year modulation.

The eccentricity of the moon's orbit has a mean value of 0.0549, more than three times greater than the eccentricity of the earth–sun orbit. Because of the effects of the sun's gravitational attraction the obliquity or inclination varies between $4°\,58'$ and $5°\,19'$, and the eccentricity varies from 0.044 to 0.067.

The lunar distance R_1, to a first approximation (Roy, 1978) is:

$$\frac{\bar{R}_1}{R_1} = (1 + e \cos (s - p) + \text{solar perturbations}) \tag{3:18}$$

where $\bar{R}_1$ is the mean lunar distance, s is the moon's geocentric *mean* ecliptic longitude which increases by $\sigma_2 = 0.5490°$ per mean solar hour and p is the longitude of lunar perigee, which rotates with an 8.85-year period.

The *true* ecliptic longitude of the moon also increases at a slightly irregular rate through the orbit as does the solar longitude:

$$\lambda_1 = s + 2e \sin (s - p) + \text{solar perturbations} \tag{3:19}$$

where λ_1 is in radians.

The right ascension of the moon is calculated from its ecliptic longitude and ecliptic latitude:

$$A_1 = \lambda_1 - \tan^2 (\varepsilon_1/2) \sin 2 \lambda_1$$

Relative to the ecliptic the moon's latitude is given by:

$$\sin (\text{ecliptic latitude}) = \sin (\lambda_1 - N) \sin (5°\,09')$$

where λ_1 is the ecliptic longitude and N is the mean longitude of the ascending node which regresses over an 18.61-year cycle.

Table 3:2 Basic periods and frequencies of astronomical motions. The mean solar day and the tropical year are used in everyday activities. Note that $\omega_0 + \omega_3 = \omega_1 + \omega_2 = \omega_s$. The ω values are usually expressed in terms of radians per unit time. The speeds in degrees per unit time are conventionally represented by the symbol σ.

	Period	Frequency		Angular speed	
		f	σ	symbol in radians	rate of change of
Mean solar day	1.00 mean solar days	1.00 cycles per mean solar day	15.0 degrees per mean solar hour	ω_0	C_s
Mean lunar day	1.0351	0.9661369	14.4921	ω_1	C_1
Sidereal month	27.3217	0.0366009	0.5490	ω_2	s
Tropical year	365.2422	0.0027379	0.0411	ω_3	h
Moon's perigee	8.85 Julian years	0.0003 0937	0.0046	ω_4	p
Regression of moon's nodes	18.61	0.0001471	0.0022	ω_5	N
Perihelion	20942	—	—	ω_6	p'

The lunar declination can be calculated from this ecliptic latitude using similar formulae.

The declination and right ascension of the moon and of the sun may all be represented as series of harmonics with different amplitudes and angular speeds. The Equilibrium Tide may also be represented as the sum of several harmonics by entering these astronomical terms into equation (3:12), as we shall see in Chapter 4.

3:3:4 Basic astronomical frequencies

The hour angle C_p for the lunar or solar Equilibrium Tide used in equation (3:10) needs further clarification. Table 3:2 summarizes the basic astronomical frequencies involved in the motions of the moon–earth–sun system. The time taken for one complete cycle of the lunar hour angle C_1 is called the mean lunar day. For the sun, one cycle of C_s is the mean solar day which we use for normal civil activities. Because of the movement of the moon and sun these days are slightly longer than the sidereal day:

$$\text{Mean solar day} \qquad 2\pi/(\omega_s - \omega_3) \equiv 2\pi/\omega_0$$
$$\text{Mean lunar day} \qquad 2\pi/(\omega_s - \omega_2) \equiv 2\pi/\omega_1$$

with the symbols as defined in Table 3:2. Hence we have:

$$\omega_s = \omega_0 + \omega_3$$

and

$$\omega_s = \omega_1 + \omega_2$$

for the frequency of sidereal earth rotation. By convention the ω speeds are in terms of radians per unit time.

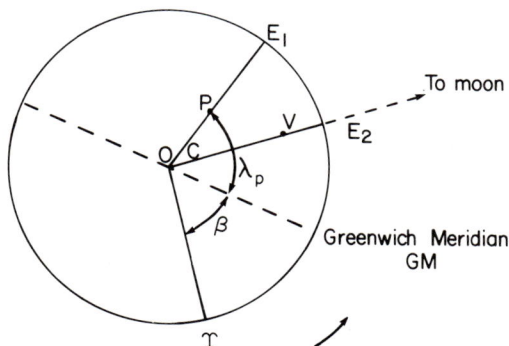

Figure 3:11   Projection of the celestial sphere on the plane of the equator to show the relationship between the hour angle C_p, the movement of the sub-lunar point, and the earth's rotation.

C_p is the angular difference between the terrestrial longitude of the point P and the longitude of the sub-lunar point V. Figure 3:11 shows a projection of the celestial sphere onto the plane of the equator, with the positions P and V as defined in Figure 3:6. All angles are related to the fixed direction of the First Point of Aries, Υ. The lunar hour angle C_1 is:

$$C_1 = E_1 \hat{O} \Upsilon - E_2 \hat{O} \Upsilon$$

$E_1 \hat{O} \Upsilon$ increases as the earth rotates on its axis, completing a revolution in absolute space in one sidereal day, 23 h 56 min, with an angular speed ω_s. It is conveniently expressed as:

$$\lambda_p + \beta = \lambda_p + \omega_s t$$

where λ_p = positive east longitude of P, and β = right ascension of the Greenwich Meridian = $\omega_s t$; here t is the sidereal time at the Greenwich Meridian, measured from Υ. Mean solar time at the Greenwich Meridan is the familiar Greenwich Mean Time, measured from the *lower transit* of the mean sun:

$$\omega_0 t + h - \pi = (\omega_0 + \omega_3) t - \pi$$

where $2\pi/\omega_0$ is the mean solar day, and $2\pi/\omega_3$ is the tropical year.

Table 3:3   Different kinds of day, month and year defined in terms of the basic frequencies of Table 3:2.

Type	Reference point	Frequency	Period (msd)
Days			
Sidereal	Fixed celestial point	$\omega_s = \omega_0 + \omega_3$	0.9973
		$\omega_s = \omega_1 + \omega_2$	
Mean solar	Solar transit	ω_0	1.0
Mean lunar	Lunar transit	ω_1	1.035
Months			
Nodical	Lunar ascending node	$\omega_2 + \omega_5$	27.2 122
Sidereal	Fixed celestial point	ω_2	27.3 217
Anomalistic	Lunar perigee	$\omega_2 - \omega_4$	27.5 546
Synodic	Lunar phases	$\omega_2 - \omega_3$; $\omega_0 - \omega_1$	29.5 307
Years			
Tropical	Υ	ω_3	365.2 422
Sidereal	Fixed celestial point	—	365.2 564
Anomalistic	Perihelion	$\omega_3 - \omega_6$	365.2 596

The angle $E_2 \hat{O} \Upsilon$ is the lunar Right Ascension A_1, which increases more slowly as the moon orbits through a period of one sidereal month with a *mean* angular speed which we call ω_2.

Combining these factors we have:

$$C_1 = \lambda_p + (\omega_0 + \omega_3)t - \pi - A_1 \qquad (3:20a)$$

Similarly, the solar hour angle may be expressed as:

$$C_s = \lambda_p + (\omega_0 + \omega_3)t - \pi - A_s \qquad (3:20b)$$

The angular speeds ω_2, ω_3, ω_4, ω_5, ω_6, in Table 3:2 are the mean rates of change with time of the astronomical coordinates s, h, p, N and p' respectively. In addition to defining three types of day, the basic astronomical frequencies in Table 3:2 may also be used to define several different kinds of months and years as listed in Table 3:3. These definitions in terms of simple sums and differences of basic astronomical frequencies illustrate the basic ideas for the harmonic expansion of the tidal potential which will be developed in Section 4:2. Although it is usual to define the astronomical speeds numerically in terms of radians per unit time, for which we use the symbol ω, tidal applications usually work in terms of degrees per unit time, for which we use the symbol σ (see Section 4:2).

3:4 Tidal patterns

The detailed descriptions of the Equilibrium Tide and of the lunar and solar motions developed in Sections 3:3:2 and 3:3:3 are necessary for a rigorous development, but several features of the observed tides plotted in Figure 1:1(a) can be explained in more general terms (Doodson and Warburg, 1941; Webb, 1976). These features include the relationship between lunar and solar declination and large diurnal tides, and the spring–neap cycle of semidiurnal amplitudes.

3:4:1 Diurnal tides

Comparing the tidal changes of sea-level plotted in Figure 1:1(a) for Karumba with the lunar changes plotted in Figure 1:1(b), we observe that maximum diurnal tidal ranges occur when the lunar declination is greatest, and that the ranges become very small when the declination is zero. This is because the effect of declination is to produce an asymmetry between the two high and the two low-water levels observed as a point P rotates on the earth within the two tidal bulges. In Figure 3:12, where these tidal bulges have been exaggerated, the point at P is experiencing a much higher Equilibrium Tidal level than it will experience half a day later when the earth's rotation has brought it to P'. The two high-water levels would be equal if P were located on the equator. However, as we shall discuss in Chapter 5, the ocean responses to tidal forcing are too complicated and not sufficiently localized for there to be zero diurnal tidal amplitudes at the equator, nor may the latitude variations of the observed tides be described in terms of Figure 3:12.

Over an 18.6 year nodal period the maximum lunar monthly declination north and south of the equator varies from 18.3° to 28.6°. There are maximum values in March 1969, November 1987 and minimum values in July 1978 and March 1996. The solar declination varies seasonally from 23.5° in June to $-23.5°$ in December. In Figure 1:1(a) the diurnal tides at Karumba approach zero amplitude when the lunar declination is zero because during the month of

March the solar declination is also zero. Similarly, the total amplitude of the diurnal forcing becomes very small in September. In other months the solar diurnal forces are significant and so the total diurnal tides will not completely disappear. The largest diurnal tides occur in June and December when the solar contribution is greatest. At times and places where the solar diurnal tides are important, if the diurnal high-water levels occur between midnight and noon during the summer, they will occur between noon and midnight during the winter because of the phase reversal of the Equilibrium solar bulges and hence the diurnal tidal forces.

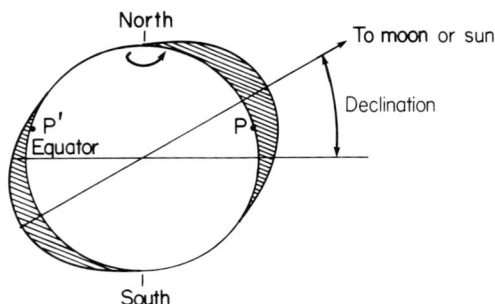

Figure 3:12   Showing how the unequal semidiurnal tides are produced when the moon or sun is north or south of the equator (diurnal tide generation).

3:4:2   Spring–neap tides

The fortnightly modulation in semidiurnal tidal amplitudes is due to the various combinations of lunar and solar semidiurnal tides. At Kilindini, Bermuda and Courtown in Figure 1:1(a), maximum ranges are seen shortly after the times of new moon and of full moon. The minimum ranges occur at first quarter and last quarter. This is because at times of spring tides the lunar and solar forces combine together, but at neap tides the lunar and solar forces are out of phase and tend to cancel. Figure 3:13 shows how the phases of the moon and the ranges of the semidiurnal tides are related. In practice the observed spring and neap tides lag the maximum and minimum of the tidal forces, usually by one or two days.

The synodic period from new moon to new moon is 29.5 days $(2\pi/(\omega_2 - \omega_3)$ or $2\pi/(\omega_0 - \omega_1))$ from the angular speeds defined in Section 3:3:4 and Tables 3:2 and 3:3, and the time from one spring tide to the next is 14.8 days. Within a lunar synodic period the two sets of spring tides are usually of different amplitudes. For example, in Figure 1:1(a) the Kilindini record shows the ranges of the first set are larger than those in the second set. This difference is due to the varying lunar distance; lunar perigee (the nearest approach of the moon to the earth) occurs on this occasion just after the new moon which accompanies the first set of spring tides (Figure 1:1(b)) so that the moon is near apogee (furthest from the earth) for the second set of spring tides. One complete cycle from perigee to perigee takes an Anomalistic month of 27.6 days. At perigee the $(1/R_1)^3$ term of equation (3:12) gives tidal forces which are 15 per cent larger than the forces when R_1 has its mean value. At apogee when the moon is at its furthest distance from the earth, the lunar forces are 15 per cent less than those for the mean value of R_1. These modulations due to distance variations are much smaller than the 46 per cent modulations in the total semidiurnal forces, produced by spring–neap variations, but their effects are nevertheless significant.

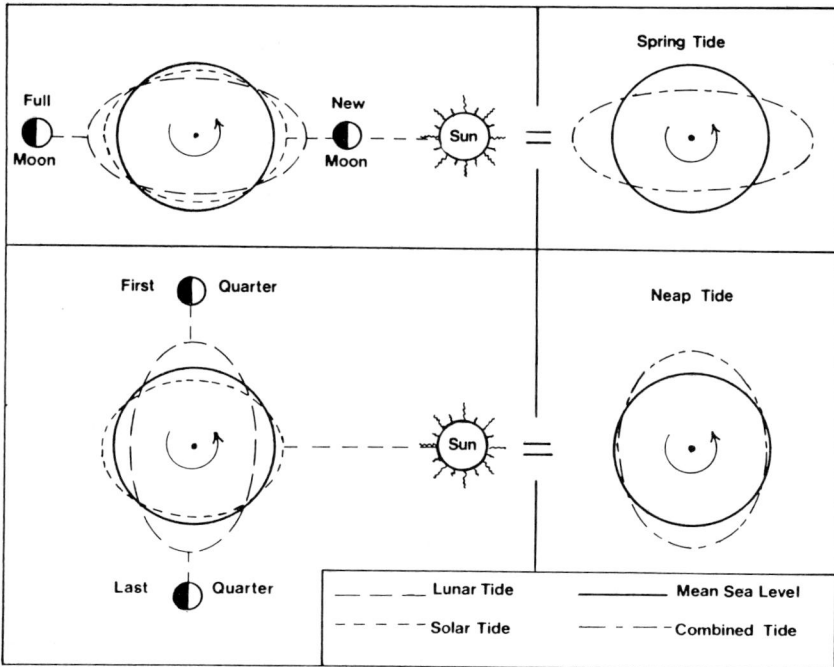

Figure 3:13   Spring–neap tidal cycles are produced by the relative motions of the moon and sun, at 14.8-day intervals. The lunar and solar Equilibrium Tides combine to produce spring tides at new and full moon and neap tides at the moon's first and last quarter. (From W. D. Forrester, 1983) (*Reproduced by permission of the Canadian Hydrographic Service.*)

A different modulation of the semidiurnal forces is produced by the varying declination of the moon and of the sun according to the $\cos^2 d$ relationship (equation (3:12)) which produces maximum forces when they are in the equatorial plane. For example, the semidiurnal lunar forces are reduced by 23 per cent when the moon has its maximum declination of 28.6°. The solar semidiurnal forces are reduced by a factor of 16 per cent in June and December when the declination reaches its maximum value of 23.5°. In March and

September near the equinoxes $\cos^2 d_s = 1$ and so the solar semidiurnal forces are maximized with the result that spring tides near the equinoxes are usually significantly larger than usual; these tides are called *equinoctial spring tides*.

3:4:3 Extreme tidal forces

By simple arguments it is evident that extreme tidal forces will occur when the moon and sun are in line with the earth and at their closest respective distances. For maximum semidiurnal tides, the moon and sun should also have zero declination. The condition of simultaneous perihelion and zero solar declination will not occur again until the year 6581; at present perihelion occurs early in January. However, there are several times when zero solar declination (the equinox), and lunar perigee occur almost simultaneously with zero lunar declination.

Table 3:4   Times of maximum semidiurnal tidal forces during the period 1980–2029 and the corresponding astronomical arguments. Only the highest value in each equinoctial period is given.

Year	Day	α (m)	Moon		Sun	
			$\overline{R}/R$	$\cos d$	$\overline{R}/R$	$\cos d$
1980	16 March	0.4 575	1.074 776	0.991 585	1.005 149	0.999 526
1984	25 Sept.	0.4 544	1.076 815	0.997 665	0.997 177	0.999 893
1993	8 March	0.4 597	1.077 784	0.998 957	1.007 454	0.996 248
1997	9 March	0.4 550	1.073 331	0.997 658	1.007 159	0.996 832
1997	17 Sept.	0.4 553	1.076 465	0.999 535	0.994 939	0.999 177
1998	28 March	0.4 588	1.076 407	1.0	1.001 968	0.998 786
2002	27 March	0.4 570	1.076 410	0.994 282	1.001 994	0.998 777
2002	4 Oct.	0.4 566	1.076 068	0.999 448	1.000 120	0.996 245
2007	18 March	0.4 552	1.072 543	0.999 951	1.004 60	0.999 905
2011	18 March	0.4 575	1.076 096	0.999 954	1.004 690	0.999 908
2015	19 March	0.4 554	1.074 894	0.999 766	1.004 339	0.999 979
2015	27 Sept.	0.4 544	1.077 101	0.999 854	0.997 805	0.999 500
2020	8 April	0.4 549	1.076 849	0.999 251	0.998 744	0.991 898
2028	11 March	0.4 559	1.073 711	0.999 987	1.006 691	0.998 039
2028	18 Sept.	0.4 540	1.075 540	0.999 193	0.995 178	0.999 538

Times of extreme semidiurnal tidal forcing are given in Table 3:4 for the period 1980–2030, together with the corresponding values of the orbital parameters for distance and for declination. These have been identified by predicting the hourly values of the Equilibrium semidiurnal tide at the Greenwich meridian, for the months of March and April, and September and October, in each year. The highest value in each period (α) was then tabulated, provided that it was not less than 0.4540 m. For comparison, the Equilibrium M_2 tidal amplitude is 0.2435 m. The highest value in the period (8 March 1993) is $1.89 H_{M_2}$, in the symbols developed in Chapter 4. Exact coincidence of other conditions with the solar equinox is not essential, as is illustrated by the values in early March in 1993 and 1997, and October 2002. The conditions in 1922 have been identified as particularly favourable for extreme semidiurnal tides (Cartwright, 1974), with New Moon occurring within one day of the equinox on 21 September. The semidiurnal amplitude, 0.4570 m was then 1.87 H_{M2}; however, larger amplitudes occurred on 13 March 1922 (0.4559 m; 1.88 H_{M2}) and on four occasions in Table 3:4.

Table 3:4 and other analyses of predicted extreme equinoctial spring tidal ranges over several years show a cycle with maximum values occurring approximately every 4.5 years. These maximum ranges occur when the time of lunar perigee (a maximum in the $(\overline{R_1}/R_1)^3$ term) corresponds with either the March or September equinox ($\cos^2 d_s = 1$), during its 8.85 years' cycle. Other periodicities in the recurrence of extreme tides can be expected as different factors in the relationship combine; for example, lunar perigee and zero lunar declination coincide every six years. Highest and lowest astronomical semidiurnal tides are often estimated by scrutiny of five years of tidal predictions. Macmillan (1966) describes a 1600-year cycle in extreme tides, but this periodicity is not supported by more elaborate analyses. For further discussions of extreme tidal forces see Cartwright (1974), Amin (1979) and Wood (1986).

It should be realized that, although maximum predicted tidal ranges usually result from extremes in the Equilibrium Tidal forcing, local ocean responses to these forces may produce local extreme tides at times other than those tabulated. Also, for maximum diurnal tides a different set of conditions would apply. Of course, the small differences between the extreme tidal predictions is of mainly academic interest as the observed extremes will also depend on the prevailing weather conditions as discussed in Chapters 6 and 8.

3:5   The geoid

If there were no tidal forcing, no differences in water density, no currents and no atmospheric forcing, the sea surface would adjust to take the form of an equipotential surface. In this condition, because the force of gravity is always perpendicular to this surface, there are no horizontal forces to disturb the water. There are many reasons, discussed in Chapter 9, why this is not the case, and why the mean sea-level surface deviates from an equipotential surface by as much as a metre. Nevertheless, an equipotential surface is still an important reference for mean sea-level studies and it is appropriate to discuss the concept further here.

This undisturbed equipotential surface is called the *geoid*. Its exact shape depends on the distribution of mass within the earth, and on the rate of rotation of the earth about its own axis (Garland, 1965; Jeffreys, 1976). Over geological time the earth has adjusted its shape to the rotation by extending the equatorial radius and reducing the polar radius. The polar flattening is computed as:

$$\frac{\text{equatorial radius} - \text{polar radius}}{\text{equatorial radius}}$$

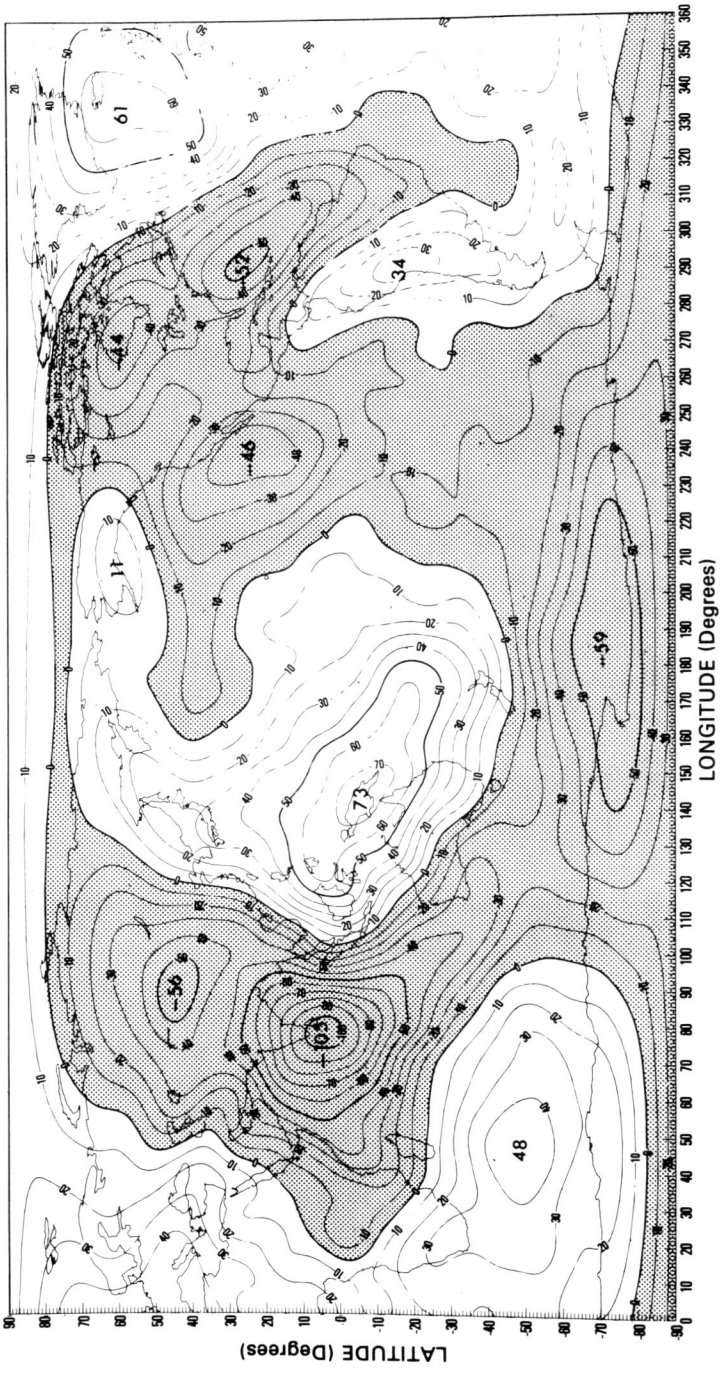

Figure 3:14 Map of the geoid surface in metres relative to a mean ellipsoid (1/298.25), computed from the GEM10 model. (copyright © 1979 the American Geophysical Union.)

which has a value close to 1/298.25. The shape generated by this rotational distortion is called an ellipsoid of revolution. If the earth consisted of matter of uniform density, an ellipsoid would be an adequate description of the geoid; however, the actual geoid has positive and negative excursions of several tens of metres from a geometric ellipsoid, due to the uneven mass distribution within the earth. Figures 3:14 and 3:15 show two representations of these excursions. The shape in Figure 3:15, showing that the actual geoid is more than 10 m higher than the ellipsoid at the North Pole and 30 m lower than the ellipsoid at the South Pole, is often described as pear-shaped. The earth is only pear-shaped relative to the ellipsoid, and is actually still convex rather than concave at the South Pole.

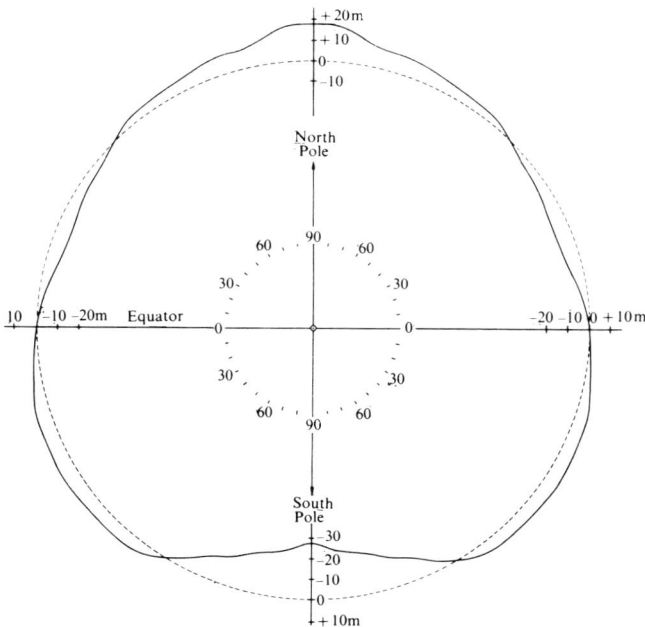

Figure 3:15   Heights of the mean meridional section (solid curve) of the geoid relative to a sphere of flattening 1/298.25 (dotted curve). (From D. G. King-Hele, C. J. Brookes and G. E. Cook (1980), *Nature*, **286**, 377–8. Copyright © 1980, Macmillan Journals Limited.)

Figure 3:14 shows a geoid depression of -105 m corresponding to a mass deficiency south of India, and geoid elevations of 73 m north of Australia, corresponding to mass excess. Interestingly the highs and lows of the geoid show no direct correlation with the major tectonic surface features; they are thought to be due to density differences deep inside the earth.

The variations of mass distribution, for example due to sea-mounts, give local variations in the mean sea-level surface, relative to a smooth surface (represented by h_3 in Figure 2:7) of several metres, which can be detected by satellite altimetry. The importance of these variations for physical oceano-graphy lies in the need to define water levels and water-level differences accurately relative to the geoid in order to know the horizontal pressure gradient forces acting on the water.

As the mass of the earth is redistributed by geological processes such as isostatic adjustment or glacial melting, the shape of the geoid gradually changes, and so too must the shape of the mean sea-level surface. Some examples of these geoid and sea-level adjustments are discussed in Chapter 10.

THE HYDRODYNAMIC EQUATIONS

The physical behaviour of the waters under the influence of tidal and other forces can be described in terms of a set of hydrodynamic equations (Lamb, 1932; Gill, 1982). These equations take slightly different forms depending upon the degree of approximation which is justified in their development. For studies of tides and other long-period movements it is usual to neglect the vertical movements and the vertical accelerations, and the water is usually assumed to have a uniform density.

* 3:6   The hydrostatic approximation

We adopt the normal Cartesian coordinate system shown in Figure 3:16 with the axes X and Y in the horizontal plane and the Z axis directed vertically upwards. The zero level for vertical displacements is taken as the long-term mean sea-level. Some authors prefer to direct Z downwards from the surface, but still define sea-level changes from the mean as being positive when upwards.

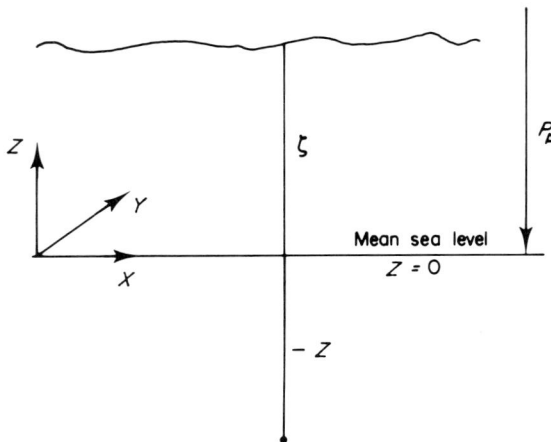

Figure 3:16   The coordinate system used for development of the hydrodynamic equations.

In our coordinate system the hydrostatic pressure at a point, depth z below the water surface, due to the combination of the atmospheric pressure P_A acting on the water surface and the weight of water above the point is:

$$P_z = P_A - \rho g (z - \zeta) \qquad (3:21)$$

where ρ is the water density, assumed constant, and ζ is the displacement of the water level from the mean. The negative sign appears because z increases upwards; as z is negative below the water surface the formula correctly represents the increase of pressure with depth. This hydrostatic formula has already been used in a simple form (equation (2:1)) to convert bottom pressures measured using the equipment described in Section 2:2:3, into changes in sea level. The assumption that this hydrostatic formula is also valid for moving fluids is called the 'hydrostatic approximation', and is true for long waves, the form in which ocean tides propagate (see Chapter 5 and Appendix 4).

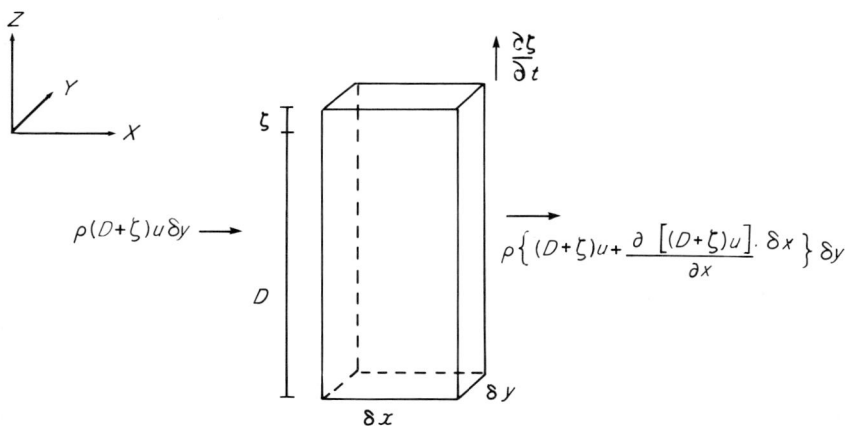

Figure 3:17   The mass and volume balance for a column of water.

* 3:7   Conservation of mass

The assumption that the density of the water is constant means that the mass and the volume of the water are equivalent parameters. Consider the mass of fluid within a column of water above an element of area which is fixed in space, where the seabed is at a depth $-D$, as in figure 3:17. The mass flux through the left-hand face of the column is $\rho(D + \zeta)u\delta y$, where u is the speed in the X-direction. Through the opposite right-hand face the flux is:

$$\rho\{(D + \zeta)u + \frac{\partial [(D + \zeta)u].}{\partial x} \delta x\} \delta y$$

The net mass flux out of the column through these two faces is the difference:

$$\rho\{\delta x \,\delta y \,\frac{\partial}{\partial x}\,[(D + \zeta)u]\}$$

Similarly along the Y-axis the net flux out of the column is:

$$\rho\delta x \,\delta y \,\frac{\partial}{\partial y}\,[(D + \zeta)v]\}$$

The sum of these two must equal the rate of change of the total mass of the column:

$$\frac{\partial}{\partial t}\,[\rho\delta x \,\delta y \,(D + \zeta)] = \rho\delta x \,\delta y \,\frac{\partial\zeta}{\partial t}$$

Hence:

$$\rho\delta x \,\delta y \,\frac{\partial\zeta}{\partial t} = -\,\rho\delta x \,\delta y \,\frac{\partial}{\partial x}\,[(D + \zeta)u] - \rho\delta x \,\delta y \,\frac{\partial}{\partial y}\,[(D + \zeta)v]$$

which reduces to:

$$\frac{\partial\zeta}{\partial t} + \frac{\partial}{\partial x}\,[(D + \zeta)u] + \frac{\partial}{\partial y}\,[(D + \zeta)v] = 0$$

In the usual case where the water depth is large compared with the displacement of the level from the mean, this becomes:

$$\boxed{\frac{\partial\zeta}{\partial t} + \frac{\partial}{\partial x}\,(Du) + \frac{\partial}{\partial y}\,(Dv) = 0} \qquad (3{:}22)$$

Where the sea-bed is flat so that the water depth is also constant, the final simplification gives:

$$\frac{\partial\zeta}{\partial t} + D\left(\frac{\partial u}{\partial x} + \frac{\partial v}{\partial y}\right) = 0$$

These mass conservation equations are also called continuity equations. They formally represent the basic fact that a net flux of water into or out of an area must be balanced by a corresponding change in the water level.

* 3:8 The horizontal momentum equations

We consider the forces acting on a Cartesian element of water with dimensions δx, δy, δz as shown in Figure 3:18. The element is again located at a fixed place on the surface of the rotating earth so that our development is according to Eulerian dynamics. The alternative of developing the equations in terms of the path of a single particle, the Lagrangian approach, would make the treatment of accelerations slightly easier, but the forces would then be much more difficult to define.

By restricting our interest to horizontal forces and accelerations, we need only consider the balance in the X and in the Y directions. Each yields a separate equation. Consider the forces and accelerations along the X-axis as shown in Figure 3:18. Three separate forces can be distinguished, the direct tidal forces, the pressure forces, and the shear forces acting on the upper and lower surfaces.

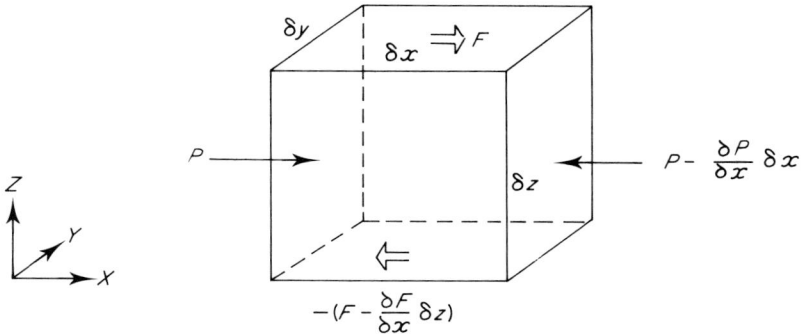

Figure 3:18   The forces acting on an element of water due to pressure and stress gradients.

From Section 3:2:1 we deduce the tidal forces in terms of the tidal gravitational potential as $-\delta\Omega/\delta x$ per unit mass. The external pressure forces on the left-hand face is (pressure × area), $P\delta x\delta y$, while the external pressure on the right-hand face is $[P + (\partial p/\partial x).\delta x]\delta x\delta y$. The resultant pressure force in the positive X direction is therefore the difference of these two opposing pressure forces:

$$-\frac{\partial P}{\partial x}\,\delta x\,\delta y\,\delta z$$

The stress force on the upper surface is $F\delta x\delta y$ in the positive X direction. The lower surface will transmit a stress $[F - (\partial F/\partial z)\cdot\delta z]\delta x\delta y$ to the element of water below, but by Newton's third law of motion there will therefore be an equal opposing stress on our Cartesian element. The resultant stress force on the element is therefore:

$$\frac{\partial F}{\partial z}\cdot\delta x\delta y\delta z.$$

Combining these three forces we have the total force per unit mass as:

$$-\frac{\partial\Omega}{\partial x} - \frac{1}{\rho}\left(\frac{\partial P}{\partial x} - \frac{\partial F}{\partial z}\right)$$

If we were considering Lagrangian dynamics on a non-rotating earth, the acceleration would be simply du/dt the so-called 'total' or 'individual' derivative. However, our fixed Eulerian frame requires additional terms to allow for the acceleration necessary to give the water within our fixed element at an instant of time, the appropriate velocity for a subsequent time and position to which it has been advected. Also, because Newton's second law applies for motions in an absolute unaccelerating coordinate system, when the law is applied to motion in a rotating earth-bound coordinate system an additional acceleration term, the Coriolis term, is required. A rigorous mathematical development of these is beyond the scope of this account, but it is possible to indicate the nature of these terms.

The required acceleration in the Eulerian system is:

$$\frac{\partial u}{\partial t} + u\,\frac{\partial u}{\partial x} + v\,\frac{\partial u}{\partial y}$$

The terms in $\partial u/\partial x$ and $\partial u/\partial y$ are called *advective accelerations* because they are related to the horizontal movement of the water.

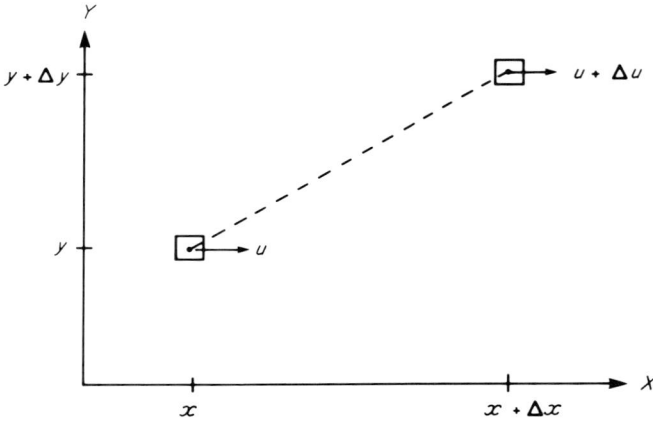

Figure 3:19 Showing the origin of the advective accelerations in the hydrodynamic equations.

Suppose the particle has a velocity component u at time t and position (x,y) in a water circulation pattern which is not changing with time. At time $t + \Delta t$ the particle will have a component $u + \Delta u$ at a location $(x + \Delta x, y + \Delta y)$, as shown in Figure 3:19. The advective acceleration in the X-direction of the particle is $\Delta u/(\text{time of travel})$. The time of travel to $x + \Delta x$ is $\Delta x/u$, which gives an acceleration $u(\Delta u/\Delta x)$. Similarly, the time of travel to $y + \Delta y$ is $\Delta y/v$ giving an acceleration $v(\Delta u/\Delta y)$. In the limit of very small elements and times these become the advective accelerations $u(\partial u/\partial x)$ and $v(\partial u/\partial y)$.

The Coriolis accelerations are introduced to allow the application of Newton's second law in a rotating coordinate system. Any particle initially projected in a particular direction in the northern hemisphere moves along a path over the earth's surface which curves to the right. In the southern hemisphere the curvature of the path is to the left. For normal daily activities such as throwing a ball, the effect is negligible because the time of flight is very much less than the period of rotation of the earth, but for tidal and other currents which persist for times which are a significant fraction of a day, the curvature cannot be neglected. The effect is represented in the force equations as an acceleration at right angles to the motion. This acceleration is proportional to the particle speed, and increases from zero at the equator to a maximum at the poles. In our coordinate system the acceleration along the X-axis is $-2\omega_s \sin\varphi\, v$, and the acceleration along the Y-axis is $2\omega_s \sin\varphi\, u$. These terms are also valid in the southern hemisphere if latitudes south of the equator are treated as negative.

A rigorous transformation of motion in one system to motion in a second system, which rotates with respect to the first, is a straightforward exercise in vector analysis. For our purposes it will be sufficient to restrict our interest to horizontal motions and to show that the above accelerations are physically reasonable. We do this by considering the components of motion in the northern hemisphere along a parallel of latitude and along a meridian of longitude.

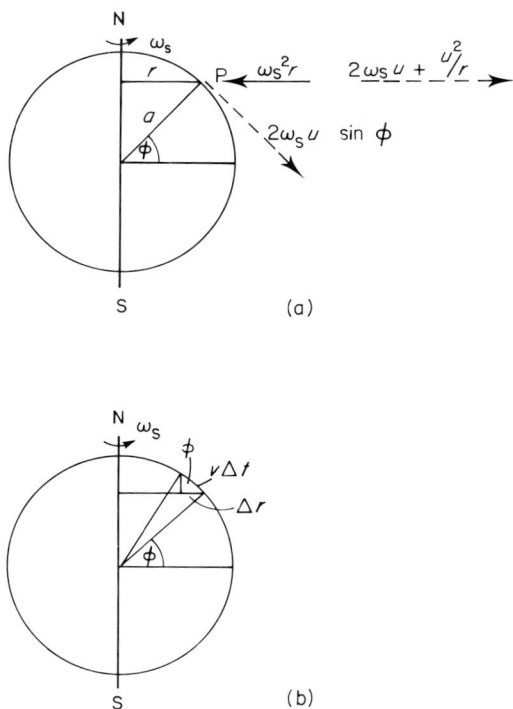

Figure 3:20 Accelerations of a particle on a rotating earth, to illustrate the origin of the Coriolis acceleration in the hydrodynamic equations (a) for motion along an east–west axis and (b) for motion along a north–south axis.

Consider first a particle P at rest with respect to the earth, which is rotating with an angular velocity ω_s as in Figure 3:20(a). The acceleration of this particle towards the axis of rotation is:

$$\frac{(\text{radial velocity})^2}{\text{radius}} = \frac{(\omega_s r)^2}{r} = \omega_s^2 r$$

If the radial velocity is increased by giving the particle an eastward velocity u over the surface of the earth, the new acceleration which the particle *should* have towards the centre becomes:

$$\frac{(\omega_s r + u)^2}{r} = \omega_s^2 r + 2\omega_s u + \frac{u^2}{r}$$

In the absence of any additional force to provide the additional central acceleration necessary to maintain the particle in equilibrium at latitude φ, the eastward moving particle experiences an apparent outward acceleration relative to the earth's axis of $(2\omega_s u + u^2/r)$. The u^2/r term is negligible for the speeds such as those of ocean currents, which are much less than the radial speed of the earth. Therefore, on the earth's surface the horizontal component of the apparent outward acceleration is directed towards the equator and has an amplitude of $2\omega_s \sin\varphi\, u$.

The Coriolis acceleration experienced by a particle due to the component of motion along the meridian of longitude can be evaluated by applying the law of conservation of angular momentum to the movement of the particle about the earth's axis. The angular momentum of the particle of unit mass at P which is stationary on the rotating earth is:

$$\text{particle velocity} \times \text{radius of motion} = \omega_s r^2$$

If the particle has a northward velocity v, the value of r decreases with time. In order to conserve the angular momentum the velocity of the particle in its orbit must increase by an amount Δu. This increase is observed in the northern hemisphere as an acceleration of the particle towards the east. To conserve angular momentum we have:

$$\omega_s r^2 = [\omega_s (r - \Delta r) + \Delta u] (r - \Delta r)$$

With rearrangement this becomes:

$$\Delta u = 2\,\omega_s\,\Delta r$$

Figure 3:20(b) shows that in the limit of very small time increments Δr is related to the northward particle displacement $v\Delta t$:

$$\sin\varphi = \frac{\Delta r}{v\Delta t}$$

Hence

$$\frac{\Delta u}{\Delta t} = 2\omega_s \sin\varphi\, v$$

which is the apparent eastward acceleration of the particle. If the particle moved towards the equator the increase in r would require a compensating decrease in the velocity, which would be apparent to an observer in the northern hemisphere as a westward acceleration.

For convenience the Coriolis parameter $2\omega_s\sin\varphi$ is usually written as f. Note that because there is no acceleration along the direction of motion of the particle, there is no net gain of particle energy. Many accounts of the movement on a rotating earth prefer to develop the idea of an apparent force at right angles to the particle motion, the Coriolis force. This places the Coriolis effects on the opposite side of the equation of motion, but of course, the dynamics are unchanged.

Although the role which Coriolis (1835) played in developing ideas of accelerations in rotating systems and their application to fluid motions on a rotating earth is acknowledged by attaching his name to the parameters, in fact the basic ideas were propounded earlier by Laplace, in connection with his study of tides in 1775.

By combining the total accelerations and the forces acting on the particle we arrive at the two basic horizontal momentum equations.

$$\frac{\partial u}{\partial t} + v\frac{\partial u}{\partial y} + u\frac{\partial u}{\partial x} - fv = -\frac{\partial \Omega}{\partial x} - \frac{1}{\rho}\left(\frac{\partial P}{\partial x} - \frac{\partial F}{\partial z}\right) \qquad (3:23)$$

$$\frac{\partial v}{\partial t} + u\frac{\partial v}{\partial x} + v\frac{\partial v}{\partial y} + fu = -\frac{\partial \Omega}{\partial y} - \frac{1}{\rho}\left(\frac{\partial P}{\partial y} - \frac{\partial G}{\partial z}\right) \qquad (3:24)$$

$$\text{time} + \text{advective} + \text{Coriolis} = \text{tidal} + \text{pressure} + \text{shear}$$
$$\text{Accelerations} = \text{Forces/Mass}$$

These two equations, together with the Continuity Equation (3:22) and the Hydrostatic Equation (3:21) will be used in subsequent chapters to describe the physical response of the oceans to the tidal and to the meteorological forces.

CHAPTER 4

Analysis and Prediction

By what astrology of fear or hope
Dare I to cast thy horoscope!
By the new moon thy life appears;

Longfellow, 'To a Child'

Tidal analysis of data collected by observations of sea levels and currents has two purposes. Firstly, a good analysis provides the basis for predicting tides at future times, a valuable aid for shipping and other operations. Secondly, the results of an analysis can be interpreted scientifically in terms of the hydrodynamics of the seas and their responses to tidal forcing. An analysis provides parameters which can be mapped to describe the tidal characteristics of a region.

The process of analysis reduces many thousands of numbers, for example a year of hourly sea levels consists of 8760 values, to a few significant numbers which contain the soul or quintessence of the record (Godin, 1972). An example of statistical tidal analysis is given in the description of sea levels in Section 1:5. In tidal analysis the aim is to reproduce significant time-stable parameters which describe the tidal regime at the place of observation. These should be in a form suitable for prediction, should be related physically to the process of tide generation and should have some regional stability. The parameters are often termed *tidal constants* on the implicit assumption that the responses of the oceans and seas to tidal forcing do not change with time. Also implicit in the use of this term is the assumption that if a sufficiently long series of levels or currents is available at a site, then a true value for each constant is obtained by analysis. In practice measurements extend over finite periods, often a year, a month or even a few days, and so the results from analysing these finite lengths of data can only approximate the true constants. The longer the period of data available for analysis, the better will be the approach to these true values.

A good system of analysis represents the data by a few significant stable numbers, but a bad analysis represents the data by a large number of parameters which cannot be related to a physical reality, and for which the values obtained depend on the time at which the observations were made. If possible, an analysis should also give some idea of the confidence which should be attributed to each tidal parameter determined. Details of analysis procedures are given in Doodson and Warburg (1941), Godin (1972), Schureman

(1976), Foreman (1977) and Franco (1981). The reader of the older texts is warned that many of the techniques designed to reduce labour of calculation in the days before digital computers are no longer necessary.

The close relationship between the movements of the moon and sun, and the observed tides, makes the lunar and solar coordinates a natural starting point for any analysis scheme. The Equilibrium Tide developed in Chapter 3 defines a tidal level at each point on the earth's surface as a function of time and latitude. The observed tides differ very markedly from the Equilibrium Tide because the oceans have complicated responses which are limited and controlled by their real depths and boundaries. In shallow water the tides are further distorted from the Equilibrium form by the loss of the energy through bottom friction and by the preservation of the continuity of water flow. Tidal behaviour in shallow water will be considered in detail in Chapter 7, but in this chapter we need to remember that any good analysis scheme must be capable of resolving these complications.

Three basic methods of tidal analysis have been developed. The first, which is now generally of only historical interest, termed the *non-harmonic method*, consists of relating high and low water times and heights directly to the phases of the moon and other astronomical parameters. The second method, which is generally used for predictions and scientific work, called *harmonic analysis*, treats the observed tides as the sum of a finite number of harmonic constituents whose angular speeds and phases are determined from the astronomical arguments. The third method develops the concepts, widely used in electronic engineering, of a frequency-dependent system *response* to a driving mechanism. For tides the driving mechanism is the Equilibrium potential. The latter two methods are special applications of the general formalisms of time series analysis and we will discuss their relative advantages and disadvantages later.

Analyses of changing sea levels, which are scalar quantities, are obviously easier to perform than those of currents, which are vectors. The development of the different techniques is first discussed in terms of sea levels; their application to currents is treated in a subsequent section. The more elaborate details of the procedures are given in separate Appendices, and the reader is referred to selected original scientific papers.

4:1   Non-harmonic methods

The simplest and oldest technique of tidal analysis is to relate the time of local semidiurnal high water to the time of lunar transit. The interval between lunar transit at new or full moon and the next high tide used to be known as the *local establishment*. Small adjustments may be made to allow for variations in the time interval through a spring–neap cycle. The *age* of the tide is an old but still useful term applied to the interval between the time of new or full moon and the time of maximum spring range, which is usually about two days. Several other non-harmonic tidal parameters such as Mean High Water Springs are defined in Chapter 1. The advantage of these non-harmonic parameters is their ease of determination and of application. Their limitations include the difficulty of using them in scientific work, and the fact that they contain insufficient information for a full tidal description and prediction scheme.

Non-harmonic methods of analysis and prediction were developed on a mathematical basis by Sir John Lubbock in 1832. Times and heights of high and low water are related to the lunar transits. Corrections are applied for the age of the tide, the lunar and solar declinations and their distances (in astronomical terms, their parallax). Further developments are still being used for official Dutch and German tidal predictions, where the method is sufficiently robust to cope with the shallow-water tidal distortions found along their coasts. Dutch predictions use separate sets of non-harmonic constants for each month of the year and for the upper and lower lunar transits.

Along the Channel and Atlantic coasts of France, a scaling factor, due to Laplace, called the Coefficient de la Marée is widely used. The scale extends over 100 points from 20 to 120, with mean equinoctial spring tides, 6.42 m at Brest, corresponding to 100. 120 corresponds to exceptional equinoctial spring tides such as those discussed in Section 3:4:3; 20 corresponds to very small neap tides.

4:2 Harmonic analysis

4:2:1 Expansion of the Equilibrium Tide

The basis of harmonic analysis is the assumption that the tidal variations can be represented by a finite number N, of harmonic terms of the form:

$$H_n \cos (\sigma_n t - g_n)$$

when H_n is an amplitude, g_n is a phase lag on the Equilibrium Tide at Greenwich (see Sections 4:2:2 and 4:2:4), and σ_n is an angular speed. For a correct mathematical development the angular speeds and phase lags should be expressed in radians. However, in tidal notation the phase lags g_n are usually expressed in degrees. We use the notation ω_n in radians per mean solar hour through most of this chapter. The angular speed in degrees per mean solar hour is denoted by $\sigma_n = 360 \, \omega_n/2\pi$. A further useful extension of the convention is to express angular speeds in cycles per unit time as f_n ($f_n = \sigma_n/360$ cycles per mean solar hour or $f_n = \sigma_n/15$ cycles per mean solar day).

The angular speeds ω_n are determined by an expansion of the Equilibrium Tide into similar harmonic terms: the speeds of these terms are found to have the general form:

$$\omega_n = i_a\omega_1 + i_b\omega_2 + i_c\omega_3 + i_d\omega_4 + i_e\omega_5 + i_f\omega_6 \tag{4:1}$$

where the values of ω_1 to ω_6 are the angular speeds already defined in Table 3:2 (i.e. the rates of change of C_1, s, h, p, N, and p'), and the coefficients i_a to i_f are small integers. The amplitudes and phases are the parameters determined by analysis which define the tidal conditions at the place of observation. The phase lags g_n are defined relative to the phase of the term in the harmonic expansion of the Equilibrium Tide which has the same angular speed, σ_n (ω_n in radian measure); by convention these phases of the Equilibrium Tide are expressed relative to the Greenwich Meridian.

If the moon were to revolve continuously in an equatorial circle around the earth then each tide would follow the previous one at intervals of 12 hours 25 minutes and each high water would have the same amplitude. If the sun also moves in an equatorial circle an additional tide of period 12 hours would be present, and the beating between those two tides would produce the spring–neap cycle, which would be completely represented by four parameters, the times of lunar and solar high waters and the amplitudes of the two tides. The real movements of the moon and sun, north and south of the equator and their varying distances and angular speeds, can be considered as the combined effects of a series of phantom satellites. These phantom satellites are of various masses and they move in various planes and at various speeds; each one moves either in an orbit parallel to the equator or stands in a fixed position amongst the stars. Each satellite therefore has a simple tide which is represented as an amplitude and a time of high water. These concepts formed the basis for the early development of Harmonic analysis by Laplace, Lord Kelvin and George Darwin (1911). The additional frequencies are also analogous to the side-bands of a modulated radio carrier frequency, as discussed below.

The first necessary step in the development of harmonic analysis is an expansion of the Equilibrium Tide into a series of harmonic terms from the expressions of equation (3:12) and the full formulae for distance, declination and hour angle, which are given in attenuated form in Section 3:3. The most complete algebraic expansions were completed by Doodson (1921) and these have been extended and adjusted by Fourier analysis of an extended time series of the Equilibrium Tide computed from the ephemerides, the lunar and solar coordinates (Cartwright and Tayler, 1971; Cartwright and Edden, 1973).

Consider first the expansion of a term of the form:

$$\gamma(1 + \alpha \cos \omega_x t) \cos \omega_y t$$

which, by the basic rules of trigonometry for combining sine and cosine terms becomes:

$$\gamma[\cos \omega_y t + \frac{\alpha}{2} \cos (\omega_x + \omega_y)t + \frac{\alpha}{2} \cos (\omega_y - \omega_x)t] \qquad (4:2)$$

The harmonic modulation in the amplitude of a constituent of speed ω_y is represented by two additional harmonic terms with speeds at the sum and difference frequencies ($\omega_y + \omega_x$) and ($\omega_y - \omega_x$).

Consider as an example the effects of the variation over a period of 27.55 days of the lunar distance R_1, and the lunar longitude. Substituting for R_1 from equation (3:18) in the semidiurnal part of the Equilibrium Tide in equation (3: 12):

$$C_2(t) = \left[\left(\frac{a}{R_1}\right)^3 \frac{3}{4} \cos^2 d_1\right] [1 + e \cos (s - p)]^3 \cos 2 C_1$$

For convenience and further simplification we write:

$$\gamma = [\left(\frac{a}{R_1}\right)^3 \tfrac{3}{4} \cos^2 d_1]$$

We make the approximation:

$$[1 + e \cos (s - p)]^3 \approx 1 + 3e \cos (s - p)$$

where terms in e^2 and above are omitted.

For the lunar longitude, from equation (3:20) we have:

$$\cos 2C_1 = \cos 2 [\omega_0 t + h - s - \pi - 2e \sin (s - p)]$$

with the approximations that $A_1 = \lambda_1$, and λ_1 is given by equation (3:19). We adopt the convention of expanding the potential for the Greenwich meridian, so $\lambda_p = 0$.

If we also write:

$$\delta = \omega_0 t + h - s - \pi$$
$$\cos 2 C_1 = \cos 2 \delta \cos [2e \sin (s - p)] + \sin 2 \delta \sin [2e \sin (s - p)]$$
$$\approx \cos 2 \delta + 2e \sin (s - p) \sin 2 \delta$$

Hence by multiplying out and neglecting the terms in e^2 we have:

$$
\begin{aligned}
C_2(t) &= \gamma [\cos 2\delta + 4e \sin (s - p) \sin 2\delta + 3e \cos (s - p) \cos 2\delta] \\
&= \gamma [\ \cos 2\delta \\
&\quad + \tfrac{4}{2} e \cos (2\delta - s + p) - \tfrac{4}{2} e \cos (2\delta + s - p) \\
&\quad + \tfrac{3}{2} e \cos (2\delta + s - p) - \tfrac{3}{2} e \cos (2\delta - s + p) \\
&= \gamma [\cos 2\delta \\
&\quad + \tfrac{7}{2} e \cos (2\delta - s + p) \\
&\quad - \tfrac{1}{2} e \cos (2\delta - s + p)
\end{aligned}
$$

Replacing the full definition of δ and writing $-\cos \beta = \cos (\beta + 180°)$

$$
\begin{aligned}
C_2(t) = \gamma [&\cos (2 \omega_0 t + 2h - 2s) \\
&+ \tfrac{7}{2} e \cos (2 \omega_0 t + 2h - 3s + p) \\
&+ \tfrac{1}{2} e \cos (2 \omega_0 t + 2h - s - p + 180°)] \quad\quad (4:3)
\end{aligned}
$$

The first term in $\cos (2\omega_0 t + 2h - 2s)$ is the main semidiurnal lunar tide due to the mean motion of the moon. By convention this tide is called M_2, where the suffix 2 denotes approximately two tides in a day. The lunar and solar mean longitudes s and h increase at rates of ω_2 and ω_3 respectively which gives $2(\omega_0 - \omega_2 + \omega_3)$ as the total speed of M_2. The values in Table 3:2 give speeds 1.9 323 cycles per mean solar day or 28.9 842° per mean solar hour. Of course, this speed is the same as $2\omega_1$, twice the speed of the mean moon, as shown in Section 3:3:4.

The second and third terms have total angular speeds:

$$2 \omega_1 - \omega_2 + \omega_4; \quad f = 1.8\ 960 \text{ cpd}; \quad \sigma = 28.4\ 397° \text{ h}^{-1}$$
$$2 \omega_1 - \omega_2 + \omega_4; \quad f = 1.9\ 686 \text{ cpd}; \quad \sigma = 29.5\ 285° \text{ h}^{-1}$$

They represent the two extra terms in equation (4:2), with the important difference that in this case the amplitudes of the terms are not equal. This is because our analysis included both the amplitude variation, which would have produced terms of equal amplitude, and the effects of the varying angular speed on the right ascension, which increase the amplitude of the term at the lower speed and decrease as the amplitude of the term at the higher speed. To emphasize their speed symmetry about the speed of M_2, these are called N_2 and L_2. According to the expansion (4:3) the relative amplitudes are ($e_1 = 0.0\,549$):

$$N_2 : M_2 : L_2 = 0.196 : 1.000 : 0.027$$

and these compare favourably with the relative amplitudes given in Table 4:1(b). The full expansion of $C_2(t)$ must also include the effects of the obliquity on the right ascension, and the effects of the declination, $\cos^2 d_1$. The approximation for the orbit parameters given in Section 3:3:3 must also be expanded to further terms. When this is done for both the moon and sun the list of harmonic constituents is very long. Nevertheless, examination of their relative amplitudes shows that in practice a limited number of harmonics are dominant. Similar expansions for the long-period and diurnal terms show that these too are dominated by a few major harmonic terms.

Table 4:1 lists the most important of these constituents, gives the appropriate values of s, h, p, N and p' and the corresponding periods and angular speeds. The amplitudes are all given relative to $M_2 = 1.0\,000$. If the coefficient γ is evaluated the M_2 amplitude is $0.9\,081$ and the principal solar semidiurnal harmonic amplitude is $0.4\,229$. Because only the relative coefficients are of further interest the normalization to H_{M2} is more convenient and is tabulated here. The different latitude variations for the three species are given in equation (3:12). The spectrum of tidal harmonics may be plotted for each species, as shown in Figure 4:1(a), to illustrate the importance of a few major tidal harmonics and their distribution into groups. At this stage in the development it is useful to reassure ourselves that further development of the Equilibrium Tide will be relevant for subsequent tidal analysis. Figure 4:1(b) shows the relationships between the Equilibrium Tide amplitudes and phases, and those obtained by harmonic analysis of observations at Newlyn. It confirms that the relationships are generally similar for constituents of similar angular speed. We will return to this similarity in Section 4:3:1.

The values of s, h, p, N and p' at any time are given in Table 4:2. Times are calculated from Greenwich midnight on 1 January 1900. Other time origins such as Greenwich noon on 31 December 1899 are sometimes used. Small secular changes in the rates of increase of these angles, involving terms in T^2 are also given (see also Cartwright (1985) for a discussion of modern time scales and their implications for tidal analysis). The rates of increase of these arguments are the angular speeds ω_2, ω_3, ω_4, ω_5, and ω_6 given in Table 3:2. In Table 4:1 the angular speed of each constituent is calculated using equation (4:1), where $i_a = 0, 1, 2$ for the long period, diurnal and semidiurnal tides. The values of i_a are said to define the tidal *species*.

Table 4:1 **Major harmonic tidal constituents** (based on Doodson, 1921; Schureman, 1976; Cartwright and Edden, 1973). The latitude variations are given in equation (3:12).

Table 4:1(a) Astronomical long-period tides, $i_a = 0$.

	Argument i_b (s)	i_c (h)	i_d (p)	i_e (N)	i_f (p')	Period (msd)	Speed f (cpd)	Speed σ (°/h)	Relative coefficient (M_2 = 1.0000)	Origin
*S_a	0	1	0	0	-1	364.96	0.0027	0.0411	0.0127	Solar annual*
*S_{sa}	0	2	0	0	0	182.70	0.0055	0.0821	0.0802	Solar semi-annual*
M_m	1	0	-1	0	0	27.55	0.0363	0.5444	0.0909	Lunar monthly
M_f	2	0	0	0	0	13.66	0.0732	1.0980	0.1723	Lunar semi-monthly

* Strongly enhanced by seasonal climate variations. For this reason the p' argument of S_a is ignored in modern tidal analysis (see text and Section 9:5:1).

Table 4:1(b) Astronomical diurnal tides; $i_a = 1$.

	Argument i_b (s)	i_c (h)	i_d (p)	i_e (N)	i_f (p')	Period (msd)	Speed f (cpd)	Speed σ (°/h)	Relative coefficient (M_2 = 1.0000)	Origin
$2Q_1$	-3	0	2	0	0	1.167	0.8570	12.8543	0.0105	Second-order elliptical lunar
σ_1	-3	2	0	0	0	1.160	0.8618	12.9271	0.0127	Lunar variation
Q_1	-2	0	1	0	0	1.120	0.8932	13.3987	0.0794	Larger elliptical lunar
ρ_1	-2	2	-1	0	0	1.113	0.8981	13.4715	0.0151	Larger evectional
O_1	-1	0	0	0	0	1.076	0.9295	13.9430	0.4151	Principal lunar
M_1 ⎧	0	0	-1	0	0	1.035	0.9658	14.4874	0.0117	Smaller elliptical lunar
⎨	0	0	0	0	0	1.035	0.9661	14.4920	0.0073	Lunar parallax
⎩	0	0	1	0	0	1.035	0.9664	14.4967	0.0326	Smaller elliptical lunar
χ_1	0	2	-1	0	0	1.030	0.9713	14.5695	0.0062	Smaller evectional
π_1	1	-3	0	0	1	1.006	0.9945	14.9179	0.0113	Larger elliptical solar
P_1	1	-2	0	0	0	1.003	0.9973	14.9589	0.1932	Principal solar
S_1	1	-1	0	0	1	1.000	1.0000	15.0000	—	Radiational
K_1 ⎧	1	0	0	0	0	0.997	1.0027	15.0411	0.3990	Principal lunar
⎩	1	0	0	0	0	0.997	1.0027	15.0411	0.1852	Principal solar

	i_b s	i_c h	i_d p	i_e N	i_f p'	Period (msd)	f (cpd)	σ (°/h)	Origin
χ_1	1	1	0	0	−1	0.995	1.0055	15.0821	Smaller elliptical solar
φ_1	1	2	0	0	0	0.992	1.0082	15.1232	Second-order solar
θ_1	2	−2	1	0	0	0.967	1.0342	15.5126	Evectional
J_1	2	0	−1	0	0	0.962	1.0390	15.5854	Elliptical lunar
OO_1	3	0	0	0	0	0.929	1.0759	16.1391	Second-order lunar

All the terms except for the parallax part of M_1 and the radiational S_1 (see Section 5:5), arise from the effects of lunar and solar declinations. The relative amplitude of the vector combination of the lunar and solar parts of K_1 is 0.5838.

Table 4:1(c) Astronomical semidiurnal tides; $i_a = 2$.

	Argument					Period	Speed		Relative coefficient	Origin
	i_b s	i_c h	i_d p	i_e N	i_f p'	(msd)	f (cpd)	σ (°/h)	($M_2 = 1.0000$)	
$2N_2$	−2	0	2	0	0	0.538	1.8597	27.8954	0.0253	Second-order elliptical lunar
μ_2	−2	2	0	0	0	0.536	1.8645	27.9682	0.0306	Variational
N_2	−1	0	1	0	0	0.527	1.8960	28.4397	0.1915	Larger elliptical lunar
ν_2	−1	2	−1	0	0	0.526	1.9008	28.5126	0.0364	Larger evectional
M_2	0	0	0	0	0	0.518	1.9322	28.9841	1.0000	Principal lunar
λ_2	1	−2	1	0	0	0.509	1.9637	29.4556	0.0074	Smaller evectional
L_2 {	1	0	−1	0	0	0.508	1.9686	29.5285	0.0283	Smaller elliptical lunar
}	1	0	1	0	0	0.508	1.9692	29.5378	0.0071	Smaller elliptical lunar
T_2	2	−3	0	0	1	0.501	1.9973	29.9589	0.0273	Larger elliptical solar
S_2	2	−2	0	0	0	0.500	2.0000	30.0000	0.4652	Principal solar
R_2	2	−1	0	0	−1	0.499	2.0027	30.0411	0.0039	Smaller elliptical solar
K_2 {	2	0	0	0	0	0.499	2.0055	30.0821	0.0865	Declinational lunar
}	2	0	0	0	0	0.499	2.0055	30.0821	0.0402	Declinational solar
M_3	0	0	0	0	0	0.345	2.8984	43.4761	0.0131	Lunar parallax ($i_a = 3$)

The relative amplitude of the vector combination of the lunar and solar parts of K_2 is 0.1266.

Table 4:2 The orbital elements used for harmonic expressions of the Equilibrium Tide. Small secular trends in the speeds are included by the T^2 terms.

Mean longitude of moon	$s = 277.02° + 481\,267.89° \cdot T + 0.0\,011° \cdot T^2$
Mean longitude of sun	$h = 280.19° + 36\,000.77° \cdot T + 0.0\,003° \cdot T^2$
Longitude of lunar perigee	$p = 334.39° + 4\,069.04° \cdot T + 0.0\,103° \cdot T^2$
Longitude of lunar ascending node	$N = 259.16° - 1\,934.14° \cdot T + 0.0\,021° \cdot T^2$
Longitude of perihelion	$p' = 281.22° + 1.72° \cdot T + 0.0\,005° \cdot T^2$

T is the time in units of a Julian century (36 525 mean solar days) from midnight at the Greenwich meridian on 0/1 January 1900.

For zero hour GMT on day D in year Y:

$$T = \frac{365 \cdot (Y - 1900) + (D - 1) + i}{36\,525}$$

where i is the integer part of $(Y - 1901)/4$.
Note that 1900 was not a leap year, but 2000 is a leap year.

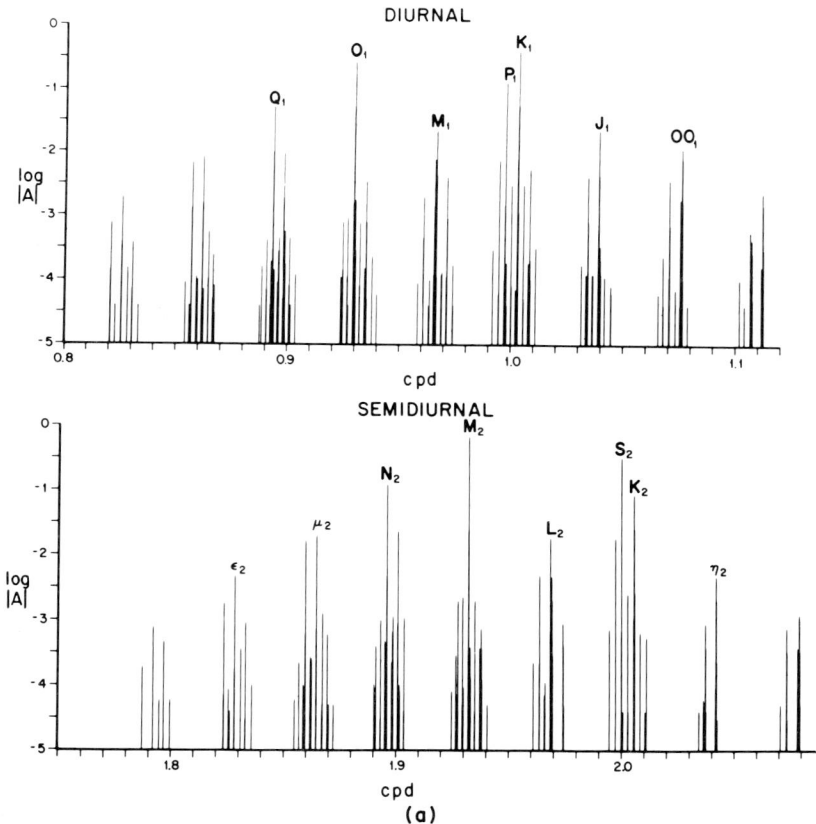

(a)

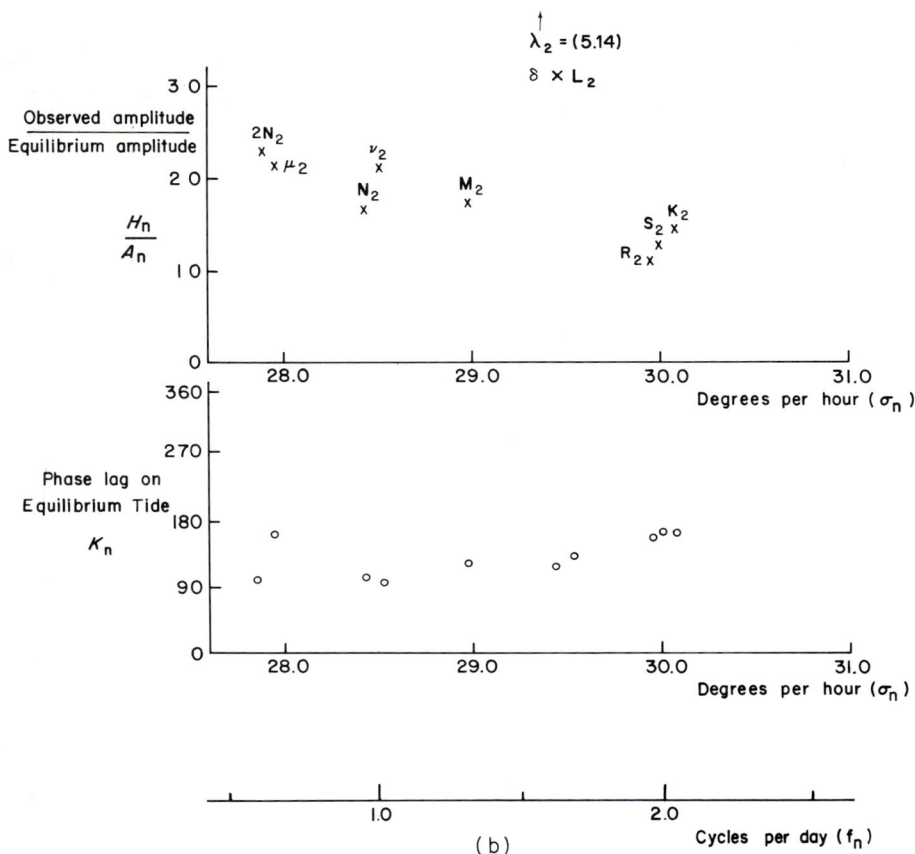

Figure 4:1 (a) The spectrum of harmonics in the Equilibrium Tide, from the expansion of Cartwright and Edden (1973). (Supplied by D. Carr Agnew.) (b) The ratios of the observed constituent amplitudes (A_n) to the Equilibrium Tidal amplitudes (H_n) for latitude 50°, and the phase lags (κ_n) in local time for the principal constituents of the semidiurnal species at Newlyn.

The energy in these species is modulated by harmonic terms involving ω_2, ω_3, etc. In the complete expansion i_b varies from -5 to $+5$ and defines the *group* within each species. Within each group the value of i_c, which also varies from -5 to $+5$ is said to define the *constituent*. In general, analysis of a month of data can only determine independently those terms which are separated by at least one unit of ω_2 (or $\partial s/\partial t$), $\Delta\sigma = 0.5\,490°$/h. Similarly, analysis of a year of data can determine harmonic constituents which differ by one unit in $\omega_3(\partial h/\partial t)$, $\Delta\sigma = 0.0\,411°$/h. The full set of integers i_a to i_f is sometimes said to define a *harmonic*, but this word is also more generally used to describe any term determined by a harmonic analysis.

Note that our equation (4:1) defines species in terms of ω_1, lunar speeds, rather than in terms of ω_0, the time in solar days. Times in lunar days are favoured by many writers including Doodson (1921) and Cartwright and

Edden (1973). The coding for M_2 is (2 0 0 0 0 0). Schureman (1976) has (2 –2 2 0 0 0) as for M_2 because he uses ω_0 instead of ω_1. Conversion between the two systems by using the relationship $\omega_1 = \omega_0 - \omega_2 + \omega_3$ is straightforward: it shows that Schureman's values of i_b will be one unit less than ours for the diurnal species, and two units less for the semidiurnal species. Also, his values of i_c are one unit more than ours for the diurnal species and two units more for the semidiurnal species. The conversion to the system in solar days is an essential step in the preparation of predictions in solar time (see Table 4:8). In Doodson's tabulations he avoided negative values of the i coefficients by adding five units to each so that his M_2 became (2 5 5 5 5 5).

The traditional alphameric labelling for the individual larger hamonics has been developed to include all the terms in our tables, but the full expansion contains hundreds of smaller unnamed terms. Strictly, these labels apply only for terms defined by the full six integers i_a to i_f, but they are also applied to the constituents determined by analysis of a year of data. The use of a suffix to denote the species is normal for all except the long-period tides, where the zero is usually omitted. The basic astronomical movements which each constituent represents are also given in Table 4:1. The elliptical terms and the declinational terms have already been discussed in a general way in Chapter 3. Variational terms are due to the changes in the sun's perturbing force on the moon's motion during a synodic month. Evection is the change in the eccentricity of the moon's orbit caused by the sun's gravitational pull.

In addition to the astronomical forces, there are a few harmonic constituents which are primarily due to periodic meteorological forcing, notably the annual and semi-annual tides and the S_2, diurnal tide. These have been called the *radiational* tides because of their relationship to the cycles in direct solar radiation.

The largest term in the long-period species is usually S_a, the annual variation in mean sea-level, which has an angular speed of $(\omega_3 - \omega_6)$, but this term is very small in the gravitational forcing. The large observed annual variations of level are due to seasonal meteorological effects. These seasonal effects have a speed of ω_3 and are of course indistinguishable from the gravitational tide in a harmonic analysis. It is standard practice to analyse for the radiational S_a term of speed ω_3; of course the frequency difference is insignificant, but it is vital to define the phase reference consistently. The gravitational semi-annual tide $S_{sa}(2\omega_3)$ is larger than the annual gravitational tide, and the observed tide also includes seasonal effects. These long-period tides are discussed in detail in Chapter 9. The terms $M_m(\omega_2 - \omega_4)$ and $M_f(2\omega_2)$, the lunar monthly and fortnightly tides are the lunar equivalent to S_a and S_{sa}, but there are no extra radiational effects at these periods.

In the diurnal species all but two small terms arise directly from the effects of lunar and solar declination. These diurnal terms have a tendency to occur in pairs which balance each other when the lunar or solar declination and hence the diurnal component amplitude is zero, as shown in Figure 1:1 for March 1981. The lunar declination is represented by the two terms $O_1(\omega_1 - \omega_2)$ and

$K_1(\omega_1 + \omega_2)$ which are in phase every 13.66 days. The K_1 $(\omega_1 + \omega_2 = \omega_0 + \omega_3)$ term also contains a part due to the solar declination which balances the P_1 solar term $(\omega_0 - \omega_3)$ at intervals of 182.55 days, the times of spring and autumn equinoxes when the solar declination d_s is zero. The constituents $Q_1(\omega_1 - 2\omega_2 + \omega_4)$, $J_1(\omega_1 + 2\omega_2 - \omega_4)$, and parts of the $M_1(\omega_1 \pm \omega_4)$ constituent are due to the changing lunar distance R_1. The M_1 term is complicated because it contains three small factors, two related to the lunar declination and varying lunar distance, and a third part which is an exact subharmonic of the principal semidiurnal lunar tide. Separation of these terms theoretically requires 8.85 years of data according to the Rayleigh criteria discussed in Section 4:2:5. The exact subharmonic $M_1(\omega_1)$, is derived by expanding the $\mathscr{P}_3$ $(\cos\varphi)$ term in equation (3:7). Physically it represents the fact that the diameter of the earth is not negligible compared with the lunar distance, so that the double bulge is not quite symmetrical. If the orbit of the moon were permanently in the plane of the equator none of the declinational diurnal tides would occur, but this small lunar diurnal tide would remain. For practical purposes it is too small to be important, but its determination is of some analytical interest (Cartwright, 1975). Turning to another small but interesting constituent, the gravitational tide producing force at S_1 is negligible, but the observed tides contain significant energy at this frequency because of the diurnal meteorological forcing, and this radiational S_1 tide may have amplitudes of a few centimetres in extreme cases (see Section 5:5). Temperature-sensitive sea-level recorders have also been known to introduce a spurious S_1 term into the data, as may an incorrectly mounted chart drum which rotates once every 24 hours.

The semidiurnal species is dominated by the principal semidiurnal lunar tide $M_2(2\omega_1)$ and the principal semidiurnal solar tide $S_2(2\omega_0)$, which are in phase at spring tides, every 14.76 days. Of the two main terms which represent the movement of the moon in its elliptical orbit, $N_2(2\omega_1 - \omega_2 + \omega_4)$, is much bigger than $L_2(2\omega_1 + \omega_2 - \omega_4)$, as already discussed. The two terms which similarly represent the effects of the varying solar distance show the same asymmetry (the calculations above can be repeated to show this) with the T_2 term $(2\omega_0 - \omega_3 + \omega_6)$, being much larger than the term at the higher frequency, $R_2(2\omega_0 + \omega_3 - \omega_6)$. The $K_2[2(\omega_1 + \omega_2) = 2(\omega_0 + \omega_3)]$ term represents the lunar and the solar declination effects. Also shown in Table 4:1(c) is the small third-diurnal harmonic $M_3(3\omega_1)$ which, like the (ω_1) part of the M_1 term arises from the asymmetry of the semidiurnal tidal bulges; this term is also obtained by expanding the $\mathscr{P}_3(\cos \varphi)$ term in equation (3:7). The L_2 constituent is interesting because it contains two terms whose angular speeds differ by $2\omega_4$. These two terms cannot be separated from analysis of only a year of data; because of this, special methods must be used to represent the slowly changing amplitude of the total L_2 constituent.

* 4:2:2 Nodal factors

Certain lunar constituents, notably L_2, are affected by the 8.85-year cycle, and all the lunar constituents are affected by the 18.6-year Nodal cycle; these modulations, which cannot be separately determined from a year of data must also be represented in some way. In the full harmonic expansion they appear as terms separated from the main term by angular speeds:

$$i_d \omega_4; \ i_e \omega_5$$

The terms that are separated by $i_f \omega_6$ may be considered constant for all practical purposes. The modulations which cannot be resolved as independent constituents by analysis of a year of data are represented in harmonic expansion by small adjustment factors f and u. Each constituent is written in terms of degrees:

$$H_n f_n \cos [\sigma_n t - g_n + (V_n + u_n)] \qquad (4:4)$$

where V_n is the phase angle at the time zero, shown in Table 4:2. The nodal terms are:

f_n the nodal factor
u_n the nodal angle

both of which are functions of N and sometimes of p. The nodal factor and the nodal angle are 1.0 and 0.0 for solar constituents.

The application of nodal terms can be illustrated by the variation of $\mathbf{M_2}$. If yearly analyses are made throughout a nodal period of 18.61 years the value of the $\mathbf{M_2}$ amplitude will increase and decrease from the mean value by about 4 per cent. The full harmonic expansion shows that in addition to the main $\mathbf{M_2}$ term $(2\omega_1)$ there are several other nearby terms; the only significant one has angular speed $(2\omega_1 + \omega_5)$ and a relative amplitude of -0.0373, the negative coefficient signifying that in the full expansion this term appears as $-\cos[(2\omega_1 + \omega_5)t]$.

Writing the total constituent as:

$$H_{\mathbf{M2}} \cos (2\omega_1 t) - \alpha H_{\mathbf{M2}} \cos (2\omega_1 t + \omega_5 t)$$

we have

$$H_{\mathbf{M2}} \{(1 - \alpha \cos \omega_5 t) \cos 2\omega_1 t + \alpha \sin \omega_5 t \sin 2\omega_1 t\}$$
$$= H_{\mathbf{M2}} \{f \cos u \cos 2\omega_1 t - f \sin u \sin 2\omega_1 t\}$$
$$= H_{\mathbf{M2}} \{f \cos (2\omega_1 t + u)\}$$

where

$$f \cos u = 1 - \alpha \cos \omega_5 t$$
$$f \sin u = -\alpha \sin \omega_5 t$$

and hence for small values of α:

$$f^2 = 1 - 2\alpha \cos \omega_5 t + \alpha^2$$
$$f \approx 1 - \alpha \cos \omega_5 t = 1 - 0.0373 \cos \omega_5 t$$

and

$$\tan u = \frac{-\alpha \sin \omega_5 t}{1 - \alpha \cos \omega_5 t}$$

(where α is in radians)

$$u \approx -\alpha \sin \omega_5 t = -2.1° \sin \omega_5 t.$$

The phase of the nodal angle $\omega_5 t$ is measured from the time when the ascending lunar node, at which the moon crosses the ecliptic from south to north, is at the First Point of Aries (Υ). Around this time the excursions of the moon north and south of the equator reach maximum declinations of 28° 36′, and the f value for M_2 has a minimum value of 0.963. After an interval of 9.3 years when the moon has minimum declination excursions of 18° 18′ the f value reaches its maximum, 1.037. If the nodal adjustments were not made, analyses of a year of data made at the time of maximum declinations would give values of H_{M2} which were 7.5 per cent lower than those which would be obtained 9.3 years later.

The factors which determine f and u for the major lunar constituents are given in Table 4:3; for the constituents K_1 and K_2 only the lunar part has a nodal modulation and this has been allowed for in the development. The terms which represent the changes in declination have the largest nodal variations. The O_1 amplitude varies by 18.7 per cent, the K_1 amplitude by 11.5 per cent and the K_2 amplitude by 28.6 per cent. Small terms in these expansions, involving arguments of $2N$ and $3N$ have been omitted.

Table 4:3   Basic nodal modulation terms for the major lunar constituents.

	f	u
M_m	$1.000 - 0.130 \cos(N)$	$0.0°$
M_f	$1.043 + 0.414 \cos(N)$	$-23.7° \sin(N)$
Q_1, O_1	$1.009 + 0.187 \cos(N)$	$10.8° \sin(N)$
K_1	$1.006 + 0.115 \cos(N)$	$-8.9° \sin(N)$
$2N_2, \mu_2, \nu_2, N_2, M_2$	$1.000 - 0.037 \cos(N)$	$-2.1° \sin(N)$
K_2	$1.024 + 0.286 \cos(N)$	$-17.7° \sin(N)$

$N = 0$ March 1969, November 1987, June 2006, etc., at which time the diurnal terms have maximum amplitudes, but M_2 is at a minimum. M_2 has maximum Equilibrium amplitudes in July 1978, March 1997, October 2015, etc.

For the constituents M_1 and L_2, the arguments also involve the longitude of lunar perigee p, which has an 8.85-year cycle. Some authorities make more elaborate expansions of the long-term modulations and include the effects of p in the factors for other constituents; these are called j and v to distinguish them from the f and u factors which are usually used. Older publications used to include expansive tables of f and u for each constituent and for each year, but modern computing procedures make this practice obsolete.

* 4:2:3 Shallow-water terms

In shallow water the progression of a tidal wave is modified by bottom friction and other physical processes which depend on the square or higher powers of the tidal amplitude itself. Figure 1:1 shows the presence of strong higher harmonics at Courtown; the double high and double low waters are due to non-linear shallow water effects. These distortions can also be expressed as simple harmonic constituents with angular speeds which are multiples, sums or differences, of the speeds of the astronomical constituents listed in Table 4:1. To illustrate this consider that shallow-water effects are proportional to the square of the tidal water level through a spring–neap cycle:

$$k'' \{H_{M_2} \cos 2 \omega_1 t + H_{S_2} \cos 2 \omega_0 t\}^2$$
$$= k'' \{\tfrac{1}{2}(H_{M_2}^2 + H_{S_2}^2) + \tfrac{1}{2}(H_{M_2})^2 \cos 4 \omega_1 t + \tfrac{1}{2}(H_{S_2})^2 \cos 4 \omega_0 t$$
$$+ H_{M_2} H_{S_2} \cos 2 (\omega_1 + \omega_0)t + H_{M_2} H_{S_2} \cos 2 (\omega_0 - \omega_1)t\}$$

where k'' is the constant of proportionality.

The relationship involving the square of the amplitude has produced additional harmonics at $4\omega_1$, and $4\omega_0$ which are called $\mathbf{M_4}$ and $\mathbf{S_4}$ because their frequencies appear in the fourth-diurnal species. The term at speed $2(\omega_1 + \omega_0)$, also in the fourth-diurnal species, is called $\mathbf{MS_4}$ to show that it originates from a combination of $\mathbf{M_2}$ and $\mathbf{S_2}$. The interaction has also produced a change in the mean sea-level, represented by the terms in $H_{M_2}^2$ and $H_{S_2}^2$, and a long-period harmonic variation with a speed equal to the difference in the speeds of the interacting constituents, $2(\omega_0 - \omega_1)$, called $\mathbf{M_{sf}}$, which has a period of 14.77 days, exactly the same as the period of the spring–neap modulations on the tidal amplitudes. The development may be continued to include interactions between other tidal harmonic constituents, and may also be extended to higher powers of interaction. Suppose, for example, that the interactive effects are related to the cube of the elevations, and that only the $\mathbf{M_2}$ constituent is present:

$$k''' (H_{M_2} \cos 2 \omega_1 t)^3 = \frac{k'''}{4} H_{M_2}^3 \{3 \cos 2 \omega_1 t + \cos 6 \omega_1 t\}$$

where k''' is the proportional constant for triple interaction.

Two non-linear terms are present. The first is at exactly the constituent speed $(2\omega_1)$, and the second is in the sixth-diurnal species. This term, which has a speed three times the speed of $\mathbf{M_2}$, is called $\mathbf{M_6}$.

In practice a whole range of extra constituents are necessary to represent distortions in shallow water. Some of the more important of these are given in Table 4:4. The nodal factors for shallow-water constituents, f and u are assumed to follow the factors for the constituents involved in their generation:

$$f (\mathbf{M_4}) = f (\mathbf{M_2}) \times f (\mathbf{M_2})$$
$$u (\mathbf{M_4}) = 2 u (\mathbf{M_2})$$

Clearly the shallow-water part of $\mathbf{M_2}$, generated by cubic interactions will have a greater nodal variation than that of the dominant astronomical part. In Table 4:4, the terms marked with asterisks are at the same speeds as pure astronomical constituents; $2MS_6$ is at the same speed as μ_2, and there is a weak astronomical harmonic constituent which coincides with M_{sf}.

Table 4:4   Some of the more important shallow-water harmonic constituents.

	Generated by	Angular speeds	°/hour (σ)	Nodal factor (f)
Long-period				
$*M_{sf}$	M_2, S_2	$M_2 - S_2$	1.0 159	$f(M_2)$
Diurnal				
MP_1	M_2, P_1	$M_2 - P_1$	14.0 252	$f(M_2)$
SO_1	S_2, O_1	$S_2 - O_1$	16.0 570	$f(O_1)$
Semidiurnal				
MNS_2	M_2, S_2, N_2	$M_2 + N_2 - S_2$	27.4 238	$f^2(M_2)$
$*2MS_2$	M_2, S_2	$2M_2 - S_2$	27.9 682	$f(M_2)$
MSN_2	M_2, S_2, N_2	$M_2 + S_2 - N_2$	30.5 444	$f^2(M_2)$
$2SM_2$	M_2, S_2	$2S_2 - M_2$	31.0 159	$f(M_2)$
Third-diurnal				
MO_3	M_2, O_1	$M_2 + O_1$	42.9 271	$f(M_2)*f(O_1)$
MK_3	M_2, K_1	$M_2 + K_1$	44.0 252	$f(M_2)*f(K_1)$
Fourth-diurnal				
MN_4	M_2, N_2	$M_2 + N_2$	57.4 238	$f^2(M_2)$
M_4	M_2	$M_2 + M_2$	57.9 682	$f^2(M_2)$
MS_4	M_2, S_2	$M_2 + S_2$	58.9 841	$f(M_2)$
MK_4	M_2, K_2	$M_2 + K_1$	59.0 662	$f(M_2)*f(K_1)$
S_4	S_2	$S_2 + S_2$	60.0 000	1.00
Sixth-diurnal				
M_6	M_2	$M_2 + M_2 + M_2$	86.9 523	$f^3(M_2)$
$2MS_6$	M_2, S_2	$2M_2 + S_2$	87.9 682	$f^2(M_2)$
Eighth-diurnal				
M_8	M_2	$4M_2$	115.9 364	$f^4(M_2)$

* Also contain a significant gravitational component ($2MS_2 \equiv \mu_2$).

Two terms which have not been included in Table 4:4 are sometimes necessary to represent a seasonal modulation in the M_2 constituent. This modulation, which is characteristic of M_2 tides in shallow-water regions, may be thought of as an interaction between M_2 and S_a, but its physical origin is more likely to be due to seasonal changes in the weather and their effects on the tides. These two terms are:

$$MA_2 \quad 2\omega_1 - \omega_3 \quad \sigma = 28.9\,430° \, h^{-1}$$
$$MB_2 \quad 2\omega_1 + \omega_3 \quad \sigma = 29.0\,252° \, h^{-1}$$

Harmonic analyses of observed tidal levels show that shallow-water terms in the fourth, sixth and higher even-order bands are usually more important than the terms in the odd-order bands (third, fifth, etc.). However, the relative importance of the shallow-water terms and the number required to represent the observed tidal variations varies considerably from region to region, depending on the physics and the severity of the interaction, as we discuss in Chapter 7.

4:2:4 Least-squares fitting procedures

In the harmonic method of analysis we fit a tidal function:

$$T(t) = Z_0 + \sum_N H_n f_n \cos [\sigma_n t - g_n + (V_n + u_n)] \qquad (4:5)$$

where the unknown parameters are Z_0 and the series (H_n, g_n). The fitting is adjusted so that $\Sigma S^2(t)$ the square of the difference between the observed and computed tidal levels:

$$S(t) = O(t) - T(t)$$

when summed over all the observed values, has its minimum value. The f_n and u_n are the nodal adjustments and the terms $\sigma_n t$ and V_n together determine the phase angle of the Equilibrium constituent. V_n is the Equilibrium phase angle for the constituent at the arbitrary time origin. The accepted convention is to take V_n as for the Greenwich Meridian, and to take t in the standard time zone of the station concerned (see Section 4:5 for conversion procedures). An older method, now obsolete for analysis and prediction, calculated the phase lags relative to the Equilibrium constituent transit at the station longitude; these phases were termed the 'kappa' phases and were denoted by κ. The main disadvantages of working in terms of local time was that for an ocean as a whole, which spans several time zones, the harmonic constants were difficult to interpret physically. Analyses of earth tides do not have the same difficulty and so the 'kappa' notation is still widely used.

The least-squares fitting procedure involves matrix algebra which is outside the scope of this account (standard subroutines are available in mathematical software packages), but schematically the equations may be written:

[observed level] = [Equilibrium Tide] [empirical constants]   (4:6)
   known   known   unknown

The tidal variation function is represented by a finite number of N harmonic constituents, depending on the length and quality of the observed data. Typically, for a year of data $N = 60$, but in shallow water more than a hundred constituents may be necessary. The choice of which constituents to include depends on the relative amplitudes in the expansion of the astronomical forcing (Table 4:1 and Figure 4:1(a)), but the Equilibrium amplitudes are not themselves involved in the computations.

The least-squares fitting is now performed rapidly by computer matrix inversion, but earlier versions of the harmonic method used tabulated filters to facilitate manual calculations of the constituents. In these older filtering techniques, gaps in the data were likely to cause difficulties, and interpolation was necessary. To overcome some of these difficulties, the Admiralty (Great Britain, 1986) developed a graphical analysis technique which allows the determination of many harmonic constituents, and has the virtue of allowing the analyst to understand the limitations of his data.

Least-squares fitting has several advantages over the older methods:

- Gaps in the data are permissible. The fitting is confined to the times when observations were taken. Analyses are possible where only daylight readings are available, or where the bottom of the range is missing (Evans and Pugh, 1982). However, care is necessary to avoid energy leaking from one harmonic to another.
- Any length of data may be treated. Usually complete months or years are analysed.
- No assumptions are made about data outside the interval to which the fit is made.
- Transient phenomena are eliminated—only variations with a coherent phase are picked out.
- Although fitting is usually applied to hourly values, analysis of observations at other time intervals is also possible.

4:2:5 Choice of constituents; related constituents

The selection of the values of σ_n, the constituents whose harmonic amplitudes and phases are to be determined in an analysis, is sometimes thought of as a black art. However, certain basic rules exist. In general the longer the period of data to be included in the analysis, the greater the number of constituents which may be independently determined.

One criterion often used is due to Rayleigh, which requires that only constituents which are separated by at least a complete period from their neighbouring constituents over the length of data available, should be included. Thus, to determine M_2 and S_2 independently in an analysis requires:

$$360/(30.0 - 28.98) \text{ hours} = 14.77 \text{ days}$$

To separate S_2 from K_2 requires 182.6 days. This minimum length of data necessary to separate a pair of constituents is called their *synodic period*. It has been argued that the Rayleigh criterion is unnecessarily restrictive where instrumental noise and the background meteorological noise are low. In practice, the Rayleigh criterion is a good guide for tidal analyses of continental shelf data from middle and high latitudes, but finer resolution is feasible in ideal conditions such as tropical oceanic sites.

When choosing which values of σ_n to include, scrutiny of tidal analyses from a nearby reference station is helpful. Where the data length is too short to separate two important constituents (separation of K_2 and S_2 is the obvious example), it is usual to relate the amplitude ratio and phase lag of the weaker term to the amplitude and phase of the stronger term. If a local reference station is not available for specifying these relationships, then their relationships in the astronomical forcing function, the Equilibrium Tide, may be used:

$$\alpha_E = \frac{\text{related amplitude}}{\text{reference amplitude}}$$

$$\beta_E = \text{related phase} - \text{reference phase} \qquad (4:7)$$

In the case of S_2 and the weaker K_2, $\alpha_E = 0.27$, and there is a zero phase lag. Table 4:5 gives a set of 8 related constituents and the Equilibrium relationship which is effective for analyses of 29 days of data around the northwest European Shelf, when included with the independently determined constituents listed alongside. For analyses of a year of data, sets of 60 or even 100 constituents are normally used. Analyses of 19 years of data may legitimately require more than 300 independent constituents.

Table 4:5 A basic set of 27 independent and 8 related harmonic constituents for the analysis of data (used for tides of the north-west European Shelf, and valid in many other areas.)

Independent constituents

Z_0	mean sea-level
M_m, M_{sf}	long-period
Q_1, O_1, M_1, K_1, J_1, OO_1	diurnal
μ_2, N_2, M_2, L_2, S_2, $2SM_2$	semidiurnal
MO_3, M_3, MK_3	third-diurnal
MN_4, M_4, SN_4, MS_4	fourth-diurnal
$2MN_6$, M_6, MSN_6, $2MS_6$, $2SM_6$	sixth-diurnal

Related constituents

	Related to	Equilibrium relationships α_E	β_E
π_1	K_1	0.019	0.0
P_1	K_1	0.331	0.0
ψ_1	K_1	0.008	0.0
φ_1	K_1	0.014	0.0
$2N_2$	N_2	0.133	0.0
ν_2	N_2	0.194	0.0
T_2	S_2	0.059	0.0
K_2	S_2	0.272	0.0

Any use of related constituents from a nearby station assumes that the relationships have some regional stability. The use of an Equilibrium relationship assumes that the ocean response to forcing at the different frequencies is the same, which is generally valid for fine nodal frequency differences. For constituents whose speeds are separated by the annual factor ω_3 or for groups separated by the monthly factor ω_2 the assumption needs more careful justification, particularly where there are extensive shallow-water distortions (see Section 4:2:3). The ocean responses to the astronomical forcing from different species are quite different as we shall discuss in Section 4:3.

In the process of least-squares harmonic analysis it is also possible to remove a constituent whose regional characteristics are known, before fitting for the independent constituents. The most probable examples for inclusion are the long-period harmonics S_a and S_{sa}. This input can improve the reliability of the final analysis.

When selecting constituents for inclusion it is also necessary to observe the Nyquist criterion, which states that no term having a period of less than twice the sampling interval may be resolved. In the usual case of hourly data sampling, this shortest period is two hours, so that resolution of M_{12} would just be possible. In practice this is not a severe restriction except in very shallow water, where sampling more often than once an hour is necessary to represent the tidal curves.

4:2:6 Harmonic equivalents of some non-harmonic terms

The harmonic constituents can be related to some of the common non-harmonic terms used to describe the tides, as defined in Section 1:4. The most useful of these terms describes the spring–neap modulations of the tidal range which, in the absence of shallow-water distortions are given by the combination of the principal lunar and principal solar semidiurnal harmonics:

$$Z_0 + H_{M2} \cos (2 \sigma_1 t - g_{M2}) + H_{S2} \cos (2 \sigma_0 t - g_{S2})$$

where the time is now taken for convenience as zero at syzygy, when the moon earth and sun are in line at new or full moon, when the gravitational forcing reaches its maximum value. We also note that $H_{M2} > H_{S2}$ and $\sigma_0 > \sigma_1$.

If we develop similar trigonometrical arguments to those used for the nodal effects in Section 4:2:2, and rewrite:

$$2 \sigma_0 t - g_{S2} = 2 \sigma_1 t - g_{M2} - \theta$$

where

$$\theta = 2(\sigma_1 - \sigma_0)t + (g_{S2} - g_{M2})$$

the combined level becomes:

$$Z_0 + (H_{M2} + H_{S2} \cos \theta) \cos (2 \sigma_1 t - g_{M2}) + H_{S2} \sin \theta \sin (2 \sigma_1 t - g_{M2})$$
$$= Z_0 + \alpha' \cos (2 \sigma_1 t - g_{M2} - \beta') \tag{4:8}$$

where

$$\alpha' = (H_{M2}^2 + 2 H_{M2} H_{S2} \cos \theta + H_{S2}^2)^{\frac{1}{2}}$$
$$\tan \beta' = (H_{S2} \sin \theta)/(H_{M2} + H_{S2} \cos \theta)$$

The maximum values of the combined amplitudes are:

$$\text{mean high-water springs} = Z_0 + (H_{M2} + H_{S2})$$
$$\text{mean low-water springs} = Z_0 - (H_{M2} + H_{S2})$$

when $\cos \theta = 1$.

The minimum values are:

$$\text{mean high-water neaps} = Z_0 + (H_{M2} - H_{S2})$$
$$\text{mean low-water neaps} = Z_0 + (H_{M2} - H_{S2})$$

when $\cos \theta = -1$

The time of maximum tidal ranges after new or full moon is given by:

$$\theta = 2(\sigma_1 - \sigma_0)t + (g_{S2} - g_{M2}) = 0$$

that is

$$t_{max} = \frac{(g_{S2} - g_{M2})}{2(\sigma_0 - \sigma_1)}$$

(4:9)

This lag usually has a positive value of several hours, which is the *age* of the tide. In shallow water there will be other constituents which modify these expressions, notably M_4 and MS_4, and so these expressions are only accurate where shallow-water distortions can be neglected.

The Mean High Water Interval between lunar transit and the next high tide in the usual case of a predominantly lunar semidiurnal tide is given by:

$$\frac{g_{M2}}{2\sigma_1}$$

where the phase lag is calculated on local time (see Section 4:5).

Highest and Lowest Astronomical Tides are not easily defined in terms of harmonic constituents (Amin, 1979). Merely to add together all the amplitudes of the harmonic constituents would give values which were too extreme, because the constituents are not physically independent. In particular the semidiurnal tides are slightly reduced when the diurnal tides are greatest ($\cos^2 d$ and $\sin 2d$ dependence). Also, the shallow water constituents conspire to reduce the extreme ranges: they represent the tendency for non-linear processes in nature to reduce extremes (see Chapter 7) by enhancing energy dissipation. The values of HAT and LAT are usually determined by examination of at least five years of predictions, or by making predictions for years of known astronomical extremes. Chart Datum is defined locally to be close to the Lowest Astronomical Tide level (see Appendix 5). One early definition which was used for Chart Datum where diurnal and semidiurnal tides are both important, as they are in parts of the Indian Ocean, was Indian Spring Low Water:

$$Z_0 - (H_{M2} + H_{S2} + H_{K1} + H_{O1})$$

The relative importance of the diurnal and semidiurnal tidal constituents is sometimes expressed in terms of a Form Factor:

$$F = \left(\frac{H_{K1} + H_{O1}}{H_{M2} + H_{S2}}\right)$$

The tides may be roughly classified as follows:

$F = 0$ to 0.25   semidiurnal form
$F = 0.25$ to 1.50   mixed, mainly semidiurnal
$F = 1.50$ to 3.00   mixed, mainly diurnal
$F = $ greater than 3.0   diurnal form

The tidal variations plotted in Figure 1:1(a) have Form Factors of 7.5 (Karumba), 1.6 (Musay'id), 0.25 (Bermuda) and 0.20 (Mombasa). The variations at Courtown which has strong shallow-water distortions are not sensibly represented by a Form Factor. A more exact definition of the tidal regime in terms of the relative importance of the diurnal and semidiurnal components must involve the ratios of the total variance in each species, but at best this type of description is only an approximate representation of the full tidal characteristics.

4:2:7   Analysis of altimeter data

The standard methods of analysing regularly sequenced values of sea-level observed over a month or a year are not applicable to the data produced by the altimetry instruments described in Section 2:2:6. Although details of the new techniques developed to deal with the along-track measurements of geocentric heights (Parke and Rao, 1983), are beyond the scope of this account, a brief summary is appropriate.

The altimeter measures its distance above the sea surface (h_1 in Figure 2:7), from which the level of the sea surface above the reference ellipsoid is calculated ($h_3 + h_4$) by knowing the satellite elevation h_2. The geocentric tide h_4 includes the ocean tide, as measured by a tide gauge, and the earth tide, which has two components, the body tide and the loading tide. The body tide is calculable from the Equilibrium Tide by multiplying by the appropriate Love number, and the loading tide can be allowed for by a small percentage reduction of the ocean tide (Woodworth and Cartwright, 1986).

For each orbit of the satellite there is an uncertainty in the satellite height of about 1 m, but this problem can be overcome very elegantly by analysing height differences along short arcs of the orbit (Cartwright and Alcock, 1981). The method consists of a fit to a set of level differences between cross-over points on the grid of track intersections, as shown in Figure 2:9. Three parameters are required to describe the sea surface height data for each point on the grid: two parameters describe the amplitude and phase variations of the semidiurnal tide and the third represents the variations of geoid height. A reliable $\mathbf{M}_2$ amplitude and phase is needed at one cross-over point in order to compute the values at all other points from these differences. The values at each point are otherwise independent of the values at each other point. An alternative analysis technique fits the altimetry data to smooth functions in space, such as the normal modes discussed in Section 5:3:2. When longer series of data are available, the method of differencing may be used to determine several more tidal parameters at each point, in addition to the $\mathbf{M}_2$ constituent.

4:3 Response analysis techniques

The basic ideas involved in Response Analysis are common to many activities. A system, sometimes called a 'black box' is subjected to an external stimulus or input. The output from the system depends on the input and the system's response to that input. The response of the system may be evaluated by comparing the input and output functions. These ideas are common in many different contexts. Financially one might ask how the economy responds to an increase in interest rates, and measure this response in terms of industrial production. In engineering the response of a bridge to various wind speeds and directions may be monitored in terms of its displacement or its vibration. Mathematical techniques for describing system responses have been developed and applied extensively in the fields of electrical and communication engineering.

In our own field, we have already considered the response of a stilling well, as measured by the levels in the well, to external waves of different amplitudes and periods (Section 2:2). For oscillations of long period (tides) the internal variations are not attenuated (the amplitude response is unity), but there may be a small phase lag. For short-period oscillations (wind waves) the stilling-well system is designed to have a very small response (the amplitude response approaches zero).

In tidal analysis we have as the input the Equilibrium Tidal potential. The tidal variations measured at a particular site may be considered as the output from the system. The system itself is the ocean, and we seek to describe its response to the gravitational forces. This treatment has the conceptual advantage of clearly separating the astronomy, the input, from the oceanography, which is the 'black box'. Although primarily intended to describe linear systems, where the output is proportional to the input, the techniques of response analysis can be extended to deal with moderately non-linear situations, such as the propagation of tides in shallow water. The ideas can also be extended to include other inputs such as the weather, in addition to the gravitational forcing.

4:3:1 The credo of smoothness

When we reduce a year of hourly sea levels to sixty or a hundred pairs of harmonic constants, it is proper to ask whether the information could be contained equally well in fewer more significant parameters. The answer is certainly 'yes'. If the ratio between the observed amplitudes and the Equilibrium Tide amplitudes are plotted as a function of the angular speed of the constituents as in Figure 4:1(b) it is clear that a pattern exists within the harmonic data, with amplification falling steadily from greater than 20 for constituents below $28.0° h^{-1}$ to around 12 for constituents near $30° h^{-1}$. The phase lags also show a systematic change with the angular speed of the constituent, increasing from near $90°$ to nearly $180°$. Nevertheless, despite the general consistency, individual constituents may deviate from the general trends of these patterns. In Figure 4:1(b) the ratios and phase lags are for the Equilibrium Tide at the Newlyn latitude and longitude, but this relationship to local coordinates is an unnecessary refinement for general response analyses.

The smooth response curves of Figure 4:1(b) make it reasonable to suppose that ocean responses to gravitational forcing do not contain very sharp resonance peaks. We have already made this assumption: that responses to gravitational forcing at adjacent frequencies will be the same, in our development of formulae for the nodal adjustments of harmonic constituents; similarly the use of Equilibrium relationships in the analysis of short periods of data makes this implicit assumption.

4:3:2 Analysis procedures

The full details of the methods of response analysis are too complicated for this discussion; an outline of the theory is given in Appendix 2, but the serious analyst should consult the original paper by Munk and Cartwright (1966), and subsequent research publications.

The gravitational potential is represented as a series of complex spherical harmonics. The two which matter are $C_2^1(t)$ and $C_2^2(t)$, which correspond to the diurnal and semidiurnal species of the harmonic development. These are equivalent to C_1 and C_2 in equation (3:12). The tidal variations at any particular time are calculated as a weighted combination of the Equilibrium potential at past, present and future times:

$$
\begin{aligned}
\mathbf{T}(t) = \; & w(-96)\, C(t-96) + w(-48)\, C(t-48) \\
& + w(0) \quad C(t) \\
& + w(48) \quad C(t+48) + w(96) \quad C(t+96)
\end{aligned}
\qquad (4:10)
$$

In practice the w and C functions are mathematically 'complex'. Five discrete values of the forcing potential taken at 48-hour intervals are usually sufficient to give a good representation of the ocean's response functions. The frequency-dependent response itself is calculated as the Fourier transform of the weights. In the simple case of no lags, where only $w(0)$ is used, the admittances $Z(\omega)$, and hence the amplitude and phase responses, are constant across the band, and independent of frequency ω. As more lags are included more detail is added to the frequency variability of the response; this detail is sometimes called the 'wigglyness'! For weaker inputs such as $C_3^1(t)$ and $C_3^2(t)$, a single unlagged weight is adequate.

Figure 4:2 shows the amplitude and phase responses for Ilfracombe and Castletownsend in the Celtic Sea, obtained by fitting response weights by a least-squares procedure to the inputs $C_2^1(t)$ and $C_2^2(t)$. The ocean responses at the semidiurnal forcing are immediately seen to be much greater than the responses to the diurnal forcing; this suggests a near-resonant response at semidiurnal frequencies, as will be discussed in Chapter 5. The increase of phase lag with frequency in the semidiurnal band is consistent with the positive age of the tide which we have already expressed in Equation (4:9) in terms of the gradient:

$$\frac{(g_{S2} - g_{M2})}{2\,(\sigma_0 - \sigma_1)}$$

For Ilfracombe the responses are reasonably smooth. Castletownsend is on the south coast of Ireland, 350 km west of Ilfracombe. Although the semidiurnal responses are similar, the diurnal responses at Castletownsend show a different pattern with a minimum in the centre of the band close to 0.98 cpd. This minimum response is due to the proximity of a diurnal amphidrome (see Chapter 5).

* 4:3:3 Non-gravitational inputs

The normal response analysis procedure deals with the linear response of the ocean to the gravitational tidal forces. However, the amplitudes and phases determined by harmonic analyses do not always fit exactly on the response

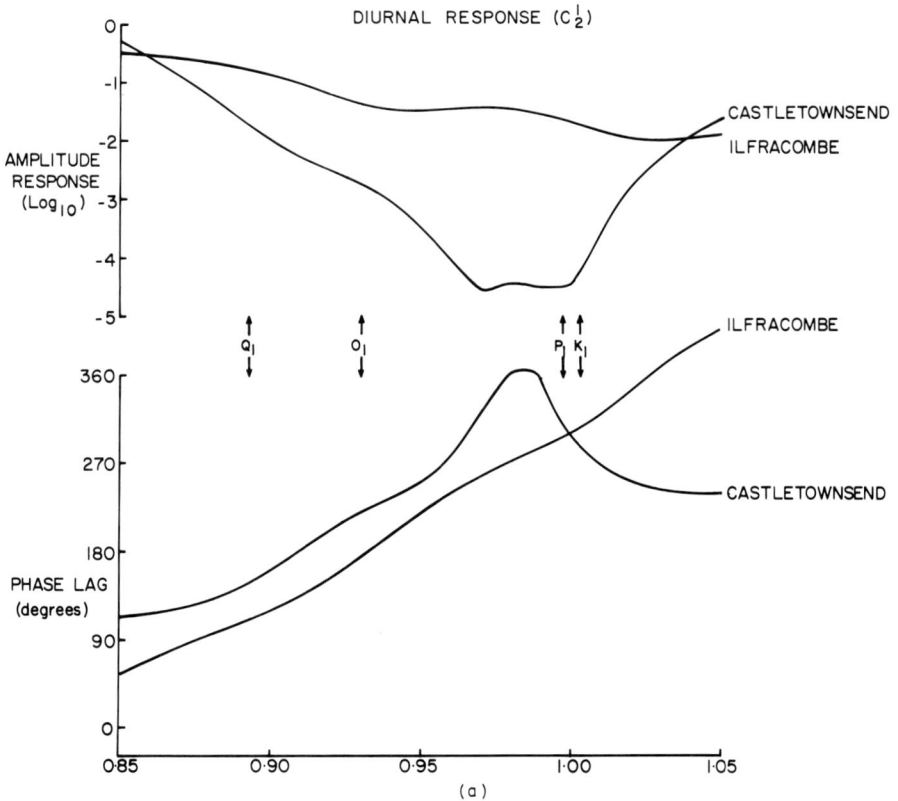

DIURNAL RESPONSE (c_2^1)

(a)

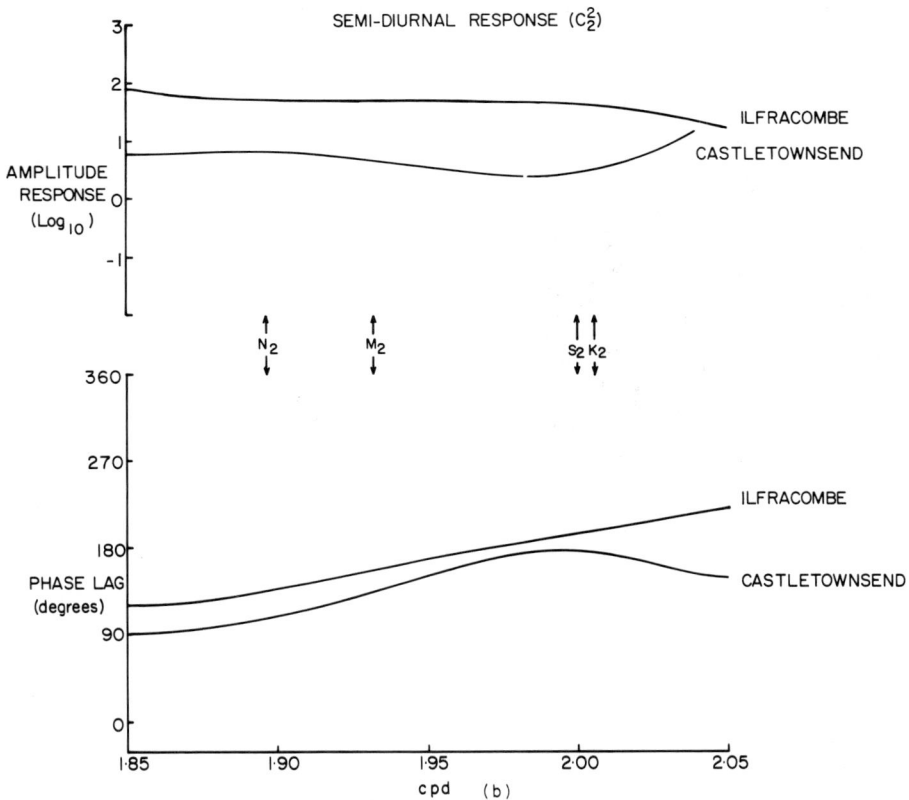

Figure 4:2   The response functions for tidal amplitudes and phases at Ilfracombe and Castletownsend on the west coast of the British Isles. The diurnal C_2^1 shows near-zero responses at K_1 for Castletownsend because of the local diurnal amphidrome.

curves: notable differences occur for μ_2, L_2 and even for M_2 and S_2. The diurnal species S_1, which has a very weak gravitational input, has an anomalously large amplitude response. The strength of the response analysis technique is that it not only identifies these anomalies, but it also allows us to investigate them in terms of additional input forcing functions. It is not possible to go into mathematical detail here, but the basic ideas behind the possible additional inputs may be outlined to emphasize the flexibility of the response approach.

Radiational inputs

Periodic influences on the oceans due to approximately regular cycles in the weather are recognizable in tidal observations as the annual term S_a and the diurnal term S_1, both of which are relatively large compared with their gravitational input. The semidiurnal term S_2 also has slightly anomalous phase and amplitude in the data from all parts of the world; response analysis separates the gravitational part from the radiational part of S_2. Physically the regular variations of tropical atmospheric pressure with a period of 12 hours and maximum values near 1000 and 2200 local time, the *atmospheric tides*, are the driving force. These pressure variations have amplitudes which vary approximately as $1.25 \cos^3\varphi$ mb (where φ is the latitude); a fuller discussion is given in Section 5:5. The physics of the annual and the diurnal constituents is related more directly to the solar radiation. Response analysis avoids the details of the physics by defining an input *radiation potential* which is calculated from the sun's coordinates (see Appendix 2). Response analyses which allow for both the gravitational and radiational inputs give a radiational component of S_2 with amplitudes 10–20 per cent of the gravitational amplitudes.

Shallow water inputs

Non-linear effects are negligible at ocean stations, but become increasingly important on the continental shelf and in regions of shallow water (Section 4:2:3). Harmonic analysis represents the distortions explicitly by introducing many additional constituents, and implicitly by absorbing them into the estimates of constituents already included. Response analysis allows us to manufacture additional inputs to represent the shallow-water interaction. These inputs are taken as the products of the local first-order linearly computed long period, diurnal and semidiurnal changes of sea-level $\zeta_0(t)$, $\zeta_1(t)$, $\zeta_2(t)$. Local values of the tide are used because non-linear effects are locally generated. The relative importance of the paired and the triple products depends on the type of local tidal regime. The $\hat{\zeta}(t)$, terms are actually mathematically complex qualities which have complex conjugates $\hat{\zeta}^*(t)$, and these are also included in the inputs. The table below gives some interactions which may be important where semidiurnal tides are dominant.

Species	Products	Notation
0	$\hat{\zeta}_2(t)\,\hat{\zeta}_2^*(t)$	I^{2-2}
1	$\hat{\zeta}_2(t)\,\hat{\zeta}_1^*(t)$	I^{2-1}
2	$\hat{\zeta}_2(t)\,\hat{\zeta}_2(t)\,\hat{\zeta}_2^*(t)$	I^{2+2-2}
3	$\hat{\zeta}_2(t)\,\hat{\zeta}_1(t)$	I^{2+1}
4	$\hat{\zeta}_2(t)\,\hat{\zeta}_2(t)$	I^{2+2}
6	$\hat{\zeta}_2(t)\,\hat{\zeta}_2(t)\,\hat{\zeta}_2(t)$	I^{2+2+2}

These interactions are neatly summarized in the I notation in column three. It is not generally necessary to compute lagged weights for these inputs, but the I^{2+2} term produces a much better fit if a lag term is introduced (Cartwright, 1968).

Standard station levels

When only short records of a few days or even months are available it is not possible to make a complete response analysis by direct reference to the gravitational potential. In harmonic analysis, several constituents are linked according to their Equilibrium Tide relationship or according to their relationships in a full analysis for a nearby standard station. The response analysis equivalent is to apply the observed or predicted tides at the standard station as an input, and to compute the response of the secondary station to this input. Only four or perhaps six weights need be computed for this local response. The full structure of the response curves is then inferred by using the characteristics of the standard station responses. This reference to a permanent local station is particularly useful for analysis of short periods of non-simultaneous bottom pressure records, because it allows the development of a pattern of regional tidal behaviour, without separate direct references to the astronomical forcing.

4:3:4 Comparisons of harmonic and response analysis

When the amplitude and phase responses at a station have been computed it is a simple matter to calculate the equivalent harmonic constituents from the harmonic equivalents in the gravitational and radiational tidal potentials. (The calculations are less straightforward for the responses to the shallow-water inputs because the individual harmonic terms have to be multiplied first.) The equivalence of the two methods also enables approximate response functions to be computed from the factors H_n/A_n and κ_n as in the plots of Figure 4:1. However, there are several other complementary aspects of the different approaches which should be appreciated.

The response concept of the oceans as a 'black box' whose physical characteristics are isolated for further study is very appealing to the physical oceanographer, and offers scope for further developments which has not yet been fully exploited. According to our criterion of a good analysis being one which represents the data with the smallest possible number of parameters, the response procedure is superior to the harmonic approach: typically the response technique can account for slightly more of the total variance than the harmonic method can accommodate, using less than half the number of harmonic parameters. Its advantages as a research tool include the identification of additional forcing inputs, and the opportunity of making subtle choices of factors for inclusion. A typical response analysis will have the gravitational input, the solar radiation input and a series of shallow-water interactive inputs. By comparison, the harmonic analysis approach offers few alternatives, apart from the addition of extra constituents, and little scope for development. The laborious adjustments for nodal variations illustrate this restriction. These variations are automatically taken care of in response analysis. In harmonic analyses the shallow-water effects and the radiation effects are automatically absorbed together into the estimates of the constituents. The penalty for the more flexible response approach is that more elaborate computations are required—for example, the shallow-water treatment requires a preliminary response analysis to calculate the linear inputs $\hat{\zeta}_1(t)$, $\hat{\zeta}_2(t)$, for input to the final analysis.

The response admittances, or more specifically their amplitude and phase responses at a particular frequency, are stable in the same way as the amplitudes and phases of the harmonic constituents are stable. Nevertheless, it is usual to present maps of tides in terms of the harmonic amplitudes and phase lags for individual constituents. They have clearly identified units of length and time which are useful for practical applications. Also, charts of harmonic constituents can be interpreted scientifically in terms of wave propagation, as discussed in Chapter 5.

4:4 Analysis of currents

Currents are intrinsically more difficult to analyse than elevations because they are vector quantities which require more parameters for a proper description (Godin, 1972; Foreman, 1977). Also, measurements of currents usually show a larger proportion of non-tidal energy than do elevations, which makes the errors in the estimated tidal components larger (Section 4:6). A third restriction is the difficulty and expense of making current measurements offshore, as described in Chapter 2. Although the techniques used for analysing levels are also available for dealing with currents, when analysing currents less stringent standards can be applied for these reasons. However, vector techniques are available for analysing currents which are not applicable to levels. Currents are traditionally measured as a speed q and a direction of flow θ clockwise from the north. Mathematically it is usual to work with Cartesian coordinates and current components in a positive north and positive east direction, v and u. The relationships are straightforward:

$$v = q \cos \theta$$
$$u = q \sin \theta$$

4:4:1 Non-harmonic methods for currents

The currents traditionally measured from an anchored ship consisted of hourly or half-hourly speeds and directions over a period of 13 or 25 hours. When plotted these would form a series of vectors as in Figure 1:2, called a central vector diagram. The vector lines may be replaced by a single dot at the outer end which marks the vector point. When several hundreds of these vector points are plotted to represent observations of currents made over some weeks, by a recording current meter (usually these meters sample every 10 or 15 minutes rather than every hour), a clear pattern emerges. An example of one of these *dot plots* is given in Figure 4:3 for 54 days of 10-minute samples of current at the Inner Dowsing light tower. The axes shown are for the components of speed in a positive north-going and positive east-going convention. At strong

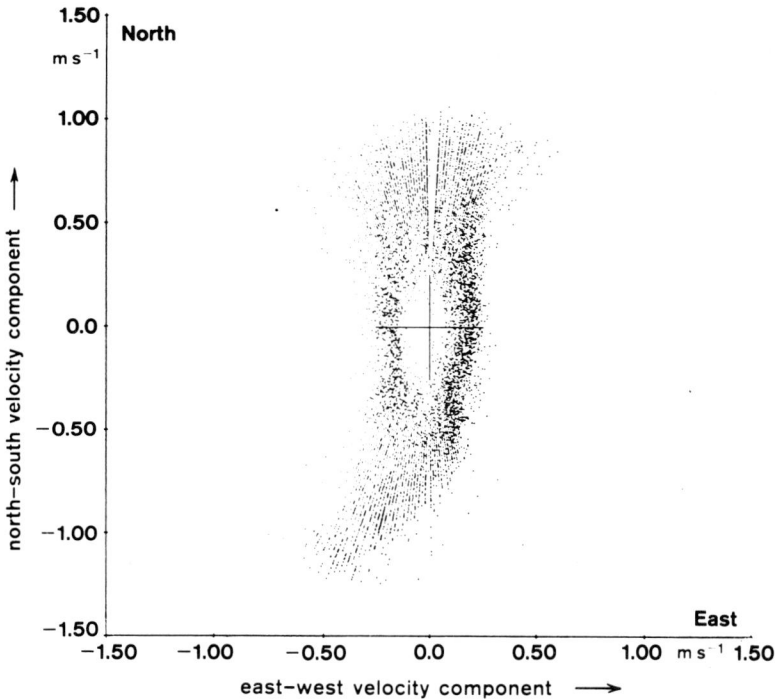

Figure 4:3 Dot-plot representation of 54 days of current measurements at the Inner Dowsing light tower. Dots away from the regular pattern show the influence of the weather on the measured currents. The central area which is nearly blank shows that zero current speeds are very rare. (From D. T. Pugh and J. M. Vassie (1976).) (*Reproduced by permission of Deutsches Hydrographisches Institut.*)

south-going spring currents, the flows are deflected more towards the west than is the case for neap currents because of the local coastal topography. The narrow blank segment at the top of the pattern is due to a dead space in the potentiometer used in the current meter to sense the direction of the flow.

A series of 25 hourly measurements of current speed and direction may be analysed by comparison with the times of high water at a nearby Standard Port as described in Table 4:6. The spring and neap ranges and the mean range of the tide for the period of the observations are also needed for scaling the current speeds. Here we are analysing 25 hours of currents at the Inner Dowsing light tower from 1800 on 1 March 1972 to 1800 on 2 March 1972. Current speeds and directions are plotted separately against time as in Figure 4:4. Note that for this analysis it is usual to measure directions in the geographical sense, clockwise from north. The observations are then joined by a series of smooth lines and the times of high water at the chosen Standard Port, in this case Immingham, are marked on the time axis. The next step is to list the Inner Dowsing current speeds and directions for the hourly times relative to the time of Immingham high water; these are read off the curves for each of the two cycles. The observed values will not be identical for each cycle because of

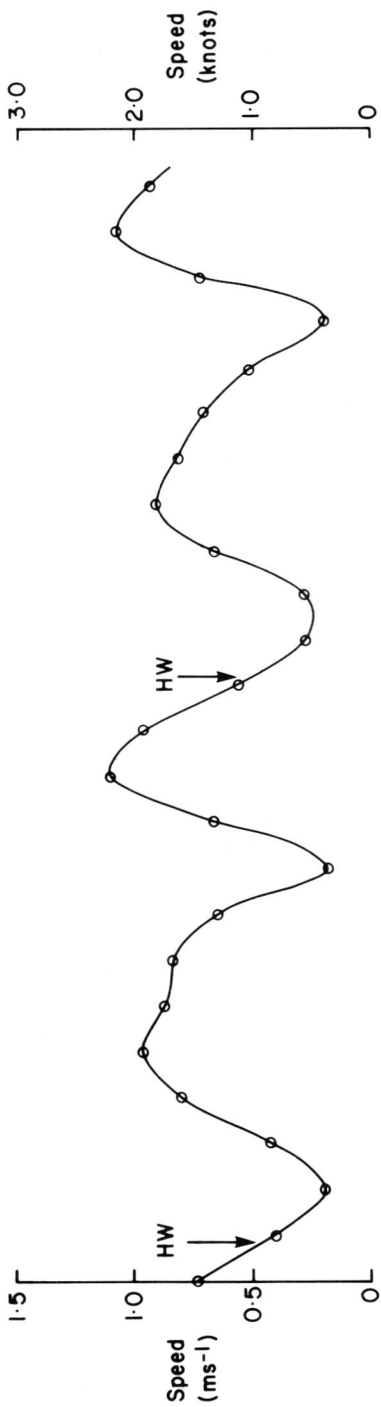

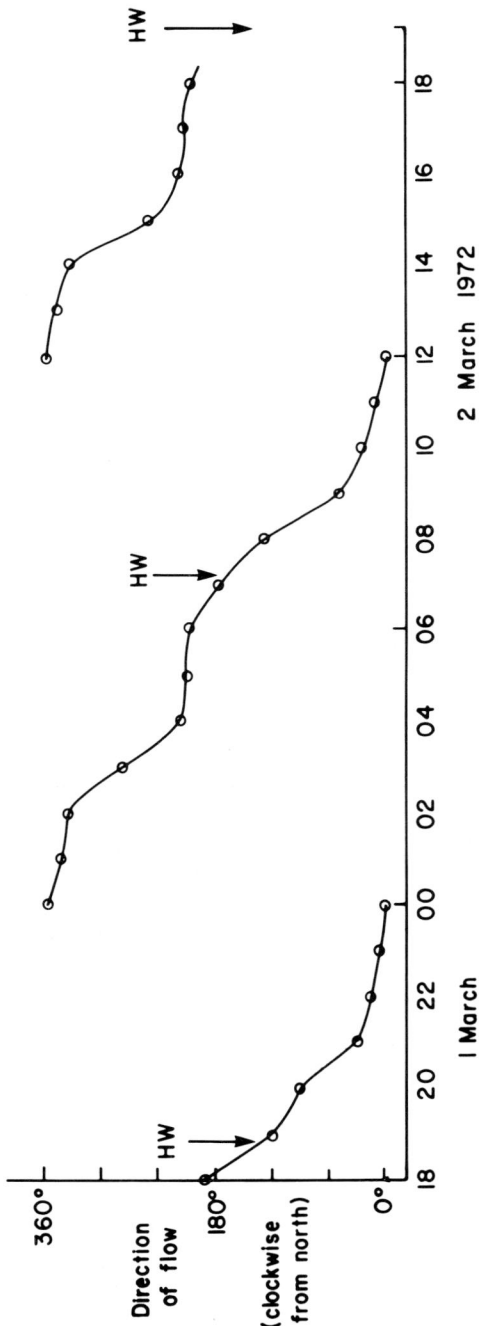

Figure 4:4 Observed current speeds and directions at the Inner Dowsing, 1/2 March 1972, used as a basis for calculating tabulated tidal currents.

Table 4:6 Calculations of tidal currents at the Inner Dowsing, 1/2 March 1972. Immingham is used as a Standard port for scaling.

Time relative to Immingham high water	Directions (degrees from north)			Speeds (m s^{-1})				
	Cycle		Mean	Cycle		Mean	Springs[a]	Neaps[b]
	1	2		1	2			
High water	130	163	147	0.48	0.50	0.49	0.51	0.25
+1 hr	94	107	101	0.21	0.20	0.21	0.22	0.11
+2	43	47	45	0.32	0.32	0.32	0.34	0.17
+3	16	18	17	0.70	0.71	0.71	0.75	0.37
+4	7	3	5	0.93	0.91	0.92	0.97	0.48
+5	355	353	354	0.91	0.79	0.85	0.89	0.44
+6	343	342	343	0.83	0.68	0.75	0.79	0.39
−6	338	341	340	0.82	0.68	0.75	0.79	0.39
−5	324	322	323	0.58	0.42	0.50	0.53	0.26
−4	257	233	245	0.21	0.20	0.21	0.22	0.11
−3	214	213	214	0.76	0.84	0.80	0.84	0.42
−2	209	210	210	1.10	1.06	1.08	1.13	0.59
−1	201	200	201	0.90	0.89	0.90	0.95	0.47

[a] Spring conversion factor $= \dfrac{\text{mean spring range at Immingham}}{\text{mean range for 1/2 March}} = \dfrac{6.4\,\text{m}}{6.1\,\text{m}} = 1.05$

[b] Neap conversion factor $= \dfrac{\text{mean neap range at Immingham}}{\text{mean range for 1/2 March}} = \dfrac{3.2\,\text{m}}{6.1\,\text{m}} = 0.52$

diurnal tidal changes, reading errors and weather effects. The mean values are then calculated for the two sets. Finally the current speeds for mean springs and mean neaps are calculated by using the adjustment factors. If residual flows are significant these should be removed before preparing Figure 4:4. The correct procedure for removing the residual flows is rather tedious because each hourly value must be resolved into two components, then the mean over 25 hours for each component calculated and removed from each hourly value prior to recomputing the hourly tidal speeds and directions from the adjusted components. In many cases, including the one discussed here, the residual currents are too weak to be significant, but in other cases they must be allowed for and stated in the final analysis.

4:4:2 Cartesian components, ellipses

If the currents are resolved into east-going and north-going components, u and v, the speeds and angles are:

$$q = (u^2 + v^2)^{\frac{1}{2}} \qquad \theta = \arctan(v/u) \text{ anticlockwise from east}$$

$$\text{or } \theta = \arctan(u/v) \text{ clockwise from north.}$$

The two components u and v can be subjected to separate harmonic analyses by least-squares fitting:

$$u(t) = U_0 + \sum_N U_n f_n \cos [\sigma_n t - g_{u_n} + (V_n + u_n)]$$

$$v(t) = V_0 + \sum_N V_n f_n \cos [\sigma_n t - g_{v_n} + (V_n + u_n)]$$

(4:11)

as for elevations (equation (4:5). The unknown parameters to be determined by the analysis are the mean current components U_0 and V_0, and two sets of harmonic constituent amplitudes and phases (U_n, g_{u_n}) (V_n, g_{v_n}). Each harmonic tidal constituent is represented by four parameters, two amplitudes and two phases.

In Appendix 3 these four parameters are shown to define an ellipse, the *current ellipse* which is traced by the tip of the current vector during one complete cycle. Several valuable properties of the constituent ellipses can be derived from the component parameters:

$$\text{current direction} = \arctan \frac{V \cos (\sigma t - g_v)}{U \cos (\sigma t - g_u)}$$

$$\text{current speed} = \{U^2 \cos^2 (\sigma t - g_u) + V^2 \cos^2 (\sigma t - g_v)\}^{\frac{1}{2}}$$

for anticlockwise rotation: $0 < g_v - g_u < \pi$
for clockwise rotation: $\pi < g_v - g_u < 2\pi$
for rectilinear flow: $g_v - g_u = 0, \pi, 2\pi \ldots$

Major and minor axes of the ellipse are calculated using equation (A3:5a) and (A3:5b). The direction of the major axis is given by (A3:8).

Although the analyses define a separate ellipse for each harmonic constituent, experience shows that those constituents which have similar angular speeds have ellipses of similar shape and orientation. This is dynamically reasonable, and corresponds to the uniform elevation responses plotted in Figure 4:1. Figure 4:5 shows the ellipses for the principal constituents in each species at the Inner Dowsing light tower. There is a steady clockwise rotation of the major axis of the ellipses as we move from diurnal currents through to the sixth-diurnal currents. Another interesting feature at this location is the sense of rotation, which is anticlockwise for all the harmonic constituents from the astronomical forces, but clockwise for those generated by shallow-water dynamics. The changes in amplitudes of the ellipses are also consistent with their different amplitudes in the astronomical forcing.

When choosing the constituents to be included in an analysis of a month of current observations, the terms most likely to be important are the same as the major terms for elevations at a nearby coastal station, and in the Equilibrium Tide. Non-resolvable constituents may be related to neighbouring constituents using the local elevation relationship, but care should be exercised when doing this for currents and elevations near an amphidrome (see Chapter 5).

Figure 4:5 Current ellipses for the principal constituents in tidal species at the Inner Dowsing light tower. Note the different scales for each species. The arrows show the sense of rotation of the current vector. (From D. T. Pugh and J. M. Vassie (1976).) (*Reproduced by permission of Deutsches Hydrographisches Institut.*)

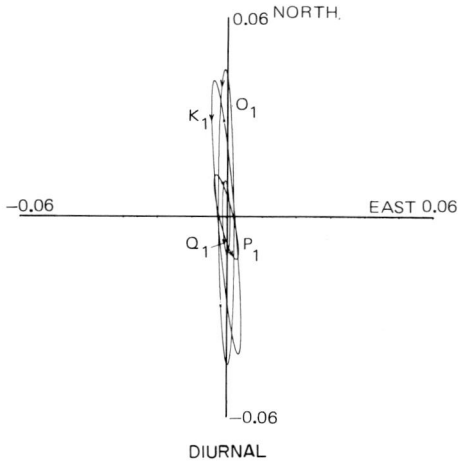

DIURNAL

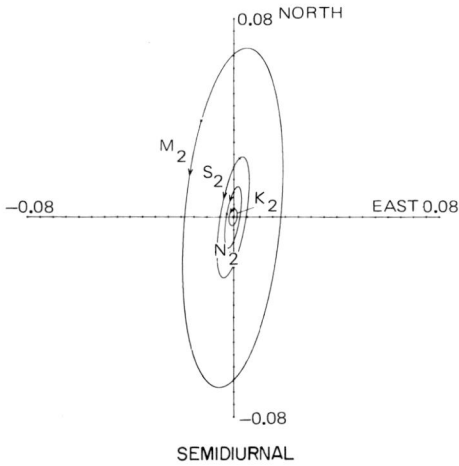

SEMIDIURNAL

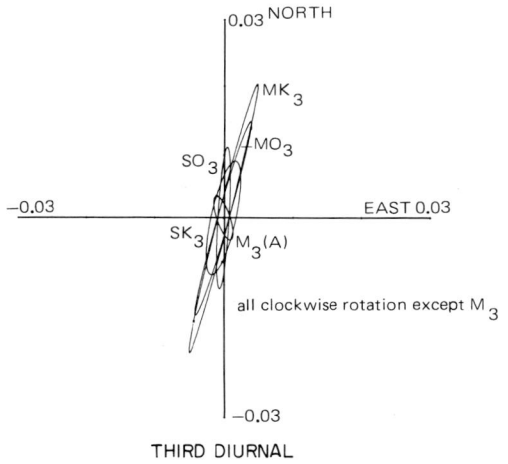

all clockwise rotation except M₃

THIRD DIURNAL

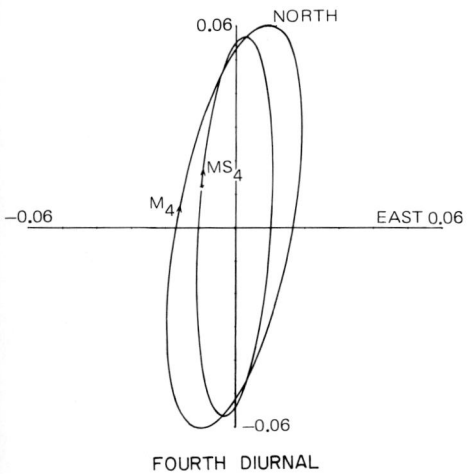

FOURTH DIURNAL

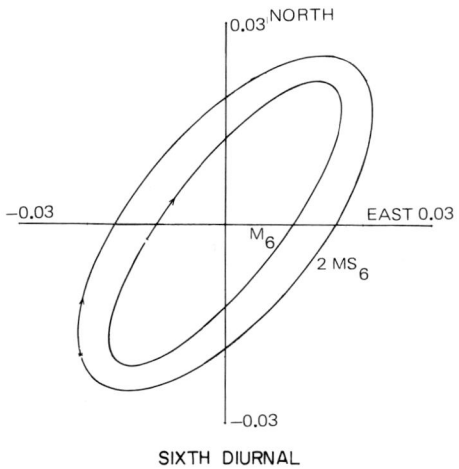

SIXTH DIURNAL

Although it is common practice to resolve the current components into north and east directions, it is sometimes preferable to resolve along the major and minor axes of the dominant flow; the major axis of flow is usually the major axis of the semidiurnal species (in Figure 4:5, by chance, the north–south axis nearly coincides with the major axis of the ellipse). Resolution along the major axis would be particularly appropriate if the currents are nearly rectilinear. Of course, all the properties of the ellipses themselves are independent of the axes chosen.

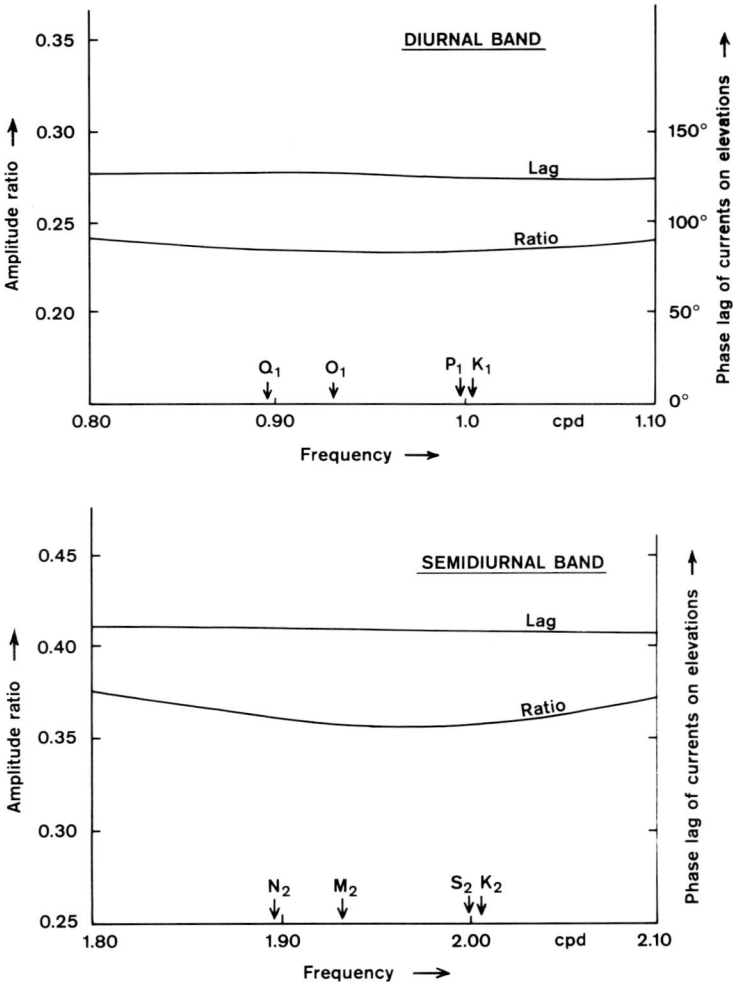

Figure 4:6 Phase and amplitude response relationship between currents and elevations at the Inner Dowsing, in the diurnal and semidiurnal tidal species. Currents in m s^{-1}, elevations in m. (From D. T. Pugh and J. M. Vassie (1976).) (*Reproduced by permission of Deutsches Hydrographisches Institut.*)

The versatility of the response analysis computer programs for interpreting tidal dynamics is shown by the curves in Figure 4:6, where the tidal elevations at the Inner Dowsing tower are used as an input and the dominant V-component of currents are the output. The variation of the phase and amplitude differences between elevations and currents are shown as a function of frequency. Such analyses can be used to determine the relative importance of progressive and standing wave elements in the tidal dynamics, and the location of amphidromes which will be discussed in Chapter 5.

4:4:3   Rotary current components

An alternative way of representing each tidal constituent ellipse, and the ellipse characteristics is to replace the amplitudes and phases of two components (U, g_u) (V, g_v) by the amplitudes and phases of two vectors each of which has a constant amplitude, which rotate at the angular speed of the constituent, but in opposite directions. When the vectors are aligned there are maximum currents, but when they are opposed the current speeds are a minimum. Further details of the calculations of the amplitudes and phases of the clockwise and anti-clockwise amplitudes and phases (Q_C, g_C) (Q_{AC}, g_{AC}) are given in Appendix 3.

The technique of analysing currents into rotary components is particularly valuable when studying inertial currents (section 6:4:6). These have a non-tidal origin, being generated by rotary winds and maintained by a balance between Coriolis and centrifugal accelerations. Their period is given by $2\omega_s \sin\varphi$, where φ is the latitude. They appear as a peak in the residual spectrum, after removing the tides from the rotary components, but only in the clockwise component in the northern hemisphere and in the anticlockwise component in the southern hemisphere.

4:5   Time zone conversion

Omitting the nodal factor, the harmonic part of a tidal constituent is:

$$\cos (\sigma_n t - G_n + V_n)$$

where σ_n is the angular speed of the nth constituent in degrees per unit time, and V_n is its phase in the Equilibrium Tide at the Greenwich Meridian at the zero time. G_n is the phase lag of the constituent in the observed tide on the Equilibrium Tide at Greenwich, in Greenwich Mean Time. If we wish to express the time in some other zone instead of GMT, we may do this, while still relating our phases to the same astronomical time zero, by rewriting:

$$t' = t - j$$

where j is the number of hours difference between the GMT value t and the zone time t'. By convention j is measured positive for zones west of Greenwich. The constituent argument is then:

$$\cos(\sigma_n t' - (G_n - j\sigma_n) + V_n)$$

or

$$\cos(\sigma_n t' - g_n + V_n)$$

where g_n is the phase lag in the new time zone:

$$g_n = G_n - j\sigma_n$$

The phase adjustment is slightly different for each constituent, even within a species. Note that where observations are supplied and predictions are required for the same time zone j, the values of g_n may be calculated in the analyses and used for prediction without any adjustment. For many purposes, such as preparing cotidal maps of oceans, the phases must all be adjusted to the same time zero, usually to the 'big G' notation. In this case the adjustment is:

$$G_n = g_n + j\sigma_n \qquad (4{:}12)$$

$$\text{Greenwich phase}$$
$$= \text{Local time zone phase}$$
$$+ \text{[Zone shift in hours, west positive]} \times \text{[constituent speed in } °\,h^{-1}]$$

Example of analysis of data from GMT + 5 (75°):

$$G_{M2} = g_{M2} + 5 \times 28.98 = g_{M2} + 145°.$$

The old convention of referring phases to local Equilibrium arguments involved redefining the astronomical zero time and hence V_n for each longitude: the λ_p factor in equation (3:20) was not kept at zero. Because of the difficulty of comparing tides at different longitudes the system is now obsolete, and these phases (called κ, the 'Kappa notation') are no longer used for marine tides. However, earth–tide analyses still favour this notation.

4:6   Accuracy of tidal parameters

It is usual to refer to harmonic tidal *constituents* rather than tidal *constants* when presenting the results of an analysis. Nevertheless, we expect these constituents to have only slight changes if a different period of data for the same place is analysed. However, analyses of individual months of tidal data from the same location invariably show small variations in the tidal parameters about some mean value. Table 4:7(a) summarizes the variability of the O_1, M_2 and M_4 harmonic constituents of elevation and current components at the Inner Dowsing light tower. In general, elevations are less variable than currents; greatest percentage variability is found in the weaker constituents of the current components. For M_2 the standard deviations of the monthly values of elevations, the V-component and the weaker U-component of current are

1.5, 8.1 and 14 per cent of the mean values of constituent amplitude in each case. Variability of the constituents from year to year is much less, but still not negligible. Table 4:7(b) summarizes the annual variability of major constituents in levels at Lerwick, Scotland over eight separate years. For M_2 the amplitude variation has a standard deviation of 0.5 per cent of the mean value.

Table 4:7(a) Stability of the amplitudes and phases of Inner Dowsing harmonic tidal constituents over one year of monthly analyses.

		Amplitude (m)		Phase (°)	
		mean	sd	mean	sd
O_1	H	0.177	0.012	114.0	7.3
	U	0.006	0.002	153.1	75.2
	V	0.045	0.008	232.2	13.3
M_2	H	1.969	0.029	162.2	1.0
	U	0.200	0.028	231.1	10.6
	V	0.693	0.056	304.0	11.3
M_4	H	0.035	0.004	242.3	8.9
	U	0.033	0.021	107.3	65.3
	V	0.065	0.012	30.1	20.0

sd = standard deviation about mean value.
H = levels in metres.
U = east-west components of current (m s^{-1}).
V = north-south component of current (m s^{-1}).
Phases in degrees relative to Equilibrium Tide at Greenwich.

Table 4:7(b) Stability of Lerwick harmonic tidal constituents from annual analyses (eight years 1959–1967, 1963 excluded).

	Amplitude (m)				Phase (°)			
	mean	sd	max	min	mean	sd	max	min
S_a	0.079	0.019	0.114	0.052	225.2	10.4	241.9	207.6
O_1	0.079	0.002	0.082	0.077	31.3	1.9	33.3	28.8
K_1	0.076	0.001	0.078	0.074	164.2	1.0	165.7	162.6
M_2	0.583	0.003	0.586	0.577	312.2	0.6	312.7	311.0
S_2	0.211	0.001	0.213	0.209	346.7	0.9	347.5	345.0
M_4	0.015	0.001	0.017	0.014	282.9	4.7	291.1	175.8

Reasons for the variability of the tidal parameters include analysis limitations due to non-tidal energy at tidal frequencies, inconsistencies in the measuring instruments and changes and real oceanographic modulations of the tidal behaviour. In the case of a harmonic tidal analysis we may estimate the confidence limits by making use of the fact that the standard error due to a random background noise of variance S^2 in the elemental band around a constituent frequency is:

Standard error in amplitude (H): $S/\sqrt{2}$
Standard error in phase (in radians): $S/H\sqrt{2}$

For a time series of length T, the elemental band has a frequency span $1/T$. However, better stability is achieved by averaging the noise level over a series of elemental bands; for example, the whole of the diurnal or of the semidiurnal frequency band might be included. If the non-tidal residual variance in the averaging band, of width $\Delta\omega$ is $S_{\Delta\omega}^2$, then the noise density is $S_{\Delta\omega}^2/\Delta\omega$. The value of S^2 to be used to calculate standard errors is given by:

$$S^2 = \frac{S_{\Delta\omega}^2}{T \cdot \Delta\omega}$$

Averaging over a whole tidal species for $S_{\Delta\omega}^2$ will tend to underestimate S^2 because it assumes a uniform noise density across the band, whereas the noise background in practice rises around the major tidal lines. For this reason some workers choose a narrower band centred on the individual constituents to compute the residual variance. In either case, a spectral analysis of the residual variance is required to determine its frequency distribution.

When the above procedures are applied to shelf tides, the standard errors computed are significantly less than the observed variations of constituent amplitudes and phases from month to month. Some of the variability is undoubtedly instrumental—for example, a chart recorder cannot normally be read to better than 2 minutes, equivalent to a phase error of $1°$ in the semidiurnal constituents, and operating errors may easily be greater than this. Nevertheless, particularly for monthly analyses, there remains a real oceanographic variability, which is not due to variations in the astronomical forcing functions, as these variations persist no matter how fully the forces are represented. In any particular region, the amplitudes and phases at several simultaneous measuring stations are found to be similarly influenced during a particular month. These regional modulations present a problem when seeking undistorted 'true' harmonic constants from 29-days or less of data at a site. One possible palliative to reduce the error is to compare the results with a 29-day analysis of simultaneous data from a nearby standard port, for which a reliable long-term mean value of the constituent has been determined from analysis of a long period of data. The 29-day constituents at the site may then be adjusted according to the ratio and phase differences of the short and long-term constituents at the Standard Port.

4.7 Tidal predictions

The standard practice is to publish high and low water times and heights in official tables, for a selected list of Standard or Reference Stations. The tables also include sets of constants for adjusting the times and heights for intermediate places, called Secondary or Subordinate Stations. Tables refer predicted levels to the local Chart Datum of soundings. To avoid any confusion due to publication of two slightly different sets of tidal predictions for a port, based on different methods, agreements exist between the national publishing authorities for exchange of predictions, each authority being responsible for ports in their country.

4:7:1   Reference or Standard stations

Predictions for the Standard or Reference Stations are prepared directly from the astronomical arguments, using sets of harmonic constituents. The procedure, which reverses the methods of harmonic analysis, is illustrated in Table 4:8 for the M_2 constituent at Newlyn for a particular time in 1993. The method may be repeated for all the other constituents previously determined by analysis, and the full set of harmonics added together to give the full prediction of levels. Figure 4:7 compares the levels predicted by M_2 alone, and by summing the full set of 100 constituents.

Table 4:8   Procedures for calculating the M_2 tidal constituent at Newlyn on 13 July 1993, by entering the appropriate values in equation (4:5). The procedure is repeated for the contributions from each other constituent.

1. From analysis $H_{M2} = 1.72$ m $g_{M2} = 135.2$
   $Z_0 = 3.15$ m above Admiralty Chart Datum

2. From Table 4:2: $s = 35.9°$
   $h = 110.9°$
   $p = 180.1°$ 0000 13 July 1993.
   $N = 250.2°$
   $p' = 282.8°$

3. From Table 4:1(c) and Section 4:2:1:
   $V_{M2} = -2s + 2h = 150.0°$ (in solar time $M_2 = 2\ -2\ 2\ 0\ 0\ 0$).
   and from Table 4:3:
   $f_{M2} = 1.000 - 0.037\cos(N) = 1.013$
   $u_{M2} = -2.1\sin(N) = 2.0°$

4. The M_2 harmonic is:
   $H_{M2}\,f_{M2}\,\cos[\omega_{M2}t - g_{M2} + (V_{M2} + u_{M2})]$
   $= 1.72 \times 1.013\,\cos[28.98°t - 135.2° + (150.0° + 2.0°)]$
   t is measured in hours from midnight.
   $= 1.74\cos(28.98°t + 16.8°)$

5. Some calculated values are:
0000	$Z_0 + 1.67 = 4.82$ m ACD
0600	$Z_0 - 1.71 = 1.44$
1200	$Z_0 + 1.73 = 4.88$
1800	$Z_0 - 1.74 = 1.41$
2400	$Z_0 + 1.72 = 4.87$

6. Figure 4:7 shows the curve through hourly values of the M_2 constituent predictions, and through the full predictions for 13 July 1993, based on the sum of 100 constituents

The times of high and low water are difficult to determine from the curves of total levels. An alternative method makes use of the turning point relationship

$$\partial T/\partial t = 0$$

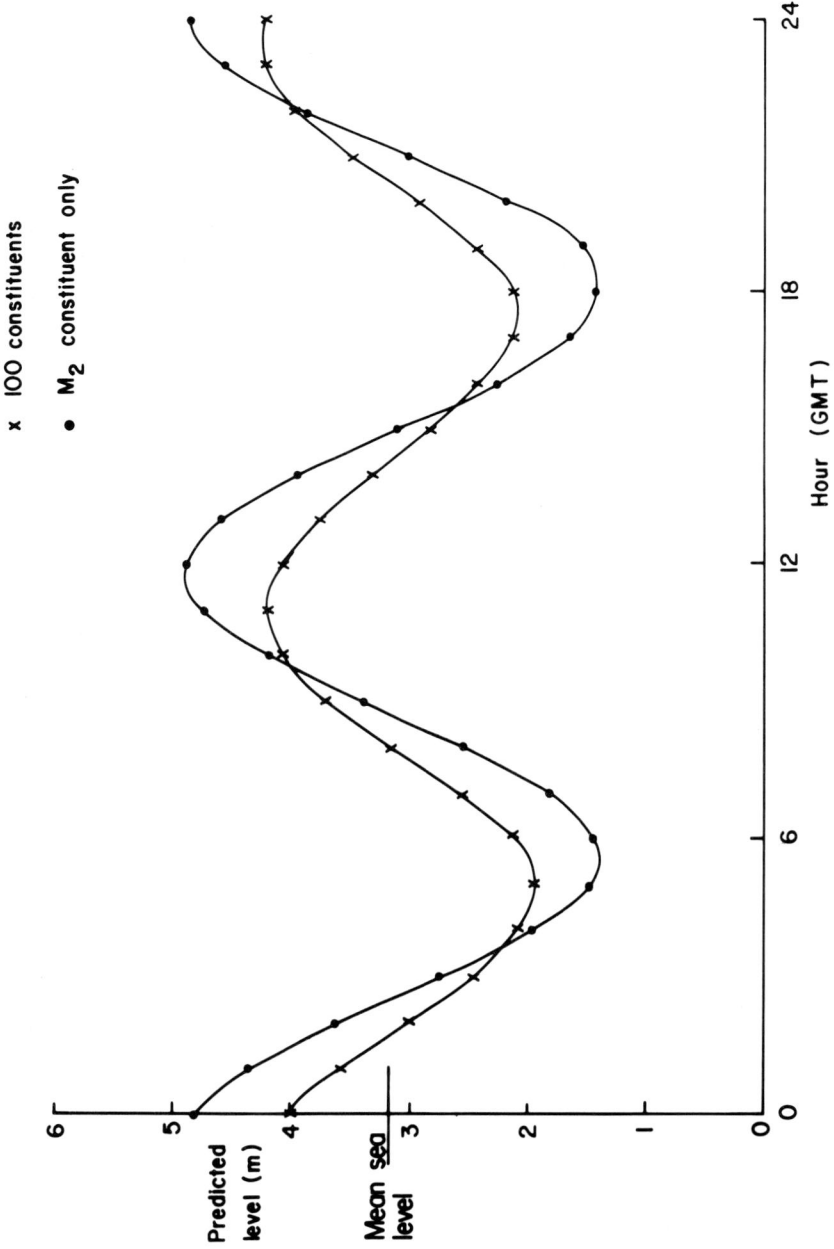

Figure 4:7 Newlyn predictions for 13 July 1993.

and since:

$$\mathbf{T} = \sum_N H_n \cos (\sigma_n t - g_n)$$

$$\frac{\partial \mathbf{T}}{\partial t} = \sum_N H_n \sigma_n \sin (\sigma_n t - g_n) \qquad (4:13)$$

the times of zero rate of change of water level, the high and low water times, can be found by calculating (4:13) for the sum of all the constituents, and finding the times of zero values. This is done numerically by computing (4:13) with the time increasing in increments of 3 hours (increments of 1 hour may be necessary where there are mixed tides, or severe shallow water distortions). If the sign of (4:13) has changed, a zero value must have occurred between the two times of calculation. The exact time of this zero is found by a sequence of calculations of (4:13) at intermediate times. Some of the more elaborate old mechanical predictors had the facility for simultaneously computing both the sin $(\sigma_n t - g_n)$ terms and the cos $(\sigma_n t - g_n)$ terms to get the best estimates of the times and the heights of the turning points.

In estuaries and regions of shallow water the harmonic method may fail to represent completely the sharp turning points, especially those of low water. The number of constituents in the higher species can be increased to improve the predictions, but eventually, more elaborate adjustments are necessary. Often these adjustments take the form of empirical corrections which are particular to each location.

Although high and low waters are the traditional method of presenting tidal predictions, for some operations it is better to have levels or currents predicted at hourly or even more frequent intervals. Again, there are problems when predicting the full curve of a severely distorted shallow-water tide.

Tidal streams are also included in Tables but usually only for rectilinear streams in straits where they are critical for planning shipping movement. Examples include the Torres Strait north of Australia, the approaches to Singapore Harbour, and the San Francisco Bay Entrance. The more common method for predicting tidal streams is to use a tabulation of current speeds and directions, which are related to local high-water levels, as in Table 4:6.

The tidal predictions are prepared about two years ahead of their effective date, to allow time for printing and publication. Although predictions could be prepared several years ahead, this is not usually done because the latest datum changes or permanent time zone changes need to be incorporated. If there are major port developments it is advisable to identify any changes which may have been inflicted on the tidal regime, by making new analyses of a year of data. Normally the tidal constituents are effectively constant over several decades, as discussed in Section 4:4.

4:7:2 Secondary or Subordinate stations

Although the published annual Tide Tables contain daily high and low water information for more than two hundred stations, these represent only a

138

fraction of the places for which tidal parameters are available. It would be impractical to prepare daily predictions directly from the astronomical arguments for all of these places. Instead, time and height differences are published which allow adjustment to the published values to give predictions to chart datum for the secondary or subsidiary stations. The Tables published by the United States National Oceanic and Atmospheric Administration contain data for more than 6000 subsidiary stations. The British Admiralty also publishes data for several thousand secondary ports.

The term secondary port does not imply secondary importance as a port and the same port may be a Standard port in one publication but a secondary port in another. The Standard port chosen may not be the nearest—the most suitable Standard will be a port which has similar tidal chracteristics. For this reason the Admiralty Tide Tables choose to refer some Japanese ports where semidiurnal tides are dominant, to the Australian port of Darwin, which has similar tides, rather than the nearer Japanese Standard ports which have mixed tides.

Secondary parameters may be based on analyses of long periods of data, but a month of observations is more usual, and normally adequate. In some remote locations observations may extend over only a few days.

Table 4:9 Comparison of secondary and full predictions for Newlyn, 13 July 1993. The differences between the secondary predictions and the full (100 harmonic constituents) predictions show the uncertainty associated with the procedure. For comparison, weather effects cause observed levels at Newlyn to differ from predicted levels by more than 0.15 m, for one hourly value in three.

	Low water		High water		Low water		High water	
Full (100) Newlyn predictions:								
	0457	1.95	1057	4.20	1730	2.10	2329	4.26
	(h)	(m)						
Secondary differences on Plymouth:								
Plymouth	0529	2.01	1146	4.29	1800	2.21	0012	4.26
(full predictions)								
ATT factors	−0022	−0.2	−0043	0.0	−0020	−0.2	−0040	0.0
Newlyn	0507	1.81	1103	4.29	1740	2.01	2332	4.26
NOS factors	−0032	0.00	−0052	0.06	−0032	−0.00	−0052	0.06
Newlyn	0457	2.01	1054	4.35	1728	2.21	2320	4.32

Admiralty Tide Tables (ATT) height factors are given only to 0.1 m.
National Ocean Service (NOS) factors are calculated as the combined differences of Newlyn and Plymouth on Queenstown, as Plymouth is not a Reference station for NOS.

The United States tables give time adjustments for high and low water, together with height adjustments and in some cases scaling factors. The British tables are more elaborate, allowing both the time and the height factors to be

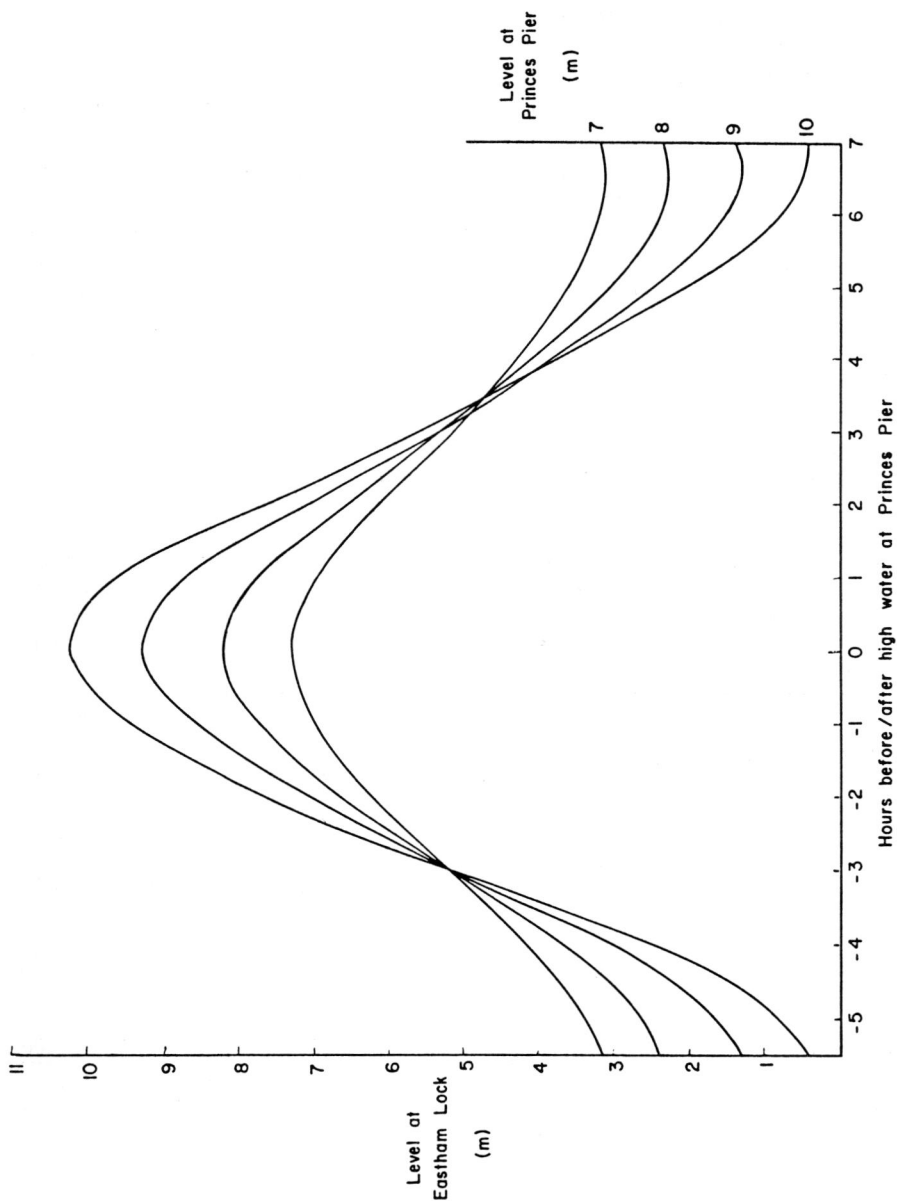

Figure 4:8 Levels at Eastham Lock, River Mersey, relative to high water times and ranges at Princes Pier, Liverpool. For example, if the predicted high-water level at Princes Pier is 8 m, then the level at Eastham Lock five hours after this time will be 3.1 m above the datum level.

adjusted between spring and neap tidal conditions; equation (4:8) may be used to justify these adjustments for a semidiurnal regime. The best results are obtained where the S_2/M_2 ratios of amplitude and phase differences are the same. In cases of extreme distortion such as Swanage or Selsey on the south coast of England (see Section 1:2 and Chapter 7) special curves must be prepared for each individual port. British tables also allow more elaborate calculations based on the harmonic constituents O_1, K_1, M_2, S_2, and adjustments for shallow-water fourth- and sixth-diurnal amplitudes and phases (f_4, F_4; f_6, F_6). Other national authorities have slightly different methods of making the adjustments and full details of the procedures are given in the individual publications. Table 4:9 shows how Newlyn high and low water parameters for a particular day can be calculated by different methods, and indicates the general agreement between them.

Another flexible method of presenting the tidal characteristics of a secondary station is shown in the reduction curves of Figure 4:8. These are prepared from a series of tabulations of observations over at least one spring–neap cycle, but better results are obtained if several cycles are included. Here the curves are plotted for different tidal ranges at the same station. Alternatively curves could be drawn for several stations in an estuary for a specified fixed range at the reference port (MHWS, for example).

Tidal stream atlases published by the British Admiralty give speeds and directions of currents at several locations for each state of the tidal elevations at Dover. Other authorities use similar techniques, and it is common to include this type of information on large-scale charts, relative to the time of high water at some local Standard Port.

4:7:3 Historical aspects

An ability to predict future events, particularly those of practical importance, must inevitably have attracted some veneration to those who practised the art. The priests of the Egyptian temples (Section 2:2) could predict the onset of the flooding of the river with their nilometers. The English cleric, the Venerable Bede (AD 673–735), who was familiar with the tides along the coast of his native Northumbria, discussed the phases of the moon and how the tides could be calculated by the 19-year lunar cycle.

By the early ninth century, tide tables and diagrams showing how neap and spring tides alternate during the month were appearing in several manuscripts (see footnote to page 126 of Jones, 1943). A later example of these is the St Albans Abbey tide table (ca 1250), which gives times of high water at London Bridge in terms of the age of the moon. Howse (1985) describes several early methods of presenting tidal information. By the seventeenth century, annual tide tables for London Bridge were produced regularly; Flamsteed, the UK Astronomer Royal, commented that his 1683 predictions were the first to give the times of two high waters each day.

The ability to predict tides was also a source of income for those who knew the secret. Tide tables for Liverpool were prepared by the Rev. George Holden from 1770, but the techniques were closely guarded family secrets. When Sir John Lubbock (Section 4:1) sought to investigate scientifically the basis on which the predictions for London were produced, he was refused access to the information. Instead Lubbock analysed several years of observations of London high waters, day and night, and by 1830, developed his own technique, the *non-harmonic* method of tide prediction. The need for observations and analyses at each port was now firmly established: new methods for automatically measuring the water levels were developed and the British Admiralty published the first official predictions (of high water times only), for four ports, Plymouth, Portsmouth, Sheerness and London Bridge, in 1833.

Lubbock's methods had great difficulty dealing with mixed tides. The method of harmonic analysis developed some years later was found to be superior; however, the harmonic method would have been difficult to apply for routine calculations without the parallel development of tide-predicting machines by Lord Kelvin and Edward Roberts in 1873. These machines, which were an early form of mechanical computer, contained a set of pulleys, one for each constituent, which rotated at different speeds scaled according to the speed of the constituent they represented. The rotations are converted into harmonic motions and summed by a means of a wire which passed over each pulley in turn. Phases and amplitudes for each constituent were set by adjusting the pulley settings for each port before running the machine. To compute high and low water times and heights for a single port for one year took about two days of work. These machines were the basis for tidal predictions until around 1965 when they were replaced by electronic computers: a year of predictions can now be calculated in a few seconds.

Many people continue to regard tidal analysis and prediction as a black art, as did people in earlier times. However, tidal analysis is neither as difficult nor as mysterious as they imagine. It is important to decide which level of complexity is appropriate for a particular application. Provided that certain basic rules relating to the data length and to the number of independent parameters demanded in an analysis are followed, satisfactory results are obtainable. Inexperienced analysts often go wrong when they ask for too much from too little data. Nevertheless, to extract the maximum tidal information from a record, for example by exploiting the complementary aspects of different analysis techniques, further experience and informed judgement are necessary.

CHAPTER 5

Tidal Dynamics

Who can say of a particular sea that it is old? Distilled by the sun, kneaded by the moon, it is renewed in a year, in a day, or in an hour.

Thomas Hardy, *The Return of the Native*

5:1 The real world

The Equilibrium Tide developed from Newton's theory of gravitation consists of two symmetrical tidal bulges, directly under and directly opposite the moon or sun. Semidiurnal tidal ranges would reach their maximum value of about 0.5 m at equatorial latitudes. The individual high water bulges would track around the earth, moving from east to west in steady progression. These characteristics are clearly not those of the observed tides.

The observed tides in the main oceans have mean ranges of about 0–1 m (amplitudes 0–0.5 m), but there are considerable variations. The times of tidal high water vary in a geographical pattern, as illustrated in Figure 5:1, which bears no relationship to the simple ideas of a double bulge. The tides spread from the oceans onto the surrounding continental shelves, where much larger ranges are observed. In some shelf seas the spring tidal ranges may exceed 10 m: the Bay of Fundy, the Bristol Channel, the Baie de Mont St Michele and the Argentine Shelf are well known examples. In the case of the North West European shelf tides approach from the Atlantic Ocean in a progression to the north and to the east, which is quite different from the Equilibrium hypothesis. Some indication of the different tidal patterns generated by the global and local ocean responses to the tidal forcing are shown in Figure 1:1. The reasons for these complicated ocean responses may be summarized as follows:

(1) Movements of water on the surface of the earth must obey the physical laws represented by the hydrodynamic equations of continuity and momentum balance: we shall see that this means they must propagate as long waves. Any propagation of a wave from east to west around the earth would be impeded by the north–south continental boundaries. The only latitudes for unimpeded circumpolar movement are around the Antarctic continent and in the Arctic basin. Even around Antarctica the connection is very restricted through the Drake Passage between Cape Horn and Graham Land.

142

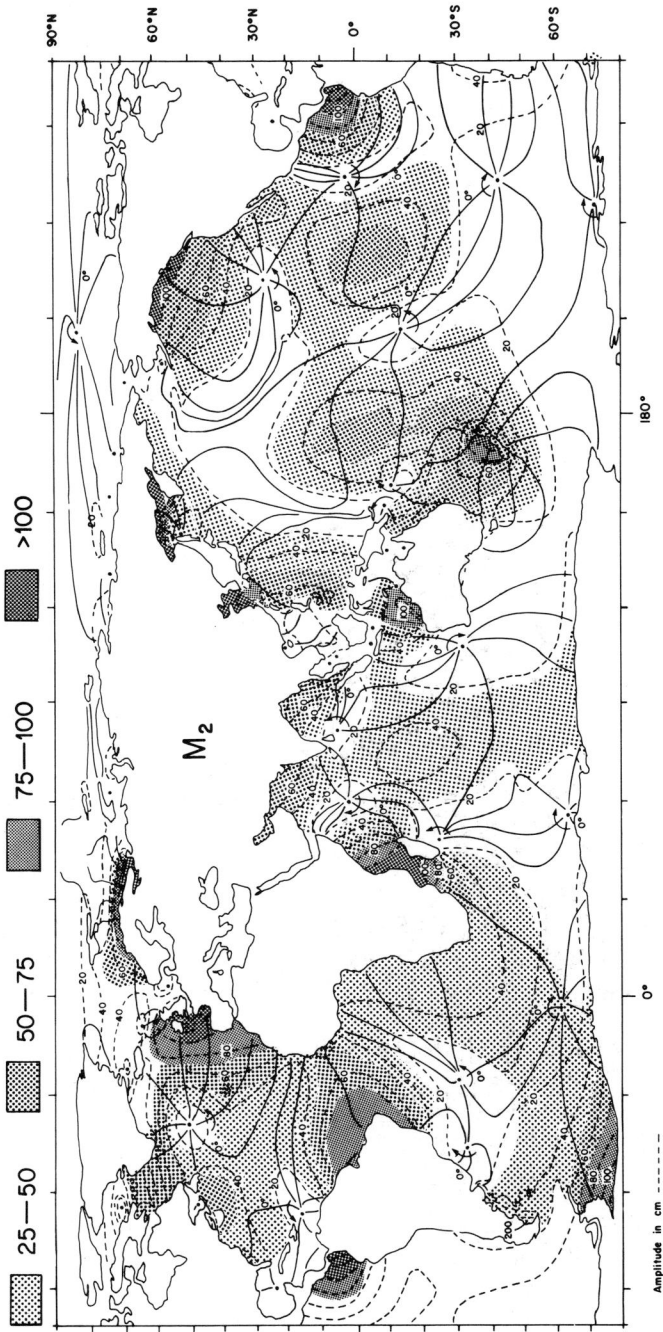

Figure 5:1   Map of lines joining places where high waters for the M_2 tide occur simultaneously, based on a numerical model. (From Schwiderski, 1979.)

(2) Long waves travel at a speed given by (water depth × gravitational acceleration)$^{\frac{1}{2}}$. Even in the absence of barriers it would be impossible for an Equilibrium Tide to keep up with the moon's tracking, because the oceans are too shallow. Taking an average depth of 4000 m the wave speed is 198 m s^{-1}, whereas at the equator the sub-lunar point travels westwards at an average speed of 450 m s^{-1}. Around Antarctica, however, at 60°S the speeds are nearly equal. At one time it was thought that the tides were generated in these latitudes, from where they spread to other areas. This was supposed to explain the age of the tide: the time between new or full moon and the maximum observed tides in northern latitudes; however, we now know that the responses are more complicated than these simple ideas suggest.

(3) The various ocean basins have their individual *natural modes* of oscillation which influence their responses to the tide generating forces. There are many resonant frequencies, but the whole global ocean system seems to be near to resonance at semidiurnal tidal frequencies and the observed tides are substantially larger than the Equilibrium Tide. The responses to forcing at diurnal tidal frequencies are much weaker, as shown in Figure 4:2 for Ilfracombe. However, the local responses of each area of the continental shelf to the driving by the ocean tides allow a different set of resonances to apply. In some cases there is a local amplification of the diurnal tides as in the Gulf of Tongking and the Gulf of Carpentaria (see Karumba in Figure 1:1 and Castletownsend in Figure 4:2).

(4) Water movements are affected by the rotation of the earth. The tendency for water movement to maintain a uniform direction in absolute space means that it performs a curved path in the rotating frame of reference within which we make observations. Alternatively, motion in a straight line on a rotating earth is curved in absolute space and must be sustained by forces at right angles to the motion. These effects are represented by the Coriolis accelerations in the hydrodynamic equations. The solutions to the equations show that certain modified forms of wave motion are possible, the most important of which have a form described as Kelvin waves.

(5) The solid earth responds elastically to the imposed tidal forces, as discussed in Section 3:2:3. The extent of this response is described in terms of Love numbers. Although the response of the solid earth to the direct tidal forcing is well described in these terms, there are local effects due to the depression of shallow-water areas and the surrounding land by the tidal loading. The tide measured at a coastal station, or by a pressure sensor on the sea-bed, is the difference between the change in the geocentric position of the sea surface, and the geocentric position of the land surface reference point. Altimetry measurements give the strictly geocentric displacements of the sea surface.

In this chapter we discuss the generation of ocean tides, their propagation onto the surrounding continental shelves, and the characteristics of shelf tides.

The development of our understanding of tides in the oceans and in continental shelf seas has been through a synthesis of theoretical ideas of wave propagation on a rotating earth and direct observations. In the first part of the chapter we concentrate on the basic physical ideas. In the second part we show how they are applied to describe the observed tidal patterns in selected areas. These regional accounts are presented in some detail because the information is not readily available in a single text. Fuller mathematical treatment of the more important ideas is given in Appendix 4 for those who require a more theoretical approach. Local harmonic tidal constants for more than 3000 coastal sites are collected and made available by the International Hydrographic Organization (1930 and on request), and details are also given in National Tide Tables.

5:2 Long-wave characteristics

Waves are commonly observed in nature. The most familiar form is probably the progression and breaking of waves on the beach. These waves, which are generated by winds acting on the sea surface, have periods from a few seconds to perhaps 15 seconds. Tidal waves have similar characteristics of propagation and individual wave crests can be followed and charted (the Venerable Bede (AD 673–735) knew of the progression of the time of high water from north to south along the east coast of Britain (Jones, 1943)). However, the tidal waves have periods of about 24 hours and 12 hours, corresponding to diurnal and semidiurnal tides and so their wavelengths are increased to many hundreds of kilometres. Waves whose wavelengths are much longer than the water depth are called *long waves* and they have some special properties.

When presenting information on tides in a region it is usual to draw maps for each individual harmonic constituent (Young, 1807; Marmer, 1928). These maps have contours joining places where high waters occur simultaneously, which are called *cotidal* lines; for a harmonic constituent this also means that low water and all other phases of the tide also occur simultaneously at all places joined by a cotidal line. The maps may also have contours joining places which have harmonic constituents of equal amplitude (also, of course, equal range) called *coamplitude* lines (or *corange* lines).

* 5:2:1 Wave propagation in a non-rotating system

We begin by summarizing some basic properties of wave motion on the surface of water in the absence of rotation. Figure 5:2 shows a disturbance travelling in the positive X direction without change of shape, in water of depth D below the undisturbed sea-level. As the wave moves past some fixed point, A, a succession of high and low sea levels will be observed. If the wave travels at a speed c and has a period T, the wave length λ is:

$$\lambda = cT$$

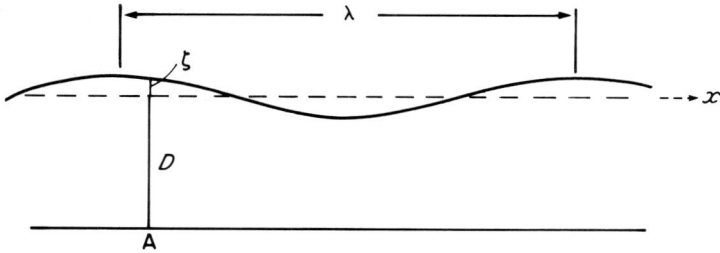

Figure 5:2 Basic parameters for a one-dimensional sea-surface disturbance travelling in the positive X direction.

It can be shown (see for example, Lamb (1932), Proudman (1953) and Appendix 4) that provided the wave amplitudes are small compared with the depth, and the depth is small compared with the wavelength, then the speed is:

$$c = (gD)^{\frac{1}{2}} \tag{5:1}$$

where g is gravitational acceleration. Similarly, the currents u are related to the instantaneous level ζ at A:

$$u = \zeta (g/D)^{\frac{1}{2}} \tag{5:2}$$

At local high water there is a maximum current in the direction of wave propagation, but at low water there is a current in the opposite direction. These results are not dependent on the exact shape of the travelling disturbance.

To derive these expressions from the continuity and horizontal momentum equations (3:22) and (3:23), the equations are reduced to a very simple form for small disturbances propagating in a deep channel:

$$\left. \begin{array}{c} \dfrac{\partial \zeta}{\partial t} + D \dfrac{\partial u}{\partial x} = 0 \\[2mm] \dfrac{\partial u}{\partial t} + g \dfrac{\partial \zeta}{\partial x} = 0 \end{array} \right\} \tag{5:3}$$

and these have a solution in ζ and u:

$$\left. \begin{array}{c} \zeta = H_0 \cos (kx - \omega t) \\[2mm] u = H_0 (g/D)^{\frac{1}{2}} \cos (kx - \omega t) \end{array} \right\} \tag{5:4}$$

which represent a free progressive harmonic wave travelling in the positive X direction, where:

$$\omega \text{ is the harmonic angular speed} = 2\pi/T$$
$$k \text{ is the wave number} = 2\pi/\lambda$$

This can be considered as an individual tidal constituent of amplitude H_0, travelling at a speed:

$$c = \frac{\omega}{k} = \frac{\lambda}{T} = (gD)^{\frac{1}{2}}$$

Long waves have the special property that the speed c is independent of the angular speed ω, and depends only on the value of g and the water depth. Any disturbance which consists of a number of separate harmonic constituents will not change its shape as it propagates—this is called *non-dispersive* propagation. Waves at tidal periods are long waves, even in the deep ocean, as shown in Table 5:1 and their propagation is non-dispersive. Wind waves, which have much shorter periods and wavelengths usually undergo *dispersive* propagation, with swell from distant storms travelling more rapidly ahead of the shorter period waves.

Table 5:1   Speeds and wavelengths of tidal waves in water
of different depths.

Depth (m)	Speed		Wavelength	
	m s^{-1}	km h^{-1}	Diurnal	Semidiurnal
4000	198	715	17 720	8 860
200	44	160	3 960	1 980
50	22	80	1 980	990
20	14	50	1 250	625

The maximum tidal currents in the direction of wave propagation occur at local high water. At local low water the currents are directed in a direction opposite to the direction of wave propagation.

Details of the behaviour of progressive waves when they encounter a step change in water depth, and of their energy fluxes are given in Appendix 4.

5:2:2   Standing waves, resonance

In the real oceans, tides cannot propagate endlessly as progressive waves. They undergo reflection at sudden changes of depth, and at the coastal boundaries. The reflected and incident waves combine together to give the observed total wave (Defant, 1961; Redfield, 1980). Consider the simplest case of a wave travelling in a long channel being reflected without loss of amplitude at a closed end. The interference between the two waves produces a fixed pattern of *standing waves* which have alternate nodes, positions where the amplitude is zero, and antinodes, positions where the amplitude is a maximum, each separated by a distance $\lambda/4$ where λ is the wavelength of the original progressive wave as illustrated in Figure 5:3. The first *node* is located at a distance $\lambda/4$ from the reflecting barrier and at this point there is no net change of water level, but the currents have their maximum amplitude. At a distance $\lambda/2$ from the barrier there is an *antinode* where the changes in level have the same range as at the reflecting barrier, and there are no currents. The absence of horizontal water movements across the antinode means that a solid impermeable boundary could be inserted into the channel at this point without affecting the oscillations once they had become established.

148

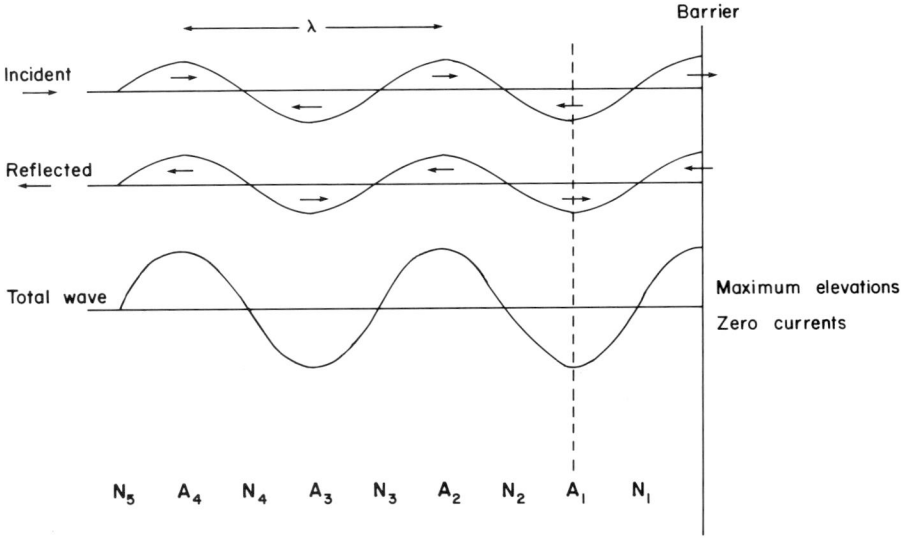

(a)

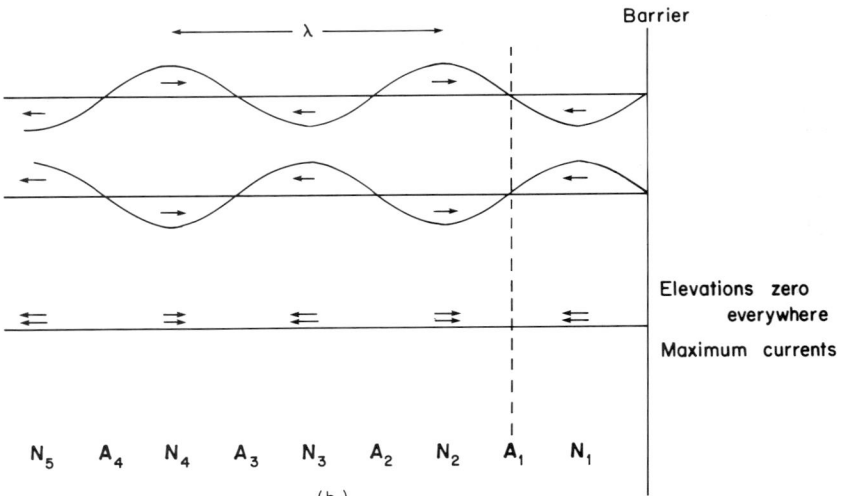

(b)

Figure 5:3 Showing how incident and reflected waves of equal amplitude produce a standing wave pattern with nodes and antinodes. (→ are currents). (a) high water at the reflecting barrier; (b) a quarter of a period later the incident wave has moved a quarter wavelength towards the barrier, and the reflected wave has moved a quarter wavelength in the opposite direction away from the barrier. After half a cycle the elevations and currents are the reverse of (a); after three-quarters of a period they are the reverse of (b).

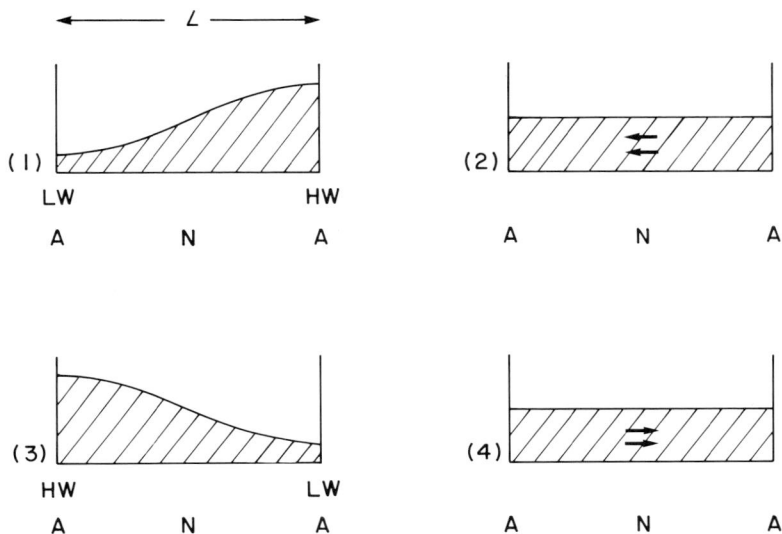

Figure 5:4   First gravitational mode for oscillations of water in a closed rectangular box, showing nodes and antinodes for levels and currents.

One of the most effective ways of describing the behaviour of standing waves is to consider the movement of water in a rectangular box of length L (Figure 5:4). This configuration may represent a domestic bath, a swimming pool, or a long narrow lake. The water movements can be likened to those of a pendulum, with the least horizontal motion at high and low water ((1) and (3)) and the maximum horizontal motion at the middle of the oscillation ((2) and (4)). There is a continuous transfer from maximum potential energy (1) through a position of zero potential energy but maximum kinetic energy (2) to a second position of maximum potential energy (3). The sequence continues through a second condition of maximum kinetic energy (4) as it returns to its initial state of maximum potential energy (1). This can be considered as two progressive waves travelling in opposite directions with perfect reflections at both ends. The natural period of oscillation of the water is the time taken for a wave to travel from one boundary and to return after reflection at the second boundary. From equation (5:1) this time is:

$$\frac{2 \times \text{box length}}{(g \times \text{water depth})^{\frac{1}{2}}} = \frac{2L}{(gD)^{\frac{1}{2}}} \tag{5:5}$$

Equation (5:5) is known as 'Merian's Formula' after its originator (Proudman, 1953). Examples of some natural periods for this half-wave oscillation are given in Table 5:2. Natural periods of long lakes may be estimated using this formula, but the observed periods will vary slightly because the depths are not uniform. For the fundamental *mode* of oscillation there is one node, and the term uninodal seiche is often applied. The next highest mode has two nodes, and so

Table 5:2 Examples of the natural period of oscillation of water bodies in the first gravitational mode, from Merian's formula (equation (5:5)).

	Length	Depth	Period
Bath	1.5 m	0.2 m	2.1 s
Swimming-pool	10	2.0	4.5
Loch Ness, Scotland	38 km	130	35 min

on. The irregular shapes of real lakes result in several natural periods of oscillation including lateral as well as longitudinal modes.

Standing waves may also occur in a box which is closed at one end but driven by oscillatory in and out currents at the other open end. The simplest case would be a box whose length was a quarter wavelength, so that the open end was at the first node, equivalent to half a box in Figure 5:4. The currents at the entrance could produce large changes of level at the head. The natural period for this type of forced oscillation is:

$$\frac{4 \times \text{box length}}{(g \times \text{water depth})^{\frac{1}{2}}} = \frac{4L}{(gD)^{\frac{1}{2}}} \tag{5:6}$$

This model of an open box approximates to the tidal behaviour of many shelf sea basins, but an exact quarter-wave dimension would be very unlikely. In reality the open boundary may lie within the node or outside the node as shown in Figure 5:5, but the probability of tidal amplification still exists. However, if the length of the basin is only a small fraction of the tidal wavelength, then the amplification will be small.

This type of oscillation will only continue for as long as the currents and elevations drive it from the open end. This forced motion is different from the natural free seiching oscillations of a body of water which can be initiated by an impulse such as a sudden squall, and which will continue until damped by frictional energy losses and imperfect reflections. Table 5:3 gives the lengths and corresponding depths of basins which would have quarter-wavelengths appropriate for the semidiurnal M_2 tide.

Table 5:3 Some lengths and depths of basins which would have quarter-wave resonance if driven by a semidiurnal M_2 tide.

Water depth (m)	Basin length (km)
4000	2200
1000	1100
200	500
100	350
50	250

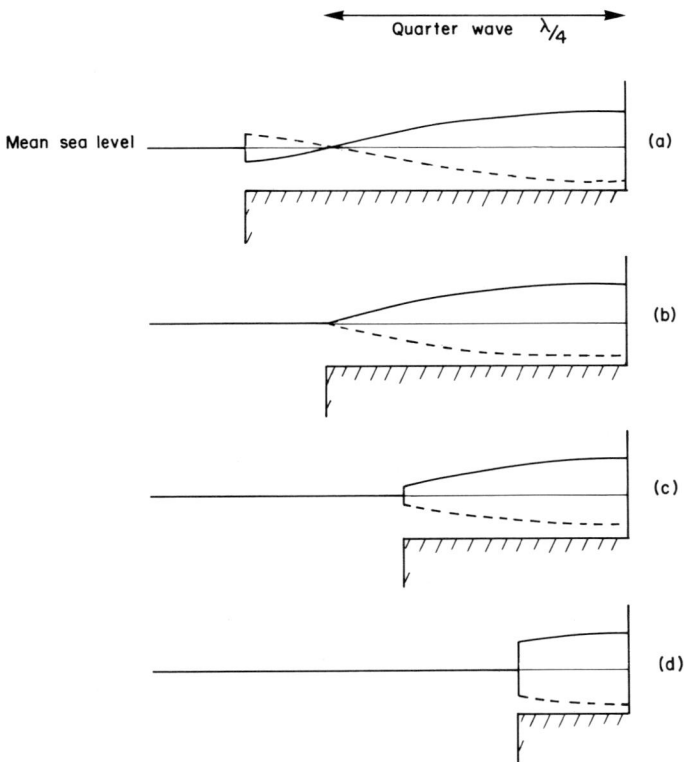

Figure 5:5 Responses of sea basins driven by tidal level changes at the open end, near to quarter-wave resonance. There is maximum amplification when the basin length corresponds to a quarter of a tidal wavelength.

Systems which are forced by oscillations close to their natural period have large amplitude responses. This *resonant* behaviour is familiar in many physical systems including electrical circuits and badly designed acoustic speakers. As we shall discuss later, the responses of oceans and many seas are close to resonance. In nature the forced resonant oscillations cannot grow indefinitely because energy losses due to friction increase more rapidly than the amplitudes of the oscillations themselves. Because of energy losses, tidal waves are not perfectly reflected at the head of a basin, which means that the reflected wave is smaller than the ingoing wave. It is easy to show that this is equivalent to a progressive wave superimposed on a standing wave (Appendix 4), with the progressive wave carrying energy to the head of the basin. Standing waves cannot transmit energy because they consist of two progressive waves of equal amplitude travelling in opposite directions.

5:2:3 Waves on a rotating earth

A long progressive wave travelling in a channel on a rotating earth behaves differently from a wave travelling along a non-rotating channel. The geostrophic forces which describe the motion in a rotating system cause a deflection of the currents towards the right of the direction of motion in the northern hemisphere. This sideways component of the flow continues until impeded by the right-hand boundary of the wave channel. The build-up of water on the right of the channel gives rise to a pressure gradient across the channel, which in turn develops until at equilibrium it balances the geostrophic force. The mathematical description of the resulting waves is given in Appendix 4. These waves were originally described by Lord Kelvin and are now named after him. Rotation influences the way in which the wave amplitude H_y decreases across the channel away from the value H_0 at the right-hand boundary (in the northern hemisphere):

$$H_y = H_0 \exp\left(-fy/c\right) \cos\left(kx - \omega t\right) \qquad (A4:15)$$

according to an exponential decay law with a scale length c/f, which is called the Rossby radius of deformation, where c is the wave speed $(gD)^{\frac{1}{2}}$, and f is the Coriolis parameter. At a distance $y = c/f$ from the boundary the amplitude has fallen to $0.37 H_0$. In this type of wave motion, the speed is the same as for a non-rotating case, and the currents are always parallel to the direction of wave propagation. Figure A4:3 shows the profile across a Kelvin wave. Note that Kelvin waves can only move along a coast in one direction. In the southern hemisphere the coast is on the left of the direction of propagation.

The case of a standing wave oscillation on a rotating earth is of special interest in tidal studies. At the head of a basin, where reflection takes place, the currents and elevations have waveforms which can only be described by complicated mathematics, but away from the boundary the tidal waves can be represented by two Kelvin waves travelling in opposite directions. Instead of oscillating about a nodal line, the wave can now be seen to rotate about a nodal point, which is called an amphidrome. Figure 5:6 shows the cotidal and coamplitude lines for a Kelvin wave reflected without energy loss at the head of a rectangular channel (Taylor, 1921). The sense of rotation of the wave around the amphidrome is anticlockwise in the northern hemisphere and clockwise in the southern hemisphere. The cotidal lines all radiate outwards from the amphidrome and the coamplitude lines form a set of nearly concentric circles with the centre at the amphidrome, at which the amplitude is zero. The amplitude is greatest around the boundary of the basin.

If the reflected Kelvin wave is weaker than the in-going Kelvin wave, then the amphidrome is displaced from the centre of the channel to the left of the direction of the in-going wave (northern hemisphere). Figure A4:3 shows this displacement increasing as the reflected Kelvin wave is made weaker (Taylor, 1919; Pugh, 1981). In a narrow channel the amphidromic point may move outside the left-hand boundary. Although the full amphidromic system shown in Figure 5:6 is not present, the cotidal lines will still focus on a point inland, which is called a virtual or *degenerate amphidrome* (see Section 7:9:2).

We are now able to apply these theoretical ideas of progressive and standing waves, resonance, Kelvin waves and amphidromic systems to describe the dynamics of the observed tides in the oceans and shelf seas.

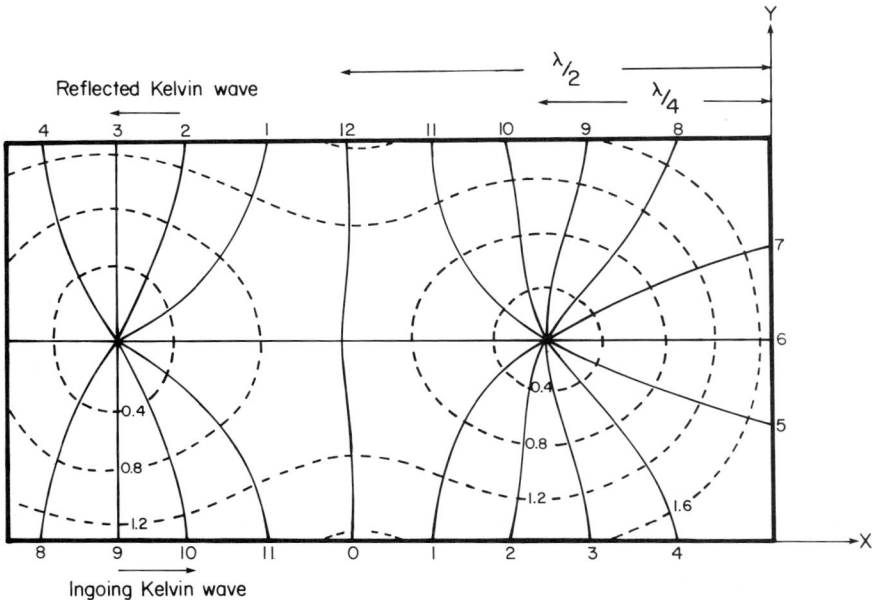

Figure 5:6   Cotidal and coamplitude lines for a Kelvin wave reflected without energy loss in a rectangular channel (after G. I. Taylor, 1921). Continuous lines are cotidal lines at intervals of 1/12 of a full cycle. Broken lines are lines of equal amplitude. Progression of the wave crests in the northern hemisphere is anti-clockwise for both the amphidromic systems shown. (*Reproduced by permission of Oxford University Press.*)

5:3   Ocean tides

Dynamically there are two essentially different types of tidal regime: in the wide and relatively deep ocean basins the observed tides are generated directly by the external gravitational forces; in the shelf seas the tides are driven by co-oscillation with the oceanic tides. The ocean response to the gravitational forcing may be described in terms of a forced linear oscillator, with weak energy dissipation. There is a flux of energy from the deep oceans on to the shelf where the energy is dissipated by bottom friction opposing the tidal currents.

This natural tidal progression from the oceans to the shelf is also followed in our discussions. Many earlier accounts have begun with the known coastal tides and extended the discussion outwards to the oceans, probably because the ocean tides were so poorly defined. Extrapolation of shore-based shelf measurements through complex coastal regions of irregular boundaries and depth, where bottom friction is important, is fraught with uncertainties. Even the measurements made at ocean islands were sparse, and sometimes not truly representative of the open ocean tidal regime: for example, tides measured in the protected waters of a coral lagoon are attenuated and lag those of the open ocean outside the lagoon. The measurement of ocean bottom pressures using modern instruments as described in Chapter 2 has enabled a proper definition of the tides at a number of selected important sites, but the procedures are too slow and expensive for general coverage. Values of open ocean tidal constants are published periodically by the International Association for the Physical Sciences of the Ocean (Cartwright and Zetler, 1985). Satellite altimetry gives the best prospects for a complete map of ocean tides by direct observation.

5:3:1   Towards tidal charts

The most direct way of preparing cotidal and coamplitude charts of ocean tides is simply to interpolate the coastal and island tidal parameters across oceans (Young, 1807; Marmer, 1928) and to supplement these where possible with data from bottom-pressure observations. With so few observations available it is advisable to constrain the charts by the known theoretical properties of long-waves propagating on a rotating earth. These properties include the previously developed ideas of progressive and standing waves, resonance, Kelvin waves, and amphidromic systems. The physical responses of the oceans to gravitational tidal forcing are still not fully understood despite valuable contributions made by recent observations, and by numerical and analytical models. It is useful to review briefly the development of scientific understanding and to place modern ideas in a historical context.

Following Newton's development of the Law of Gravitation and of an Equilibrium Tide, Laplace established in 1775 the hydrodynamic equations of continuity and momentum for a fluid on a rotating earth similar to those given in Chapter 3, including the rotation term later credited to Coriolis. Laplace assumed a spherical earth with a geocentric gravitational field uniform both horizontally and with time, a rigid ocean bottom and a shallow ocean which allowed Coriolis accelerations associated with vertical movements and with the horizontal component of the earth's rotation to be neglected. The consequences of neglecting vertical accelerations have been carefully considered by subsequent theoretical work which shows that for the vertically stratified ocean the approximation is justified. Using these equations Laplace considered the types of wave motion possible on a thin sheet of fluid of uniform depth on a rotating sphere and showed that the response to the tide generating forces was critically dependent on the depth. For depths which exceed a critical value the driven tide is in phase with the gravitational forcing and is said to be *direct*, but for depths less than the critical value the driven tide has the opposite sign to the forcing and is said to be inverted. For the special case of zonal canals the equation is the same as equation (6:11) for the forcing of sea-level changes by a moving atmospheric pressure disturbance. The dynamics of propagation in zonal canals, which had previously been considered in 1740 by Bernoulli was further investigated in 1845 by Airy, who introduced energy losses through a linear friction term and considered the resulting distortions to the progressive wave.

A completely different approach was taken by Harris (1897–1907) who pictured the global tides as a mosaic of standing wave systems, each with natural periods equal to that of the tides. Although severely criticized for its lack of mathematical rigour and its neglect of the earth's rotation, Harris's techniques produced charts which were generally compatible with the available observations and a considerable improvement on those previously drawn. His emphasis on the influence of resonance was an important contribution to modern ideas of near-resonant ocean response to semidiurnal forcing.

Attempts to solve Laplace's tidal equations for ocean basins of different geometrical shapes and depths have served to demonstrate the types of motion which may occur and to investigate the sensitivity of the ocean tides to parameters such as basin width and depth (Proudman 1925, 1942), but the real oceans are too complex for detailed realistic solutions. Numerical solutions of the equations by powerful computers can give better fits to all of the observations (Parke, 1982; Schwiderski, 1983). The principle is straightforward: the oceans are divided into a network of component areas sufficiently small to define their shape and depth variations. Some compromise is necessary between a very detailed resolution and the computing power available, but this is not a severe limitation using modern computers: a grid consisting of $1°$ by $1°$ elements is easily possible. Energy dissipation can be introduced by radiating energy through the boundaries, by bottom friction or by a form of eddy-viscosity. Both the solid earth tides and the ocean loading effects can be included, as can the gravitational self-attraction of the tidal bulges. Figure 5:1 is an example of an M_2 chart produced in this way, with the frictional parameters adjusted to give a satisfactory fit where observations are available.

A different approach, concerned with physical interpretation of the observed tides, begins by numerically computing the normal modes of oscillation of the oceans, more realistically than Harris's imposed resonances. Treating the ocean tides as a forced linear oscillating system with weak dissipation, the observed tides must be represented as a linear combination of these modes (Platzman, 1983). The contribution of each mode to the synthesis depends on the closeness of its frequency to that of the tide and to its spatial coherence with the Equilibrium Tidal potential. Ten modes can account for 87 per cent of the energy in the M_2 tide; the most energetic mode, which accounted for 19 per cent of this, was dominant in the South Atlantic. For the K_1 tide, which has a period of 23.93 hours, 93 per cent of the energy is accounted for by the first ten modes; remarkably, the most energetic mode, a combined fundamental basin oscillation of the Pacific Ocean and a fundamental Kelvin wave around Antarctica of period 28.7 hours, accounts for 61 per cent of the tidal energy. The diurnal responses are weaker because the normal modes are at periods further removed from the periods of the tidal forcing than are the semidiurnal normal modes.

5:3:2 General features

Figure 5:1 shows a global chart of the principal lunar semidiurnal tidal constituent M_2, produced numerically by constraining the lines to satisfy the hydrodynamic equations as far as possible between the observation stations. The most obvious feature is the large number of amphidromes. As a general rule these conform to the expected behaviour for Kelvin wave propagation, with anticlockwise rotation in the northern hemisphere, and clockwise rotation in the southern hemisphere. Some caution is necessary: for example, the M_2 system west of Africa in the south Atlantic Ocean rotates anticlockwise, showing that other types of wave dynamics are also involved. Similar charts have been prepared for all the other major tidal constituents, and each of these contains many amphidromes. The amplitudes of the semidiurnal tides are significantly greater than the amplitudes of the diurnal tides because the oceans have a near-resonant response to forcing at semidiurnal frequencies. The possibility of a near-resonant response for semidiurnal tides is confirmed by rough estimates using Merian's formula (equation (5:5)): in 4000 m depth the progressive wave speed of $198 \, \mathrm{m \, s}^{-1}$ and a typical ocean length scale of 10 000 km gives a period of 14 hours. Examples of the relative importance of the diurnal and semidiurnal tides are given in Table 1:1, in the Form Factors of Section 4:2:6, and in the Ilfracombe response curves of Figure 4:2. For progressive tidal waves to develop fully as Kelvin waves, the channel in which they propagate must be wide compared with the Rossby radius (c/f) (see equation (A4:15)). At $45°$ latitude, the Rossby radius in 4000 m depth is 1900 km, but this increases as the latitude decreases (Table 5:4). The Pacific is clearly wide enough for full development, but the Atlantic may be too narrow.

Ocean tidal currents are very weak. For a progressive wave of 0.5 m amplitude in 4000 m depth, equation (5:2) gives a speed of $0.025 \, \mathrm{m \, s}^{-1}$, which is small compared with currents generated by other forces. Internal tides (Section 5:6) give currents at tidal periods with greater speeds, but these currents are irregular and lack the coherence of shelf tidal currents.

In the **Pacific**, as elsewhere, semidiurnal tides are dominant, but in the north diurnal tides are also well developed. The semidiurnal amphidrome at $25°$ N, $135°$ W has been firmly established by observations, as has the amphidrome near $25°$ S, $150°$ W. This amphidrome gives very small M_2 tides in the vicinity of the Society Isles. Because the S_2 amphidrome is not identically placed, the S_2 tides are dominant at some sites. Where this occurs the semidiurnal tides have high and low water at the same time every day. This is a very local phenomenon, but something similar is found along the coast of southern Australia from western Tasmania to Cape Leeuwin, where the M_2 and the S_2 constituents have similar amplitudes. At Thevenard, for example, for part of the spring–neap cycle the times of high water are similar for several successive days. This type of behaviour can be analysed in terms of the equations developed in Section 4:2:6 for the combination of two harmonic constituents.

The semidiurnal ocean tides progress around New Zealand as around an amphidrome, but in an anticlockwise sense, which is not as expected in the southern hemisphere, and the dynamics cannot be those of cancelling Kelvin waves because the amplitudes are greatest at the coast (Bye and Heath, 1975). The reason for this behaviour is probably the trapping of the tidal waves by the topography of the surrounding continental slope.

Changes in bottom topography have a strong influence on the tidal propagation. As the tide approaches the Hawaiian Islands the wave is diffracted by the submarine ridge which extends over 500 km, rather than by the individual islands; this gives a focusing of cotidal lines from a more extensive region, as illustrated for a coastal headland in Figure A4:4. The semidiurnal tides at Honolulu have a negative age, with maximum amplitudes 14 hours ahead of maximum gravitational forcing, which is unusual as the age of the tide generally has a lag of one or two days.

The semidiurnal tides in the northern **Indian Ocean** have two distinct regimes. The Arabian Sea is broad enough for the standing wave system to develop an amphidrome; however, this amphidrome, which is situated close to the equator, cannot be described simply in terms of Kelvin waves. The entrance to the Bay of Bengal is too narrow for an amphidromic system to develop. The wave propagates to the north along the west coast of Sumatra and Thailand and also along the east coast of Sri Lanka, where the range is low and there is a tendency towards a degenerate amphidrome. In the south central Indian Ocean there is an extensive region of large semidiurnal tides over which the phases change only slowly. Dynamically this phenomenon, which corresponds to an antinode in Figure 5:3, is called an *anti-amphidrome*. A similar semidiurnal system is shown in the south central Pacific Ocean. Large tidal ranges are also observed between the island of Madagascar and the African mainland, in the Mozambique Channel, because of a similar double standing wave system.

In the **Southern Ocean**, there is a circumpolar zonal channel around 60° S. Early scientific ideas supposed that this would allow a resonant response to the tidal forces: for a circumference of 20 000 km and a wave travelling in 4000 km water depth, simple theory gives a complete cycle in 28 hours, and allows resonances at harmonics of this. Whewell argued in 1833 that the directly driven tides of the Southern Ocean were the source of Atlantic tides, which were 'derivative' tides. The observed and computed tides of the Southern Ocean do show a general westward propagation, particularly for the diurnal tides, but the amplitudes are not particularly large. The narrowness of the Drake Passage between South America and Graham Land has a strong restricting influence on the pattern of wave propagation.

Whewell argued further that the one to two day age of the semidiurnal tide in the North Atlantic was a consequence of the time taken for the wave to propagate from the Southern Ocean. Measurements made in June 1835 on both sides of the Atlantic for a fortnight at 28 places, caused Whewell (1836) to review the earlier ideas. The observed diurnal tides were particularly difficult to reconcile with their generation in the distant Southern Ocean because the maximum inequality in successive semidiurnal high-water levels occurred two or three days later on the European side than on the American side. Further, simultaneous observations at the Cape of Good Hope near the supposed tidal source showed maximum inequalities simultaneous with those at Spain and Portugal. With no prospect of the necessary mid-ocean observations, Whewell abandoned his plan to draw cotidal lines for the oceans, and prepared a chart only for the north-west European continental shelf.

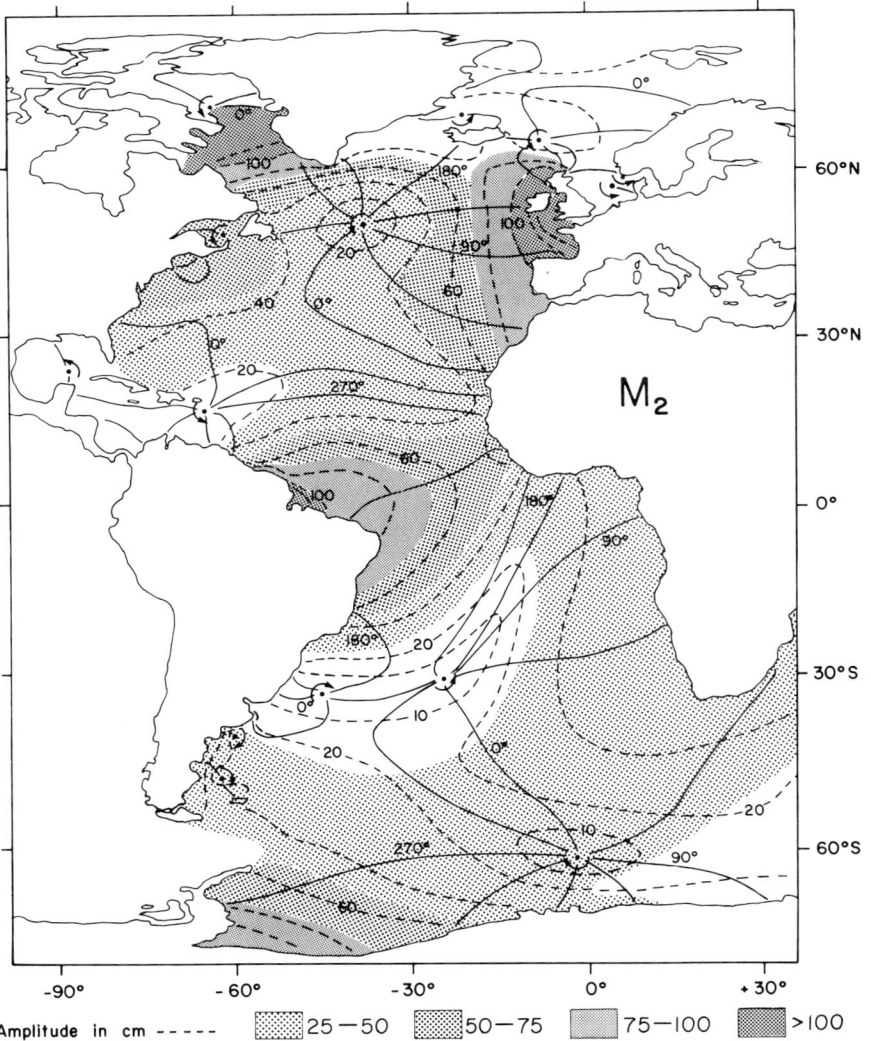

Figure 5:7 Cotidal and coamplitude charts of semidiurnal (M_2) tide of the Atlantic Ocean, based on a numerical model. (From Schwiderski, 1979.)

More recently, the **Atlantic tides** have been systematically observed and mapped by the United Kingdom Institute of Oceanographic Sciences (Cartwright *et al.*, 1980). Figure 5:7 shows the detailed chart, based on a numerical model (Schwiderski, 1979), of the principal lunar semidiurnal M_2 tide. The systematic northward progression of the semidiurnal phases along the coast of Brazil and West Africa is well established, but there is some uncertainty about the existence of amphidromes in the central South Atlantic around 35° S. The small observed amplitudes suggest a tendency for an amphidromic system to develop, but the geometry of amphidromic systems requires *two*, rotating in opposite senses, to satisfy the coastal observations. The anticlockwise one nearer the African coast would have the wrong sense of rotation for the southern hemisphere, so that complicated dynamics are implied. The ranges are relatively large near the equator and the phases nearly constant over an extensive area, high waters occurring along the whole coast of northern Brazil from 35° W to 60° W within an hour, behaviour consistent with standing wave dynamics. Further north, around 20° N, smaller amplitudes and a rapid northward increase of phase show a tendency for an amphidrome to develop.

The most fully developed semidiurnal amphidrome is located near 50° N, 39° W. The tidal waves appear to travel around the position in a form which approximates to a Kelvin wave, from Portugal along the edge of the north-west European continental shelf towards Iceland, and thence west and south past Greenland to Newfoundland. There is a considerable leakage of energy to the surrounding continental shelves and to the Arctic Ocean, so the wave which is reflected in a southerly direction is much weaker than the wave travelling northward along the European coast. Subsidiary anticlockwise amphidromic systems are formed between the Faeroe Islands and Iceland and between Iceland and Greenland. This results in a complete circulation of the semidiurnal phase around Iceland in a clockwise sense, similar to that observed around New Zealand, but for reasons which involve the dynamics of Arctic tides rather than local topographic trapping alone.

As in other oceans, for constituents in the same tidal species the tidal charts show broad similarities, but in the Atlantic Ocean there are also significant differences. The ratio between the M_2 and S_2 amplitudes, 0.46 in the Equilibrium Tide, falls in the North Atlantic to 0.22 at Bermuda. This relative suppression of the principal solar semidiurnal tide S_2 extends over a very large area of the North Atlantic and is observed on both the American and European coasts.

Along the Atlantic coast of North America from Nova Scotia to Florida the ocean tides are nearly in phase, consistent with standing-wave dynamics along the north-west to south-east axis; superimposed on this is a slow progression of phase towards the south. The semidiurnal tides of the Gulf of Mexico and the Caribbean have small amplitudes. An anticlockwise amphidromic system is apparently developed in the Gulf of Mexico for the semidiurnal tides.

The diurnal tides of the Gulf of Mexico are larger than the semidiurnal tides because of a local resonant response but the diurnal tides for the Atlantic Ocean as a whole are generally weaker than in the other oceans. They can be described in terms of amphidromic systems, a clockwise system in the south Atlantic and an anticlockwise system in the north Atlantic, consistent with

Kelvin wave dynamics. Although the link with Antarctic tides (Gill, 1979) is important for tides in the Atlantic (but not in the simple way originally proposed by Whewell), it is clear that the role of direct gravitational forcing within the Atlantic is also important.

5:3:3   Tides in enclosed seas

Basins of oceanic depths such as the Mediterranean Sea and the Red Sea, which connect to the oceans through narrow entrances, have small tidal ranges. The dimensions of the basins are too restricting for the direct tidal forces to have much effect, and the areas of the entrances are too small for sufficient oceanic tidal energy to enter (Section A4:1a) to compensate for the energy losses which would be associated with large tidal amplitudes.

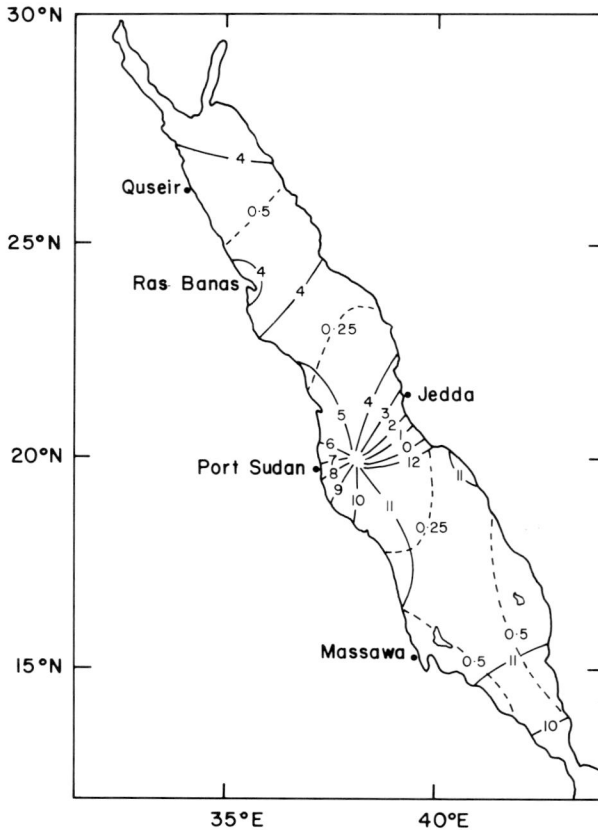

Figure 5:8 **M$_2$** tide of the Red Sea, based on observations. (*Reproduced by permission of Deutsches Hydrographisches Institut.*)

The semidiurnal tides of the **Red Sea** are of interest because they are closely represented by a standing wave having a single central node (Figure 5:8) (Defant, 1961; Deutsches Hydrographisches Institut, 1963). Careful analysis shows that there is a progression of the wave in the expected anticlockwise sense around a central amphidrome. From Merian's formula (equation (5:5)) the period of oscillation of the Red Sea would be approximately 12.8 hours, for a depth of 500 m and a length of 1600 km, the distance between Kamaran and Shadwan, the antinodes of the main oscillation. Because of its long narrow shape and steep sides, the Red Sea has been used to test dynamical theories of tides, including early numerical solutions of the equations of motion.

The tides of the **Mediterranean** are also weak. Essentially there are two basins, separated by the Sicilian Channel and the Straits of Messina. The tides of the western basin are strongly influenced by the Atlantic tides which penetrate through the Straits of Gibraltar, and a semidiurnal amphidrome is developed near the line of zero longitude. There is a second amphidromic region near the boundary between the two basins, and a third in the eastern basin south of Greece. Because the connection with the Atlantic Ocean through the Straits of Gibraltar is so restricted, the influence of direct gravitational forcing within the Mediterranean is probably of comparable importance to the external forcing. The diurnal tides of the Adriatic are relatively large because a natural oscillation is excited by the Mediterranean tides at the southern entrance. A simple semidiurnal amphidromic system similar to that in the Red Sea is also developed. The effect of the standing oscillations is to produce large tides at the northern end, in the vicinity of Venice.

Although the tidal amplitudes are generally small in the Mediterranean, strong tidal currents occur where separate basins having different tidal regimes are connected by a narrow channel. The currents in the Straits of Messina may exceed $2 \, \mathrm{m \, s^{-1}}$ because of the strong gradients on the sea surface. Even stronger currents are observed through the Euripus between mainland Greece and the island of Euboea; tidal currents under the bridge at Khalkis can exceed $3 \, \mathrm{m \, s^{-1}}$ at spring tides.

The **Arctic Ocean** has depths in excess of 5000 m, but also contains the world's most extensive shelf region. Its tides are driven by the Atlantic tides, through the connection between Scandinavia and Greenland (Zetler, 1986), which is small compared with the ocean area. Figure 5:9 shows the cotidal and corange lines for the M_2 tide. The wave enters primarily through the Greenland Sea to the west of Spitzbergen. As it propagates northwards it decreases in amplitude as it circles an anticlockwise amphidrome located near 81° 30′N, 133° W in the deep waters of the Canadian Basin. Part of the wave propagates onto the broad shallow shelf of the East Siberian Sea where its energy is dissipated. Tides on the North Siberian Shelf were described by Sverdrup (1927) in terms of a particular type of non-coastal wave, now called Sverdrup waves (see also Platzman, 1971; Gill, 1982). A secondary branch of the M_2 tide enters the Arctic between Spitzbergen and Norway; this produces high tides in the southern Barents Sea and around Novaya Zemlya. The permanent ice covering is thought to have little influence on the observed tides, but where tidal currents diverge on the shallow shelf seas, they can cause a periodic opening and closing of the cracks in the ice cover (Kowalik, 1981).

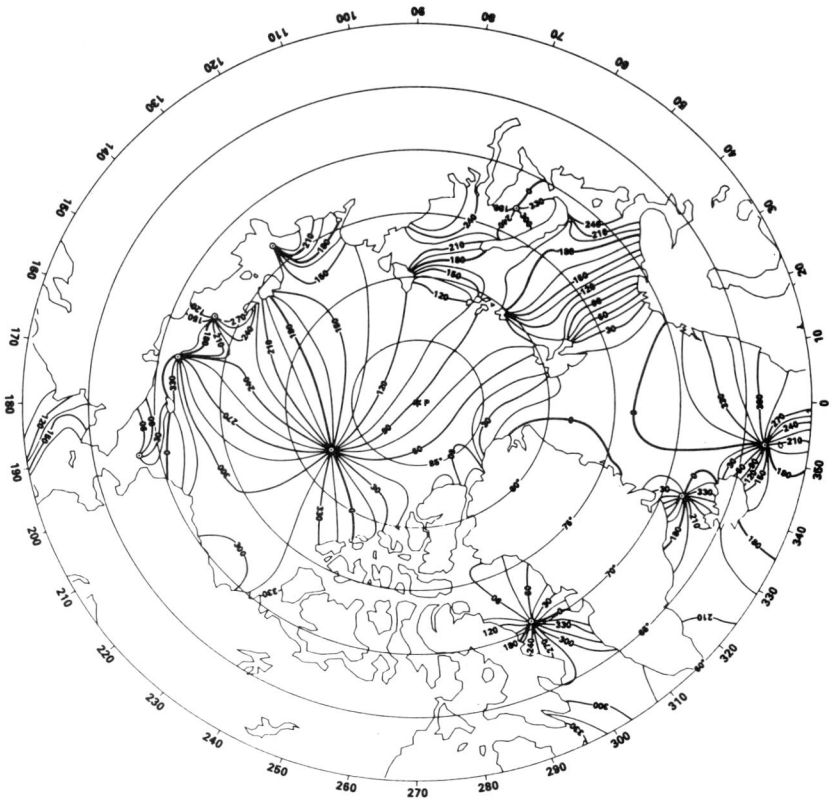

ARCTIC COTIDAL MAP OF M₂ OCEAN TIDE
GREENWICH PHASES δ IN DEGREES
30° ≈ 1 HOUR
⊙ AMPHIDROMES   ✳ P:NORTH POLE

(a)

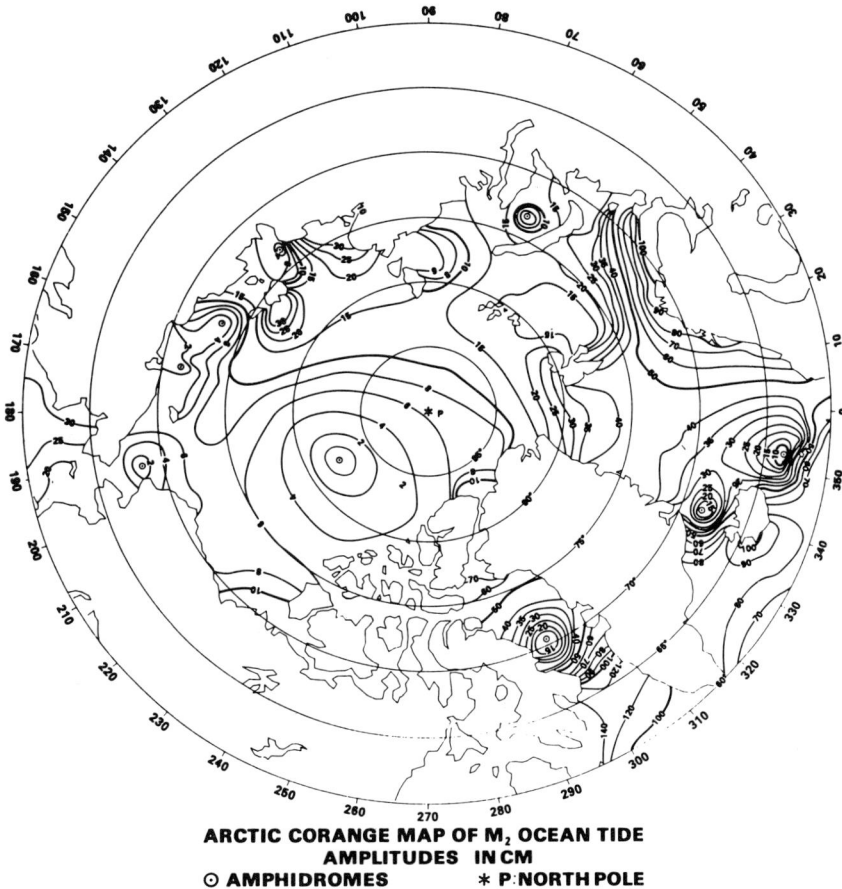

ARCTIC CORANGE MAP OF M₂ OCEAN TIDE
AMPLITUDES IN CM
⊙ AMPHIDROMES   ＊ P:NORTH POLE

(b)

Figure 5:9 M_2 tide of the Arctic Ocean based on a numerical model. (From Schwiderski, 1979.)

5:4   Shelf tides

5:4:1   Transition from the ocean to the shelf

The process of transition of the oceanic tidal wave onto the relatively shallow continental shelves has not been observed in detail, but the general effects of the decrease in depth are the same as for any other kind of long wave. As the depth decreases, so does the wave speed c, as a result of which the wave is refracted so that the crests tend to align themselves parallel to the bathymetry contours.

Suppose that the boundary is represented by a step discontinuity in depth from 4000 m to 200 m. For a plane wave approaching an infinitely long ocean shelf boundary at an angle, the law of refraction applies, the angles of incidence and refraction being related by:

$$\frac{\sin\,(\text{incident angle})}{\sin\,(\text{refracted angle})} = \frac{\text{ocean wave speed}}{\text{shelf wave speed}} \approx \left(\frac{4000\,g}{200\,g}\right)^{\frac{1}{2}} = 4.5$$

In the absence of other effects such as the earth's rotation, a wave approaching nearly parallel to the shelf edge (incident angle $90°$) would have a refracted angle of $13°$—the direction of wave propagation is turned through $77°$.

A normally incident long wave will be partly reflected back to the ocean and partly transmitted onto the shelf. Of course the real continental margin will have a different response to that of the theoretical step, but the relative theoretical amplitudes are instructive. For an incident wave of amplitude 1.0 m the reflected wave has a theoretical amplitude of 0.64 m and the transmitted wave has an amplitude of 1.64 m (Appendix 4:1b). The reason for this can be argued by energy flux considerations. In Appendix 4:1a the energy transmitted in unit time by a progressive wave is shown to be:

$$\tfrac{1}{2}\rho g H_0^2 (gD)^{\frac{1}{2}}$$

per unit distance along the wavefront. In the absence of reflection and energy losses due to bottom friction, the quantity:

$$(\text{wave amplitude})^2 \times (\text{water depth})^{\frac{1}{2}}$$

must remain constant as the wave moves into shallower water. If the depth is reduced by a factor of 20 and there is total energy transmission, the wave amplitude increases by a factor of $20^{\frac{1}{4}} = 2.1$. For the step model of the boundary between the ocean and the shelf, 60 per cent of the incident energy is transmitted and 40 per cent is reflected, but the transmitted energy is still enough to give larger amplitudes in the shallower water by a factor of 1.64.

The tidal currents on the shelf are also enhanced. Applying equation (5:2), which relates currents to wave amplitude and water depth, the ocean wave of amplitude 1.0 m in 4000 m depth has current speeds of $0.05\,\text{m s}^{-1}$ whereas the transmitted wave of amplitude 1.64 m in 200 m depth has current speeds of $0.36\,\text{m s}^{-1}$. The current speeds are increased by a greater factor than the increase in the elevation amplitudes.

Waves incident at an angle to the actual sloping topography of the continental margins will undergo continuous refraction. Reflected waves may also be refracted back again giving rise to a succession of topographically trapped edge waves travelling along the ocean–shelf interface. For waves travelling with the shallow water to their right (northern hemisphere) trapping is also possible because of the earth's rotation. The tides around New Zealand have been explained in terms of weakly trapped semidiurnal waves (Bye and Heath, 1975).

The relative importance of the different kinds of hydrodynamic waves which may exist at the continental margin will depend on the particular circumstances. One well-studied example is the northward progression of the semidiurnal tidal phases ($140\,\mathrm{m\,s^{-1}}$) along the Californian coast of the western United States, where the continental shelf is narrow (Munk *et al.*, 1970). This progression can be described as the superposition of a Kelvin-like edge wave, a free Poincaré wave (another type of long wave which can propagate on the rotating earth: Appendix 4) and a directly forced wave, with amplitudes at the coast of $0.54 : 0.16 : 0.04$ m. Phase progression is consistent with the semidiurnal amphidrome located at 25° N, 135° W. Diurnal tides may also be explained in terms of these three wave types. They have a faster ($214\,\mathrm{m\,s^{-1}}$) speed and a southern progression; their component wave amplitudes are $0.21 : 0.24 : 0.09$ m at the coast.

The importance of the earth's rotation in the generation and propagation of different types of waves, and the different responses to the diurnal and semidiurnal tides is illustrated by the behaviour of sea levels and currents off the west coast of Scotland. The observed semidiurnal levels and currents are scaled according to expected Kelvin wave dynamics, but the diurnal waves have very large currents compared with the small amplitudes of the vertical diurnal tides. The diurnal tides include a continental shelf wave, constrained to travel parallel to the shelf edge with shallow water to the right in the northern hemisphere; because these waves are only possible for waves whose period exceeds the inertial period due to the earth's rotation (14.3 hours at 57° N), they are impossible for the semidiurnal tides at that latitude. Similarly enhanced diurnal currents due to non-divergent shelf waves are observed on the continental shelf off Vancouver Island, British Columbia, and on the shelf between Nova Scotia and Cape Hatteras.

5:4:2 Some examples of shelf tides

The patterns of tidal waves on the continental shelf are scaled down as the wave speeds are reduced. Table 5:1 gives the wavelengths for typical depths. The Rossby radius, which is also reduced in the same proportion, varies as shown in Table 5:4. In the very shallow water depths (typically less than 20 m) there will be strong tidal currents and substantial energy losses due to bottom friction; these severe non-linear distortions are discussed in Chapter 7. For average shelf depths the waves are strongly influenced by linear Kelvin wave dynamics and by basin resonances. Energy is propagated to the shallow regions where it is dissipated. In this discussion it will be possible to describe only a few representative cases of shelf tidal behaviour.

(a) *The tides of the north-west European continental shelf* have been mapped in detail as shown in Figure 5:10(a) and (b) (see also Proudman and Doodson, 1924; Great Britain, 1940; Oberkommando der Kreigsmarine, 1942; Sager and Sammler, 1975; Huntley, 1980; Howarth and Pugh, 1983). These charts have

Table 5:4 Rossby radius in km for different water depths and latitudes. A Kelvin wave amplitude is reduced by e^{-1} ($= 0.36$) in a distance of one Rossby radius.

	Rossby radius $= (gD)^{\frac{1}{2}}/f$ Latitude (°)				
	10	30	50	70	90
$f \times 10^{-5}$	2.53	7.29	11.2	13.7	14.6
Water depth (m)					
4000	7820	2720	1770	1450	1360
200	1750	610	395	325	305
50	875	305	200	160	150
20	550	190	125	102	96

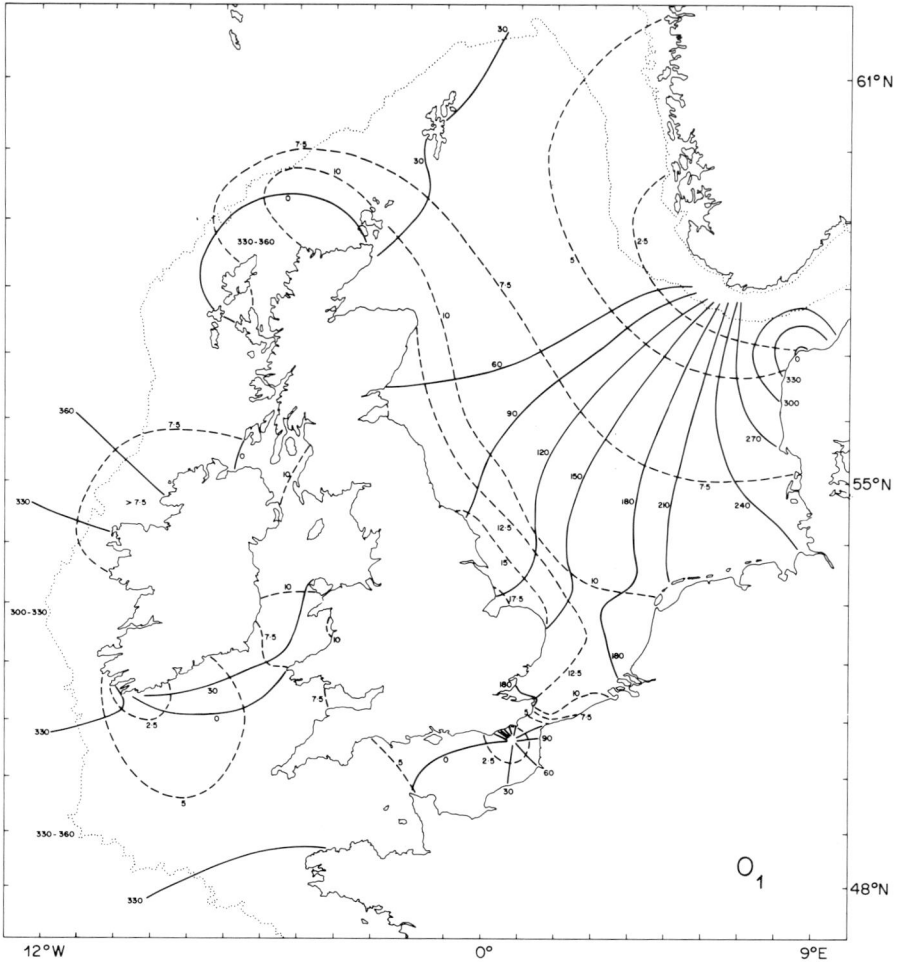

(a)

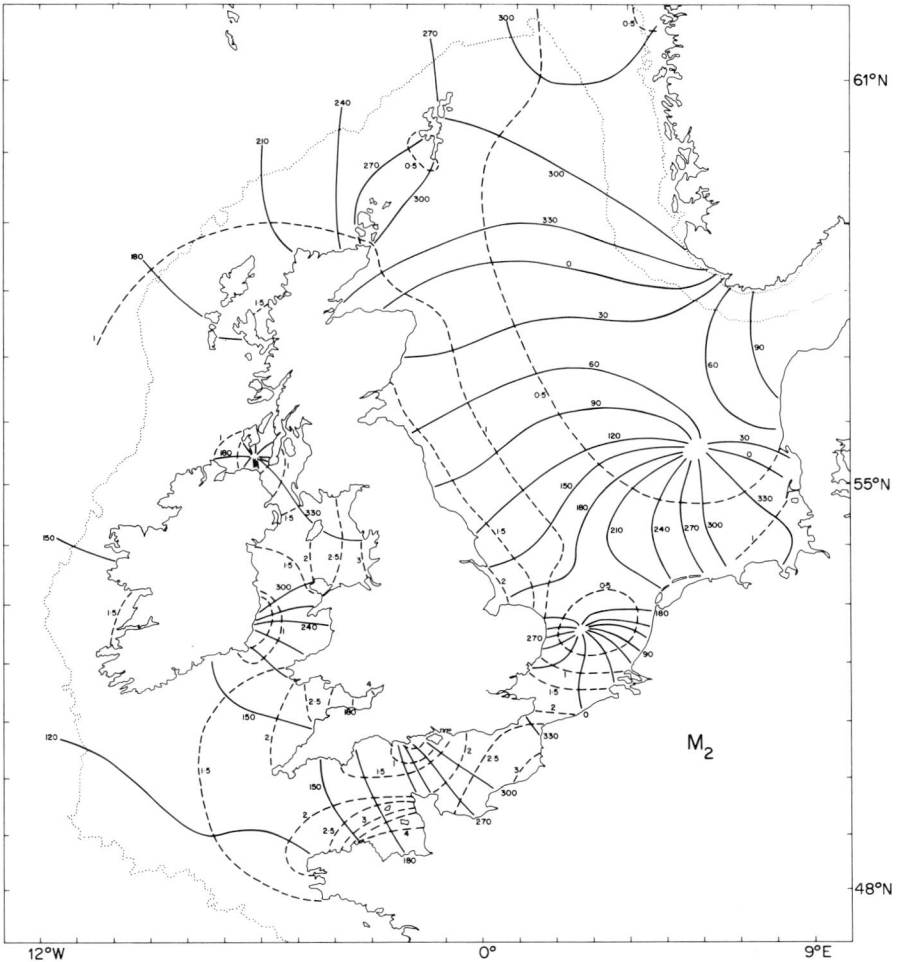

Figure 5:10 O_1 and M_2 tides of the north-west European continental shelf from observations (from Howarth and Pugh, 1983) (*Reproduced by permission of Elsevier.*)

been produced from coastal sea-level observations, measurements of offshore bottom pressures and offshore current measurements combined hydrodynamically as described in Appendix A4:4. The Atlantic semidiurnal Kelvin wave travels from south to north. Energy is transmitted across the shelf edge into the Celtic Sea between Brittany and southern Ireland; this wave then propagates into the English Channel where some energy leaks into the southern North Sea, and into the Irish Sea and the Bristol Channel. The Atlantic wave progresses northwards, taking five hours to travel from the Celtic Sea to the Shetlands.

The diurnal tidal progression along the shelf break is not so simple because the phase increases and decreases several times rather than increasing regularly. The semidiurnal wave is partly diffracted around the north of Scotland where it turns to the east and to the south into the North Sea.

The semidiurnal tides of the North Sea (which is broad compared with the Rossby radius) consist of two complete amphidromic systems and a third, probably degenerate, system which has its centre in southern Norway. The largest amplitudes occur where the Kelvin south-travelling wave moves along the British coast. Coamplitude lines are parallel to the coast, whereas cotidal lines are orthogonal. The identification of the amphidrome in the southern North Sea is of some historic interest: it was originally proposed by Whewell in 1836, and confirmed in 1840 by Captain Hewett, who measured an almost constant depth throughout a tidal cycle from a moored boat (Figure 5:11). Subsequent further measurements and theoretical developments (see Appendix 4 and Proudman and Doodson, 1924) have refined the original charts.

The amphidromic system shown in Figure 5:10(b) may be compared with Taylor's theoretical amphidrome in Figure 5:6. However, although the southern amphidrome is located near the centre of the sea, progressive weakening of the reflected north-going Kelvin-type wave places the second and third amphidromes further and further to the east of the central axis. Indeed, even the central position of the southern amphidrome is probably partly due to an enhancement of the reflected wave by a north-going wave entering through the Dover Straits.

The English Channel and the Irish Sea are relatively narrow in terms of the Rossby radius. They respond similarly to the incoming wave from the Celtic Sea. The wave takes about seven hours to travel from the shelf edge to the head of the Irish Sea, and a similar time to reach the Dover Straits. The wave which travels along the English Channel reaches the Dover Straits one complete cycle earlier than the wave which has travelled the greater distance around Scotland and through the North Sea. The large tidal amplitudes in the Dover Straits (greater than 2.0 m) are due to these two meeting waves combining to give an anti-amphidrome.

The English Channel has a response similar to that of a half-wave resonator (equation (5:5)), with a nodal line between the Isle of Wight and Cherbourg. Tidal levels at the Dover Straits have the opposite phase to those at the shelf edge. The large amplitudes on the French coast (3.69 m at St Malo near La Rance tidal power station, for the M_2 constituent) are due to Kelvin-wave dynamics and local standing wave resonance. Because of frictional dissipation and leakage of energy into the southern North Sea, the reflected tidal wave is much weaker than the ingoing wave. A full amphidromic system cannot develop because the Channel is too narrow; instead, there is a clustering of cotidal lines towards a degenerate amphidrome located some 25 km inland of the English coast.

The large tidal amplitudes in the upper reaches of the Bristol Channel ($H_{M_2} = 4.25$ m at Avonmouth) appear to be due to quarter-wave resonance

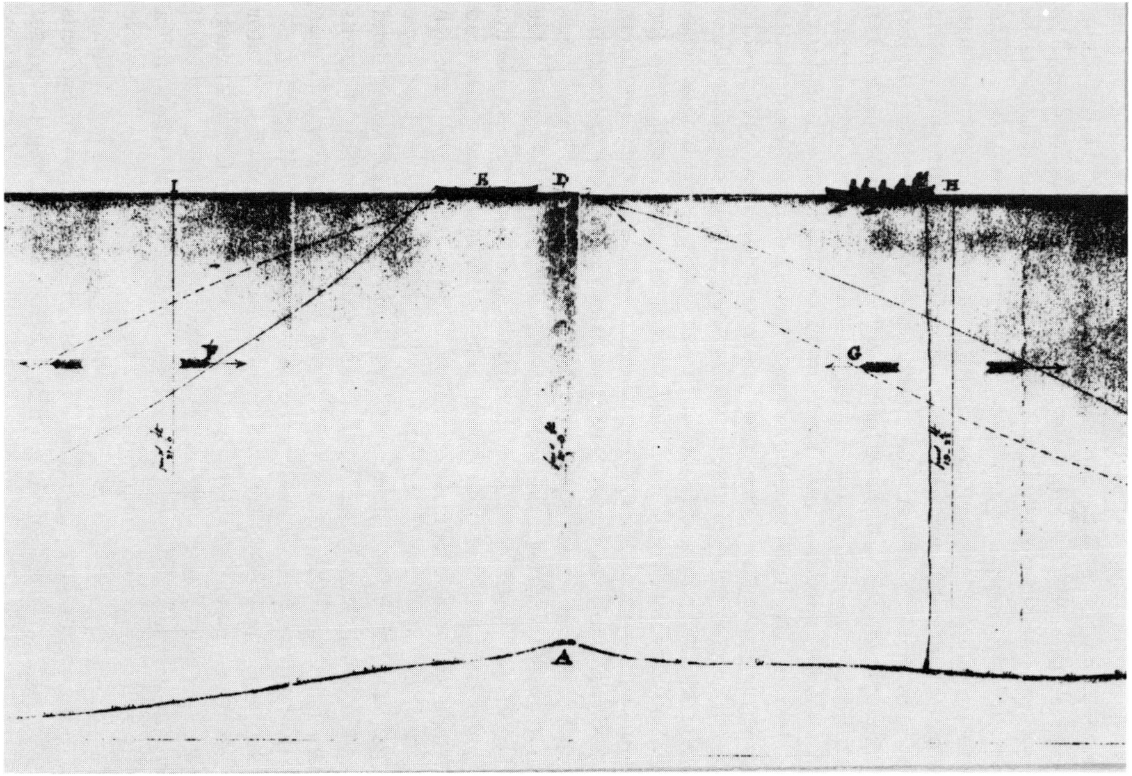

Figure 5:11 Part of the diagram prepared by Captain W. Hewett R.N., *H.M.S. Fairy*, illustrating his method of obtaining tidal observations in August 1840 from an anchored boat in the southern North Sea. His observations confirmed the existence of the tidal amphidrome proposed by Whewelll from theoretical arguments. Shortly after these observations were made *H.M.S. Fairy* was lost with all hands in the great gale of 13 November 1840. This copy is provided by the Hydrographic Department of the Admiralty.

(equation (5:6)); the quarter-wave resonant period of the Celtic Sea and Bristol Channel together has been estimated at slightly greater than 12 hours, close to but less than the period of M_2.

The standing wave response of the Irish Sea is similar to that of the English Channel. In this case the nodal line is located between Wales and Ireland, with a degenerate M_2 amphidrome located some 10 km inland from the Irish coast (Taylor, 1919; Robinson, 1979; Pugh, 1981). Energy is transmitted through this nodal region into the northern part of the Irish Sea. Over the northern part the tidal times are nearly simultaneous, characteristic of a standing wave. The strength of the reflected wave returning through St George's Channel varies throughout the spring/neap cycle because of the varying frictional energy losses, as we discuss in Chapter 7. A small amount of energy enters the Irish Sea through the narrow North Channel and there is a further amphidrome between Ireland and the Scottish island of Islay.

The diurnal tides (Figure 5:10(a)), represented by the O_1 chart, progress anticlockwise around the North Sea in one cycle with an amphidrome which may be degenerate, close to the shore of southern Norway. There is little evidence of shelf amplification or resonance for the diurnal tides; their maximum amplitude does not exceed 0.20 m, which is only three times the shelf edge value. There is an amphidrome west of the Dover Straits and another, slightly degenerate, in south-west Ireland. The response curves for Castletownsend (Figure 4:2), located close to the amphidrome, show its effects very clearly: diurnal amplitude responses have a sharp minimum value. This degenerate amphidrome is located half of the diurnal tidal wavelength from the head of the Irish Sea. Throughout the Celtic Sea, the Bristol Channel and English Channel, as well as in the Irish Sea, the diurnal tide behaves as a standing wave, but without any tendency to resonance.

(b) *The semidiurnal tides of the Yellow Sea*, shown in 5:12, may be compared with those of the North Sea: the main basin has three amphidromes, progressively displaced from the central axis. The wave which enters from the East China Sea travels in an anticlockwise sense as a Kelvin wave (Larsen *et al.*, 1985). The largest amplitudes are found along the coast of Korea. The returning Kelvin wave which travels south along the coast of China is much weakened by energy losses to Po Hai basin and to Korea Bay. As a result, the amphidromes are progressively nearer the Chinese coast and the third amphidrome near Shanghai may be degenerate.

(c) The nearly simultaneous semidiurnal tides observed over more than 1000 km along the *east coast of the United States* between Long Island and Florida have been explained by Redfield (1958) in terms of an Atlantic Ocean tide which is nearly simultaneous along the shelf edge. Standing waves develop across the shelf, the coastal amplitude being greatest where the shelf is widest, as shown in Figure 5:13. The smallest ranges occur near Cape Hatteras where the shelf is narrowest. Extrapolating the values in Figure 5:13 to zero shelf width

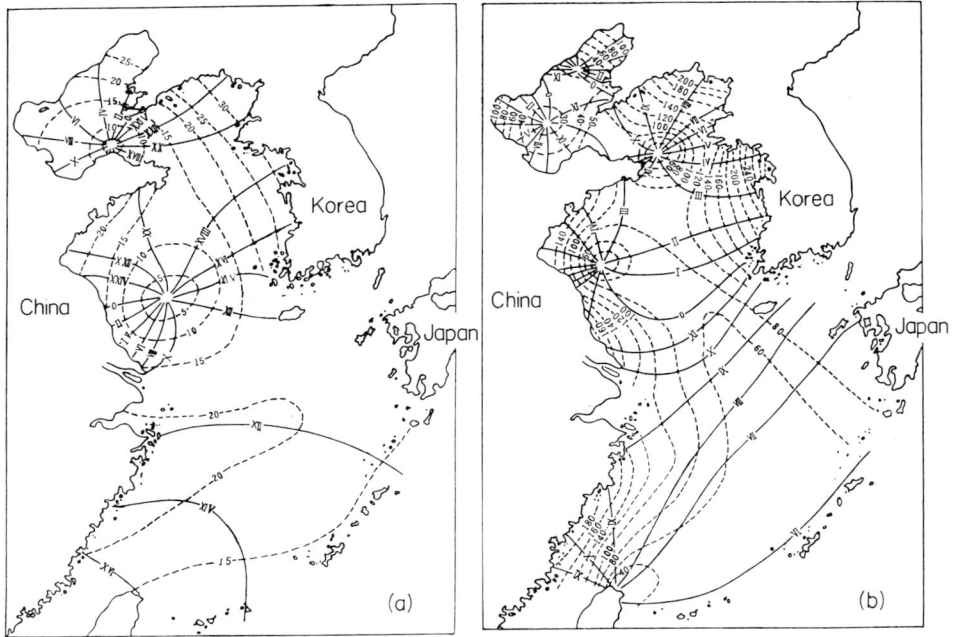

Figure 5:12 (a) O_1 tide of the Yellow Sea, from a numerical model. (b) M_2 tide. (From Shen Yujiang, Numerical computation of tides in East China Sea, *Collected Oceanic Works*, **4**, 36–44 (1981).)

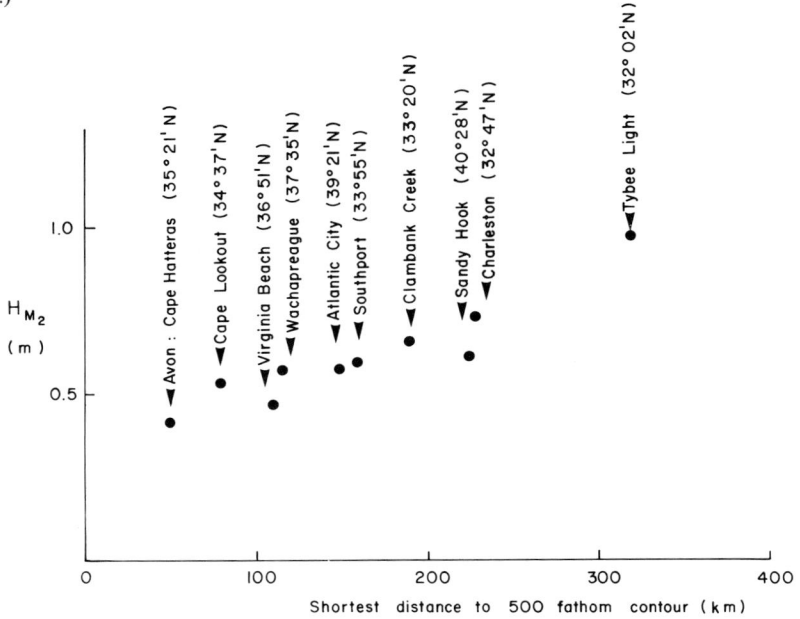

Figure 5:13 The relationship between the M_2 coastal amplitude and the shortest distance to the 500-fathom depth contour, along the east coast of the United States. The increase of amplitude with shelf width is consistent with standing wave oscillations at right angles to the coast.

172

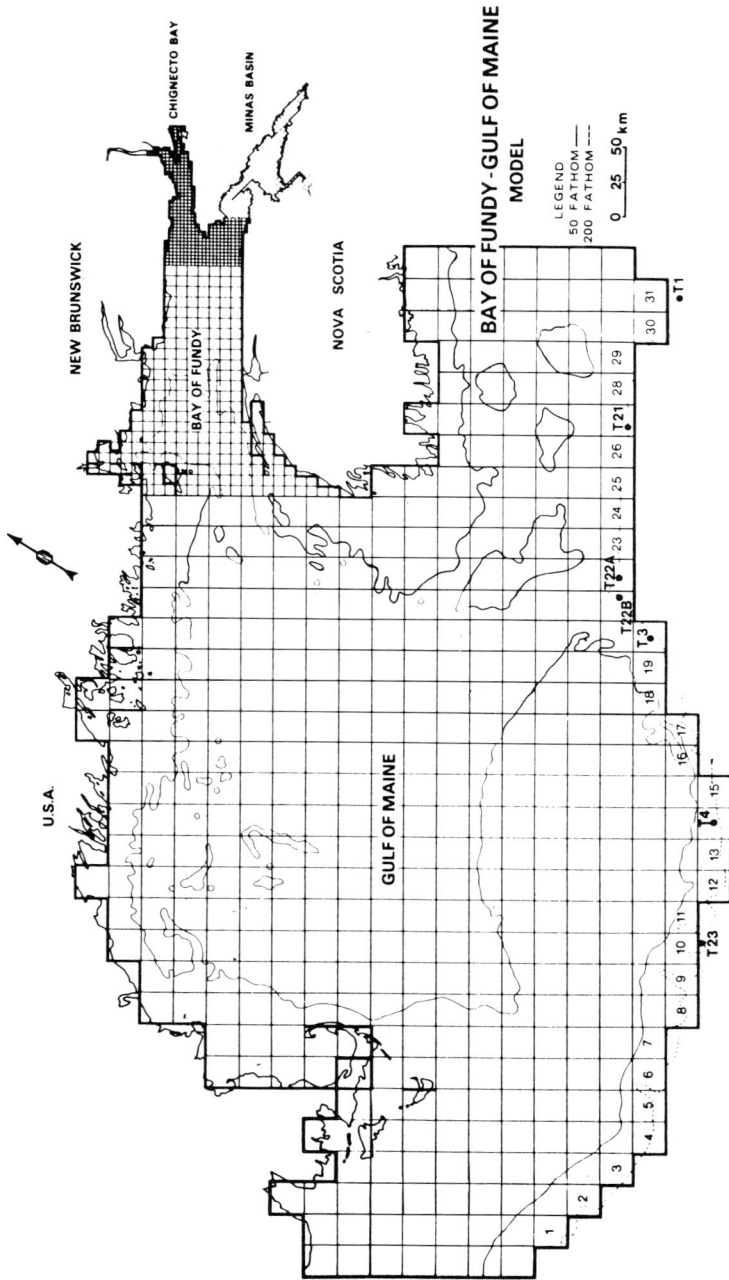

(a)

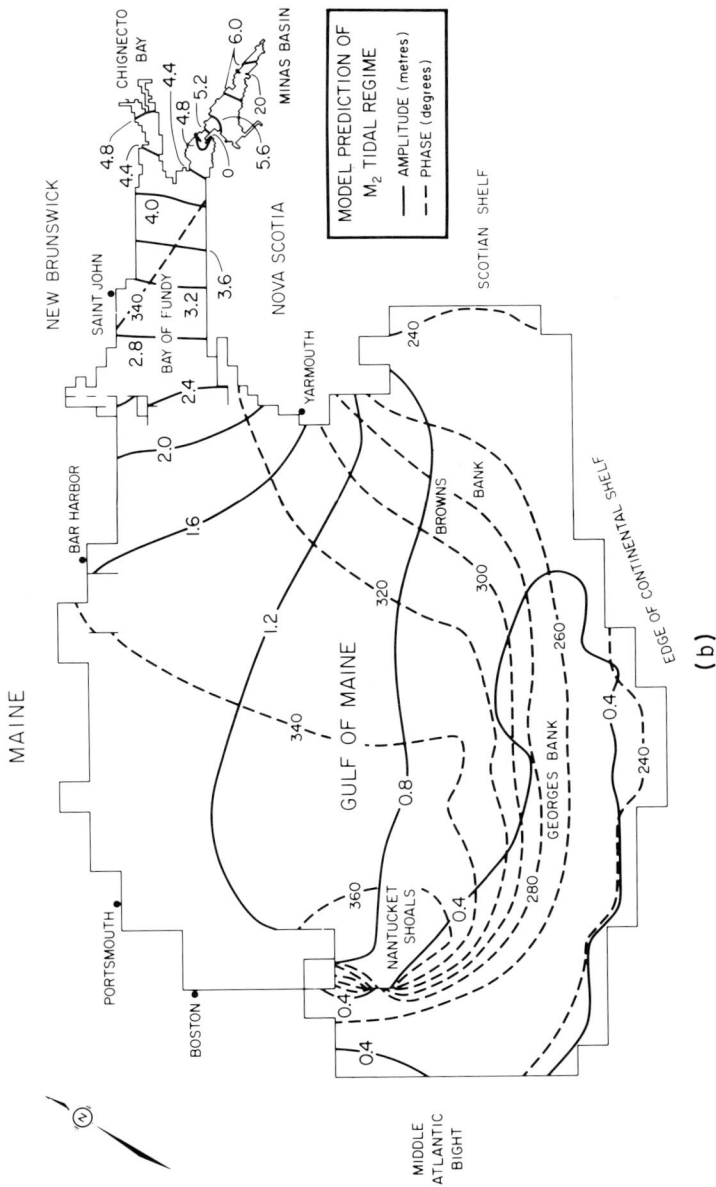

Figure 5:14 Numerical model description of the tidal regime of the Bay of Fundy and Gulf of Maine, (a) the grid of different mesh sizes used in the computations (from Garrett, 1984). (*Reproduced by permission of Pergamon Books Ltd.*) and (b) the computed cotidal and coamplitude lines (supplied by D. A. Greenberg).

suggests an oceanic M_2 amplitude of about 0.4 m. The shelf is too narrow for any resonant responses to develop. Redfield found that the reflected wave was slightly weaker than the incident wave because of energy dissipation at the coast. As a result there is a small progressive wave component, with coastal tides occurring up to an hour later than those at the shelf edge where the shelf is wide, off Charleston.

The tides of Long Island Sound (Marmer, 1926; Redfield, 1980) are dominated by a standing wave oscillation which is driven from its eastern entrance. However, there is also a tidal flow which enters through the East River. The effect of this flow from the west into the Sound is to shift the antinode where maximum tidal amplitudes occur slightly to the east of the head of the Sound. Near to Glen Cove the spring tidal range exceeds 2.6 m.

Further north, the tidal system which develops in the Gulf of Maine and the Bay of Fundy consists of near-resonant oscillations which produce the world's greatest tidal amplitudes, in the Minas Basin. Figure 5:14 shows how the M_2 amplitudes increase from less than 0.5 m at the shelf edge to more than 5.6 m at Burncoat Head. It has been estimated that the natural quarter-wave period of this system is close to 13.3 hours, which explains the large near resonant response to semidiurnal tides. Proposals to exploit the very large tidal ranges by building tidal power stations would reduce the effective length of the system, so bringing the natural period (equation (5:6)) even closer to the 12.4 hours of the M_2 tide! For tidal systems near resonance, even small physical changes can produce large changes in tidal range (Ku *et al.*, 1985). Estimates of the effect of 'tuning' the tidal responses are uncertain because the enhanced dissipation, vital for estimating responses close to resonance, is also uncertain.

A full semidiurnal amphidromic system is developed in the Gulf of St Lawrence and a further small system appears in Northumberland Strait between Prince Edward Island and New Brunswick (Godin, 1980). The tides of the Canadian Arctic are complicated by the many islands and connecting straits. A local resonance produces very large semidiurnal amplitudes in Ungava Bay (4.36 m for M_2 at Leaf Basin). In Baffin Bay a full anticlockwise semidiurnal amphidrome has been mapped; the diurnal tides appear to behave as a standing wave with amplitudes of 0.3 m at the northern end. The semidiurnal tides of Hudson Bay, which are driven through Hudson Strait, show a complex of several amphidromes, with amplitudes generally less than 0.3 m. However, the diurnal tidal pattern, which consists of a single anticlockwise amphidromic system is particularly simple.

(d) *The Persian Gulf* is a shallow sea which has mixed diurnal and semidiurnal tides (Figure 5:15(a)). It is a largely enclosed basin with only a limited

Figure 5:15 O_1 and M_2 tides of the Persian Gulf, based on Admiralty Chart 5081. (Crown Copyright. Reproduced from Admiralty Chart 5081 with the permission of the Controller of Her Majesty's Stationery Office.)

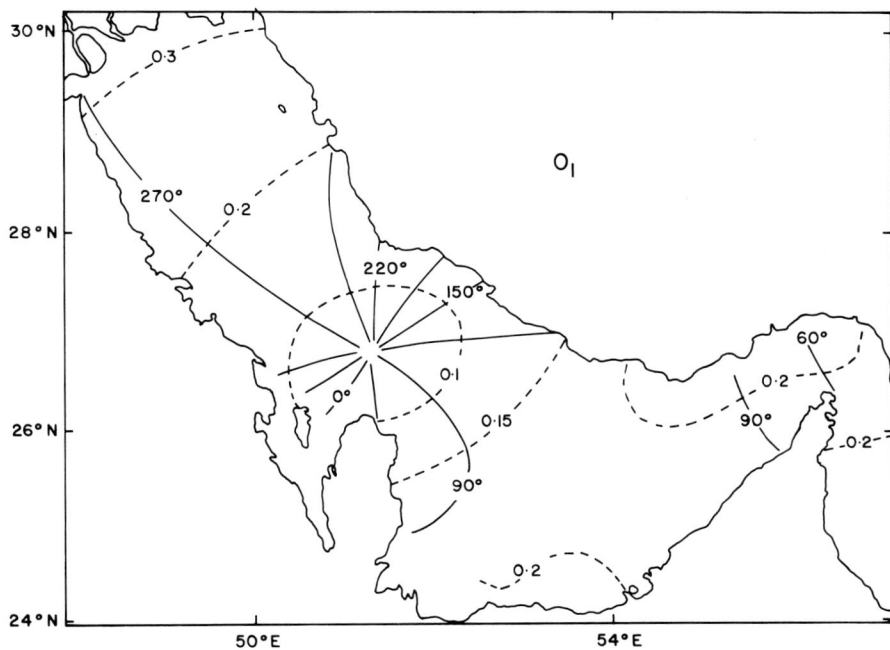

(a)

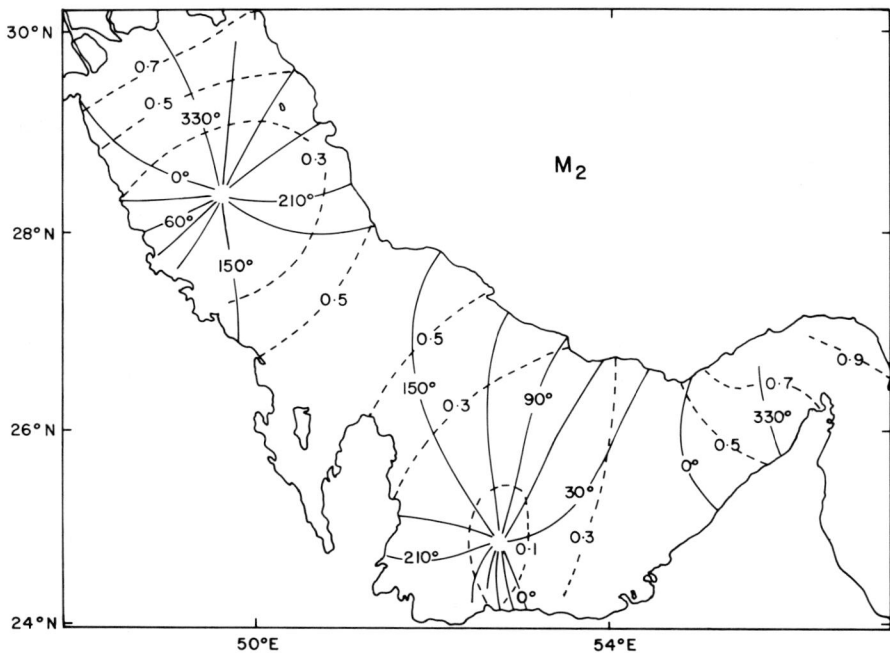

(b)

connection to the Indian Ocean through the Strait of Hormuz. Along the major north-west to south-east axis it has a length of about 850 km. The average depth is approximately 50 m, giving a resonant period near 21 hours according to equation (5:5). The Rossby radius at $27°$ N (c/f) is 335 km, comparable with the basin width. As a result the response to the diurnal forcing through the Strait of Hormuz is a single half-wave basin oscillation with an anticlockwise amphidrome (Figure 5:15(b)). The semidiurnal tides develop two anticlockwise amphidromic systems, with a node or anti-amphidrome in the middle of the basin. Near the centre of the basin the changes in tidal level are predominantly semidiurnal, whereas near the semidiurnal amphidromes they are mainly diurnal. At the northwest and southeast ends of the basin the tidal levels have mixed diurnal and semidiurnal characteristics.

These discussions of observed tides have concentrated on the semidiurnal and diurnal responses. Although the third-diurnal forcing in the Equilibrium Tide is weak, a local resonance on the shelf off Brazil gives M_3 amplitudes in excess of 0.10 m at Paranagua. Similarly, there is also a tendency for resonant M_3 amplification on the European continental shelf between Ireland and Scotland.

5:4:3 Shelf tidal currents

Tidal currents are often termed *tidal streams* by hydrographic authorities. These currents are more variable from place to place than changes of tidal levels because they are sensitive to changes of depth, and to the influences of coastal embayments and headlands (Howarth, 1982). However, at any particular location, the tidal currents may be measured, analysed and predicted in the same way as levels. Total currents may be analysed and plotted as detailed in section 4:4 and Figure 1:2. More elaborately, they may be broken down into their harmonic constituent ellipses or rotary components as described in Sections 4:4:2 and 4:4:3, Appendix 3 and Figure 4:5. Tidal currents show little vertical structure, except near the sea-bed (see Chapter 7) or in stratified estuaries, and so the concept of a depth mean tidal current is valid except in very shallow water.

For a progressive wave the currents, which are a maximum in the direction of wave propagation at local high water, are related to the wave amplitude by equation (5:2). A theoretical progressive or Kelvin wave has rectilinear currents. In reality rectilinear currents are only found near to steep coasts or in narrow channels. In the North Sea, for example (Pugh and Vassie, 1976; Davies and Furnes, 1980), in addition to the Kelvin wave which propagates southward along the British coast there is a standing wave with a current component at right angles to the coast whose amplitude is typically 15 per cent of the amplitude of the parallel component. The phase relationship between currents for a station in this area are shown in Figure 4:6. For a south-going progressive wave the currents (positive to the north) should lag the elevations by $180°$

rather than 150° observed in the semidiurnal band. In this case maximum currents in the direction of wave propagation occur one hour before local high water.

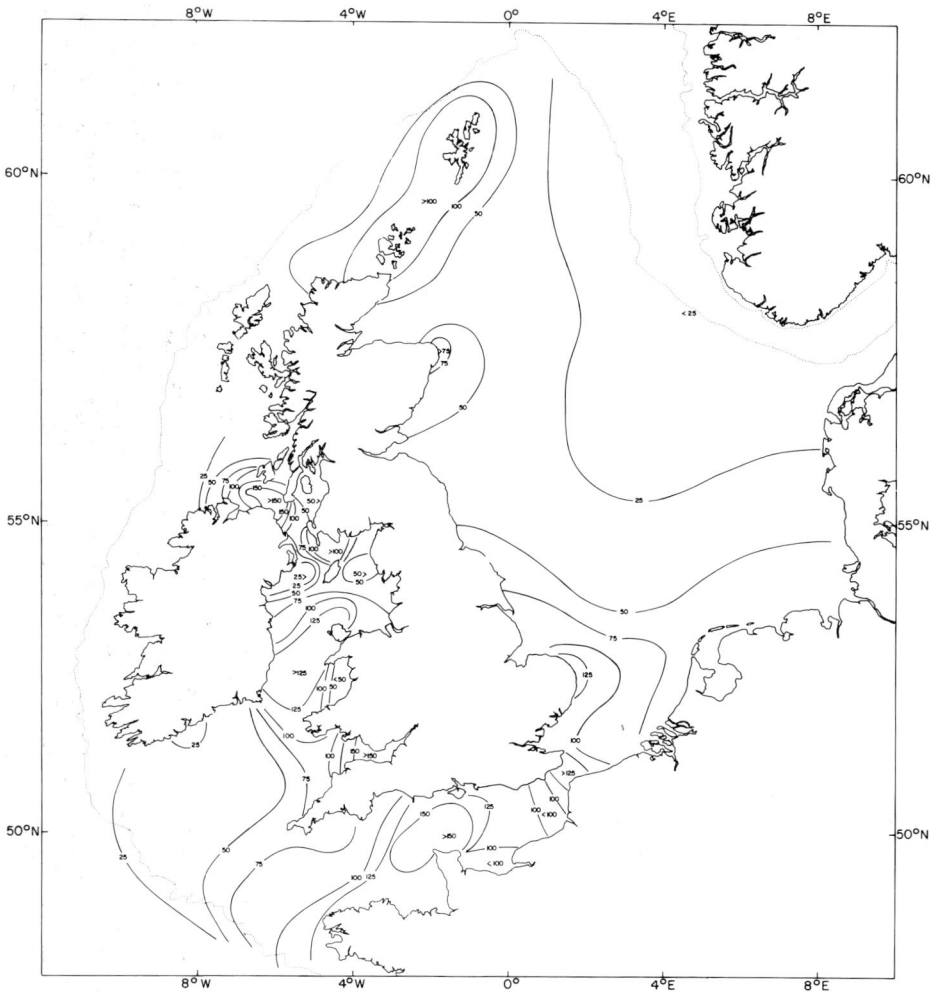

Figure 5:16 Maps of mean spring near-surface tidal current amplitudes on the north-west European continental shelf (cm s^{-1}). (*Reproduced by permission of Elsevier.*)

In a pure standing wave system, currents are again rectilinear with maximum amplitudes near the nodes or amphidromes. In practice strong currents are observed near amphidromes such as in the southern North Sea, and the southern entrance to the Irish Sea. Along the Atlantic coast of the United States the standing wave has maximum tidal currents at the shelf edge, directed towards and away from the coast.

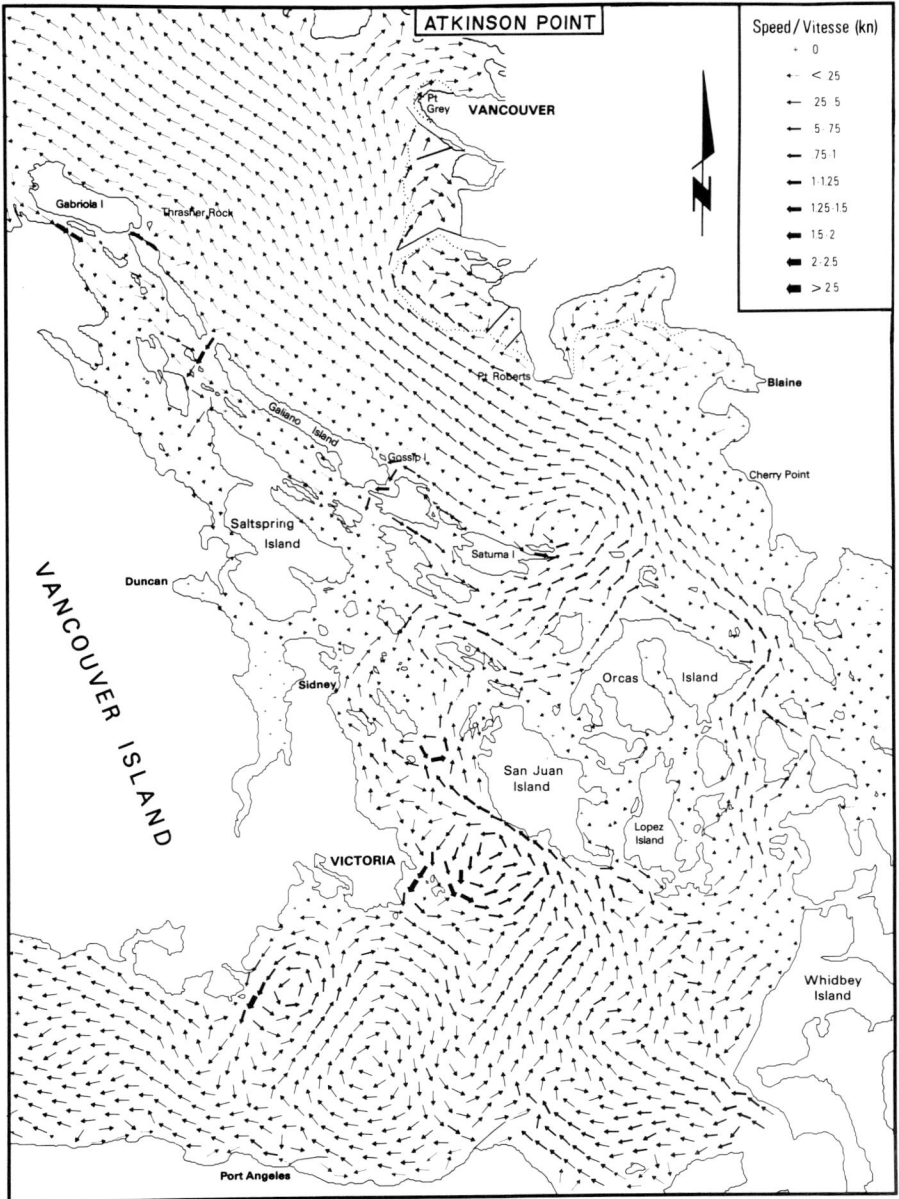

Figure 5:17 Computed tidal predictions of currents in the Juan de Fuca Strait, British Columbia. Values are predicted for one hour before high water (4.5 m) at Atkinson Point. Note the complicated flow patterns through the islands and the development of near-shore gyres. Predicted currents through Seymour Narrows at the north of Vancouver Island can exceed $8.0\,\mathrm{m\,s}^{-1}$. (*Reproduced by permission of Canadian Hydrographic Department.*)

Figure 5:16 shows the mean spring near-surface tidal currents on the northwest European shelf. The strongest currents are observed in the English Channel and Dover Straits, in the Irish Sea and north of Ireland. Currents in excess of 1.0 m s^{-1} are also observed in the channels between Scotland and the Orkney and Shetland Islands as the Atlantic tide enters the North Sea. In the Pentland Firth the spring currents exceed 4.0 m s^{-1}. Currents in narrow straits, such as these between two tidal regimes, are controlled by the balance between pressure head and friction, according to the laws of open channel hydraulics; these are discussed in Chapter 7. Figure 5:17 shows the current system which develops between a complex of islands in the Juan de Fuca Strait, British Columbia; the power of the numerical modelling technique for producing these complicated charts is evident.

The ratios between the amplitudes of different current constituents are usually fairly stable over large regions; for example, the S_2/M_2 current amplitude ratio over the northwest European shelf is close to 0.35 except near amphidromes. The ratio between diurnal and semidiurnal currents is usually stable, and close to the ratio in the elevations. However, there are places where these ratios become anomalous. In the centre of the Persian Gulf (Figure 5:15) the proximity of a diurnal amphidrome and a semidiurnal anti-amphidrome means that the currents are predominantly diurnal whereas the elevation changes are predominantly semidiurnal. A similar juxtaposition of a diurnal amphidrome and a semidiurnal anti-amphidrome gives diurnal currents and semidiurnal elevation changes for Singapore at the southern end of the Malacca Strait.

The sense of rotation, clockwise or anticlockwise of a current ellipse is controlled by many factors and there are no simple rules to decide which effects will be most important. In some cases the rotation sense may be different at the top and bottom of the water column (see Section 7:4). In the oceans, well away from the influences of the coast, the direct tidal forcing and the Coriolis accelerations both act to induce circulation of the semidiurnal current ellipses in a clockwise sense in the northern hemisphere and in an anticlockwise sense in the southern hemisphere. On the continental shelf the sense of rotation is usually controlled by the bathymetry and by coastal wave reflections. The theoretical amphidromic system shown in Figure 5:6 has anticlockwise circulation of current ellipses near the reflecting boundary in the northern hemisphere and anticlockwise circulation in the southern hemisphere.

Another important influence on the sense of ellipse rotation near a coastline is the presence of shelving beach or an embayment, as illustrated in Figure 5:18. Consider a progressive wave travelling with the coast on its right. At high water (1) the flow is directed parallel to the coast as shown. At midwater level on the falling tide there is no contribution to the flow from the wave progression, but the fall of water level in the embayment produces an offshore flow. At low water (3) the progressive wave gives a flow as shown, parallel to the coast, but as levels in the embayment are not changing, there is no component towards the coast. At midwater on the rising tide (4) the progressive wave has no

direction of wave propagation ⟹

midtide, falling
↑ (2)

low water (3) ←———————— ————————→ (1) high water

(4) ↓

midtide,
rising

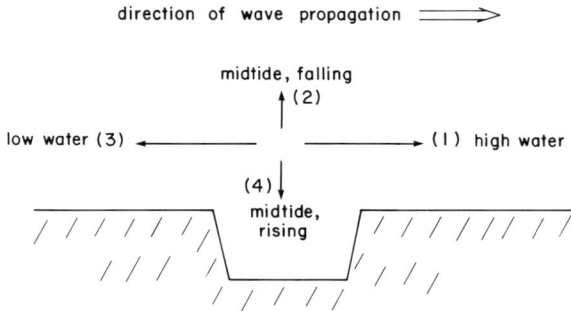

Figure 5:18   The development of an anticlockwise current vector rotation for a progressive wave with an embayment or a shelving coast to the right.

current, but the rapid increase of level in the embayment is fed by an onshore flow. The net result of this cycle is an anticlockwise sense of rotation for the current ellipse. Of course, this applies in both hemispheres as it has nothing to do with the earth's rotation. However, progressive waves with a coast to the right are more probable in the northern hemisphere. In a basin or channel which is too narrow for a full amphidromic system to develop, a progressive wave will also have a coast to the left of its direction of propagation and the embayments in this coast will influence the currents to rotate in a clockwise sense.

5:5   Radiational tides

In addition to the ocean responses to gravitational forcing, there is another type of forcing, radiational forcing, and the dynamics of the ocean response to this are poorly understood. Detailed harmonic tidal analyses, usually of a year of data, show a constituent of 24 hours period S_1 (speed ω_0) which is much larger than in the Equilibrium Tide (the true S_1 does not exist in the Equilibrium Tide, but there is a finite term with a speed of $(\omega_0 + \omega_6)$ which cannot be separated). In addition, the amplitudes and phases of the S_2 constituent are anomalous compared with the responses to the gravitational forcing at other adjacent frequencies, for example N_2 and M_2. These apparent anomalies are due to additional, regular, forcing by the weather, as outlined in Section 4:3:3; so too are the large seasonal variations of mean sea-level discussed in Chapter 9. These inputs are defined in terms of a radiation potential (Appendix 2), but the response of the oceans to these forces is not very well understood. The S_1 constituent is probably generated by local diurnal land/sea winds, and the radiational S_2 constituent is driven by the twelve-hour oscillation of air pressure in the tropics.

At Mombasa on the east coast of Africa, the S_1 term has an amplitude of 0.027 m; the land/sea diurnal winds have a harmonic amplitude of 2.0 m s^{-1}, with maximum onshore winds at 1530 hours local time. Maximum S_1 levels occur 6 hours later, just after the wind reverses and begins to blow offshore, which suggests a piling-up of the water rather than an immediate response to an imposed wind stress. Large S_1 tidal constituents are usually only found in tropical regions where the diurnal heating cycle is strong. Around the United States the S_1 amplitude is usually small, close to 0.01 m. Instrument errors may also produce an S_1 constituent in a harmonic analysis (see Section 4:2:1).

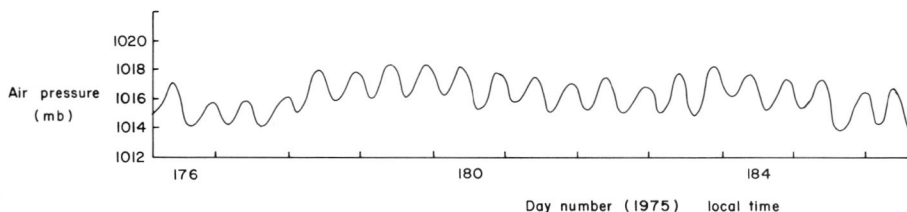

Figure 5:19   Air pressure variations at Aldabra Island, showing the variations at the S_2 tidal frequency, typical of equatorial latitudes.

Figure 5:19 shows the variations of air pressure at Aldabra Island in the west central Indian Ocean (Pugh, 1979); it is clear that these are dominated by a 12-hour variation. In the tropics variations of pressure at low frequencies from day to day are much less than at higher latitudes. In the upper atmosphere the diurnal heating cycle gives rise to diurnal pressure waves, but the dynamic structure of the atmosphere causes the semidiurnal harmonic to dominate at sea-level with an observed equatorial amplitude of 1.25 mb. By the usual inverted barometer arguments, the low atmospheric pressures at approximately 0400 and 1600 local time are equivalent to a maximum Equilibrium forcing of 0.0 125 m at the equator. The Equilibrium gravitational forcing at S_2 has a maximum of 0.164 m, at 0000 and 1200, local time. If the ocean response to the two forcing systems is the same, the radiational S_2 should have an amplitude 0.076 that of the gravitational S_2, and should lag it by 4 hours, or 120°. This is approximately true for tropical stations; the ratios and phase lags observed at three Indian Ocean stations were:

Mombasa	0.061	90°
Aldabra	0.073	125°
Mahé, Seychelles	0.066	116°

However, this good agreement for tropical stations conceals an anomalous larger S_2 radiational response at higher latitudes. The gravitational Equilibrium Tide decreases as $\cos^2\varphi$, where φ is the latitude, but the atmospheric pressure tide decreases as $\cos^3\varphi$ (Chapman and Lindzen, 1970; Forbes and Garrett, 1979). In theory the atmospheric ratio should also decrease at higher latitudes, but in practice the radiational tide becomes more important. At Bermuda the ratio is 0.10, but on both the east and west coasts of the United

States, the average ratio is 0.16. In the Celtic Sea and the English Channel the mean ratio is 0.15. The reason for this enhancement of the radiational constituent is not understood (see Cartwright (1968); and Zetler (1971)).

5:6 Internal tides

The propagation of waves at the sea surface depends on the density difference between water and the air, and on the gravitational acceleration. The air density is small enough to be ignored in the usual theoretical development. Waves may also propagate along density gradients within the ocean, but in this case the density of both layers must be taken into account. The speed of propagation of a long wave in a sea at the interface between two layers of density ρ_1 and ρ_2, the upper layer having a thickness D_1, and the lower layer having a thickness D_2, is given by:

$$C_1 = \left[\left(\frac{\rho_2 - \rho_1}{\rho_2} \right) g \left(\frac{D_1 D_2}{D_1 + D_2} \right) \right]^{\frac{1}{2}}$$

If the thickness of the bottom layer is much greater than that of the surface layer, this reduces to

$$C_1 = \left[\left(\frac{\rho_2 - \rho_1}{\rho_2} \right) g D_1 \right]^{\frac{1}{2}}$$

which is similar to the formula for the speed of a long surface wave, but with the gravitational acceleration reduced by a factor $[(\rho_2 - \rho_1)/\rho_2]$. Typical speeds for internal waves are 1.0 m s^{-1}, much lower than for surface waves because of the weaker restoring forces. However, because of their larger amplitudes, the currents associated with the wave may be quite large. Internal waves may also propagate where there is a continuous variation of density with depth. In this case a whole series of possible modes of wave propagation exists. Usually for internal tides only the lowest few modes are important. The lowest mode, which represents periodic vertical movement of the density interface, has currents flowing in opposite directions above and below the interface for each phase of the wave cycle. Maximum currents are found at the sea surface and sea-bed. At the surface they may exceed those of the surface tidal wave. Internal waves are termed baroclinic motions whereas currents due to surface gravity long waves, which are the same at each depth below the surface, are termed barotropic motions.

The generation of internal tides is ascribed to the interaction between the barotropic currents and the bottom topography in areas of critical gradients such as near the shelf edge. Very near to the generation locations the internal tides may be coherent with the barotropic tides; however, because they propagate outwards through density fields which are highly variable in both space and time, the observed internal tides generally show complicated and irregular fluctuations (Wunsch, 1975; Hendry, 1977).

Observations of currents and temperatures in stratified seas often show variations of tidal frequencies which are not coherent with the tidal forcing. Analyses of these fail to identify a stable phase and amplitude, but spring to neap modulations are sometimes apparent. In spectral terms they have a narrow band rather than a line spectrum. It has been suggested that the rise in the level of incoherent energy near to the frequencies of the major tidal constituents in the spectrum of residuals, may be partly due to the second-order effect of incoherent internal tides on the surface levels. The ratio of surface displacement to maximum internal displacement is approximately $[(\rho_2 - \rho_1)/\rho_2]$ in the simplest case, with similar scaling for more complicated modes. Thus an internal tide of 10 m amplitude would have a typical effect on surface levels of less than 0.01 m; nevertheless, these can be detected in detailed analysis of surface levels.

CHAPTER 6

Storm Surges

Mere anarchy is loosed upon the world,
The blood-dimmed tide is loosed, and everywhere
The ceremony of innocence is drowned;

W. B. Yeats, 'The Second Coming'.*

6:1   Weather effects

The regular tidal movements of the seas are continuously modified to a greater or lesser extent by the effects of the weather. Exchange of energy between the atmosphere and the oceans occurs at all space and time scales, from the generation of short-period wind-waves to the amelioration of climatic extremes by the poleward transfer of heat and by the thermal inertia of the oceans. The intermediate weather-driven movements with periods from several minutes to several days are the concern of this chapter. More detailed reviews, which emphasize the considerable developments in these studies since the advent of modern computers are given by Heaps (1967 and 1983) and Jelesnianski (1978).

Even the most carefully prepared tidal predictions of sea-level or current variations differ from those actually observed, because of the weather effects. The relative importance of tidal and non-tidal movements depends on the time of year and on the local bathymetry. Meteorological disturbances are greatest in winter, and have greatest effect where they act on shallow seas. The total level can give rise to serious coastal flooding when severe storms acting on an area of shallow water produce high levels which coincide with high water on spring tides (Lamb, 1980; Murty, 1984; Wood, 1986).

Where the surrounding land is both low-lying and densely populated the inundations can result in human disasters of the greatest magnitude: Table 6:1 lists examples of such disasters. Patterns of geological coastline development and sediment deposition often operate to produce flat low-lying land adjacent to extensive shallow seas. The northern Bay of Bengal, where huge volumes of sediment are deposited from the River Ganges, is an outstanding example. Such low-lying lands are very fertile, which encourages intensive settlement. Once flooded by sea water, previously fertile lands are unsuitable for growing

---

* Reprinted by permission of A. P. Watt Ltd. on behalf of Michael B. Yeats and Macmillan London Ltd.

crops for several years because of the saline deposits which remain after the floods have receded. Loss of life during flooding can be minimized by moving people and animals to higher ground if enough warning is given, but damage to property and crops is inevitable unless permanent protection barriers have been built. Aspects of the design of permanent defences are discussed in Chapter 8.

Figure 6:1 Extensive sea flooding at Sea Palling on the Norfolk coast, during the North Sea surge of February 1953.

Table 6:1 Estimated results of some historical storm surge events.

Date	Region	Maximum surge level	Lives lost
November 1218	Zuider Zee	?	100 000
1864, 1876	Bangladesh	?	250 000
September 1900	Galveston, Texas	4.5 m	6 000
January–February 1953	Southern North Sea	3.0 m	2 000
March 1962	Atlantic Coast, USA	2.0 m	32
November 1970	Bangladesh	9.0 m	500 000

Figure 6:2 Hurricane Carol (1954) at Rhode Island's Edgewood Yacht Club (supplied by C. P. Jelesnianski).

The term *storm surge* is normally reserved for the excess sea levels generated by a severe storm. The non-tidal residual may be defined by rearrangement of equation (1:2):

$$S(t) = X(t) - Z_0(t) - T(t) \qquad (6:1)$$

as the difference between the observed and predicted levels. This definition is adequate for most purposes but requires a further term to allow for interaction between tide and surge where this is important (see Section 7:8). The non-tidal residual is alternatively called the *non-tidal component*, or the *meteorological residual*, or the *set-up*. Hydrodynamically the term *surge* implies a sudden movement of water which is quickly generated but which is soon over. Alternative popular descriptions of these severe flooding events include *freak tide*, *storm tide*, and *tidal wave*, none of which is valid in exact scientific usage.

No two surge events are exactly alike because small variations in weather patterns may produce quite different responses in a body of water, particularly where there is a tendency for local water-mass resonances and oscillations. Physically the atmosphere acts on the sea in two distinctly different ways.

Changes in atmospheric pressure produce changes in the forces acting vertically on the sea surface which are felt immediately at all depths. Also, forces due to wind stress are generated at and parallel to the sea surface; the extent to which they are felt at depths below the surface is determined by the length of time for which they act and by the density stratification of the water column, which controls the downward transfer of momentum. Usually in any particular storm the effects of winds and air pressures cannot be separately identified.

The effects of tropical storms and of extratropical storms have very different characteristics, as illustrated by the two examples in Figure 1:3, and so a clear distinction between them is usually made. Tropical storms are usually small and very intense. They are generated at sea, from where they move in a relatively unpredictable way until they meet the coast. Here they produce exceptionally high flood levels within a confined region of perhaps tens of kilometres. Tropical storms are known variously as hurricanes (USA), cyclones (India), typhoons (Japan), willi-willies (Australia) and baguios (Philippines). Because of their compact nature, the maximum flood levels generated by such storms are unlikely to occur in the vicinity of any of the sea-level recorders on a normally distributed network: even if they do, there is a good chance that the recorder itself will be overwhelmed. However, the effects may be more extensive when a tropical storm tracks parallel to the coast.

The 1962 USA Ash Wednesday storm which caused damage from North Carolina to New York, strictly classified as a winter frontal cyclone, and known locally as a northeaster, was an example of an extratropical storm. Extratropical storms extend over hundreds of kilometres around the central region of low atmospheric pressure and are usually relatively slow moving. They affect large areas of coast over periods which may extend to several days. Because they are slower moving and cover much larger areas than tropical storms, the extreme sea-level events are likely to be detected at several recorders. Pressure and wind effects may be equally important, whereas for tropical storms wind stress effects are usually dominant. Along the east coast of the USA extratropical storms are generally more important to the north of Cape Hatteras, whereas tropical hurricane storms are most important to the south.

Storm surges in the waters around Britain are due to extratropical weather patterns, but the responses of semi-enclosed seas to the weather forcing are far from uniform. There is, for example, a tendency for North Sea surges to persist for more than one tidal cycle whereas surges on the west coast are often of much shorter duration. The more extensive spatial scales and longer periods of extratropical storms mean that the effects of the earth's rotation, represented by Coriolis forces, are more important in determining the seas' dynamical response; so too are the natural resonant periods of the seas and basins themselves.

Several possible physical responses may be modelled by analytical solutions to the hydrodynamic equations, but a full description of the complicated local effects is best achieved by numerical modelling techniques. Coupled with numerical models of atmospheric changes these models are used as the basis of flood warning schemes which allow people in the areas at risk to take precautions.

Although it is usual to consider surges only in terms of the extreme high water levels they generate at the coast, the extreme currents which are generated off-shore are also important for the design of structures such as oil rigs and sea-bed pipelines. Extreme negative surges may also be generated by storms and these too have economic significance for the safe navigation of large tankers in shallow water. Both negative and positive surges may be generated by the same storm at different stages of its progression. For example, large positive surges in the North Sea are often preceded by negative surges a day or so before.

6:2 Statistics of meteorological residuals

If the time series $S(t)$ of the hourly residuals is computed according to equation (6:1), several useful statistics may be derived. The standard deviation of $S(t)$ from the mean value of zero, varies from values of a few centimetres at tropical oceanic islands, to tens of centimetres in areas of extensive shallow water subjected to stormy weather. Table 6:2 shows that Mahé in the Seychelles has a very low standard deviation (0.05 m) whereas Southend in the southern North Sea has a relatively high value (0.25 m). These figures may be used to derive confidence limits for tidal predictions, but are not 'errors' in the predictions in the usual scientific sense.

Table 6:2   Standard deviations of non-tidal meteorological residuals at four representative sites.

Location	Standard deviation	Description
Mahé, Seychelles	0.05 m	Tropical ocean area
Newlyn, Cornwall	0.15	Shallow-water region on wide shelf
Lerwick, Shetlands	0.13	Stormy high latitude area near shelf edge
Southend, southern North Sea	0.23	Extensive area of very shallow water

The spectrum of the residuals from a year of sea-level measurements at the Inner Dowsing light tower is shown in Figure 6:3. Although the tidal variations have been removed in the analysis there are still peaks of energy at the tidal frequencies, due to small timing errors in the gauge and weak interaction between the tides and the surges. It is interesting to note that the diurnal band has no significant residual energy. Similar spectra for the u and v components of current also show peaks of residual energy in the even tidal bands.

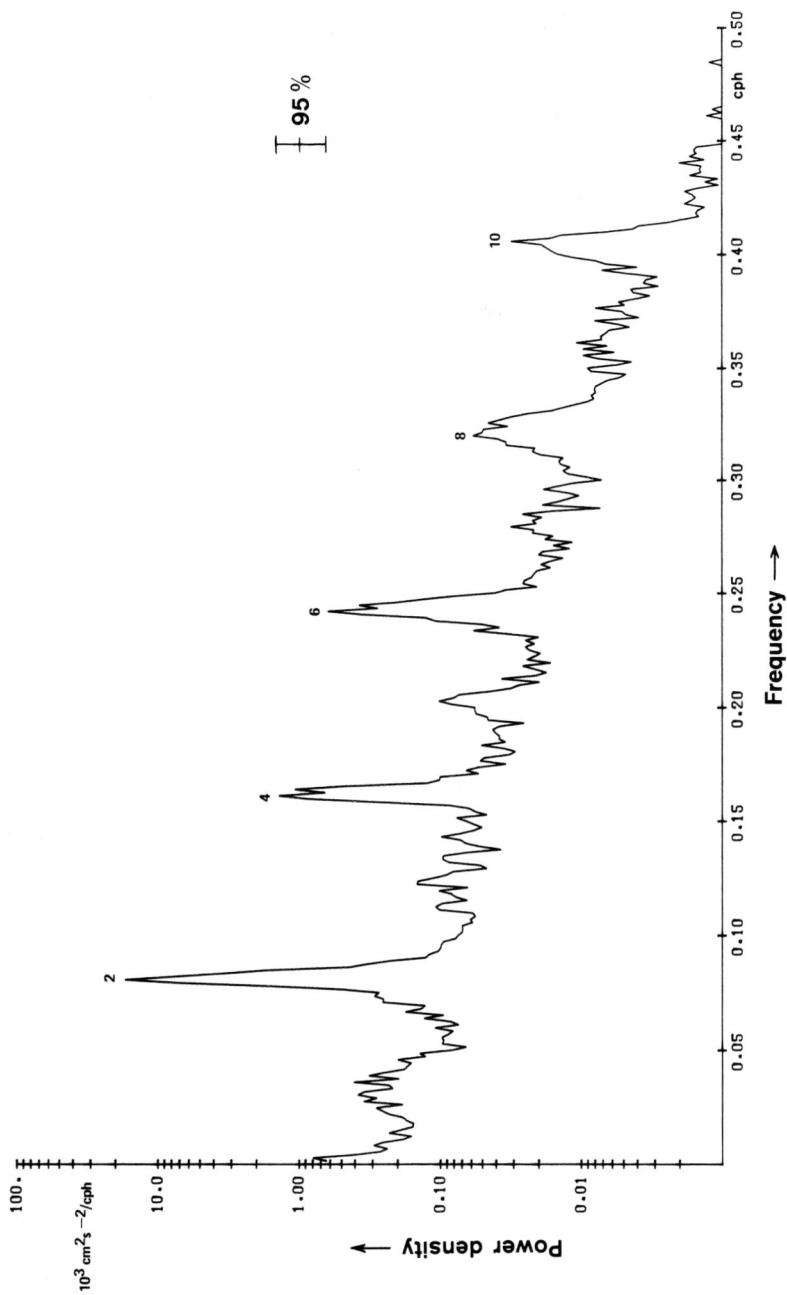

Figure 6:3 Spectrum of a year of hourly sea-level residuals at the Inner Dowsing light tower, showing residual energy in the tidal species. (From D. T. Pugh and J. M. Vassie (1976).) (*Reproduced by permission of Deutsches Hydrographisches Institut.*)

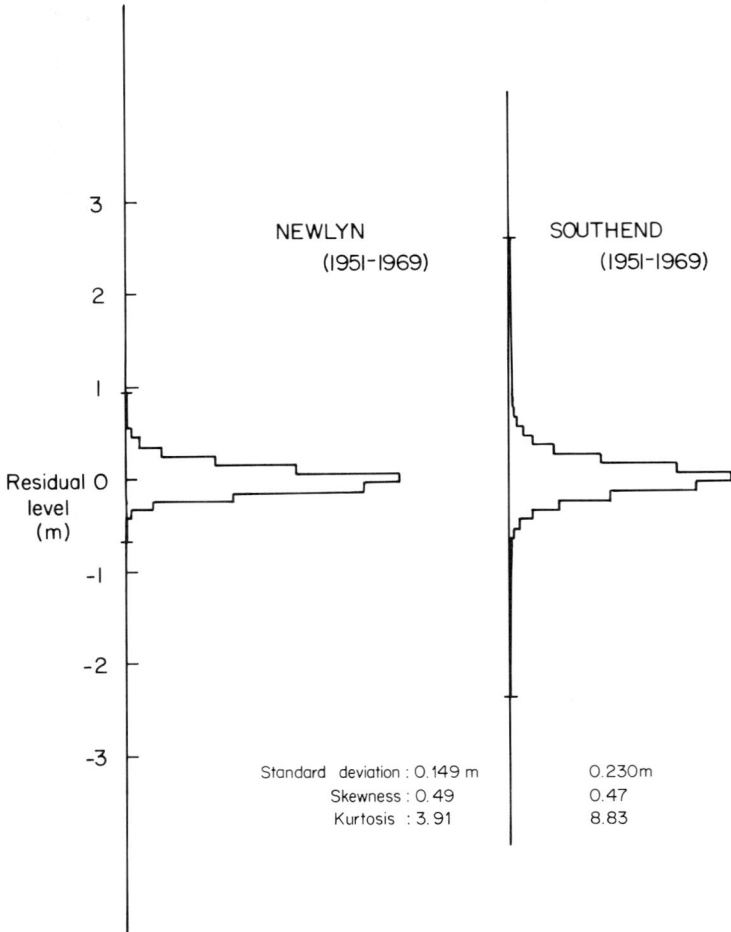

Figure 6:4   The frequency distribution of residual levels at Newlyn and Southend.

The frequency distribution of residual levels at Newlyn and at Southend are plotted in Figure 6:4, in a similar way to the frequency distribution of tidal levels shown for Newlyn and Karumba in Figure 1:4. The relative importance of tides and surges has been summarized in terms of their variance in Table 1:1. To a first approximation the distributions of the residuals at both Newlyn and Southend have the classic bell-shaped appearance of the normal or Gaussian distribution, which would be expected statistically for the sum of a large number of random variables.

The formal mathematical definition of the normal distribution is:

$$D(S) = \left[\frac{1}{\sigma_s \sqrt{2\pi}} \right] \exp\left(-S^2/2\sigma_s^2 \right) \qquad (6:2)$$

where $D(s)$ is the probability density of a residual level S, σ_s is the standard deviation of the residuals, and S is measured about a zero mean value.

There are, however, important differences between this theoretical normal distribution, and the observed frequency distribution. The residual distributions have extended tails for both positive and negative events and these tails include the major surge events. The distribution tails are more extensive at Southend than at Newlyn. There is also a tendency for large positive residuals to occur more frequently than large negative residuals. This asymmetry is summarized in Table 6:3 for Newlyn and Southend. As a general rule for most British ports, positive surges in excess of 5 standard deviations ($5\sigma_s$) occur somewhat less frequently than once a year, whereas negative surges in excess of 4 standard deviations ($-4\sigma_s$) occur on average less than once every two years. Exceptions are ports in regions of severe shallow-water hydrodynamic distortions, such as the southern North Sea.

Table 6:3 Frequency and persistence of extreme surge events at Newlyn and Southend in terms of the local standard deviations of the non-tidal residuals.

	Duration (h)	Events less than			Events greater than			
		$-6\sigma_s$	$-5\sigma_s$	$-4\sigma_s$	$4\sigma_s$	$5\sigma_s$	$6\sigma_s$	$7\sigma_s$
Newlyn (1951–69)								
$\sigma_s = 0.15\,\text{m}$	1–4			4	28	5		
	5–12				10	1		
	12+							
Southend (1951–69)								
$\sigma_s = 0.23\,\text{m}$	1–4	12	33	93	140	65	30	17
	5–12	2	6	17	35	13	4	2
	12+				2	1	1	1

For studies of surges generated by extratropical storms it is often convenient to eliminate variations at frequencies above the diurnal tidal band. Figure 1:3(a) shows a low-pass filtered surge computed by applying the X_0 filter (see Appendix 1). The sub-tidal region of the sea-level frequency spectrum between the diurnal tidal band and the monthly mean values is dominated by weather effects which are often coherent over wide areas. Figure 6:5 shows the sub-tidal variations computed using a 72-hour filter (Appendix 1) for several sites (including Newlyn) around the Celtic Sea in the South-west Approaches to the British Isles. The sea-level changes are highly coherent with each other and also with the inverse of the low-frequency air pressure variations, plotted at the bottom of the diagram. Casual inspection of Figure 6:5 suggests that the activity has a preferred period of about four days, which might be attributable either to a peak in the atmospheric forcing or to a natural response of the shelf seas. However, a detailed analysis of 61 years of Newlyn levels (Figure 6:6) shows that no such peak exists. The variations in sea-level have a uniform spectrum with a gradual increase of energy as the very low frequencies are approached.

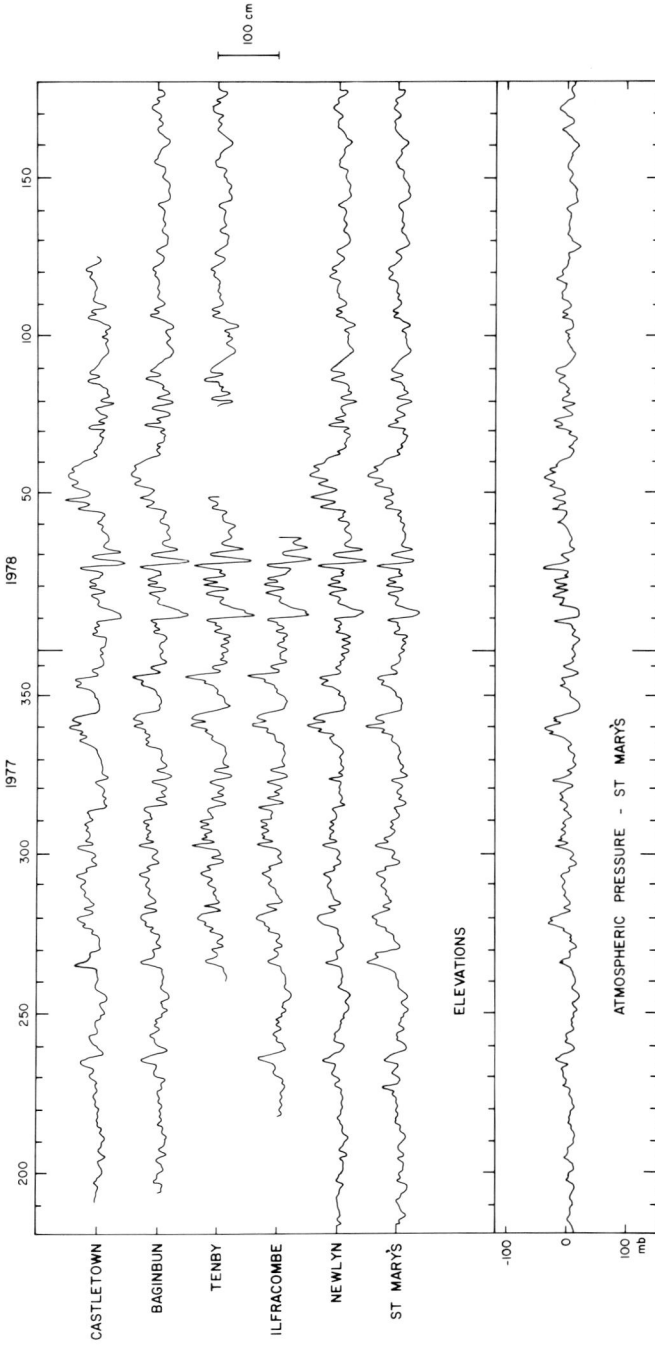

Figure 6:5 Sub-tidal variations of the sea-level around the Celtic Sea, computed using a 72-hour filter, compared with air pressures at St Mary's. The air pressures are plotted with negative values at the top. (*Reproduced by permission of Pergamon Books Ltd.*)

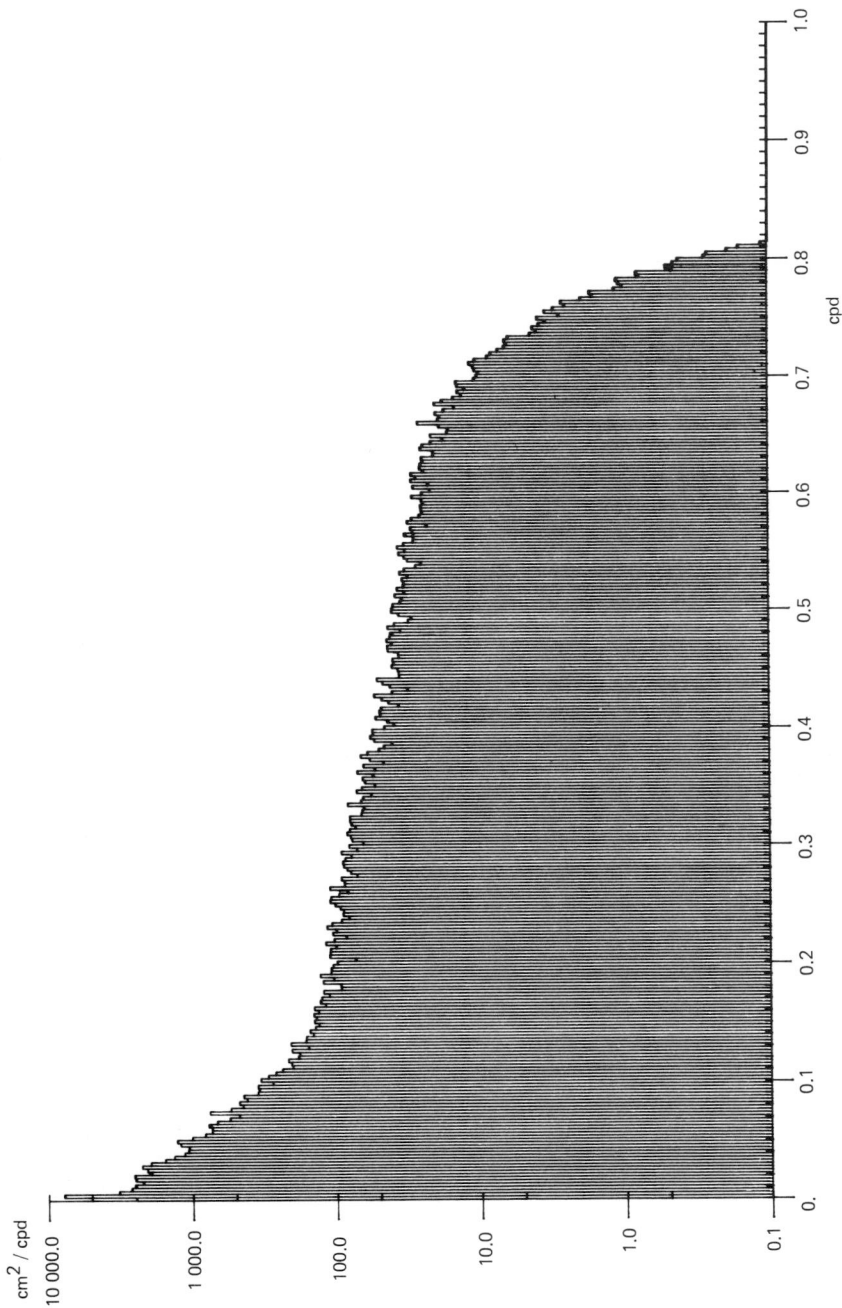

Figure 6:6 Detailed spectrum of 61 years of Newlyn sub-tidal levels, computed using the 168-hour filter. There is no evidence for resonant peaks. (*Reproduced by permission of Pergamon Books Ltd.*)

NEWLYN: MONTHLY SURGE PROBABILITY (1951–1968)

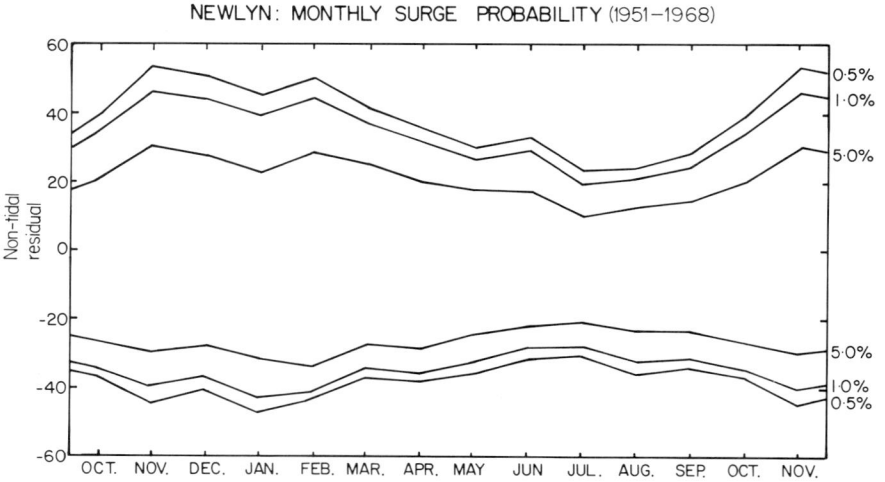

Figure 6:7 Seasonal variations of residual distributions at Newlyn (1951–1969). The percentages are of values which exceed the plotted levels.

The strong seasonal cycle in the weather is also present in the non-tidal residuals; at Newlyn (Figure 6:7) the most severe events tend to occur during the period November to February, while May to September are the quiet months. However, as the Fastnet storm of early August 1979 showed, severe storms can also occur during the late summer period.

The statistical description of sea-level variations due to tropical storms is much less satisfactory because the events are so local that they are unlikely to be properly represented in the period of measurements made at a single site. Wherever an extreme flooding event has occurred it is possible to estimate the maximum levels from the damage caused, but these observations are too scattered and unreliable to form the basis for estimating the probability of future floods. As an example of the extreme effects of these storms, hurricane Camille, in August 1969, produced a local surge of 7.5 m near Gulfport, Mississippi.

* 6:3   Responses to atmospheric pressure

The inverse relationship between sea-level and atmospheric pressure, clearly seen in Figure 6:5, can be modelled theoretically. Suppose that the sea has reached an equilibrium condition in response to an applied atmospheric pressure field so that there are no currents. Equation (3:23) becomes:

$$\left(\frac{\partial P}{\partial x}\right) = 0$$

(6:3)

with a similar expression for $(\partial P/\partial y)$ from equation (3:24). The pressure is given by the sum of the sea-level and the atmospheric pressure terms, in equation (3:21):

$$P = P_A - \rho g(z - \zeta) \qquad (6:4)$$

Differentiating with respect to x:

$$\frac{\partial P_A}{\partial x} + \rho g \frac{\partial \zeta}{\partial x} = 0$$

from which it follows that for values of x and y:

$$P_A + \rho g \zeta = \text{constant}$$

For local variations ΔP_A about the mean atmospheric pressure over the oceans, the level of the sea surface will change relative to the mean level according to:

$$\Delta \zeta = -\frac{\Delta P_A}{\rho g} \qquad (6:5)$$

If the mean atmospheric pressure over the oceans is constant then equation (6:5) also applies for local changes of sea-level with time. This is generally a good assumption but for seasonal changes of air pressure and sea-level some adjustment is necessary to allow for a shift of the global air mass towards Siberia (see Section 9:5:1).

Taking values of sea-water density $\rho = 1026\,\text{kg m}^{-3}$ and $g = 9.80\,\text{m s}^{-2}$:

$$\Delta \zeta = -0.993\,\Delta P_A$$

where ζ is in centimetres and ΔP_A is in millibars. An increase in atmospheric pressure of one millibar will produce a decrease in sea-level of one centimetre. This response of sea-level is called the *inverted barometer* effect. If sea levels have fully adjusted to atmospheric pressure changes, this compensation means that observed bottom pressures are unaffected by atmospheric pressure changes. During a typical year extratropical atmospheric pressures may vary between values of 980 mb and 1030 mb. Compared with a Standard Atmosphere of 1013 mb, this implies a range of static sea levels between $+0.33\,\text{m}$ and $-0.17\,\text{m}$.

Atmospheric pressures in tropical regions have a much smaller range, their chief characteristic being a 12-hour cycle with amplitudes around 1 millibar with maximum pressures at 1000 and 2200 hours local time; the inverted barometer response to these pressure cycles is a local radiational tide at the same frequency as the gravitational solar semidiurnal tide, S_2, with a minimum at 1000 and 2200 hours local time.

The exact inverted barometer response is seldom found in practice. One reason for this is the dynamic response of the shallower waters of the continental shelf to the movement of the atmospheric pressure field. A simple but elegant theoretical model of the response may be derived from the hydrodynamic equations for a shelf of constant depth (Proudman, 1953). For a disturbance travelling in the X-direction, omitting the advective, Coriolis, and surface and bottom stress terms, equation (3:23) becomes:

$$\frac{\partial u}{\partial t} = -\frac{1}{\rho}\left(\frac{\partial P}{\partial x}\right)$$

Substituting for P from equation (3:21) and differentiating with respect to x:

$$\frac{\partial u}{\partial t} = -\frac{1}{\rho}\frac{\partial P_A}{\partial x} - g\frac{\partial \zeta}{\partial x} \tag{6:6}$$

A general form for an atmospheric pressure disturbance propagating in a positive X-direction may be written:

$$\Delta P_A = A(x - C_A t) \tag{6:7}$$

where A is a general, but physically possible function, and C_A is the speed of propagation of the disturbance. We seek a solution for the resulting sea-level disturbance of the form:

$$\Delta \zeta = \alpha A(x - C_A t) \tag{6:8}$$

which satisfies both equation (6:6) and the Continuity equation (3:22):

$$\frac{\partial \zeta}{\partial t} + D\frac{\partial u}{\partial x} = 0 \tag{6:9}$$

where D is the constant water depth. Substituting from (6:8) in (6:9) and rearranging:

$$\frac{\partial u}{\partial x} = \frac{\alpha C_A}{D} A'(x - C_A t)$$

where $A'(x - C_A t)$ is the derivative of the function A with respect to the argument $(x - C_A t)$. This may be integrated to give:

$$u = \frac{\alpha C_A}{D} A(x - C_A t) \tag{6:10}$$

the constant term being assumed to be zero to satisfy the expected condition of no flow well away from the disturbance, where ΔP_A is zero.

Substituting into equation (6:6) from (6:7), (6:8) and (6:10) we obtain:

$$-\alpha\frac{C_A{}^2}{D} A'(x - C_A t) = -\frac{1}{\rho}A'(x - C_A t) - g\alpha A'(x - c_A t)$$

and hence:

$$\alpha = -\left[\rho g\left(1 - \frac{C_A^2}{gD}\right)\right]^{-1}$$

so that the level is:

$$\Delta \zeta = -(\Delta P_A)\Big/\left[\rho g\left(1 - \frac{C_A^2}{gD}\right)\right] \tag{6:11}$$

The negative sign expresses the inverted barometer response, which is now enhanced by the factor $[1 - (C_A^2/gD)]^{-1}$. This response may be written:

$$\text{Dynamic sea-level response} = \frac{\text{static sea-level response}}{(1 - C_A^2/gD)}$$

If the atmospheric pressure disturbance is stationary, $C_A = 0$, which gives the static response. The term C_A^2/gD is the square of the ratio between the speed of the disturbance and the speed of a free progressive long wave. As C_A increases from zero the response is amplified. If the two are equal there is a theoretically infinite response. However, this resonance would be damped by friction in a real situation. If the disturbance speed becomes very large compared with the speed of a free wave then the response becomes small again. The amplification factor for a disturbance moving at 20 km per hour over water of 50 m depth is 1.07, but if the depth falls to 25 m the factor increases to 1.14. The resonant condition for a disturbance travelling at a speed of 20 km per hour requires a very shallow depth of 3.0 m.

It is rarely possible to identify separately the effects of the wind stresses, as they are associated with the effects of atmospheric pressure disturbances. One of the earliest sets of observations, by Sir James Clark Ross, spending the 1848–9 winter in the Canadian Arctic while searching for the lost Franklin Expedition, showed a response very close to the theoretical static response, for an ice-covered sea (Ross, 1854). At Newlyn, for the data plotted in Figure 6:5, the sea-level response was -0.84 cm for each millibar of atmospheric pressure increase, but analysis of the response at different frequencies showed values of -1.10 cm mb^{-1} at very low frequencies, and a minimum response of -0.70 cm mb^{-1} at 0.2 cycles per day. These variations are due to air pressure and the wind effects being related to each other during the passage of air-pressure fields over the Celtic Sea.

* 6:4 Responses to wind stress

6:4:1 Stress laws

When two layers of moving fluid are in contact, energy and momentum are transferred from the more rapidly moving layer to the slower layer. The physics of this transfer process is very complicated; however, basic functional relationships can be combined with empirical constants to give useful formulae. This section outlines these developments, but the 'proofs' should not be considered rigorous. More detailed accounts are given by Proudman (1953), Csanady (1982), Gill (1982) and Bowden (1983).

The drag or stress τ_s on the sea surface due to the wind, measured as the horizontal force per unit area, might reasonably be expected to depend on both the wind speed W and on the air density ρ_a. A useful relationship can be derived by assuming that the relationship has the form:

$$\tau_s = C_D \, \rho_A^{\,i} \, W^j$$

where C_D is a dimensionless drag coefficient and i and j are the power laws of the relationship. Using the method of dimensional analysis to equate the powers of mass, length and time:

$$[ML^{-1}T^{-2}] = [ML^{-3}]^i \, [LT^{-1}]^j$$

gives values of $i = 1$ and $j = 2$. Hence

$$\tau_s = C_D \rho_A W^2 \text{ per unit area} \qquad (6:12)$$

The value of C_D depends on the level above the sea surface at which the wind speed is measured. Conventionally a 10 m level is chosen. Experiments have shown that the drag increases slightly more rapidly than the square of the wind speed. This may be accounted for by an increase of C_D, and justified in terms of an increased surface roughness with increasing wave heights. A further refinement would allow the increase in C_D to lag behind the increase in wind speed to allow for the time taken for the rougher seas to develop. Acceptable values for C_D (Smith and Banke, 1975) are given by:

$$10^3 \, C_D = 0.63 + 0.066 \, W_{10} \;(2.5 \, \mathrm{m\,s^{-1}} < W < 21 \, \mathrm{m\,s^{-1}})$$

The wind stress on the sea surface is in the direction of the wind. It may be resolved into two orthogonal components:

$$F_s = C_D \rho_A |W| W \cos \theta$$
$$\qquad (6:13)$$
$$G_s = C_D \rho_A |W| W \sin \theta$$

in the usual Cartesian system, with θ the direction to which the wind blows, measured anticlockwise from east. The modulus of the wind speed is necessary to give both negative and positive components of the stress. Note that, because of the square-law relationship, it is not correct to resolve the wind into two components and then to compute the stresses as the square of these wind components.

6:4:2 Wind set-up

The steady-state effect of the wind stress on the slope of the sea surface for a wind blowing along a narrow channel of constant depth is found from equation (3:23) by balancing the pressure and stress forces:

$$\frac{\partial P}{\partial x} = \frac{\partial F}{\partial z} \qquad (6:14)$$

Assuming that the stress is independent of depth this depth-averaged stress is:

$$\frac{1}{D}\int_{-D}^{0} \frac{\partial F}{\partial z}\,dz = \frac{1}{D}(F_s - F_B) \tag{6:15}$$

where F_S and F_B are the surface and bottom stresses.

Substituting into equation (6:14) from (3:21) and (6.15) gives:

$$\frac{\partial \zeta}{\partial x} = \frac{F_s - F_B}{g\rho D} \tag{6:16}$$

For a surface wind stress, derived from equation (6:13):

$$\frac{\partial \zeta}{\partial x} = \frac{C_D\,\rho_A\,W^2}{g\rho D}$$

This formula makes the important point that the effect of winds on sea levels increases inversely with the water depth and will be most important when the wind blows over extensive regions of shallow water. For a Strong Gale (Beaufort Force Nine, $22\,\mathrm{m\,s^{-1}}$) blowing over 200 km of water which has a depth of 30 m (approximately the dimensions of the southern North Sea) the increase in level would be 0.85 m. If the wind speed increased to Storm (Beaufort Force Eleven, $30\,\mathrm{m\,s^{-1}}$) the level of increase would be 1.60 m.

6:4:3   Current profiles

In practice a wind stress on the sea surface produces a variation of current speed with depth. Laboratory experiments, numerical models and observations at sea have shown that the water at the surface is driven at approximately 3 per cent of the wind speed at $z = 10\,\mathrm{m}$. These experiments give a wide range of results; see for example Bye (1965), Wu (1975) and Davies and Flather (1987). Below the surface the speeds decrease with depth as the stress is transmitted from layer to layer. By following dimensional arguments it is possible to derive physically reasonable formulae to describe the behaviour: for these arguments, the shear, or rate of change of speed with depth, is considered to depend on the surface wind stress, the water density (assumed constant), and the depth itself:

$$\frac{\partial u}{\partial z} = \left(\frac{1}{\kappa}\right)\tau_s^{\,i}\,\rho^{\,j}\,z^{\,l}$$

where κ is a dimensionless constant, called the von Karman constant, and i, j and l are the power laws of the relationship. Hence:

$$[LT^{-1}/L] = [ML^{-1}T^{-2}]^i\,[ML^{-3}]^j\,[L]^l$$

so that $i = \tfrac{1}{2}$, $j = -\tfrac{1}{2}$ and $l = -1$.

The shear becomes:

$$\frac{\partial u}{\partial z} = \frac{1}{\kappa z}\left(\frac{\tau_s}{\rho}\right)^{\frac{1}{2}} \tag{6:17}$$

which may be written in terms of a *friction velocity u**:

$$\frac{\partial u}{\partial z} = \frac{u^*}{\kappa z}$$

(6:18)

Substituting for τ_s from equation (6:12) in (6:17) the frictional velocity is seen to increase roughly in proportion to the wind speed. By experiment, a value of $0.0012\ W_{10}$ (where W_{10} is the wind speed at 10 m above the sea surface) has been found appropriate for a wide range of conditions. Integrating equation (6:18) to a depth z from the surface is not possible because equation (6:17) is not valid at $z = 0$ and so a small *roughness length*, z_0, is defined, having values between -0.5 mm and -1.5 mm. With this condition equation (6:18) integrates to give:

$$u_z = u_0 - \frac{u^*}{\kappa} \ln\left(\frac{z}{z_0}\right)$$

(6:19)

showing a logarithmic decrease of velocity with depth, since z is negative (this is why we define negative roughness lengths). Profiles for two different wind speeds are shown in Figure 6:8, using a roughness length of -1.5 mm, a value of 0.4 for the von Karman constant, and computing u_0 as 3 per cent of W_{10} the ten-metre wind speed. These currents, calculated using physically reasonable recipes, are useful for practical applications, but they are very difficult to verify by direct measurement, especially close to the surface in the presence of waves. Note that this pragmatic treatment has assumed a constant density, which may be inappropriate if there is a strong surface heating, and has also assumed that there is no earth rotation.

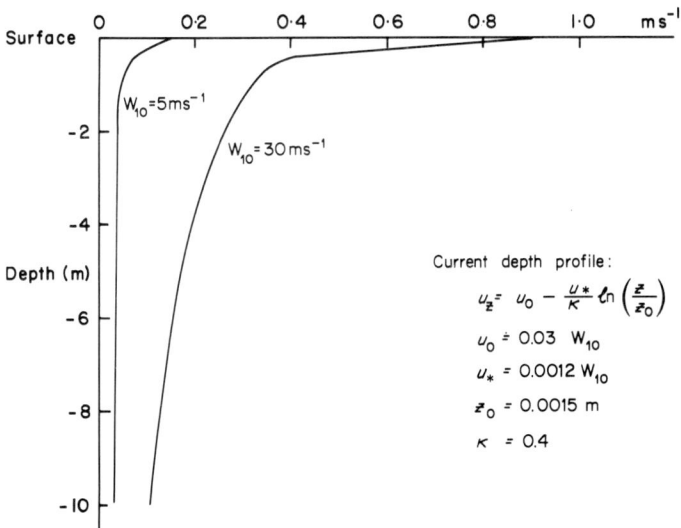

Figure 6:8   Current depth profiles for the logarithmic law.

6:4:4   Ekman transport

The Norwegian scientist and explorer, Fridtjof Nansen, observed during the 1893 to 1896 polar-ice drift of the *Fram* that the ice moved in a direction to the right of the wind and not in the direction of the wind itself. Ekman explained this in 1902 in terms of the hydrodynamic equations. Suppose that the wind is blowing over water well away from the coast, which is deep enough for the influence of bottom friction to be negligible. Also suppose that the water density is constant with depth and that the wind/water system has reached its steady-state condition. Finally, suppose that there are no horizontal pressure gradients due, for example, to sea surface gradients. In this limiting case, equations (3:23) and (3:24), applied for a wind and wind stress acting along the positive X-direction, reduce to:

$$-fv = -\frac{1}{\rho}\frac{\partial F}{\partial z} \qquad fu = -\frac{1}{\rho}\frac{\partial G}{\partial z}$$

If we integrate these equations through from the bottom of the wind-driven layer to the surface, recognizing that the stress at the bottom of this layer will be zero:

$$-f\overline{V} = F_s/\rho \qquad f\overline{U} = 0$$

where F_s is the surface wind stress and $\overline{U}$ and $\overline{V}$ are the volume transports per metre along the X and Y axes, with units of $m^2\,s^{-1}$. These equations have the solution:

$$\overline{U} = 0 \qquad \overline{V} = -F_s/f\rho \qquad (6:20)$$

The net transport is along the negative Y-direction, i.e. to the right of the wind stress in the northern hemisphere (where f is positive), and to the left in the southern hemisphere. This is called the *Ekman volume transport* per metre of section, and is a function of the wind stress and the latitude. For a relatively strong wind speed of $15\,m\,s^{-1}$, at $45°$ latitude the wind-driven layer transport can be estimated from equation (6:12) and the appropriate value of C_D, as $4.4\,m^2\,s^{-1}$. For the more normal speed of $5\,m\,s^{-1}$, the transport is $0.29\,m^2\,s^{-1}$. Nearer the equator the corresponding transports are greater as f decreases, but the theory breaks down at the equator itself where infinite volume transports are predicted.

The depth-averaged transport calculated from equation (6:20) tells us nothing about the details of the variation of current directions with depth. More detailed mathematical solutions indicate a surface flow at $45°$ to the direction of the wind stress; this surface flow in turn drives the lower layers in a direction which is further rotated, and at slower speeds. This progressive rotation and reduction of speed with depth is called the *Ekman spiral*, the net effect of which is the transport of water at right angles to the wind stress.

Although Ekman's theory has allowed powerful insight into the relationships between global wind fields and ocean circulation, the simplifying assumptions are so limiting that the full ideal Ekman spiral is unlikely to be observed in practice. The time needed to establish a dynamic equilibrium may be several days, whereas the wind is likely to change over several hours. Vertical and horizontal density gradients will give further distortions. For continental shelf seas, the important assumptions about no coastal interference or bottom friction influence are seldom valid. Some effects of the coastal influences are described in the following section. The effects of bottom friction are considered in Chapter 7, but it is relevant to consider here the depths to which a wind-driven Ekman layer might penetrate. This is defined as the depth D_E at which the progressive rotation results in flow in the opposite direction to the surface wind stress:

$$D_E = \pi \left(\frac{2A_z}{\rho f} \right)^{\frac{1}{2}}$$

(6:21)

f is the Coriolis parameter, ρ is the water density, and A_z is a quantity called the vertical eddy coefficient of viscosity, which represents the efficiency with which the water exchanges momentum vertically from layer to layer, the 'stickiness'. It has units of $\text{kg m}^{-1}\text{s}^{-1}$. (A_z/ρ) is called the kinematic eddy viscosity.

The shear stress within the water column is related to the velocity gradient by the eddy coefficient of viscosity defined in the following way:

$$F_{xz} = A_z \frac{\partial u}{\partial z}$$

(6:22)

A_z varies rapidly with depth near the surface and bottom boundary layers and near vertical density gradients. It should be realized that the previous descriptions of logarithmic boundary layers in Section 6:4:3 imply non-constant values of A_z: an understanding and coherent description of the variation of the parameter A_z with depth requires a full understanding of the basic physics of air–sea interaction.

However, for a theoretical discussion of Ekman dynamics, equation (6:21) is applied for a value of A_z which is constant at all depths. For average winds and tidal conditions (tidal currents enhance the turbulence which transfers the momentum) a typical value of A_z might be $40 \text{ kg m}^{-1}\text{s}^{-1}$, whereas for stronger winds of 20 m s^{-1}, or more the value might be as high as $400 \text{ kg m}^{-1}\text{s}^{-1}$. At a latitude of $45°$ the corresponding values of D_E from equation (6:21) are 87 m and 270 m; thus, for strong winds the depth of the Ekman layer is comparable to the depth of the continental shelves.

If the water is too shallow for the full Ekman spiral to develop, the net transport due to the wind will be at an angle of less than $90°$ to the wind direction. Another significant factor in shallow water is the turbulence due to bottom friction, which will increase A_z, and hence reduce the angle between the wind and the net transport.

6:4:5 Alongshore winds

When a component of the wind stress acts parallel to a coastline the onset of

Ekman transport is followed by the development of water-level differences which have their own influence on the water movements, so distorting the simple Ekman transport dynamics.

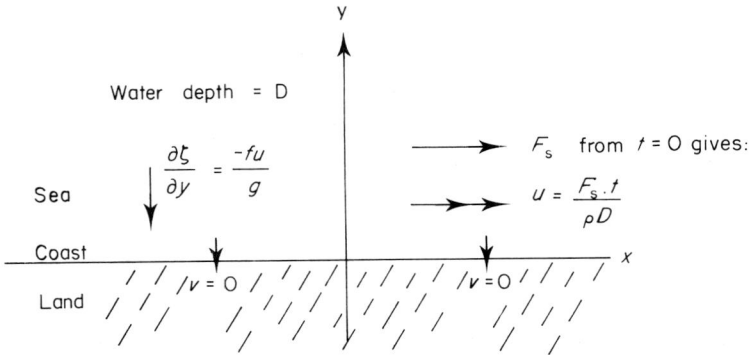

Figure 6:9   Coordinates used in developing formulae for the response of a coastal sea to a longshore wind stress.

Suppose that a semi-infinite sea of depth D exists for all values of y greater than zero (Figure 6:9), and that a uniform wind stress F is applied to the surface of this sea from time $t = 0$. In the absence of any changes with x, and in the vicinity of the coast where the condition of no cross-boundary flow ($v = 0$) applies, equations (3:23) and (3:24) become:

$$\frac{\partial u}{\partial t} = \left(\frac{1}{\rho}\right)\left(\frac{\partial F}{\partial z}\right); fu = -\left(\frac{1}{\rho}\right)\left(\frac{\partial P}{\partial y}\right)$$

Integrating these from $-D$ to the surface, assuming u is constant with depth, and substituting from equation (3:21) for $\partial P/\partial y$:

$$D\left(\frac{\partial u}{\partial t}\right) = \frac{F_s}{\rho}; fu = -g\left(\frac{\partial \zeta}{\partial y}\right)$$

which leads directly to:

$$u = \frac{F_s t}{\rho D} \tag{6:23}$$

$$\frac{\partial \zeta}{\partial y} = -\frac{fu}{g} \tag{6:24}$$

The first of these relationships shows the current parallel to the shore increasing steadily with time at a rate inversely proportional to the depth of water to be moved, as would also be the case in the absence of the earth's rotation; in practice this current will eventually be limited by bottom friction. The second equation shows a sea-level gradient normal to the coast which is in geostrophic equilibrium with the current u at all times; corresponding to this equilibrium is a linear increase of coastal sea levels with time. Both current flows and sea-level changes are greatest when the water is shallow.

The net result of the applied wind stress is a flow of water in the direction of the wind, but we can now appreciate that this apparently simple intuitive direct relationship is really more complicated: it results from the condition of no Ekman transport across the coastline. The conservation of mass requires a steady increase of coastal levels in response to Ekman transport and this increase develops in geostrophic equilibrium with the long-shore current. This sea-level change is called the *locally* generated surge, to distinguish it from surges propagating freely as progressive waves which have travelled from external areas of generation.

6:4:6 Inertial currents

The full solution for water movements in response to an imposed wind stress shows that the current u and the sea surface gradient decrease exponentially away from the coast:

$$u = \frac{F_s t}{\rho D} \exp(-y/R) \tag{6:25}$$

where R is the Rossby radius, defined as:

$$R = (gD)^{\frac{1}{2}}/f \tag{6:26}$$

At 45° latitude in water of 50 m depth this scale length is 215 km; for other examples see Section 5:2:3 and Table 5:4.

Well away from the coast the full solution contains an important oscillatory part which has a period not far removed from that of the tides. To illustrate this type of motion suppose that previous driving forces have now stopped and that the water remains in motion after this. Because of the absence of a coast there are no pressure gradients. The equations (3:23), (3:24) and (3:22) become:

$$\frac{\partial u}{\partial t} - fv = 0$$

$$\frac{\partial v}{\partial t} + fu = 0$$

$$D\left(\frac{\partial u}{\partial x} + \frac{\partial v}{\partial y}\right) = 0$$

and these have the solution:

$$u = q \cos ft$$

$$v = -q \sin ft \tag{6:27}$$

where the coordinate axes are chosen to give the maximum current q along the X axis at $t = 0$. The water masses perform circular motions with a period $2\pi/f$.

This is 17 hours at 45° latitude, and 14.6 hours at 55° latitude. Near 30° and 70° latitude, the inertial motions fall within the diurnal and semidiurnal tidal species. The radius of the circular motions, which depends on the water speed is q/f. For a speed of $0.2 \, \mathrm{m \, s^{-1}}$, typical of inertial motions on the northwest European shelf, the radius at 55° N is 1.7 km. Inertial currents are most effectively produced by an impulse of winds blowing for not more than half of an inertial period, acting on a relatively thin surface layer above a sharp thermocline, so that the masses of water to be accelerated and the dissipating forces are small. In these circumstances inertial oscillations may persist for several days after their generation and be advected to other regions. Gradually, as they decay, the speeds and radius of the motions in the inertial currents diminish. The method of rotary analysis of currents described in Section 4:4:3 is particularly effective for identifying inertial currents (see also Appendix 3).

6:5   Numerical modelling of surges

Although the simple analytical solutions, described previously for the response of a sea to winds and atmospheric pressure changes, can help towards a physical understanding of the processes involved, they cannot approach a full description of how a real sea responds to a real imposed weather field. Real seas have irregular boundaries and variable depths. To reproduce the dynamics of real seas we need to include non-linear surface and bottom stresses, and to make allowances for changing depths and boundary positions as the sea levels change.

The importance of being able to model the complicated responses of coastal shelf seas to observed and forecast weather patterns, in order to give advanced warning of coastal flooding, has encouraged the development of several different types of numerical simulations on computers (Jelesnianski, 1978; Flather, 1979; 1981). Before numerical models became sufficiently reliable, warnings of impending flooding were given on the basis of empirical formulae which related the meteorology to observed sea-level changes; some aspects of these models are discussed in Section 6:10 (see also, WMO, 1978).

Numerical models were first developed for describing the flow in rivers and the concept can usefully be explained in these terms. The river can be divided into a series of sections, each with a mean depth, breadth and bottom roughness. The change in level of the uppermost section can be computed from the equation of continuity and the flow difference between its upstream and downstream boundaries. The simultaneous changes of level for all other of the river sections can be similarly calculated if the flows are known. After a small time interval of level adjustment, the one-dimensional momentum equation is used to calculate the flow between each section, on the basis of the previously calculated level differences and appropriate bottom friction. After a second small time interval the newly computed flows are used to calculate again the changes in level for each river section. Provided the lengths of the river sections and the time steps are small enough, this repeated cycle of calculating levels and their flows can give a river behaviour which closely approximates the real situation. If the input flow from upstream is now rapidly increased because of heavy rainfall, the model is capable of predicting the adjustment of the river to the new input. A one-dimensional river model can also be driven by periodic tidal changes of level at its mouth.

Such models were first applied to tidal rivers, including the River Thames, to calculate likely flood conditions for different tidal amplitudes and fresh-water discharge rates. Similarly, tidal models of long narrow seas, such as the Red Sea, were successfully developed, but there are few seas where a one-dimensional model is appropriate. The extension of one-dimensional models to the representation of two-dimensional sea areas became possible with the advent of digital computers.

The equations to be solved are those developed in a finite-difference form in Chapter 3, which in the limiting case of very small elements become the differential equations of momentum and continuity (3:23), (3:24), (3:22). To model the characteristics of a sea, the dimensions of each rectangular element must have scales of several kilometres: the smaller the individual elements, the closer is the numerical approximation to the real situation. The spatial resolution is limited by the size, speed and running cost of the computer. Halving the dimensions of each element in a two-dimensional model increases the computing roughly by a factor of eight; the number of elements is increased fourfold, and the number of iterations is doubled because the time step must be halved to satisfy a stability condition:

$$\Delta x > \Delta t \, (gD)^{\frac{1}{2}} \qquad\qquad (6:28)$$

This stability condition ensures, for example, that a free progressive wave travelling with speed $(gD)^{\frac{1}{2}}$ will not have moved more than a grid box dimension Δx during an iteration time step Δt.

Tropical and extratropical storm surges are usually modelled in slightly different ways. Figure 6.10 shows a finite-difference mesh which has been used to model the behaviour of the northwest European continental shelf in response to extratropical weather patterns (Flather, 1979). The resolution is $\frac{1}{2}°$ in longitude and $\frac{1}{3}°$ in latitude, giving approximately 30 km between calculation points. The sea model is driven by winds and air pressures derived from an atmospheric numerical model which is being run simultaneously by the Meteorological Office for weather forecasting. The crosses shown in Figure 6:10 mark the points where the meteorological model, which is three-dimensional with ten vertical layers (since increased to fifteen), computes the forecast values of pressure and winds. The spacing is larger than for the sea grid because of the larger spatial scales of normal atmospheric phenomena.

The conditions at the boundaries of the model must be specified with care. At the coastline the flow must be parallel to the boundary. For the open sea boundaries at the edge of the shelf and at the entrance to the Baltic Sea, the relationships between currents and levels are prescribed to allow energy to enter or leave the area. The main loss of energy from the system takes place at the sea-bed through bottom friction. The energy losses due to bottom friction increase as the cubic power of the current speed, as discussed in Section 7:9. Because the total current consists of both surge and tidal components, running the model for either surge or tide alone does not reproduce the real situation. For forecasting surges the model is run by driving it at the ocean boundaries with tidal changes of level; the air pressures and winds are then imposed to give the total effects. A time step of 3 minutes between iterations satisfies condition (6:28) for even the deepest grid boxes. The weather forecast is available at hourly intervals, so interpolation is necessary at intermediate times. Surge forecasts are prepared for the next 30-hour period, with continual updating as further weather forecasts become available.

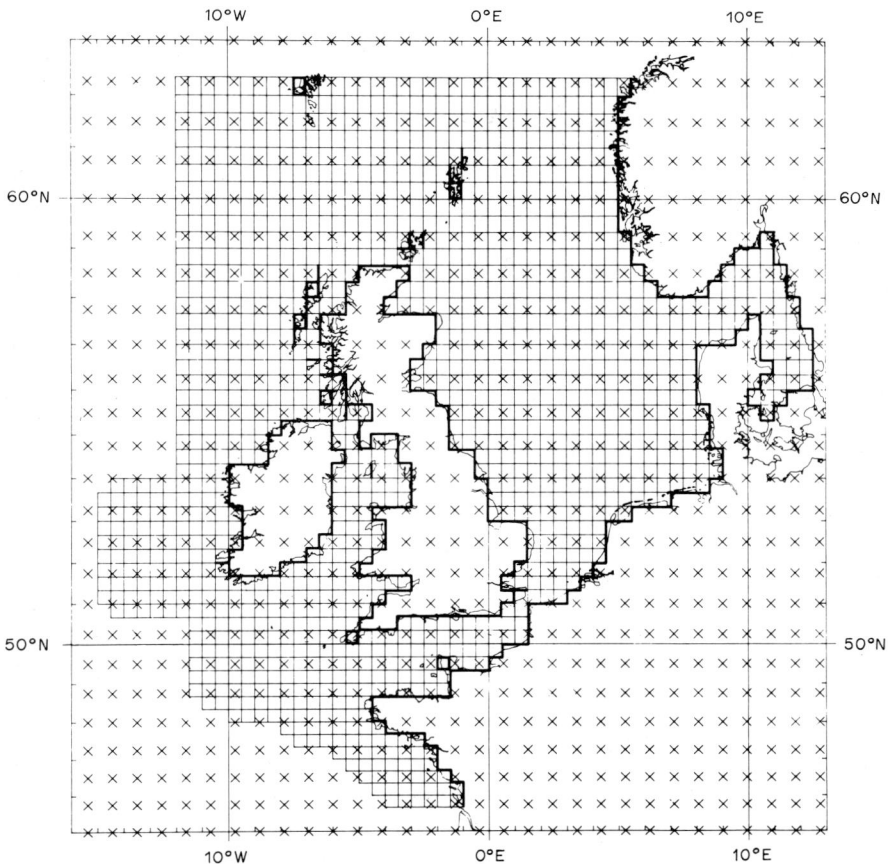

Figure 6:10   Finite-difference mesh of the north-west European shelf model with grid points (x) of the 10-level model of the atmosphere. (*Reproduced by permission of Elsevier.*)

208

The significance of these surge forecasts is illustrated in Figure 6:11 which compares the observed and 12-hour forecast levels for a major event in the North Sea in January 1978. Along the northeast coast of England this surge, which coincided with a high spring tide, gave total levels comparable with the 1953 event (Table 6:1). The circled observations indicate the observed surge residuals at the time of predicted tidal high water. At Southend these residuals clearly show a lower surge level at times of tidal high water, because of the interaction between tide and surge, but this interaction is much weaker at Walton, 50 km to the north (see Section 7:8).

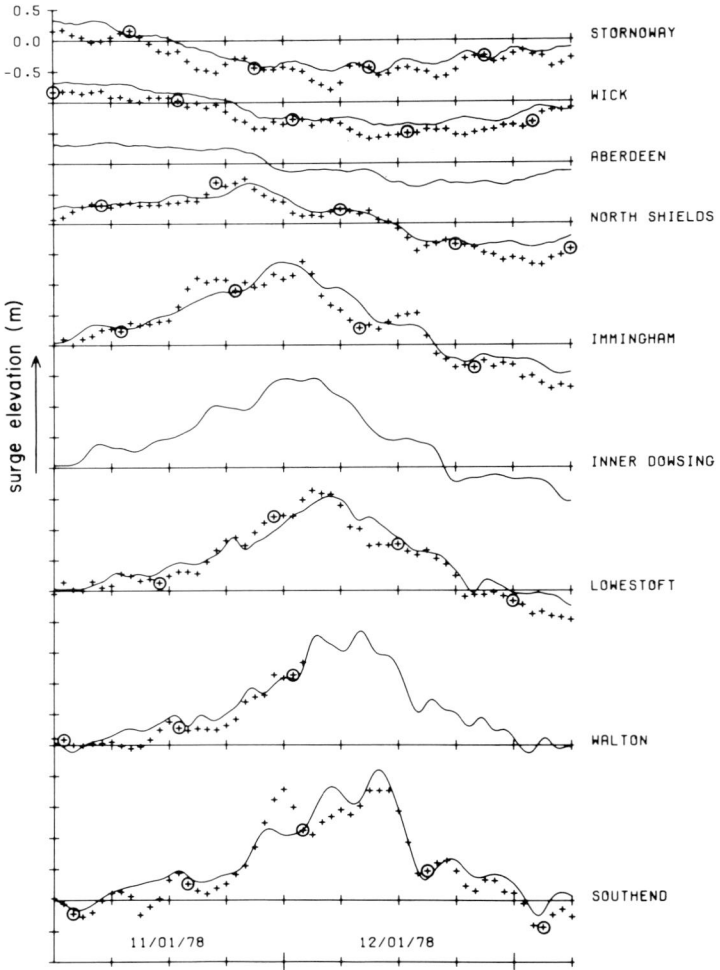

Figure 6:11 Computed and observed surge levels in the North Sea for the storm of 11–12 January 1978; ($\oplus$ are times of local high water). (*Reproduced by permission of Elsevier.*)

The propagation of the surge from north to south, which is typical of North Sea surges, allows reliable warnings of flooding to be given several hours ahead for southern coasts. In other seas where the surge development is more rapid, surge forecasts, although still valid, are less reliable. The accuracy of the weather forecasts is a critical requirement: the success of the North Sea surge models is also due to the importance of large-scale meteorological disturbances, which are well represented in the atmospheric forecasting model. The increase of wind stress as the square of the wind speed makes models very sensitive to errors in the wind forecasts. Further difficulties occur when there is intense local activity on a scale too small to be resolved by the grid size of the meteorological model. In shallow coastal bays and estuaries the accuracy of the forecasts may be improved by having a model with a finer mesh size nested inside the larger regional model.

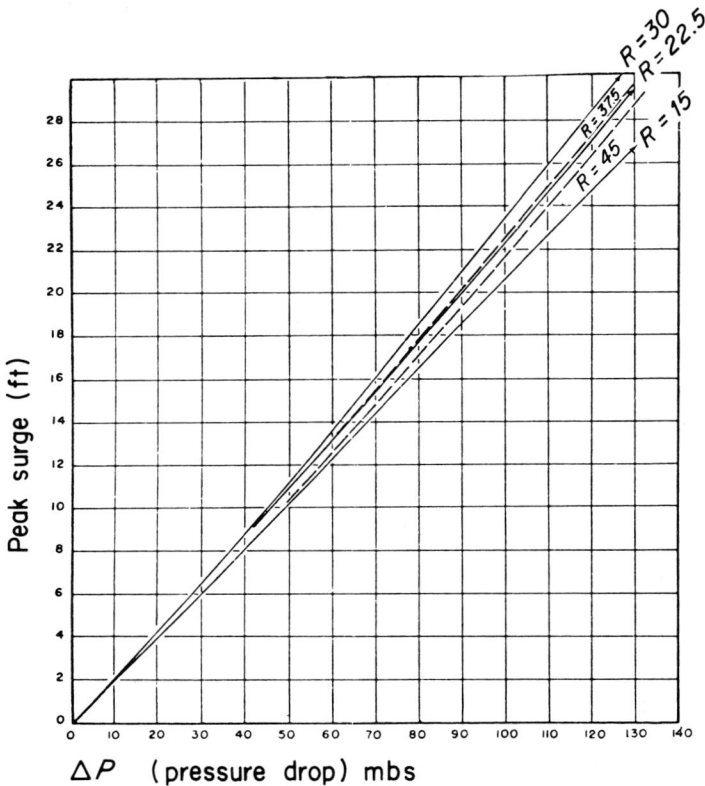

Figure 6:12 Nomogram for first estimates of peak tropical storm surge level on an open coast as a function of pressure drop to hurricane centre, and radius of maximum winds (R in statute miles). The curves are computed for standard conditions, as outlined in the text. (*Reproduced by permission of the World Meterorological Organization.*)

210

The reduced scale of tropical storms requires a smaller grid for their effects to be resolved by a shelf model. For operational forecasting of the effects of hurricanes approaching the low-lying eastern coast of the United States, the National Weather Service developed a depth-integrated finite-difference model called SPLASH (Special Program to List Amplitudes of Surges from Hurricanes), (Jelesnianski, 1972; 1978; WMO, 1978). This model covers 600 miles (960 km) of coastline from the Mexican–USA border to New England. Coastal irregularities are partly accommodated by shearing or off-setting the grid normal to the shore. The model has been used to determine the sensitivity to a standard distribution of pressures and winds on the basis of a specified pressure drop to the centre of the storm, and the estimated distance from this centre to the radius of maximum winds. Typical values are from 50–100 mb and from 20–80 km. Also specified are the tracking direction and speed of the storm centre, and the estimated landfall point. The responses of the sea to a complete range of pressure drops and radii have been numerically computed for a straight coastline in which the depth profile seawards is one-dimensional. The depth at the coast is 4.5 m and the slope is 0.5 m km^{-1}, which is a representative mean for all coastlines in the region. A standard storm motion of 7 m s^{-1} along a track normal to the coast is also assumed. Figure 6:12 shows the nomogram of peak surges for different pressure drops and for different radii. Use of this nomogram gives only a crude first estimate of the maximum surge level, and further adjustments are necessary for track speed, direction and landfall. It shows, however, that the major factor is the pressure difference between the outside ambient air pressure and the low pressure at the hurricane centre. This is because the rate of the pressure drop indicates the wind speed, which is the physical factor directly responsible.

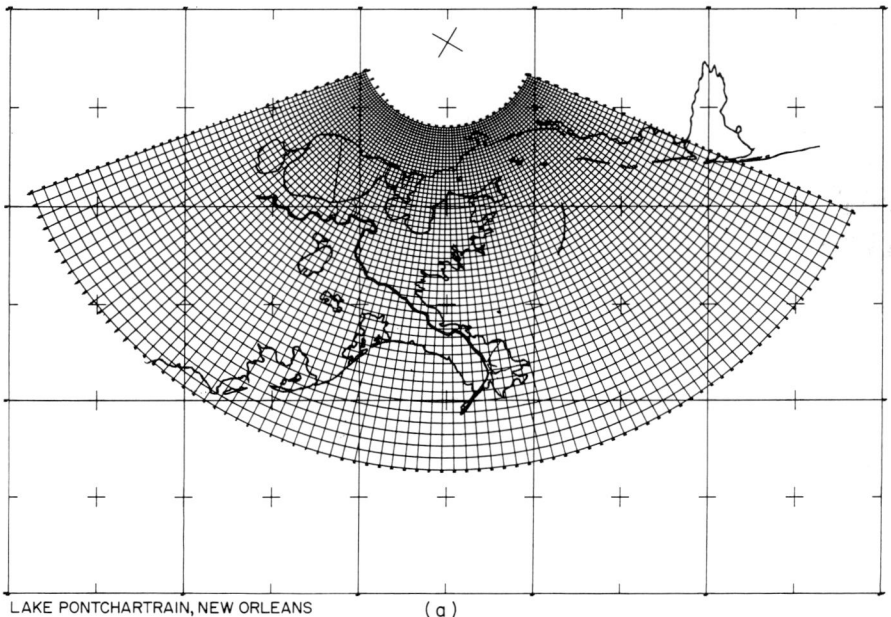

LAKE PONTCHARTRAIN, NEW ORLEANS   (a)

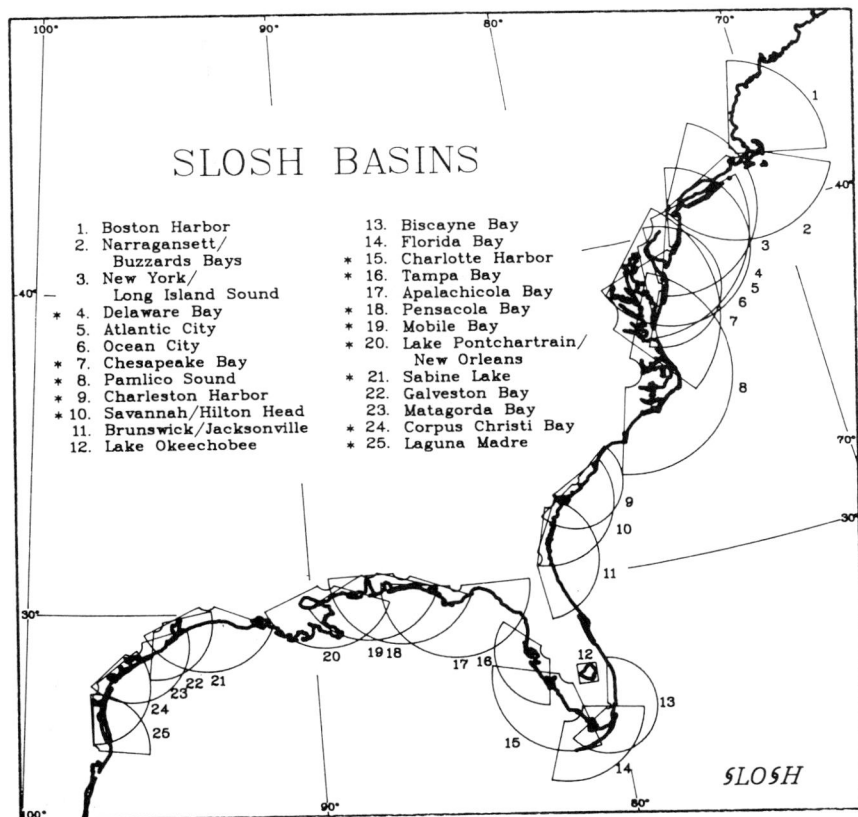

SLOSH BASINS

	1.	Boston Harbor	
	2.	Narragansett/	
		Buzzards Bays	
	3.	New York/	
		Long Island Sound	
*	4.	Delaware Bay	
	5.	Atlantic City	
	6.	Ocean City	
*	7.	Chesapeake Bay	
*	8.	Pamlico Sound	
*	9.	Charleston Harbor	
*	10.	Savannah/Hilton Head	
	11.	Brunswick/Jacksonville	
	12.	Lake Okeechobee	
	13.	Biscayne Bay	
	14.	Florida Bay	
*	15.	Charlotte Harbor	
*	16.	Tampa Bay	
	17.	Apalachicola Bay	
*	18.	Pensacola Bay	
*	19.	Mobile Bay	
*	20.	Lake Pontchartrain/	
		New Orleans	
*	21.	Sabine Lake	
	22.	Galveston Bay	
	23.	Matagorda Bay	
*	24.	Corpus Christi Bay	
*	25.	Laguna Madre	

SLOSH

(b)

Figure 6:13   (a) An example of the polar grid used by SLOSH, allowing fine resolution in the area of most interest. The area covered is Lake Pontchartrain/New Orleans (basin 20). (b) Coverage of SLOSH basins on the United States Gulf of Mexico and Atlantic coasts. (Supplied by C. P. Jelesnianski.)

SPLASH was designed for efficient operation in real time. An alternative model called SSURGE is used by the US Army Coastal Engineering Research Center to calculate statistics for the design of coastal protection works and in coastal zone management. The grid sizes are irregular, and the model has to be fitted in each case to the coastal area of interest. There are several other refinements, but the basic equations are still depth-integrated.

More recently the National Weather Service has extended the SPLASH model to compute Sea, Lake and Overland Surges from Hurricanes (SLOSH) (Jelesnianski *et al.*, 1984; Jarvinen and Lawrence, 1985; Shaffer *et al.*, 1986). The model, which is also depth-integrated, uses a polar coordinate grid scheme to allow high resolution in the coastal areas of interest. Figure 6:13(a) shows the grid used to represent the Lake Pontchartrain/New Orleans area; at each grid point a value of terrain height or water depth is supplied. The model allows for the overtopping of barriers, including dunes, levees and reefs, and for channel and river flow. Astronomical tides are not included. SLOSH is applied to a series of regions, termed basins, along the east coast; Figure 6:13(b) shows 25 operational basins used by the National Weather Service. Winds for the model operation are formed by specifying the hurricane's central pressure and its radius of maximum wind, in a similar way to SPLASH. In addition to its use as a forecast model (see Section 6:10), SLOSH has also been used as a tool to identify areas vulnerable to flooding. This is done by allowing several (typically 300) storms to impact on an area; each storm has a different intensity, size and landfall point, and moves along likely tracks. The computed responses allow statistics of coastal flooding to be assembled, and enable evacuation procedures to be designed.

Further developments of shelf-sea models include three-dimensional representation to give depth variations of currents, and computations which include the density-driven circulation of shelf seas. Although discussion has concentrated on the modelling of surges, tidal movements may also be modelled if the tidal variations at the boundaries are known. In shallow water, interactions between the major tidal constituents make it meaningless to compute the responses separately for each tidal harmonic; instead, it is essential that they are computed using a full series of tidal variations as input at the model boundaries. The development of new and better models gives great scope for mathematical ingenuity and sophistication. The physical oceanographer is able to apply these numerical models in three different ways. The primary application is for the prediction of events, either operationally (Section 6:10), or statistically for design purposes as discussed in section 8:3:4. Models may also be used to make correct hydrodynamic interpolations between sets of sparse observations, such as in the construction of cotidal charts. A third application is in the testing of physical or analytical models by simulating the conditions they specify in a controlled way, which is not possible in the real world.

6:6   Some regional examples of surge development

So far we have considered the statistics of surge residuals, some simple analytical solutions to the meteorological forcing, and the techniques available for detailed numerical simulation of these responses. It is now appropriate to consider some examples from a few well studied regions showing the physical processes which combine to give large surges, and to describe briefly the factors which are likely to be important elsewhere (see Murty (1984) for a comprehensive account).

The North Sea

The North Sea, between Britain and northern Europe, has been described as a splendid sea for storm surges (Heaps, 1983). It is open to the North Atlantic ocean in the north so that the extratropical storms which travel across this entrance from west to east are able to set the water in motion with very little resistance from bottom friction. Surges are generated by winds acting over the shelf to the north and northwest of Scotland, and by pressure gradients travelling from the deep Atlantic to shallow shelf waters. When these water movements propagate into the North Sea they are affected by the earth's rotation and by the shallowing water as they approach the narrowing region to the south. These disturbances are sometimes called *external* surges to distinguish them from the movement and changes of level brought about by the wind acting on the sea surface within the North Sea, which are called *internal* surges.

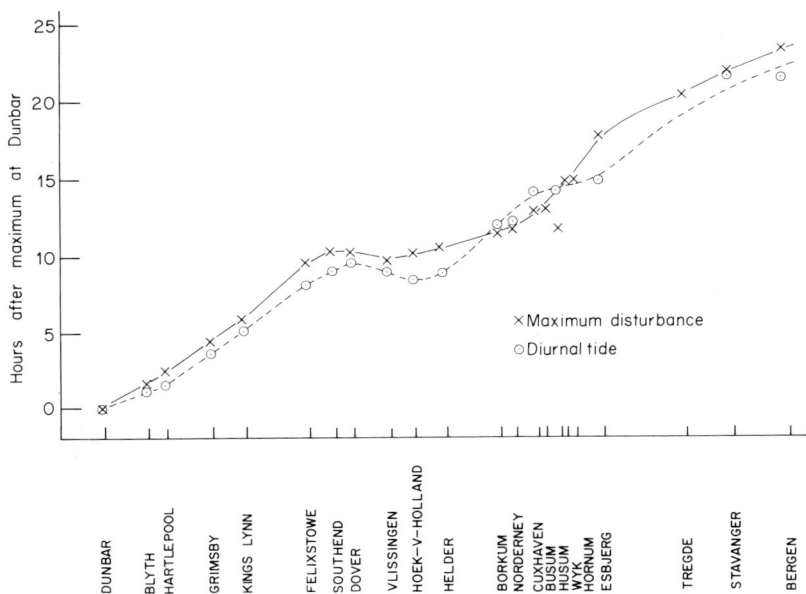

Figure 6:14 Propagation of the maximum surge disturbance at coastal gauges during the storm of 7 January 1949, compared with the phases of the regular diurnal tide, showing the anticlockwise progression of both around the North Sea. (*Reproduced by permission of the Royal Society.*)

The gradual subsidence of the southern North Sea (see Sections 9:5 and 10:5) makes the low-lying land of southeast England, including London, and Holland, increasingly vulnerable to flooding from these surges. Major floods have been recorded throughout history, the most spectacular in recent times being that of 31 January–1 February 1953, as a result of which sea defences including the Thames Barrier and the Delta defences in the Netherlands, and elaborate early-warning systems have been established (see Section 6:10). Because external surges travel like the tides from north to south as Kelvin waves along the coast of Scotland and England, reliable warnings are possible.

Some of the energy in the surge may escape through the Straits of Dover but, as for the tides, most of the wave energy is reflected and travels northwards along the European coast. The rate of surge progression, which is controlled by the water depth according to $c = (gD)^{\frac{1}{2}}$, can be compared with the speed of travel of the diurnal tide in Figure 6:14: the time to travel between Dunbar and Southend is about 9 hours.

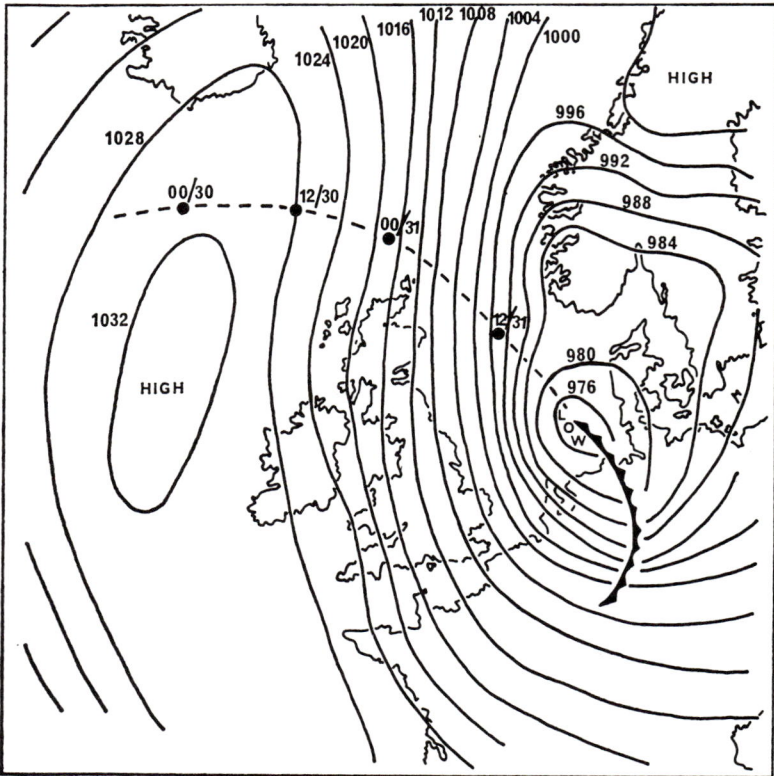

Figure 6:15 Synoptic weather chart for 0000 hours on 1 February 1953, the time of maximum surge levels in the Thames Estuary. The track of the depression is shown by the broken lines. (*Reproduced by permission of Allen & Unwin.*)

The 1953 surge and that of January 1978 (Figure 6:11) were dominated by the internal effects of strong winds blowing over the North Sea from the north. The 1953 storm conditions are shown in Figure 6:15; in the absence of geostrophic effects, steady-state wind set-up alone would give an increase in sea-level of more than 2 m in the south. For alongshore winds with geostrophic effects, equation (6:23) predicts an alongshore current which increases steadily with time. For winds of 25 m s^{-1} the increase of current speed in water of 50 m is 0.13 m s^{-1} for each hour. After 6 hours the surface gradient near the shore and at right angles to it, given by equation (6:24) is 0.01 m per km, showing that geostrophic effects contribute significantly to the coastal surges observed. Once generated, these internal surges also show a southward progression.

The depression track plotted in Figure 6:15 shows that on 30 January the strong winds from the south, located to the east of the depression, were acting to drive water out of the North Sea, giving negative surge levels. The return of this water to the North Sea produces an effect, sometimes called the *return* surge, which adds to the surge effects of the winds from the north which reach their peak about a day later, when the depression has tracked further to the east of the North Sea.

West coast of the British Isles

The dynamic characteristics of surges on the west coast of the British Isles are different from those in the North Sea although the standard deviations of the residuals are comparable. The sub-tidal oscillations shown in Figure 6:5 form a continuous weather-driven variation of the sea-level in the Celtic Sea, which is highly correlated with the barometric pressure variations according to the inverted barometer relationship (Pugh and Thompson, 1986). Those variations which are not accounted for by an inverted barometer response can be correlated with the local winds. The most effective wind direction for producing large surges is from the south and southeast, in accordance with the theory of Ekman transport to the right of the wind, and the corresponding build up of coastal sea levels. Observations of bottom pressures in the Atlantic Ocean show only weak correlation with surge pressures on the shelf, confirming the results of numerical models, which indicate that external surge propagation is not an important factor here, unlike in the North Sea.

The second difference between west coast surges and those observed in the North Sea is related to the near-resonant response of the Irish Sea and Bristol Channel to tidal forcing from the Atlantic. It appears that these resonant modes are also responsive to meteorological forcing at similar frequencies, resulting in short-lived intense surges which are quickly generated and which decay during a single semidiurnal tidal cycle. A spectacular example of this type of surge occurred at Avonmouth in March 1947, when a surge of 3.54 m on a falling tide reversed the normal tidal fall of sea-level for several hours.

The normal tracks of the depressions responsible for large surges in this region are to the south of those which are critical for the North Sea. They travel from the southwest over Ireland (Lennon, 1963a), at speeds close to 75 km h^{-1}. As they approach the UK, the winds over the Celtic Sea and the English Channel are from south to north, giving Ekman transport onto the shelf. This build-up of levels is then enhanced by the winds in the right rear quarter of the depression blowing directly from the southwest as the depression centre tracks over Ireland. The most effective speed of depression movement for generating large surges, 75 km h^{-1}, is too slow to be due to hurricane-type resonance, which would require speeds (equation (6:11)) of 140 km h^{-1} for the typical shelf depths of 150 m; it is more likely to be related to the way in which winds and pressures act together to excite natural modes of oscillation of the shelf waters. Nevertheless, serious coastal flooding can be caused by weather conditions which fail to satisfy these general criteria. One example occurred on 13 December 1981, when a slower moving depression and an associated frontal system gave a sudden alteration of winds from the south to the west over the Bristol Channel a few hours before high water on a large spring tide, which resulted in severe flooding in the upper parts of the Bristol Channel.

Surges in the English Channel are usually smaller than elsewhere in the region, but occasionally leakage of surges from the southern North Sea through the Straits of Dover, following the English coast under the influence of geostrophic forces, causes coastal flooding.

Atlantic coast of North America

Around the Gulf of Mexico and along the Atlantic coast from Florida to Cape Hatteras the greatest risk of flooding comes from tropical storms—hurricanes—which originate in the tropical Atlantic Ocean from where they travel in a westerly direction until they reach the West Indies. Here many of them turn northward towards the coast of the USA. Their greatest effects on sea levels are confined to within a few tens of kilometres of the point at which they hit the coast. Hurricanes sometimes travel as far north as New England but the areas most at risk are further south, where the danger of flooding is also greater because the coastline is low-lying. Gradually, as sea levels increase relative to the land by some 0.3 m per century, these risks are becoming greater.

Further north the surges due to extratropical storms are dominant (Redfield, 1980). The winds are less extreme, but the effects are more widely spread over hundreds of kilometres. The Ash Wednesday Storm of March 1962 caused flooding and coastal damage from North Carolina to New York. An intense depression developed from 5–7 March some 500 km east of Cape Hatteras. The strong winds from the northeast and east associated with this depression were experienced along the whole eastern seaboard; a combination of on-shore wind set-up, Ekman transport towards the coast due to alongshore winds, and spring tides resulted in some of the worst flooding ever recorded. Most of the direct destruction was due to severe beach erosion with breaking waves acting on top of the extreme sea levels.

The close correlation between alongshore winds and coastal sea levels (Section 6:4:5) is clearly shown in Figure 6:16, which relates sea levels on the coast of Nova Scotia to the Sable Island winds. Winds blowing to the west caused high coastal sea levels, with very little evident phase lag. Winds to the east give negative surge residuals.

The Bay of Bengal

The coasts of India and Bangladesh which surround the Bay of Bengal are very vulnerable to severe flooding due to tropical cyclones (Murty *et al.*, 1986). These cyclones usually originate in the southern parts of the Bay or in the Andaman Sea, from where they move towards the west before curving to the

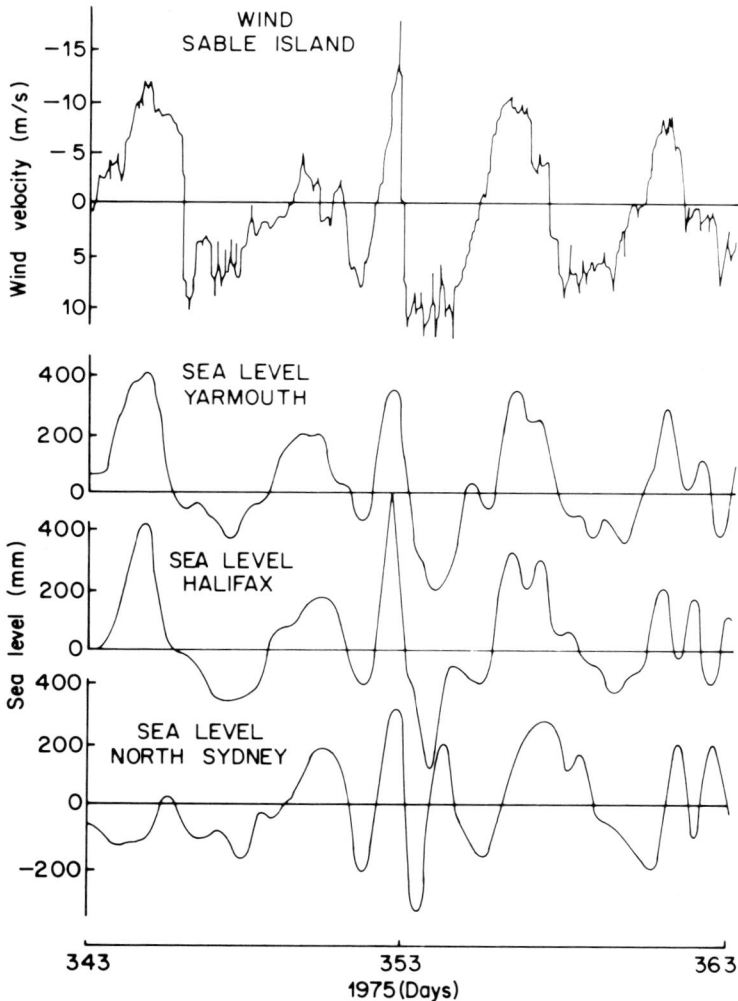

Figure 6:16 Longshore winds and residual sea levels along the Nova Scotia coast (from Sandstrom, 1980). (Copyright © 1980 the American Geophysical Union.)

north and north-east. Hurricane resonance (Section 6:3) may be significant over the narrow shelf which surrounds the normal ocean depths of the Bay, but the major factors for producing surges are atmospheric pressures and winds. In the north, where the shelf is 300 km wide, a severe cyclone in November 1970 struck the coast north of Chittagong and produced surge levels of more than 9.0 m. The fact that the coastline has a right-angle turn near Chittagong produces maximum surge levels higher than the same storm would produce moving perpendicular to a straight coast. Numerical models have confirmed that the maximum level is due to a direct surge contribution and to a contribution reflected from the neighbouring coast. Numerical modelling of the surge which hit the Orissa coast of India (Figure 6:17) has shown a positive surge of more than 4.5 m to the right of the track due to onshore winds, while a negative surge in excess of 4 m developed to the left of the track, where the winds were offshore.

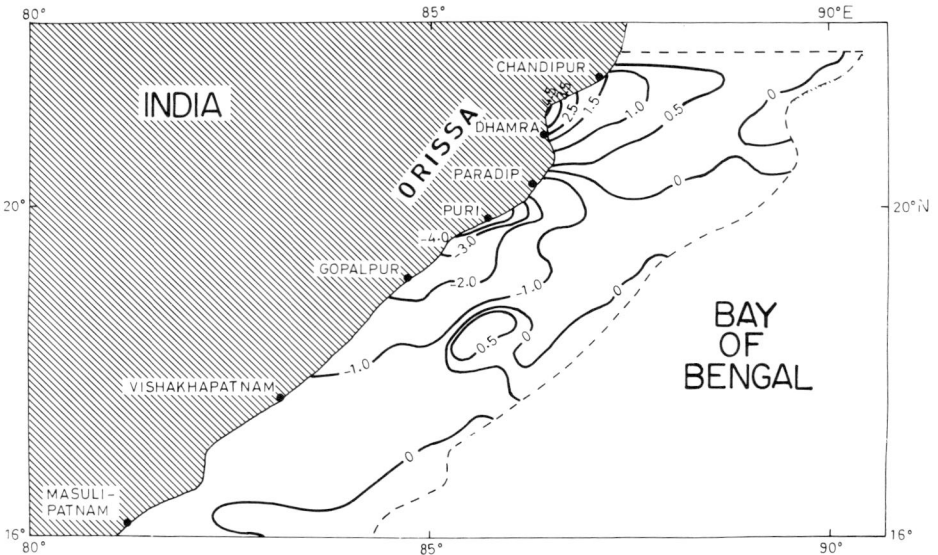

Figure 6:17   Contours of equal sea-surface elevation for the Indian coastal regions of Orissa and north Andhra Pradesh in the cyclone of 3 June 1982. Surge levels, computed by a numerical model, correspond to 30 minutes after landfall of the storm, 17 km north of Paradip. (Supplied by B. Johns.)

Japan

Several thousand people were killed in September 1959 when typhoon Vera struck Japan, producing a peak surge of 3.6 m at Nagoya on the south coast of Honshu. The outer coast of Japan has a very narrow continental shelf, so that the typhoons have their greatest effect on individual bays. The wind effects are less important here than the atmospheric pressure changes. If the direction and speed of a typhoon produces a resonant response at the natural period of the bay, then the effects can be particularly severe. Oscillations of the waters of a bay persist for several hours once they are set in motion. Computations have shown that maximum surges in Tokyo Bay would be generated by a cyclone tracking approximately 60 km north or east of the Bay, when amplitudes of 2.0 m or more are possible. In this particular case the computations showed that neither the speed of the tracking nor the existence of resonant responses was critical. Further west, Osaka Bay and the Inland Sea between Honshu and Shikoku suffer severe typhoon-generated surges which exceed 2.0 m on several occasions. The maximum surge occurs somewhat to the west of the point where the typhoon crosses the coast, and is largely independent of local topography. Observations have shown that the levels begin to rise about 20 hours before the time of the nearest approach of a typhoon and return to normal some 10 hours later.

Figure 6:18   Venice is often flooded by a combination of high tides and surges. The surges have a characteristic 22-hour period due to oscillations of the Adriatic Sea. (Supplied by P. A. Pirazzoli.)

The Adriatic

The increasing frequency with which the city of Venice, at the head of the Adriatic Sea, is subjected to flooding by high sea levels has attracted much attention in recent years (Figure 6:18). These flooding events happen more often than in historical times because of a gradual increase of mean sea-level, relative to the land, of between 3 and 5 mm per year (see Section 10:7). Storm surges in the Adriatic Sea are most effectively generated by pulses of strong winds from the south-east, directed along its length. These winds are associated with depressions which move eastwards, from their region of development over the Ligurian Sea to the north-west of Italy. Observations of the surges produced at Venice show oscillations with periods near 22 hours which persist for many days. These oscillations, which are due to excitation of the fundamental longitudinal oscillation of the Adriatic Sea, have maximum amplitudes sometimes in excess of 1 m. One of the difficulties in forecasting these surges is the poor estimation of surface wind speeds over the Adriatic; these winds show considerable differences from one place to another due to the effects of the surrounding mountain ranges. Oscillations with shorter periods which coincide with higher harmonics of the fundamental 22-hour oscillation are also likely to be generated by this varying wind stress, and those with periods close to 11 hours and 7 hours are also observed.

The Baltic Sea

The Baltic has a very small tidal range due to its limited connection to the North Sea through the Kattegat and Skagerrak, but it is subjected to severe extratropical storms which generate large surges. One of these raised sea levels by 4 m at Leningrad in 1924. Several early attempts to evaluate the wind drag coefficients C_D in expressions such as equation (6:12) were based on observations in the Baltic. An interesting series of observations of the effects of winds on sea-level gradients along the Gulf of Bothnia, the northernmost part of the Baltic Sea, showed that a value of $C_D = 0.0024$ was appropriate for strong winds during the ice-free summer months (consistent with the theories given in Section 6:4:2); however, as the Gulf became covered with ice during the winter, the effective drag coefficient gradually reduced and approached zero when the whole Gulf was covered with fast ice. Complete ice cover is rare because strong winds tend to break the ice, but when there is continuous cover, the wind stresses can be balanced by horizontal coastal thrusts transmitted through the rigid ice-sheet. The effects of seasonal changes of ice cover are also significant for surges in the Beaufort Sea in the Canadian Arctic. Surges are reduced by ice cover, positive surges being more strongly damped than negative surges. Ice cover can also affect the natural periods of free oscillations of the shallow bays in the Southern Beaufort Sea, so that resonant amplification of surges can be quite different between winter and summer. The observations made by Sir James Clark Ross (Section 6:3) through the Canadian Arctic winter, which confirmed the inverted barometer response of sea levels to atmospheric pressure changes, were made during continuous ice cover, so that the observations were not complicated by the effects of wind stresses.

General

In addition to those areas described, many other regions are vulnerable to surges generated by tropical storms. These include the coasts of southern China and Hong Kong, the Philippines, Indonesia, Northern Australia and the

Queensland coast. In all cases the surge levels and the damage caused by the storms are very sensitive to the direction and speed of their progress, so that the results of no two storms are alike. Local resonances can give further large differences in the effects on sea-level over short distances. For example, the surge in Hong Kong harbour due to a typhoon in September 1962 was 1.8 m high, but within the channels and islands a much higher surge level of 3.2 m was recorded.

6:7   Seiches

Any body of water has a set of natural periods of oscillation at which it is easy to set up motions, which are called seiches (Proudman, 1953). The periods of these seiches depend on the horizontal dimensions and the depth of the water, and range from less than a second for a tea-cup to many hours for the seas and oceans. The oceans of the world are particularly responsive to semidiurnal tidal forcing because the period of the forcing coincides with one of their natural periods of oscillation. The importance of near-resonant conditions for generating large surges has already been discussed for the Bristol Channel, for the Adriatic Sea and for the coast of Japan.

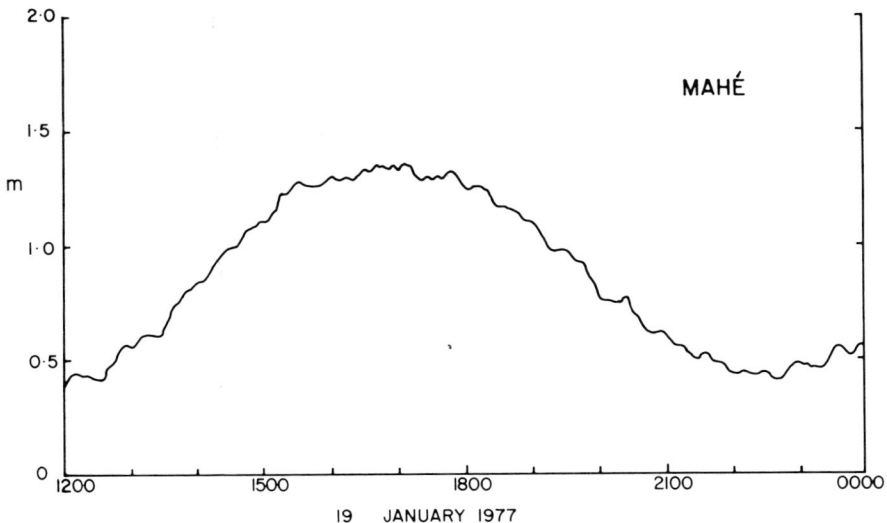

Figure 6:19   A portion of the sea-level record from Mahé, Seychelles, showing seiches of 50-minute and 12-minute periods. (*Reproduced by permission of Pergamon Books Ltd.*)

Many records from the tide gauge charts show a pattern of small oscillations imposed on the main tidal signal, similar to the pattern shown in Figure 6:19 for Mahé in the Seychelles. In this case a detailed analysis shows that two periods of oscillation are present, one near to ten minutes and one near to fifty minutes. It is not possible to relate these periods directly to any particular mode of oscillation of the surrounding shallow water, but the energy of the 10-minute oscillations is directly related to wind stress. Similar oscillations are observed for gauges around Britain, particularly for those around the northwest and the north of Scotland. Elsewhere, San Francisco Bay has oscillations with a period of about 45 minutes. Once established, these seiches often persist for several hours, showing that they are only slightly damped by friction.

The first scientific investigations of seiche phenomena were made on Lake Geneva during the latter years of the nineteenth century. Historical records of these seiches include an account of water oscillations of more than a metre in the year 1600, but fortunately amplitudes are usually much smaller than this. Using simple equipment to damp out wind waves and ripples, the Swiss naturalist Forel was able to observe oscillations as small as a millimetre, which had periods close to 74 minutes.

For an enclosed body of water the seiche motion can be considered as a standing wave with a node of no vertical motion at the centre, and maximum vertical displacement alternately at opposite ends (see Section 5:2 and Figure 5:4). The natural period of a basin is the time taken for a wave to travel from one end of the basin to the far end, and to return, after being reflected. The natural period given by Merian's formula is:

$$\frac{2L}{(gD)^{\frac{1}{2}}} \tag{5:5}$$

for this simple oscillation with only one node, the wavelength being twice the length of the basin. Applying this formula for Lake Geneva, taking $L = 70$ km and $D = 160$ m gives 58.9 minutes for the basic period; the difference between this value and the observed value of 74 minutes is due to the shape and varying depth of the lake. The two ends of the lake must always be antinodes of the oscillations, but harmonics of the basic period for oscillations having two, three or more nodes are possible, in which case the periods are:

$$\frac{2L}{n\,(gD)^{\frac{1}{2}}} \tag{6:29}$$

with n as the number of nodes.

Calculations of natural periods for harbours and bays is also possible, but in this case the node is found to be near to the open entrance, and the antinode is located at the closed end (see Figure 5:5). This gives a natural period:

$$\frac{4L}{(gD)^{\frac{1}{2}}}$$

with the length of the bay being one quarter of a wavelength. Because this configuration lacks the second reflecting boundary necessary for sustained oscillations, it is less important for seiches than for the tidal resonances described in Section 5:2:2. The nature and persistence of seiches depends critically on the boundaries of the water mass, and the period can change with the height of the tide, where this is significant compared with the mean water depth D. Seiche periods are then longest at the low tidal levels, in accordance with equation (5:5), and shortest at high water. In many cases oscillations persist where there is no obvious reflector to return the outgoing waves, for example, along a gently curving bay or straight coast. One possible explanation is the presence of a submerged sandbank or ridge offshore. A sharp increase in depth offshore is also capable of reflecting an incident wave (see Section 5:4:1 and Appendix 4:1b).

Other forms of wave motion which would appear as wave oscillations having periods of several minutes are also possible. Edge waves, which propagate along the coast under the influence of bottom topography, would give sea-level oscillations, although it is not strictly correct to call these seiches. Another possibility is for a group of islands and headlands to be spaced so that they trap the edge waves, causing them to resonate in a standing wave pattern. Similar trapped waves have been calculated and observed to propagate around islands, with a period controlled by the island radius and the offshore sea-bed topography.

Waves with periods of several minutes, intermediate between those of wind waves (seconds) and tides (hours), are of practical importance because their period happens to coincide with that of some man-made systems. Early designs of systems for extracting wave power off the west coast of Scotland were sensitive to longshore oscillations. Although the amplitudes of harbour seiches may not exceed a few centimetres, the associated currents may be relatively strong because of the short periods in which water transfers take place. Ships moored in harbours have suffered damage because their natural period of yaw coincided with that of the harbour oscillation. Seiches can also give rise to rapid currents in and out of a narrow harbour entrance, as the harbour levels adjust over short periods.

The physical forces responsible for triggering the oscillations usually take the form of an impulse such as that due to a sudden squall of wind. During times of high wind-waves oscillations may also be driven by surf-beat, in which energy is transferred to longer periods. Around the Pacific Ocean seiches are often triggered by the impulse of tsunamis, generated by submarine earthquakes. In the next two sections we consider these in more detail.

6:8 Tsunamis

Tsunamis are wave events generated by seismic activity and as such fall outside the two principal categories of forces responsible for sea motions: tides and the weather (Murty, 1977; Loomis, 1978). An alternative name is *seismic sea waves*. The popular description of them as 'tidal waves' is a misnomer because they lack the regularity associated with tides. Virtually all tsunamis are generated by submarine earthquakes, but landslides into the sea, and submarine slumping (of, for example, sediments on the continental slope) may occasionally be responsible. There are three distinct aspects of tsunamis which may be considered: their generation by earthquakes, their propagation in deep water, and their behaviour where they impinge on coasts and the surrounding regions of shallow water.

Not all earthquakes produce tsunamis. The important element appears to be a vertical crustal movement which displaces the sea-bed. After the sea-bed is displaced, a tsunami is generated by the horizontal pressure gradients in the water acting as a restoring force. The wave characteristics will depend on the amplitude of the displacement and the dimensions of the sea-bed involved. Horizontal displacements of the sea-bed will be relatively ineffective for producing tsunami because water is not displaced vertically. Direct observations of sea-bed displacements and their relation to tsunamis are not possible because these events are rare and inaccessible. However, computer simulations of several possible earthquake models of the 1964 Niigata event (see Figure 10:17) showed that a 6 m displacement with a dip angle of 45° along a 55 km fault length was consistent with the arrival times and amplitudes of the tsunami observed along the coast of Honshu. It has been estimated that the energy in a tsunami is from 0.01 to 0.10 of the total original seismic energy.

Tsunamis which propagate across deep water typically have wave periods of ten minutes or longer, which means that their wavelength is long compared with the water depth: their speed of propagation is therefore given by $(gD)^{\frac{1}{2}}$. For a wave of period 10 minutes and a water depth of 4000 m the speed is $720 \, \text{km h}^{-1}$ and the wavelength is 120 km. Amplitudes in deep water are small, probably not more than 1 m. The waves pass unnoticed by ships at sea. Travel times for tsunamis between gauges in the Pacific Ocean confirm that the long-wave speed formula is valid, and this behaviour provides a basis for giving warnings to places further away from the source.

Although the times of arrival of a tsunami can be predicted accurately, the amplitude of the wave which hits a particular length of coast is much less certain. This is because in shallow coastal waters, in addition to the normal amplification of the wave as it slows down (see Section 5:3:1), the wave undergoes reflections and refraction. Often the first wave is not the biggest. Sometimes the first arrival of the disturbance is seen as a recession of the sea which exposes levels which have not been seen before. People who follow the water out to inspect the sea-bed are then overwhelmed by the arrival of the main wave as a 'wall' of water 2 m or more high. Depending on the nature of the original earthquake, the first arrival may also be a sudden rise of sea-level: Figure 6:20 shows positive first arrivals of a tsunami at three gauges in the Hawaiian Islands on 23 May 1960. The oscillations which are superimposed on the normal tide have different periods at each station, showing that they are local seiche motions initiated by the tsunami disturbance, rather than periods imposed by the tsunami itself. Once established these motions may continue for several days. Indeed, the term *tsunami* derives from the Japanese for 'harbour wave', which is a form that they often take around the Japanese coast.

The Pacific Ocean and its coastlines are the most vulnerable to tsunamis because of the seismically active surrounding plate boundaries. The tsunamis shown in Figure 6:20 were generated by an earthquake along the coast of South Chile some 15 hours earlier. It took a further 7 hours to reach Japan, where more than 100 people were drowned. In Chile itself the tsunami death toll was over 900 people.

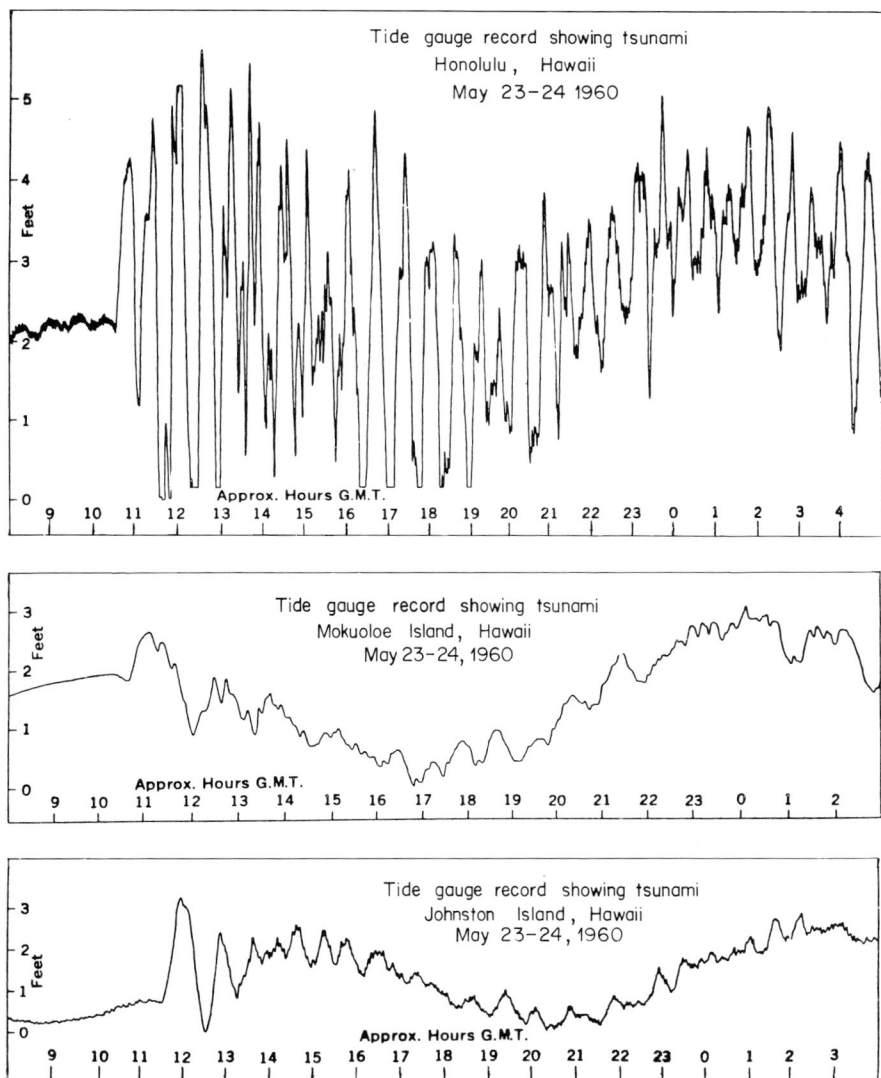

Figure 6:20   Sample tsunami records from three tide gauges in the Hawaiian Islands, 23–24 May 1960. (From J. M. Symons and B. D. Zetler (1960). 'The Tsunami of May 22, 1960 as Recorded at Tide Stations.' United States Department of Commerce, Coast and Geodetic Survey. Unpublished manuscript.)

Hawaii experiences a significant tsunami on average once every seven years. An earthquake some 200 km off the coast of north-east Japan resulted in over 17 000 people being drowned in 1896. Although most tsunamis which affect Japan arrive on the Pacific coast, serious damage and loss of life sometimes occur due to tsunamis on the west coast. The Niigata earthquake and tsunami of 1964 which has already been discussed is an example. More recently, over 100 people were drowned in 1983 following an earthquake of magnitude 7.7 on the Richter scale in the Sea of Japan, when a tsunami hit the coast around Akita in northern Honshu; three large waves were separated by 15-minute intervals. An indication that an earthquake might occur had been given by a local mean sea-level rise of 0.04 m in the previous 12 years (see Section 10:7).

Tsunamis are not restricted to the Pacific Ocean, but they are much rarer around the Indian and Atlantic Oceans. The Krakatoa explosion of 27 August 1883 was perhaps the most severe in recent history; more than 36 000 people were reported drowned by the tsunami which it generated in the Sunda Strait between Java and Sumatra. Its effects were observed not only all around the Indian Ocean, but as far as South Georgia and Tierra del Fuego in the South Atlantic Ocean. In the North Atlantic Ocean the most devastating recorded tsunami followed the Lisbon earthquake of Saturday, 1 November 1755. The destruction of the earthquake was followed by three enormous waves which overflowed the banks of the River Tagus. British lakes and harbours were observed to seiche within a few minutes of the earthquake due to the direct seismic wave, but the main tsunami, travelling as a long progressive wave, arrived on the coasts of the south-west of England and Ireland later in the afternoon. At Newlyn the levels rose by more than 3 m in 10 minutes. The very rare and unpredictable nature of earthquakes and their associated tsunami makes it very difficult to include them in probability estimates of extreme levels (Chapter 8) because they are unlikely to be included in the period of observations.

6:9 Wave set-up; surf-beat

The influence of wind waves on measurements of still water level was discussed in Section 2:2. Careful design of the measuring system is necessary to avoid non-linear responses to the waves, which appear as apparent changes of mean sea-level. Wind waves also react in a non-linear way with the shallow near-shore waters, producing changes in mean sea-level and oscillations with periods of minutes, known as *wave set-up* and *surf-beat*. It is appropriate to describe these briefly, but a full discussion is outside the scope of this book (James, 1983).

The phenomena associated with waves approaching the shore have been described in terms of a *radiation stress*, a concept developed by Longuet-Higgins and Stewart (1964). It is defined as the excess flow of momentum towards the shore due to the presence of waves. For waves normal to the shore, where the depth is small compared with the wavelength, this flux is:

$$S_w = \tfrac{3}{4} \rho g a_w^2$$

where a_w is the local wave amplitude, g is gravitational acceleration and ρ is the water density. Changes in the radiation stress are balanced by changes in the potential energy per unit area of sea, which in turn means changes in the mean level ζ. If the beach slope is sufficiently gentle for there to be no wave reflection, it is possible to calculate the changes of mean water level associated with changes in the momentum flux as the waves travel from deep water. Outside the line of breaking waves the mean level is depressed by an amount:

$$\bar{\zeta} = -\frac{a_w^2}{4D} \qquad (6{:}30)$$

where D is the local depth. This *set down* becomes greater as the water shallows, reaching a maximum value just before the waves break.

Experiments and theory show that swell waves tend to break when the water depth is approximately 2.6 times the wave amplitude. Inside the line of breakers the wave energy decreases shorewards because of dissipation, leading to a decrease in the radiation stress. After breaking, the wave amplitudes remain proportional to the mean water depth:

$$a_w = \alpha D \qquad (6{:}31)$$

where α is between 0.3 and 0.4 for spilling breakers; a_w is taken as half of the total wave height, as the waves are not now sinusoidal. If the coordinate system is chosen with the positive X-direction offshore the gradient of the mean surface can be shown to be:

$$\frac{\partial \bar{\zeta}}{\partial x} = -\tfrac{3}{2}\, \alpha^2 \frac{\mathrm{d}D}{\mathrm{d}x} \qquad (6{:}32)$$

There is an increase of mean level $\bar{\zeta}$ as the waves travel towards the beach, starting from the point where the waves break.

As an example, consider waves of 3 m amplitude on a beach with a steady slope of $\tfrac{1}{30}$. According to the conditions given above, these waves would tend to break 230 m offshore in a water depth of 8 m. The set-up gradient of (6:32) gives a total increase around 1.5 m from this point to the shoreline. Just outside the line of breakers there would be a set down of 0.29 m according to equation (6:30).

Wave amplitudes vary locally because of reflections and refraction. It is also well known that they change with time, larger waves appearing in groups. The mean level $\bar{\zeta}$ will therefore be a complicated surface which changes over periods of minutes as the wave amplitudes change, and so the position of wave breaking is shifted nearer and further away from the shore.

These low-frequency changes called *surf-beats* will be seen on a tide gauge as intermediate 'waves' with periods of minutes, though they are not propagating waves in the usual sense. They may also be observed by pressure recorders mounted offshore on the sea-bed. Careful experiments have shown a correlation between low-frequency waves propagating offshore, and the amplitudes of the increasing swell, with a time delay corresponding with the time for the swell to travel to the breaker zone, and for a generated long wave to be propagated back as a free wave. These surf beats may be refracted by the offshore topography to propagate along the coast as edge waves. Where the coast has irregularities the edge waves may be reflected, establishing a pattern of trapped edge waves at some period which is a characteristic of the coast.

Set-up may also occur in harbours, which is unfortunate for sea-level measurements as these are often chosen for permanent sea-level recorders. As waves spread into a harbour from a narrow entrance, their amplitude decreases (Thompson and Hamon, 1980). Radiation stress arguments show that at distances within the harbour greater than a few times the width of the entrance, the mean levels are actually higher than those at the mouth by an amount:

$$\bar{\zeta} = \tfrac{3}{4}\frac{a_w^2}{D}$$

where a_w is now the wave amplitude at the entrance and D is the depth inside the harbour. If $D = 10\,\text{m}$ and $a_w = 1.0\,\text{m}$ the theoretical set-up is $0.075\,\text{m}$. Because the set-up is greater for lower mean water depths, the apparent amplitude of a measured tide can be reduced if waves increase measured low tide levels more than they increase high tide levels, perhaps by a few centimetres. Adjustment is not instantaneous as a few hours are necessary for the water transfer needed for full set-up to develop.

Set-up due to waves is hard to distinguish from the larger scale surge effects of wind on the sea surface, described in Section 6:4. Attempts to measure the drag coefficient C_D using equations such as (6:13) and (6:16) are also liable to error because the true regional gradient on the sea surface is distorted by local set-up due to waves at the measuring site.

6:10   Flood warning systems

Coastal flooding by the sea from time to time is inevitable. Some coasts may be protected by suitable high walls (see Chapter 8), but these are expensive to build, and are only justified where large populations are at risk. An alternative is to give warnings of impending flooding sufficiently early for people, their animals, and other valuables to move to higher levels. The essential components of these systems are a means of detecting and predicting potential flooding, and a reliable procedure for transmitting these warnings to the public in the areas at risk.

For the North Sea, a Storm Tide Warning Service (STWS) operates from the Meteorological Office at Bracknell throughout the months from August to April. Numerical model forecasts of the weather are used to drive the numerical model of the continental shelf as described in Section 6:5, to give estimated coastal levels 36 hours ahead. There is a continual process of updating the forecasts as the times of potential flooding approach, using the latest weather forecasts and observations of the surges as they develop and progress from north to south along the coast. The coast is sectioned into five Divisions, each of which has a reference port. Local authorities in each Division nominate a Danger Level for their reference port. The STWS issues an alert to the local police if a Danger Level is likely to be reached; the police in turn consult with the local Water Authorities to decide from local knowledge what warnings are necessary. About 4 hours before the maximum level is reached a second message is issued. This may be a full Danger Warning, an Alert Confirmed if the forecast levels may approach the Danger Level, or a Cancellation if there is no longer any danger. In a typical season some 70–100 Alerts will be issued, but only about five of these will be followed by Danger Warnings. This type of procedure is also used to decide when London should be protected by closing the Thames Barrier (Figure 6:21) (Horner, 1985).

Figure 6:21 The Thames Barrier at Woolwich, which is raised in advance of large surges to protect London. (Copyright © Thames Water.)

Warnings of hurricane surges along the east coast of the United States are issued by the National Weather Service using the SLOSH model previously described. Weather parameters are difficult to obtain for hurricanes which develop and travel out at sea; in addition to coastal and ship reports, satellites, coastal radar and special reconnaissance aircraft are also used to estimate the central pressure, the radius to maximum winds and the tracking of an approaching tropical storm. Six-hourly hurricane parameters are supplied to the model, but the reliability of the model is limited by the forecasts of the hurricane positions. At present National Weather Service forecasts of hurricane position have an average error of $\pm$ 100 miles for a 24-hour forecast. Several runs for different possible tracks are examined to get an overview of the potential surge flooding from a threatening hurricane.

Warnings of tsunami approach are issued by the Tsunami Warning System in Hawaii. When an earthquake of sufficient magnitude to generate a tsunami occurs in the Pacific Ocean area its exact location is determined by a network of seismograph stations from the arrival times of the seismic waves. The network of sea-level stations is then alerted to watch for a tsunami. Only those earthquakes which cause a strong vertical movement of the sea floor generate tsunamis, but these cannot be identified directly from the seismic signals. The first positive indications of a tsunami are usually obtained from the gauges nearest to the earthquake epicentre. If a tsunami is observed, all other countries in the System are given immediate warning from Hawaii. If no tsunami is observed the warning alert is cancelled.

It is important that flooding warnings are only issued for serious events. Too many near-misses and cancelled warnings reduce public confidence with the result that genuine warnings are ignored. But the obvious consequences of delaying before issuing a warning are far more severe. The officer responsible for initiating the warning procedures must have reliable weather forecasts, and a proven model of the behaviour of the sea under the influence of these weather conditions, to avoid both these difficulties.

CHAPTER 7

Shallow-water Dynamics

God does not care about our mathematical difficulties. He integrates empirically.

Albert Einstein. In L. Infield, *Quest*.

7:1 Introduction—some observations

Chapter 5 discussed how the amplitudes of the tidal waves which are generated in the deep oceans increase when they spread onto the shallow surrounding continental shelves. On the shelves the characteristics of these waves are altered by other processes including standing-wave generation and local resonances. In this chapter we consider the more extreme distortions which occur as the waves propagate into the shallower coastal waters and the rivers. The behaviour of these distorted tides is very important for near-shore human activities such as coastal navigation, and also for sedimentation and for biological dynamics, as we shall see in later chapters.

It is useful to identify three separate physical factors which may contribute to the distortions. Although the tidal waves still satisfy the criteria for long waves, that is, they have wavelengths which are much longer than the water depth, in shallow water the amplitudes of the waves become a significant fraction of the total water depth. Secondly, the stronger currents which develop in the shallow water are resisted by the drag due to bottom friction, a process which eventually removes much of the propagating tidal energy, and reduces the wave amplitudes. A further distorting factor is the influence of topography. Irregular coastlines and varying depths impose complicated tidal current patterns; where currents take curved paths, there must be associated surface gradients to provide the necessary cross-stream accelerations. Exact mathematical solutions for the complicated combinations of these factors which apply in particular shallow-water conditions are seldom possible. Even so, it is instructive to look separately at the three factors in terms of the hydrodynamic equations, as we shall see in Section 7:3; however, that section may be omitted by readers who do not require a mathematical interpretation.

Examination of sea-level records from shallow-water locations normally shows that the interval from low to high water is shorter than the interval from high to low water: the rise time is more rapid than the fall. Offshore the flood currents are stronger than the ebb currents. This effect is seen in Figure 2:13 for Liverpool, where the tides are distorted in passing through the shallow waters of the Irish Sea and the estuary of the River Mersey. In these circumstances, simple predictions give times of high water which are later than observed, and times of low water which are earlier.

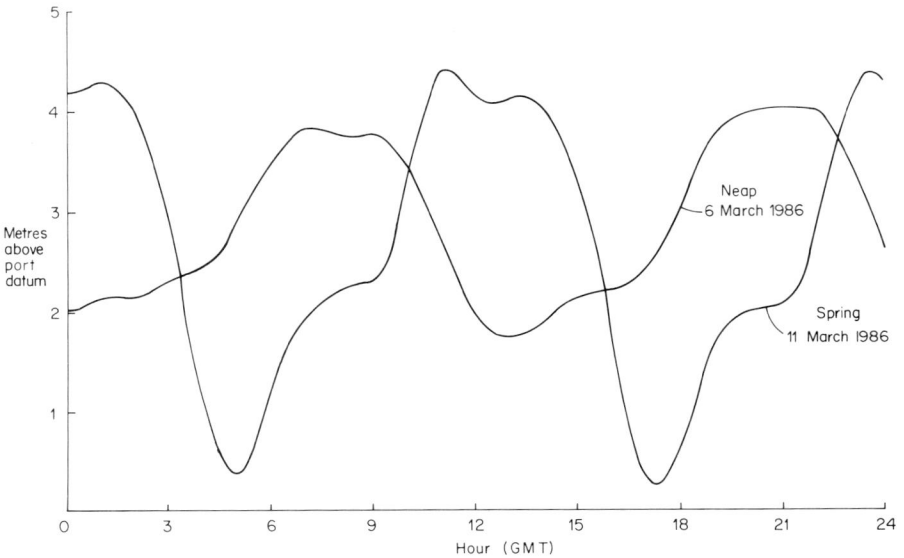

Figure 7:1 Typical curves for spring and neap tidal water level changes at Southampton showing the double high waters and the increased distortion at spring tides. (From data supplied by British Transport Docks Board, Southampton.)

In some shallow-water areas, the distortion takes the form of a double high water or a double low water (Doodson and Warburg, 1941). Figure 7:1 shows perhaps the best known example of this, the tidal variations at Southampton on the south coast of England. During spring tides, following low water the water level rises, but there is then a slackening of the tidal stream and a water level stand for a further two hours before the final rapid rise to high water, over the next three hours. The slackening effect is known locally as the 'young flood stand'. The flood and the double high water last approximately nine hours, leaving only 3 hours for the tidal ebb, which is therefore associated with very strong ebb currents. Explanations of the double high water in terms of the travel of two separate M_2 progressive waves around the Isle of Wight are wrong: Newton was among the earliest to point out that, from basic trigonometry, adding two harmonic waves which have the same period but different phases produces a single combined wave of the same period and of intermediate phase. No extra maxima or minima are created by this superposition. Observations at other places along the coast of the English Channel show that the phenomena are more extensive and that the effects are not limited to South-ampton Water. At Portland, 90 km to the west of Southampton, there are sometimes double low water levels at spring tides.

Elsewhere, double high waters are also observed at Falmouth in Buzzards Bay, Massachusetts, and at the Hook of Holland and Den Helder on the Dutch coast, and probably at Marion in Buzzards Bay, Massachusetts (Redfield, 1980).

7:2 Higher harmonics

The distortions of the normal harmonic variations of tidal levels and currents can be represented by the addition of higher harmonics, as was discussed in Chapter 4. Consider the simplest case of M_2 and its first harmonic, M_4. The phase of the higher harmonic relative to the basic wave controls the shape of the total curve, as is shown in Figure 7:2.

Rapid rise followed by a slow fall of level (Figure 7:2(a)) corresponds to the case of the M_4 phase leading the M_2 phase by $\varphi = 90°$, at $t = 0$. The case of the slow rise and rapid fall (Figure 7:2(c)) corresponds to the M_4 phase lagging the M_2 phase by $\varphi = 90°$ at $t = 0$. Double high water may occur (Figure 7:2(b)) when the M_4 phase is $\varphi = 180°$ different from the M_2 phase at $t = 0$, and the corresponding case of double low water (Figure 7:2(d)) occurs when $\varphi = 0°$, the phases of M_2 and M_4 are the same at $t = 0$. More generally, at all times, the phase relationships are defined as in Figure 7:2, in terms of $\varphi = 2g_{M2} - g_{M4}$.

In the case plotted the M_4 amplitude is set at 0.4 times the M_2 amplitude. For a double high water or a double low water, there is a minimum M_4 amplitude which can produce the effect. Consider the total water level at $t = 0$ in Figure 7:2(b):

$$H_{M2} \cos 2\omega_1 t - H_{M4} \cos 4\omega_1 t = H_{M2} - H_{M4}$$

If this is a minimum value between two maxima, then the amplitude at time Δt later or earlier must be greater. Using the trigonometric expansion $\cos\beta = (1 - \beta^2)$ for small β, this new level is:

$$H_{M2} (1 - 4\omega_1^2 \Delta t^2) - H_{M4} (1 - 16\omega_1^2 \Delta t^2)$$

which only exceeds the previous value if

$$4 H_{M4} > H_{M2}$$

Similar arguments may be used to show that a double high water involving only M_6 requires

$$9 H_{M6} > H_{M2}$$

Double high or low waters are more likely to occur where the M_2 amplitude is low, as is the case in the vicinity of Southampton because of the degenerate amphidromic system found in this region of the English Channel (see Section 5: 4:2 and Figure 5:10). In Figure 1:1(a) there are several additional turning points

$2g_{M_2} - g_{M_4} = -90°$
fast rise
slow fall
high water earlier, higher;
low water later, lower.

(a)

$2g_{M_2} - g_{M_4} = 180°$
double high water;
low water lower.

(b)

$2g_{M_2} - g_{M_4} = 90°$
slow rise
fast rise
high water late, higher;
low water early, lower.

(c)

$2g_{M_2} - g_{M_4} = 0°$
double low water;
high water higher.

(d)

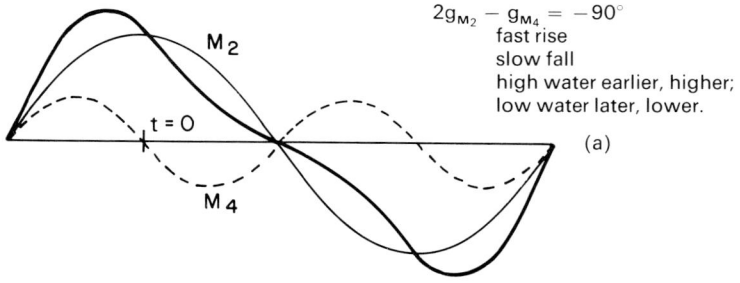

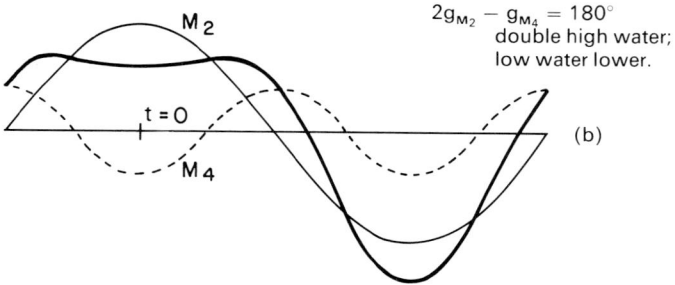

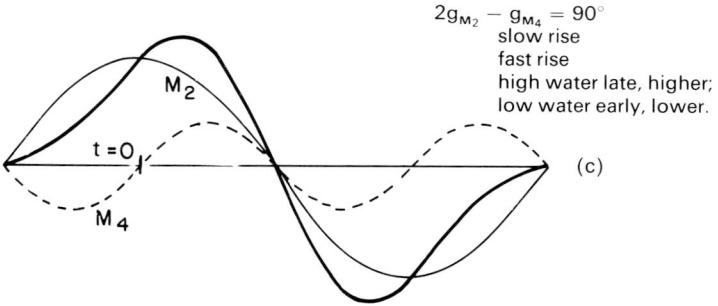

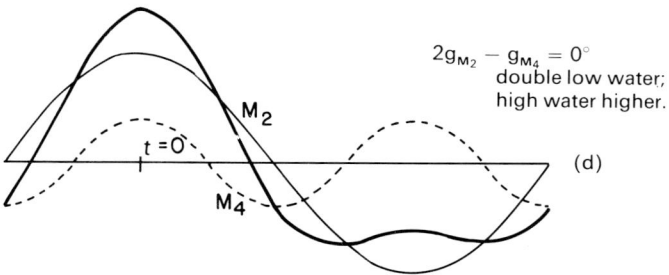

Figure 7:2 Showing how the shape of the composite sea-level curve is controlled by the relationship between the M_2 and M_4 phases. ($H_{M_4} = 0.4\ H_{M_2}$.)

during neap tides when the semidiurnal amplitude at Courtown becomes very small. At Southampton neither M_4 nor M_6 is large enough on its own to produce the double high waters: rather, it is a combination of several shallow-water harmonic terms which represent the observed distortions.

In many cases it is possible to draw cotidal and coamplitude charts for the shallow-water constituents. For the sea around Britain the M_4 amplitude is comparable with the amplitude of the diurnal constituents, in the range 0.05–0.15 m. The amplitude of M_6 is smaller, rarely exceeding 0.10 m except in the vicinity of Southampton Water and the Isle of Wight where it reaches 0.20 m. Higher harmonics than M_4 are only significant in estuaries and restricted local areas.

Figure 7:3   A chart of M_4 for the seas around the British Isles. There is insufficient data to draw full cotidal and coamplitude lines north of 53° N; the arrows show the progression of the phases along the coast. (From Howarth and Pugh, 1983). (Reproduced by permission of Elsevier.)

236

Figure 7:3 shows a cotidal chart for the M_4 constituent around the British Isles. South of 53.5° N it has been possible to produce cotidal and coamplitude lines, but north of 53.5° N there are too few observations, particularly offshore, and the amplitudes are too small for a satisfactory determination of the lines. However, the direction of phase propagation, which is consistent along the coast, is shown by arrows. The sense of phase propagation of the shallow-water constituent waves towards the ocean, and the increase from very small (< 0.01 m) amplitudes at the edge of the continental shelf to maximum values in excess of 0.40 m in local areas such as the Severn Estuary and the Gulf of St Malo are consistent with their being generated in the shallow water. Note that there are clearly defined amphidromes in the southern North Sea, off south-east Ireland, and a well developed pattern of double amphidromes in the English Channel. There is also an anti-amphidrome with M_4 amplitudes in excess of 0.3 m in the vicinity of the Dover Straits.

In the semidiurnal tidal regime the odd shallow-water harmonics (M_5, M_7, M_9, etc.) are usually very small, but where mixed tides occur, so that diurnal and semidiurnal tides can interact, the fifth-diurnal tides can be appreciable: an example is found at Anchorage in Alaska.

The observed tides, as calculated by the harmonic analysis, are usually slightly smaller and lag by a few minutes on the linear gravitational tides computed by the response technique. For example, at Newlyn, a one-year harmonic analysis of the actual M_2 tide gave an amplitude and a phase of 1.700 m and 135.5°, compared with the linear gravitational response tide of 1.872 m and 131.3°. Physically the difference results from the effects of the finite M_2 amplitude on its own propagation characteristics in shallow water—in a sense M_2 is self-limiting; wave amplitudes are reduced and the arrival time is delayed. In terms of triple interaction, this is represented by the notation of Chapter 4 as I^{2+2-2}, and may be considered as a shallow-water harmonic component of the total M_2 constituent. At Newlyn this M_2 non-linear tide has an amplitude of 0.179 m and a phase of 276°. Without this ability to compare results from the two different types of analysis, the compound nature of the observed M_2 constituent would not be apparent.

*7:3 Hydrodynamic processes

7:3:1 Non-linear interactions

In Chapter 4 we considered how dynamic processes which depend on the square or higher powers of the currents and elevations can lead to higher harmonics of the basic tidal frequencies: at spring tides, as shown in Figure 7:1, the larger amplitudes result in greater distortions of the basic astronomical harmonics. An alternative to the analysis of long periods of data for several harmonics, which are separated from each other by frequency increments equivalent to a month (groups) or a year (constituents), is to analyse single days of data for the harmonic present in each *species*. These daily harmonics may be called $\mathbf{D}_1$, $\mathbf{D}_2$, $\mathbf{D}_4$, etc. by analogue with the usual notation for the naming of harmonic constituents. Modulations of these $\mathbf{D}_n$ terms will take place over the spring–neap cycle, over a month and over a year, in the same way as the many lunar constituents in more detailed analyses vary over 18.6 years. Although not suitable for prediction purposes, daily analysis can give some insight into the physics of non-linear dynamics, which is not possible by other means. These daily analyses are particularly helpful where non-linear physical processes are involved, as these processes generate many additional harmonic constituents.

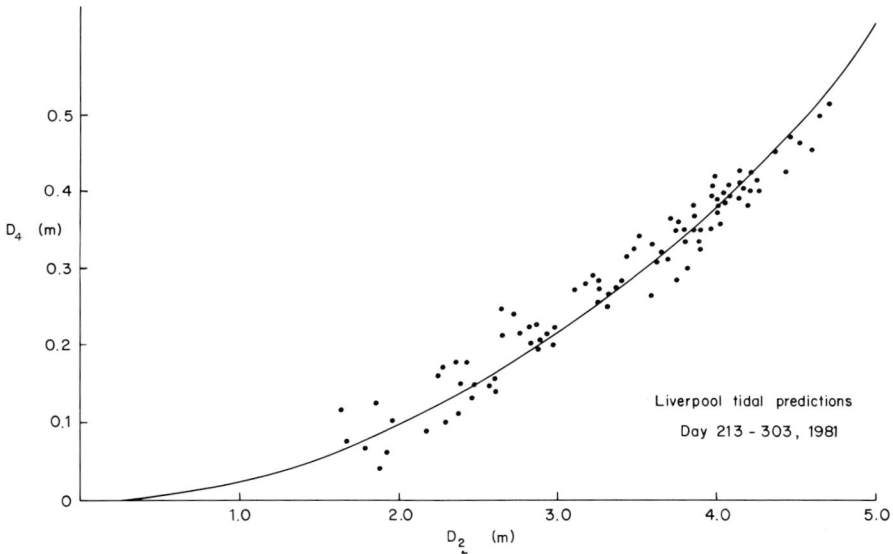

Figure 7:4   The amplitude of the daily fourth-diurnal tide, $\mathbf{D}_4$, as a function of the amplitude of the composite semidiurnal tide, $\mathbf{D}_2$, at Liverpool.

Figure 7:4 shows the variation of the $\mathbf{D}_4$ amplitude at Liverpool, England, as a function of the $\mathbf{D}_2$ amplitude. $\mathbf{D}_4$ increases rapidly, at a rate which lies between the square and the cube of $\mathbf{D}_2$. Section 4:2:3 shows how the combination of $\mathbf{M}_2$ and $\mathbf{S}_2$ leads to terms in $\mathbf{M}_4$, $\mathbf{MS}_4$ etc. if a squared law of interaction applies, whereas further higher harmonics are generated by higher powers of interaction. In practice the relationships are more complicated because there are several types of non-linear activity which influence the dynamics; bottom friction, effects due to the water depth being comparable to the tidal amplitude, and curvature imposed on the flow patterns by the topography.

The total effect of these several factors can only be modelled fully by applying numerical techniques for each particular case. Nevertheless, it is useful to examine some simple solutions of the hydrodynamic equations for each separate process. In the case of one-dimensional dynamics, such as might approximate to the situation for a narrow channel, or estuary, equations (3:22), (3:23) and (3:24) become:

$$\frac{\partial \zeta}{\partial t} + \frac{\partial \{(D + \zeta)u\}}{\partial x} = 0 \tag{7:1}$$

$$\frac{\partial u}{\partial t} + u \frac{\partial u}{\partial x} = -g \frac{\partial \zeta}{\partial x} + \frac{1}{\rho} \frac{F_b}{(D + \zeta)} \tag{7:2}$$

where direct tidal forcing is considered negligible for the local dynamics, and F_b is the force on the water due to bottom friction. On the basis of these simplified equations we may consider the three shallow-water effects in turn. The mathematical aspects of the next three sections may again be omitted by those who require only a more general description.

7:3:2 Effects of bottom friction

The effects of bottom friction are to oppose the flow and to remove energy from the motion. Experiments and theoretical dimensional analysis arguments such as those given in Section 6:4:1 support a relationship between the drag and the current speed of the form:

$$\tau_b = -C_D \rho \, \mathbf{q}|\mathbf{q}| \tag{7:3}$$

where τ_b is the bottom stress, C_D is a dimensionless drag coefficient, ρ is the water density, and $\mathbf{q}$ is the total current. In a similar way to the relationship between wind speed and the stress on the sea surface, the value of C_D depends on the level above the sea-bed at which the current is measured; usually this level is taken as 1 m, and for this case the value of C_D may lie between 0.0 015 and 0.0 025. The two components of bottom stress are:

$$F_b = -C_D \rho \, |\mathbf{q}|\mathbf{q} \cos \theta = -C_D \rho q u$$
$$G_b = -C_D \rho \, |\mathbf{q}|\mathbf{q} \sin \theta = -C_D \rho q v$$

in the usual Cartesian system. In the case of flow in a channel this reduces to:

$$F_b = -C_D \rho |u|u$$

Suppose that the currents vary harmonically as $u = U_0 \cos \sigma t$. The $u|u|$ term may then be expanded mathematically as a cosine Fourier series which contains only odd harmonics:

$$u|u| = U_0^2 (a_1 \cos \sigma t + a_3 \cos 3\sigma t + a_5 \cos 5\sigma t + \cdots)$$

where $a_1 = 8/3\pi$, $a_3 = 8/15\pi$, etc. (Gallagher and Munk, 1971). This shows that tidal currents in a channel which vary as $U_{M2} \cos 2\omega_1 t$ will result in frictional forces which include terms in $\cos 6\omega_1 t$, $\cos 10\omega_1 t$ etc. In the more general case of two-dimensional flow, both even and odd harmonics of the basic tidal frequencies are present in the frictional resistance terms.

7:3:3 Finite water depth

Most people are familiar with the behaviour of wind waves as they approach the shore. Gradually the wavefront steepens until eventually the wave breaks. Similar behaviour occurs with tidal waves, but the initial distortion is not obvious to the casual observer because the wavelengths and periods are so much greater. Figure 7:5 shows an exaggerated profile of a wave moving in the direction of increasing X. At any particular location the rise and fall of the water as the wave passes will not take equal times. The rate of rise of the water level is more rapid than the rate of fall. This difference becomes greater as the wave progresses (X increases), a phenomenon which is commonly observed in nature.

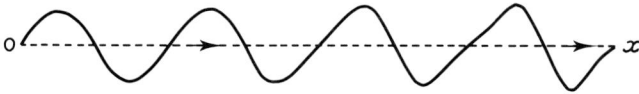

Figure 7:5   Distortion of a progressive wave travelling in shallow water up a channel in the positive X-direction. Note the shorter rise time as the wave passes a fixed observer, as the distortion increases.

Approximate arguments suggest that since the wave speed decreases as the water depth decreases, the troughs of the waves will tend to be overtaken by the crests, so giving rise to the asymmetry. Previously we have shown that the wave speed (c) where the depth is large compared with the amplitude, and the wavelength is long compared with the depth, is given by $c = (gD)^{\frac{1}{2}}$. Where the wave amplitude is comparable with the total depth a more exact relationship is given by $c = [g(D + \frac{3}{2}\zeta)]^{\frac{1}{2}}$ as will be shown in Section 7:7, equation (7:15); the differences between wave crest and wave trough speeds given by these more elaborate formulae enhance the above general argument for the development of asymmetry. However, a more elaborate theoretical analysis of the dynamics proceeds from equations (7:1) and (7:2); omitting the effects of friction and making the partial differentiation in equation (7:1):

$$\frac{\partial \zeta}{\partial t} + (D + \zeta)\frac{\partial u}{\partial x} + \frac{\partial \zeta}{\partial x} = 0$$

$$(7:4)$$

$$\frac{\partial u}{\partial t} + u\frac{\partial u}{\partial x} = -g\frac{\partial \zeta}{\partial x}$$

If these two equations are compared with equations (5:3), we see that a first order solution, ignoring for the moment the additional non-linear term, is given by equations (5:4) Bowden (1983):

$$\zeta_1 \approx H_0 \cos(kx - \sigma t)$$
$$u_1 \approx H_0 (g/D)^{\frac{1}{2}} \cos(kx - \sigma t)$$

Substituting these first order solutions into the non-linear extensions of equation (7:4) and retaining terms up to H_0^2, a more exact solution is obtained. For sea levels this gives:

$$\zeta = H_0 \cos(kx - \sigma t) - \tfrac{3}{4} \frac{kx\,H_0^2}{D} \sin 2(kx - \sigma t) + \cdots \qquad (7:5)$$

The second term is a harmonic of the original wave. Its amplitude increases as the inverse of the depth, linearly as the distance increases along the channel, and as the square of the amplitude of the original wave. Further refinement would lead to solutions containing higher harmonics. Relating this basic example to an input $\mathbf{M}_2$ tidal wave shows that an $\mathbf{M}_4$ component would be generated as the wave progressed into shallow water.

The curves shown in Figure 7:5 plot the sum of the two terms of equation (7:5). The peculiar features seen evolving on the back of the distorted wave develop into secondary maximum and minimum values after further propagation. Although it would be tempting to use these characteristics to account theoretically for tidal double high and double low waters, they are actually due to stopping the full solution to equations (7:4) at the second term in the theoretical analysis, rather than to the physical properties of the wave propagation (Lamb, 1932).

7:3:4 Flow curvature

The effects of variations of bottom topography and of coastal configurations on tidal currents have been discussed briefly in Section 5:4:3 for the cases of shelving beaches and embayments. The tidal currents tend to follow the contours of the coast. The forces which change the directions of current flow along the coast are given by the pressure gradients across the directions of the stream lines. In Figure 7:6 the arrows on the broken lines show the downward slope of the sea surface to give this gradient. Assuming that the sea surface well away from the coast is level, then the sea levels at the headland will be slightly lower and the levels in the bays will be slightly higher than those off-shore. When the tidal flows reverse during the second half of the cycle, the same gradients are again generated to maintain the curvature along the coast.

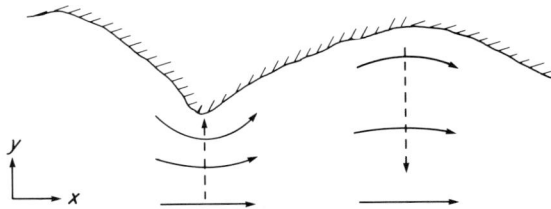

Figure 7:6 Curvature of streamlines for tidal flows near a coast (solid arrows). The surface gradients which produce the curvature are represented by the dashed arrows, which point down the slope. The situation is reversed half a tidal cycle later.

Because these gradients are in the same sense for both the flood and the ebb flows, the result is a relative depression of sea-level at the headlands twice in a tidal cycle. This would appear as an enhanced M_4 constituent in a harmonic analysis of the headland sea levels. The significance of coastal curvature for generating shallow-water waves which propagate outwards to the open sea has not been fully investigated. Some important aspects of reversing tidal flows for the generation of residual circulation will be discussed in Section 7:6.

The gradient which is necessary to produce the flow curvature can be calculated in terms of the current speed u and the radius of curvature of the flow (there is a direct analogue with the forces which keep a satellite in orbit):

$$g \frac{\partial \zeta}{\partial y} = - \frac{u^2}{r}$$

(7:6)

where r is the radius of curvature of the flow. Of course, the same result can be obtained directly by manipulation of the hydrodynamic equations of Chapter 3. For a radius of curvature of 1 km and a current speed of $1 \, \mathrm{m \, s^{-1}}$, the gradient is 0.1 m in 1 km. The effects on sea-level at the coast will depend on the distance over which this curvature of flow and the corresponding gradient are maintained. Calculations with numerical models of the flow around Portland Bill which protrudes into the English Channel show that this curvature effect is responsible for an M_4 amplitude of 0.10 m, in addition to the normal M_4 amplitude in the region (Pingree and Maddock, 1978). Because headlands are sharper coastal features than bays, the flow curvature and associated surface gradients are more severe. As a result, gauges located in broad bays will measure levels which are more representative of open-sea conditions than those located near headlands.

Curvature of streamlines near the coast also affects the coastal mean levels over a spring–neap cycle, with lower headland levels at springs. In harmonic terms this 14.8-day variation will be represented by an enhanced local M_{sf} constituent. Mean sea levels over longer periods are also affected with levels at headlands permanently lower than those in bays. The mean sea-level at the end of Portland Bill is estimated to be 0.06 m lower than the general level along the coast (Pingree, 1978).

The relative importance of the different physical processes will depend on the local coastal conditions. In the English Channel studies using numerical models (Pingree and Maddock, 1978) have shown that for M_4 the constraint of continuity of water flow is the most important distorting factor, and that bottom friction plays a relatively minor role. The curvature is important in local areas such as near to Cap de la Hague on the French coast, where the direction of the strong tidal streams is changing rapidly.

7:4 Bottom friction effects on current profiles

Tidal currents are driven locally by gradients which are established by sea surface slopes and these pressure-generated forces act uniformly through the water depth. However, at the sea-bed the opposing forces due to bottom friction act to reduce currents. The flow at the sea-bed is important for many reasons including the movement of sediment and the installation of cables and pipelines. An understanding of the influence of bottom friction on the currents at higher levels, called the *tidal current profile*, is necessary for studies of mass transport and dispersion. These profiles are a subject of active scientific research (Prandle, 1982; Soulsby, 1983). The water layer which is affected by the physical effects of the sea-bed is called the *benthic boundary layer*.

Observations of currents near the bottom are made by mounting instruments on a stable framework. Usually, tidal analysis of these currents shows that their phases are in advance of those in the main body of water. This may be explained in general terms: as the sea-level gradients reverse at low water, the existing flow continues until its momentum has been overcome by the reversed tidal pressure gradient, and by the drag of bottom friction. At the bottom the friction overcomes the momentum very shortly after the pressure gradient reverses, and flow in the reverse direction begins with only a short delay. At higher levels in the water column the reversal takes longer because the effects of bottom friction are weaker and are not felt immediately. Typically, changes of bottom currents lead those in the upper layers by 20–40 minutes.

Immediately above the sea-bed, under conditions of steady flow, there is a layer where the shear stress is constant. By similar arguments to those developed in Section 6:4:3 for surface currents generated by wind stress, this shear stress is:

$$\frac{\partial u}{\partial z} = \frac{\kappa}{z} \left(\frac{\tau_b}{\rho} \right)^{\frac{1}{2}} \tag{7:7}$$

where κ is the von Karman constant, τ_b is the bottom stress and ρ is the water density. Further, we may define a friction velocity $u_* = (\tau_b/\rho)^{\frac{1}{2}}$, and express the current speed at a level z' above the sea-bed (note that for convenience our normal sea surface datum is replaced here by a sea-bed zero level):

$$u_{z'} = \frac{u^*}{\kappa} \ln \left(\frac{z'}{z_0} \right) \tag{7:8}$$

where z_0 is a surface roughness length factor. Typical values of z_0 vary from 0.02 cm (mud) through 0.3 cm (gravels) to 0.6 cm (rippled sand). Experiment shows that the von Karman constant κ has a value close to 0.4, as is also the case for wind drag on the sea surface. Typically u_* is between 3 and 7 per cent of the undisturbed current speed. Although the constant stress layer is theoretically only about 1 m thick, the layer within which the current profile can be represented by a logarithmic expression like equation (7:8) usually extends several metres above the sea-bed. One way of estimating the mean stress on the sea-bed is to fit a logarithmic profile to a series of current measurements at different heights within this layer. However, measurements of bottom stress by other methods have shown that the stress is not steady because bottom turbulence results in intermittent bursts of stress on the sea-bed. The stress during these bursts may exceed the mean stress by a factor of 10 or occasionally by as much as 30 or more. These bursts last for perhaps 5 or 10 seconds and are repeated at intervals of 20 to 100 seconds. Observations have shown marked differences in the turbulence on the accelerating and decelerating phases of the cyclical tidal currents.

In the main body of water above the logarithmic layer, the flow gradually increases in a manner which may be represented by a power law of the form:

$$u_{z'} = u_s \left(\frac{z'}{D} \right)^{1/m} \qquad (7:9)$$

Observations suggest values of m between 5 and 7. u_s is the undistorted surface flow. The total thickness of the benthic layer may be calculated in a manner analogous to equation (6:21):

$$D_B = \pi \left(\frac{2 A_z}{\rho \omega} \right)^{\frac{1}{2}}$$

where A_z is the vertical eddy coefficient of viscosity and ω is the angular speed of the harmonic tidal constituent. Typical values of D_B, the thickness of the benthic boundary layer, lie between 50 m and 100 m, which means that the effects of the bottom friction extend to the surface over many parts of the continental shelf. Integration of equation (7:9) from the bottom to the surface gives an estimate of the relationship between the surface current and the depth averaged current $\bar{u}$:

$$\bar{u} = \frac{m}{m+1} u_s \equiv \alpha_m u_s \qquad (7:10)$$

For the experimentally determined range of values of m (5 to 7), the corresponding range for α_m is 0.83 to 0.88 respectively. Further algebra shows that the value of z' at which $u_{z'} = \bar{u}$ is close to 0.4 D. This suggests that if a single current meter is to estimate the mean tidal water transport, it should be deployed at a level 40 per cent of the total depth above the sea-bed (Muir Wood and Flemming, 1981).

These are only rough guides for estimating current profiles and the effects of bottom friction: other factors which may be important are the presence of sand waves and other substantial depth irregularities, and stratification of the water column. Within the upper layers, the effects of the earth's rotation must also be considered, but there is insufficient time for full Ekman dynamics to develop in oscillating tidal flows. One consequence of the earth's rotation is an enhancement of the anticlockwise component of current rotation (northern hemisphere) relative to the clockwise component. In some cases the normal sense of ellipse rotation (see Section 5:4:3) may be reversed from a clockwise sense at the surface to an anticlockwise sense at the sea-bed. As a general rule there is little change of current direction with depth for tidal currents in water less than 50 m deep, but as the above discussion has shown, this is not invariably the case.

7:5   Currents in channels

Strong tidal currents are often found in straits where different tidal range and time conditions prevail at the two ends. These strong currents are driven by the pressure head generated by the differences in levels acting across a short distance. This distinguishes them from the currents due to tidal wave propagation discussed in Chapter 5. Slack water with zero currents in the channel may occur at times other than those of high or low water at either end. These channels are often important for ship passage, and so an understanding of their currents is essential. Figure 5:17 shows one representation for currents in the Juan de Fuca Strait, British Columbia. Channel currents are commonly called *hydraulic currents*. Similar currents are also found through narrow or shallow entrances to bays, fjords and harbours. Some of the more spectacular examples of channel currents include flows through the Straits of Messina between Italy and Sicily, flow between the Indonesian islands and the reversing falls in St John River, New Brunswick (Forrester, 1983). Similar reversing falls, the Falls of Lora, are found at the entrance to Loch Etive, Scotland.

For these channels, which are short compared with a tidal wavelength, the hydraulic currents are essentially a balance between the pressure head and the bottom friction which opposes the flow (Wilcox, 1958). Retaining only these terms in equation (7:2) and using the quadratic friction law developed in Section 7:3:2 we have for the condition of dynamic equilibrium:

$$g \frac{\partial \zeta}{\partial x} = \frac{C_D}{D} u|u|$$

and integrating along the channel of length L:

$$u = \left\{ \frac{D_{Dg}}{C_D L} (\zeta_1 - \zeta_2) \right\}^{\frac{1}{2}} \qquad (7:11)$$

with proper regard for the initially arbitrary convention for positive and negative flow. Of course the values of ζ_1 and ζ_2, the levels at the ends, may be expressed as the sum of separate sets of harmonic constituents. The constant term within the brackets may be refined by using more elaborate friction laws and by allowing for channel depth variations (Pillsbury, 1956; Dronkers, 1964). In view of the many complicated physical processes such as boundary turbulence, channel meanderings and end effects which are incorporated, a more direct approach is to calibrate any particular channel by measuring flows over a few tidal cycles and plotting them as a function of the square root of the water head difference.

The flow through the Cape Cod Canal has been analysed and predicted in terms of the tidal level differences (Wilcox, 1958). Figure 7:7 shows the close agreement between the currents predicted and those observed over a particular 12-hour period. Analysis and prediction of currents through straits using observations of levels at the two ends avoids the need for an expensive and difficult series of measurements of the currents over a period sufficiently long

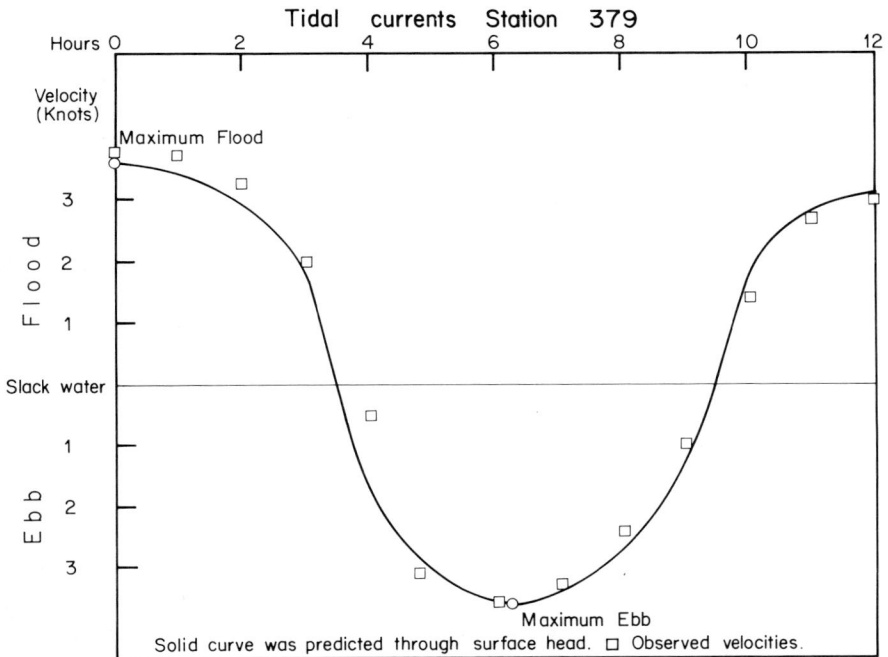

Figure 7:7 Comparison between the observed and the calculated currents in the Cape Cod canal, based on the water-level differences between the two ends. (*Reproduced by permission of the American Society of Civil Engineers.*)

for independent harmonic analysis; instead, only a few· days of current observations are required.

The very strong currents associated with channel flows sometimes give rise to strong eddies or vortices which occur at fixed times in the tidal cycle. Some, such as the Maelstrom in the channel between Moskenesoy and Mosken in northern Norway, and the Charybdis and Scilla whirlpools associated with the flow through the Straits of Messina are more celebrated in literature than they are in oceanography. The graphic descriptions given by Edgar Allan Poe of 'A descent into the Maelstrom':

> speeding dizzily round and round with a swaying and sweltering motion, and sending forth to the winds an appalling voice, half shriek, half roar, such as not even the mighty cataract of Niagara ever lifts up in its agony to Heaven

contrasts with the Admiralty Sailing Directions, which report that '... rumour has greatly exaggerated the importance of the Maelstrom', and add 'The current attains its greatest velocity, which may be estimated at about 6 knots, during westerly gales in winter'. These Directions clearly emphasize the importance of weather effects, both distant and local, for the strength of currents in channels. Contrary to popular legend, the eddy and the strait in which it develops are not bottomless: detailed surveys show a relatively shallow maximum depth of 36 m (Mooney, 1980).

Another well known example of eddies associated with straits occurs between the islands of Jura and Scarba to the west of Scotland where the tidal currents reach speeds in excess of $4 \, \text{m s}^{-1}$ at spring tides; strong eddies form in the Gulf of Corryvreckan (Gaelic: cauldron of the spotted seas) particularly when westward currents run at spring tides. Of the two legendary whirlpools associated with the flow through the Straits of Messina, only the eddy off Charybdis is now significant, but there is a third, important eddy which is generated at the entrance to Messina harbour. Because the water here is often stratified, as the Ionian water coming from the south is more saline than the Tyrrhenian Sea water to the north, the total pattern of circulation and mixing is exceedingly complicated.

7:6 Residual tidal currents

If the currents measured at an offshore station are averaged over several tidal cycles, a residual water movement is usually apparent. This mean flow past a fixed point is sometimes called the Eulerian residual circulation to distinguish it from the long-term movement in space of a body of water or of a drifting buoy (the Lagrangian residual). Residual circulation may be driven by density gradients, by wind stress or by tidal movements. Residual circulation requires detailed mathematical analysis for a full treatment, and as such is beyond the scope of this book (Robinson, 1983; Pingree and Maddock, 1985). Instead we will confine ourselves to a description of some of the main features of the tidally driven residual flows, which have been confirmed by observation and by detailed numerical models.

Tidally driven residual currents are usually one or two orders of magnitude less than the tidal currents themselves (typically a few centimetres per second compared with tidal currents of perhaps a metre per second), but they are important because their persistence may allow them to dominate the overall distribution and transport of characteristic water properties such as temperature and salinity. Essentially they are generated by interaction of the tidal currents with coastal features and with bottom topography. In the simplest case of a progressive tidal wave which has an associated current which is a simple harmonic, $U_0 \cos \omega t$, the transport at $t = 0$ in the direction of wave propagation is $U_0 (D + H_0)$ per metre of section, where H_0 is the wave amplitude. At low water the reduced return flow is $- U_0 (D - H_0)$. Over a tidal cycle there is a net transport of water in the direction of wave propagation. In more complicated cases the net transport depends on the phase relationship between currents and elevations in the two components of tidal flow.

Tidal residuals associated with coastal features and bottom topography can be described in terms of the vorticity or rotational qualities of a body of water.

Like momentum and energy, vorticity is a physical quantity which is conserved for any water mass in the absence of external forces. To illustrate the significance of vorticity, consider a uniform current flowing parallel to the coast (Figure 7:8(a)) which encounters a headland. The currents near the end of the headland will be much stronger than those further offshore, and will be opposed by a much stronger frictional resistance as a consequence, particularly if the water shallows near shore. This increased friction imparts an anticlockwise vorticity to the current and this vorticity is then advected with the water to the bay behind the headland. Further away from the headland the parallel flow is probably re-established, but in the bay an anticlockwise eddy will persist for a large part of the tidal cycle. Similarly when the tide turns (Figure 7:8(b)) a clockwise eddy will be established by the reversed currents in the bay on the other side of the headland because of the clockwise vorticity imparted to the water as it again passes the headland. Averaged over a tidal cycle there is a net clockwise flow in the bay to the west of the headland, and a net anticlockwise flow in the bay to the east (Mardell and Pingree, 1981). Observations of currents around Portland Bill in the English Channel (Figure 7:9) show this pattern of residual flows along the coast towards the tip of the promontory on both sides. This kind of persistent coastal residual has important implications for the design and location of sewage discharge systems, radioactive or other waste disposal, and as we discuss in section 10:3, for the movement of coastal sediment.

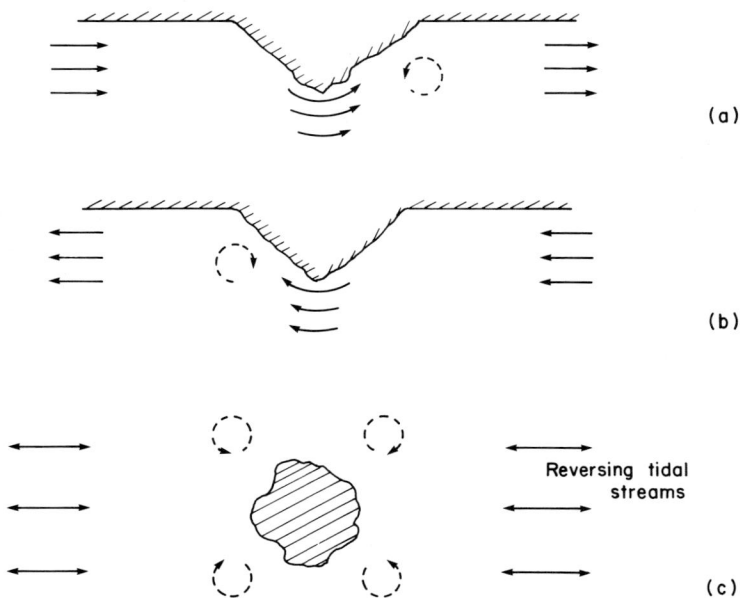

Figure 7:8 Eddies produced by headlands and islands which impede tidal currents: (a) residual eddy generated by a headland, (b) the corresponding eddy generated when the stream reverses, (c) the pattern of four eddies generated by an island in a tidal stream.

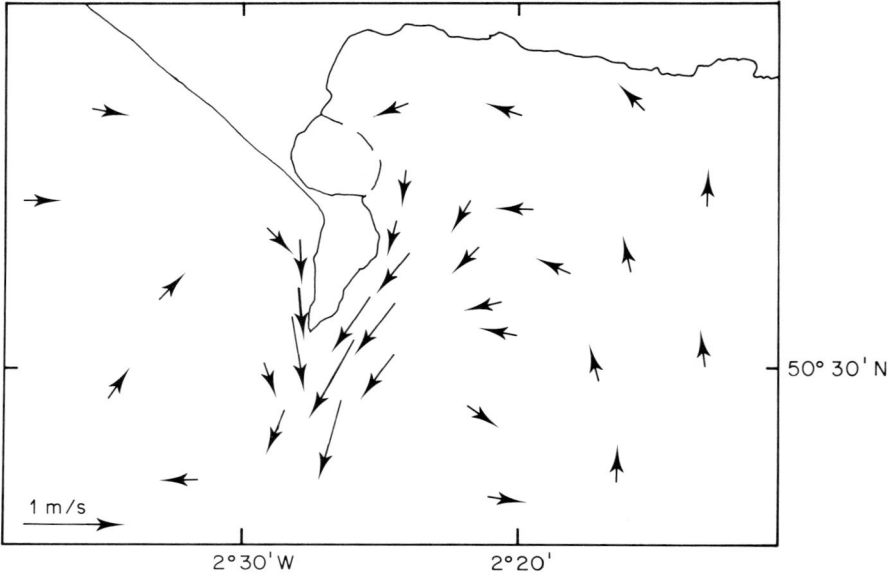

Figure 7:9 Observed headland and island eddies (a) tidal mean of hourly stream values from observations off Portland Bill in the English Channel (from Robinson, 1983). (*Reproduced by permission of Elsevier.*)

In the case of the flow past islands it is possible for four eddies to form, as shown in Figure 7:8(c). It is often a matter of local observation and knowledge that circulation in bays and around islands is directed for several more hours in one direction than in the other through a tidal cycle. This too is because of locally generated coastal vorticity. In these circumstances effective navigation is a matter of expert local knowledge, which is far more valuable than the dogma of tidal scientists!

Tidal flows also induce residual circulation around sandbanks because of the depth variations (Howarth and Huthnance, 1984; Pattiaratchi *et al.*, 1986). In the northern hemisphere this circulation is observed to be in a clockwise sense. It may be explained in terms of vorticity generation and friction effects. A typical series of sandbanks is situated off the Norfolk coast in the North Sea: their length is more than 40 km whereas their width is about 2 km, their separation averages 9 km, and they rise to within 10 m of the mean sea-level surface in water of average depth 40 m.

Bottom friction can explain a clockwise residual circulation if the axis of the sandbank is aligned to the left of the axis of the tidal streams (Robinson, 1983) (Figure 7:10). Clockwise vorticity is induced as water travels onto the bank because of the greater drag of the frictional resistance in shallow water. An anticlockwise vorticity is given to water leaving the bank, which enhances the net clockwise circulation around it before being advected away. Of course, alignment in the opposite sense would induce anticlockwise circulation, but this case is not usual for reasons which probably relate to the additional effects of vorticity due to the earth's rotation.

The influence of the earth's rotation is due to the fact that any body of water has a vorticity by virtue of its position on the rotating earth. This vorticity is zero at the equator and a maximum at high latitudes. In the northern hemisphere this vorticity is anticlockwise. A mass of water which is advected over a sandbank spreads out and covers a larger area. For the potential vorticity of this mass of water to be conserved, the anticlockwise vorticity relative to the earth must be reduced; this reduction is observed as an increase in the clockwise vorticity of the water over the sandbank, which generates clockwise residual circulation. It may be supposed that the stability of the sandbank is enhanced when the frictional and earth vorticity effects are acting together to produce a strong clockwise circulation around them. This argument suggests that in the southern hemisphere sandbanks will have their major axis rotated to the right of the tidal stream axis.

The relationship between tidal currents and sediment dynamics is discussed further in Chapter 10; the interaction, which can be very complicated, is an active subject for scientific research (Dyer, 1986).

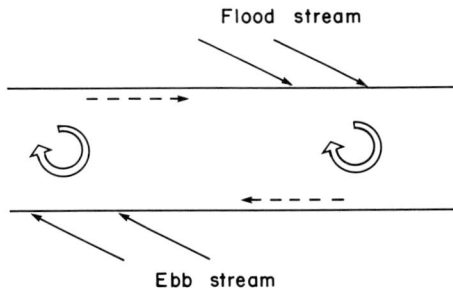

Flood stream

Ebb stream

Figure 7:10 Showing how clockwise residual circulation (— — →) is produced by bottom friction when a sandbank axis is aligned to the left of the axis of the tidal streams. This is the usual case for tidal sandbanks in the northern hemisphere.

7:7 Tides in rivers, bores

Many important coastal towns are situated on estuaries or towards the highest navigable point of major rivers. Although the distortions of the tides which occur in these narrowing and shallowing regions are both unusual and extreme, they are of considerable practical interest. The ultimate and most spectacular distortion of the tide is the generation of a tidal bore, but there are other important distortions and constraints which are observed before this is reached (Abbott, 1956; LeBlond, 1978; Lynch, 1982).

The gradual narrowing of an estuary tends to increase the tidal amplitudes; a progressive wave conveys energy at a rate proportional to the square of the amplitude (Appendix 4) and proportional to the width of the wavefront. If energy is being transmitted at a constant rate to an area of dissipation, as in the upper reaches of an estuary, then a reduction of the width of the wavefront must result in an increase in the wave amplitude:

$$H_0^2 \propto \frac{1}{\text{width}}$$

Of course, this relationship will not hold strictly because energy dissipation is not confined to the upper reaches, and will occur to some extent along the whole length of an estuary.

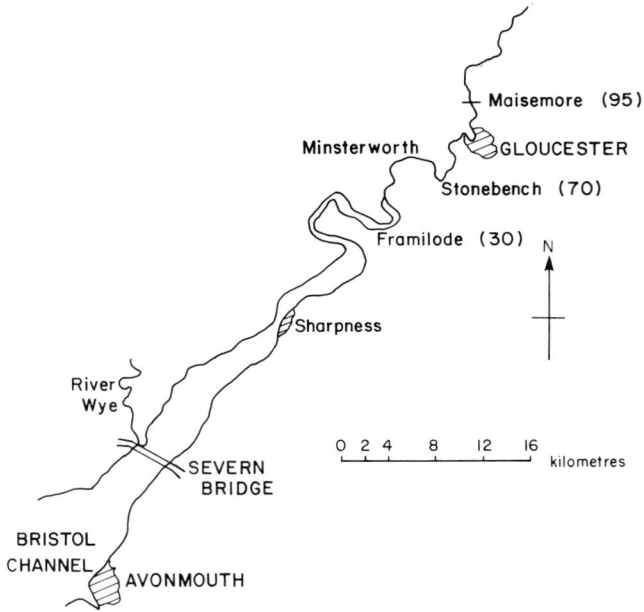

(a)

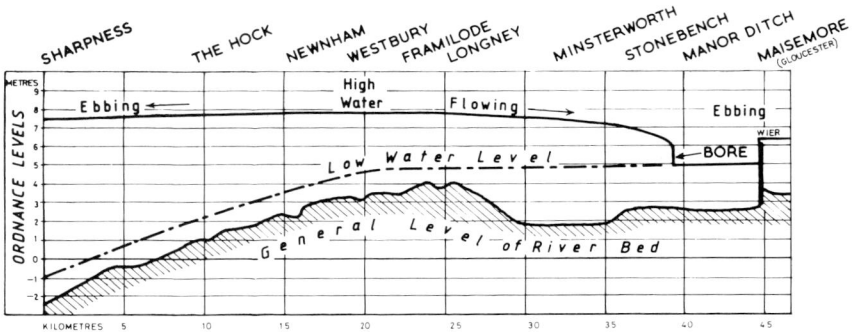

(b)

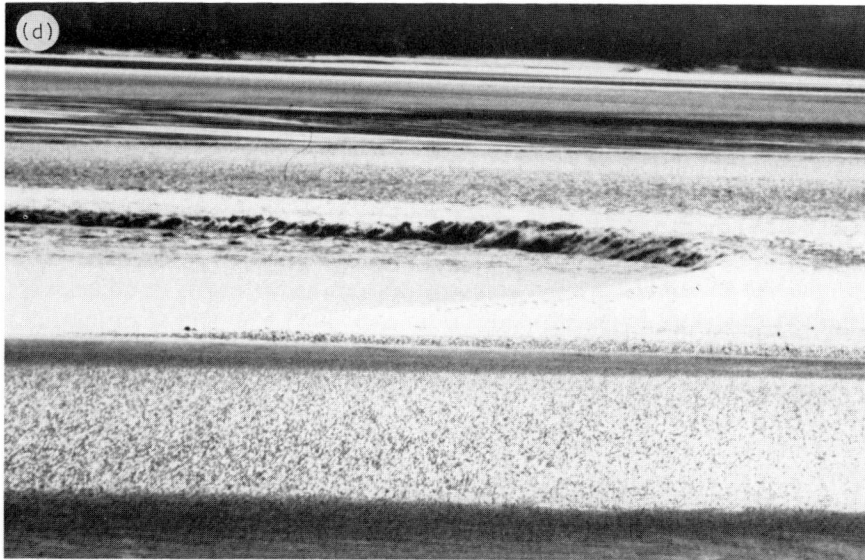

Figure 7:11 (a) The lower reaches of the River Severn showing the region where the bore develops. The figures in brackets are the approximate arrival times of the bore in minutes after high water at Avonmouth. (b) Profile of a tide of 9 m on the Sharpness gauge, showing the ebb and flow situation as the bore is approaching Gloucester. (From Rowbotham, 1983; Copyright © F. W. Rowbotham) (c) A historic view of the bore as it approaches Gloucester, September, 1921. (From Rowbotham, 1983. Copyright © F. W. Rowbotham.) (d) The bore on Turnagain Arm, off Cook Inlet, Alaska, 21 August 1986. (Photograph by J. Eric Jones.)

Frictional resistance to flow becomes the dominant physical process in the upper reaches of estuaries, over most of the tidal cycle. The times taken for the water to drain away under gravity after high water become longer than the tidal periods themselves with the result that long-term pumping-up of mean water levels occurs through the period of spring tides. In contrast to normal coastal locations, where the lowest levels occur at low water on spring tides, in the upper reaches of shallow rivers the lowest levels occur at neap tides. In the St Lawrence River in Canada, this happens upstream of Cap de la Roche. Similar low-water effects are found in the upper reaches of the Severn Estuary in Britain, and in the Seine in France. In terms of harmonic analysis the variations are represented by a large amplitude for the M_{sf} constituent with a maximum constituent level at springs and a minimum level at neaps (LeBlond, 1979).

The most spectacular form of tidal distortion, the tidal bore, is seen in only a few rivers. The basic requirement for its appearance is a large tidal range, but this condition alone is not sufficient. The main characteristic of a bore is the very rapid rise in level as its front advances past an observer; from the river bank it often appears as a breaking wall of water a metre or so high which advances upstream at speeds up to 20 km h^{-1} or greater.

Bores are known by many other names in the regions where they occur. The bore in the Amazon is called the pororoca, in the French Seine, mascaret, and in the English River Trent, aegir or eagre. The famous bore in the Tsientang river in China is reported to have a height in excess of 3 m and to advance at speeds in excess of 25 km h^{-1}. Other less spectacular bores are found in the Petitcodiac and the Salmon rivers which flow into the Bay of Fundy (Forrester, 1983), in the Hooghly River in India, the River Indus in Pakistan, the Colorado River in Mexico, the Turnagain Arm, Cook Inlet, Alaska, and the Victoria River, Australia.

The bore in the River Severn which occurs on about 130 days per year will be described in some detail (Rowbotham, 1983) because it shows the characteristics and the variability of many other bores. In the Severn as in other rivers, the largest bores are found around the times of high equinoctial spring tides; local opinion sometimes supposes that the largest bores occur at Easter, and this is often so because Easter Day is defined in the Christian calendar as the first Sunday after the full moon which happens upon or next after 21 March. Figure 7:11(a) shows the length of the river from where the bore begins to its upper limit. Between the Severn Bridge and Sharpness there may be intermittent suggestions that a bore is developing as a thin white line along a wavefront a few centimetres high, but these lines are seldom sustained. Above Sharpness on spring tides the bore establishes itself and moves upstream at speeds of about 10 km h^{-1}. Between Framilode and Stonebench it reaches its maximum height, typically more than 1 m and exceptionally in excess of 2 m in midstream. Greater heights are observed at the banks particularly on outside bends, and along these banks, the bore breaks continuously. Along this length the speed increases to perhaps 20 km h^{-1}. Thereafter the bore loses height and speed until its character is finally lost at Maisemore Weir. The exact character

of the Severn bore depends on the height of the high-water level which drives it:
a low barometric pressure and a strong southwest wind which give a positive
surge on top of a spring tide are most effective for producing big bores. The
volume of river discharge is also critical; high river flows, associated with levels
0.6 m higher than normal at Gloucester produce the biggest effects. However,
very high rates of river flow tend to reduce the size of the bore, while increasing
its speed, because of the greater river depth. Careful analysis sometimes shows
that the bore is slightly bigger for the tides immediately before the maximum
spring tide, presumably because there is still some component of residual tidal
flow in the subsequent river discharge, corresponding to the pumping and $\mathbf{M}_{sf}$
enhancement described above. The normal shape for the bore front is a steep
but not breaking wave; this is followed by a steady increase of several metres in
the water level. The water has already begun to ebb in the lower river before the
bore reaches the weir at Maisemore (Figure 7:11(b)). Only when conditions of
river discharge and sea levels are extremely favourable does the bore travel
upstream as a simple breaking wall of water across the whole width of the river.
Locally, changes in the morphology of the river banks change the character of
the bore over periods of several years and even from tide to tide. Figure 7:11(c)
shows a view of the bore taken in September 1921, as it approaches Gloucester.
Figure 7:11(d) shows a view of the bore in Turnagain Arm off Cook Inlet,
Alaska.

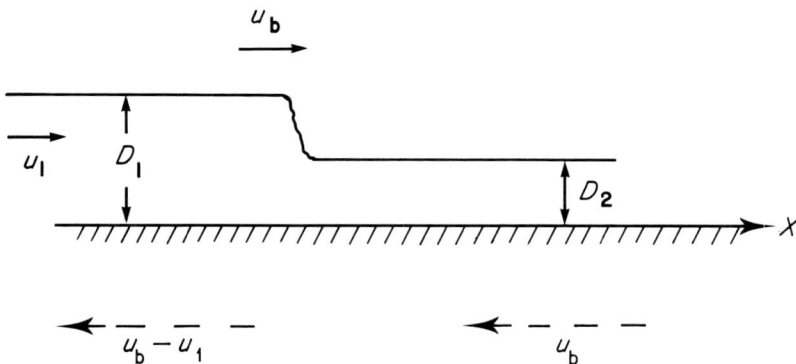

Figure 7:12   Basic configuration for calculation of the approximate speed of a bore. The bore is
travelling upsteam in the positive x direction.

The speed of advance of the wavefront of a bore may be calculated using the
following approximate theory (Lamb, 1932; Tricker, 1964). Suppose that the
bore is travelling upstream in the positive X direction with a speed u_b (Figure
7:12). Before the arrival of the bore, the water in the river has negligible speed
and the total depth is D_2. Behind the bore the water speed is u_1 upstream, and
the total depth is D_1. In order to apply the conditions of continuity of flow and
of conservation of particle energy on either side of the bore, we 'freeze' the

motion by imposing an opposite velocity of $-u_b$ on the process. For continuity, the rate of flow of water through sections across the channel ahead of and behind the bore must be equal:

$$(u_b - u_1)\, D_1 = u_b\, D_2 \tag{7:12}$$

Similarly, assuming no energy loss, and equating the kinetic energy of a particle ahead of the bore with the sum of the kinetic energy and potential energy behind the bore:

$$\tfrac{1}{2}(u_b - u_1)^2 + g(D_1 - D_2) = \tfrac{1}{2}u_b^2 \tag{7:13}$$

Expanding (7:13) and substituting from (7:12) for u_1:

$$g(D_1 - D_2) = \frac{u_b^2(D_1 - D_2)}{D_1} - \tfrac{1}{2}\frac{u_b^2(D_1 - D_2)^2}{D_1^2}$$

which yields, after some rearranging:

$$u_b^2 = \frac{2gD_1^2}{(D_1 + D_2)} \tag{7:14}$$

Replacing D_1 by $D_2 + h_b$, where h_b is the height of the bore:

$$u_b^2 = \frac{2g\,(D_2 + h_b)^2}{(2D_2 + h_b)}$$

In the case of $h_b \ll D_2$ the bore velocity is $(gD_2)^{\frac{1}{2}}$. If $h_b \gg D_2$ the theoretical velocity is $(2gh_b)^{\frac{1}{2}}$. In the more usual case of a bore height h_b less than but not negligible compared with the river depth D_2, we may expand this expression, eliminating terms in $(h_b/D_2)^2$:

$$u_b^2 = g(D_2 + \tfrac{3}{2}h_b) \tag{7:15}$$

The bore speed is greater than the speed of a wave travelling in water of depth D_1. For a bore of height 1 m travelling into water of 3 m depth, the formula gives a speed of 6.6 m s^{-1} or 24 km h^{-1}. If the river depth is reduced to 2 m the speed reduces to 21 km h^{-1}.

Hydraulic engineers have studied bore-type movement along open channels for irrigation and other applications, and their conclusions may be applied to tidal bores. Essentially the bore travels as a solitary wave, and is analogous to a sonic boom where an air pressure disturbance is forced to travel faster than the speed of sound. The character of a bore may be described in terms of the hydraulic dimensionless Froude number, the ratio of the speed of a body of water to the speed of a free wave travelling through the water (Henderson, 1964). The Froude number is related to the water depths in front of and behind the bore:

$$F_R^2 = \tfrac{1}{2}\frac{D_1}{D_2}\left(\frac{D_1}{D_2} + 1\right)$$

The larger the ratio D_1/D_2, and the greater the corresponding Froude number, the stronger is the hydraulic jump of the bore, and the more spectacular its behaviour.

If $F_R > 1$ the flow is *supercritical* but where $F_R < 1.7$ the energy difference is too small for a proper bore to develop: instead there is a train of standing waves, with some backward energy radiation downstream by the waves. In this regime the roughness of the river bed is an important factor, with a breaking wave more likely to occur when the bottom is too smooth to absorb energy by friction losses. For the range $1.7 < F_R < 2.5$ the front breaks, but is relatively quiet. Within the range $2.5 < F_R < 4.5$ the upstream flow penetrates the turbulent front as a jet oscillating between the river bed and the surface; surface waves of large amplitude, which persist for a considerable distance downstream are also generated. Between 4.5 and 9.0 the jump is strong, and free from waves because of the effective dissipation of energy; the bore moves in a jerky way with high-speed jets shooting ahead of the main turbulent front. Across a river the effective Froude number will be greatest at the banks and less for the deeper central parts, so that the bore at the river bank will be higher and stronger than the bore at the centre, where a more undular wave character develops.

Formula (7:15) for the bore speed up a river is only approximate because it makes several assumptions, including the assumption that there is no energy loss within the bore. In fact, there must be significant energy losses due to the turbulence and to the bottom friction. This energy loss increases approximately as $(D_1 - D_2)^3$. As energy is dissipated the speed of the bore is reduced until at an advanced stage of decay it may eventually be stopped and even carried back downstream by the river flow. Bores are essentially due to instabilities in the hydraulics of tidal flow, and are therefore very sensitive to changes in river morphology. Increasing the river depth by dredging, and speeding the river flow by building embankments both reduce the chances of a bore developing.

While there are reasons for supposing that the bore in Tsientang, China has become more spectacular over the centuries, in other rivers there have been deliberate attempts to reduce bores, because of the damage which they can cause to river banks. The pattern of silting and dredging at the entrance to the River Seine has reduced the size of the Mascaret. Similarly, the burro, the bore in the Colorado river, has been progressively reduced by siltation and land drainage schemes.

7:8   Tide–surge interaction

The shallow-water dynamic processes which cause interaction between the different tidal constituents, producing higher and lower harmonics, also cause tidal and surge components of the sea levels and currents to interact. Suppose, for example, that there is a process which depends on the square of the total sea-level:

$$\zeta^2 = (\mathbf{T} + \mathbf{S})^2 = \mathbf{T}^2 + \mathbf{S}^2 + 2\mathbf{TS}$$

The **TS** term represents the interaction between tides and surges. In practice the interactions are difficult to describe in terms of analytical models, and the full details are best calculated by numerical models.

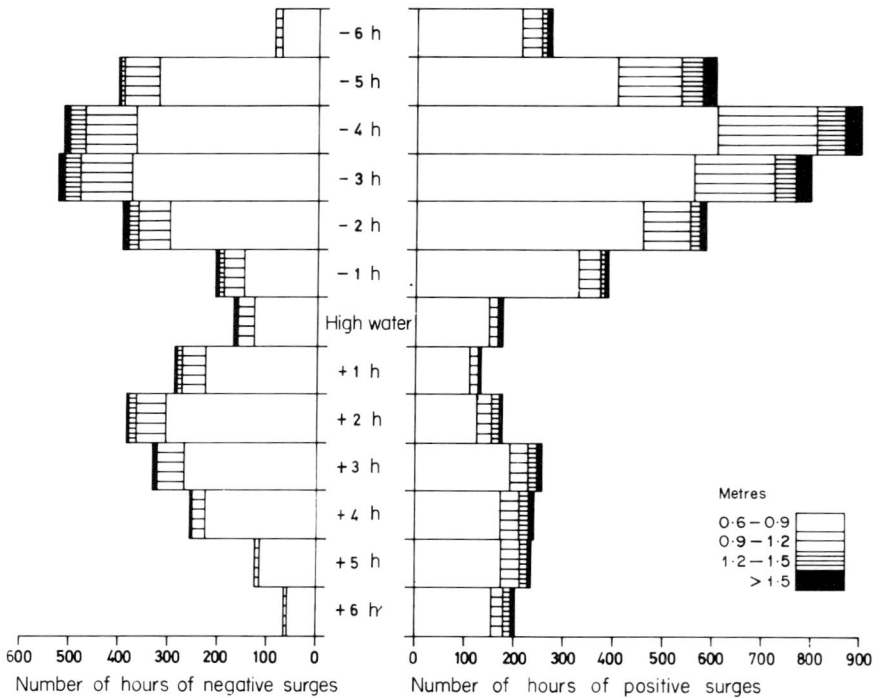

Figure 7:13   The distribution of large positive and large negative surges at Southend relative to the time of high water 1929–1969, showing that tide–surge interaction causes large surges to avoid the times of tidal high water.

Tide–surge interaction is very important because it is most apparent in shallow-water areas where large surges may be generated. Perhaps the best known example is found in the southern North Sea and River Thames (Prandle and Wolf, 1978), where the pattern of the interaction causes positive surge peaks to avoid times of tidal high water (see Figure 6:11). Positive surges are most likely to occur on the rising tide (Figure 7:13). Negative surge peaks also tend to avoid high water, but their probability distribution has peaks on both the rising and falling tide. Application of the joint-probability techniques described in Chapter 8, first with independent and then with tidally-dependent surge statistics shows that the interaction reduces the 100-year return level at

Southend by 0.5 m, which is highly significant for the design of coastal defences against flooding (Pugh and Vassie, 1980).

The flood-warning system which alerts authorities to imminent flooding must allow for the tide–surge interaction. To do so, the numerical model on which forecasts are based must include both the tides and observed surges as driving inputs. If the model were driven by surges alone the forecast levels would be wrong. Numerical models have been used to investigate the processes responsible for this interaction, and in the case of the southern North Sea they show that the primary factor is the bottom friction, which increases as the square of the current speed. This factor is dominant in the southern North Sea because of the strong currents.

There are several different ways of presenting the tide and surge statistics to represent the interaction. Figure 7:13 shows one example, where surge amplitudes are plotted against tidal phase. An alternative way of presenting the data is to calculate the standard deviation of the surges for different tidal levels; at Southend the standard deviation of the surges at mid-tide is 0.27 m compared with values of 0.19 m and 0.18 m at mean high-water and mean low-water spring tidal levels. Another form of presentation which is adequate for a first scrutiny of data is to plot each observed surge level against the corresponding tidal level: any variation of the surge distribution with the tidal height is an indication of interaction.

In Section 4:6, we discussed the variability from month to month of the tidal constituents, and showed that there is usually a consistent regional pattern in these variations. Shallow-water interactions between tides and surges are likely to be responsible for these irregular variations, as a result of more energy being lost from tides and surges travelling together, than would be lost if they were travelling separately (Amin, 1982). There is a seasonal modulation of the M_2 constituent amplitude which varies between 1 and 2 per cent in the North Sea (Pugh and Vassie, 1976). Part of these modulations is directly due to astronomical effects such as the influence of the sun on the lunar orbit, but the rest of the variability is probably due to tide–surge interaction. Because surges are bigger in winter, the winter amplitudes of M_2 should be less than the amplitudes obtained by analysis of data collected during the quieter summer months, and the observations are consistent with this expected behaviour.

7:9   Energy budgets

The energy lost by the tides through bottom friction in shallow water is converted to turbulence, and eventually generates a small amount of heat. The original source of this energy is the dynamics of the earth–moon system and, over geological time, gradual but fundamental changes have occurred in the system because of these steady losses of energy (Goldreich, 1972; Brosche and Sundermann, 1982; Webb, 1982a, 1982b).

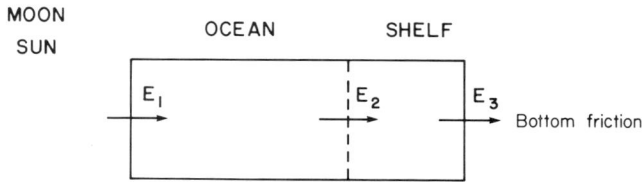

Figure 7:14 Scheme of energy flux from the astronomical forcing of the oceans, to the continental shelves, and eventual dissipation by bottom friction in shallow water.

In Chapter 5 we discussed the development of tidal oscillations in the major ocean basins and their propagation into the surrounding shallow shelf regions. The tidal energy also propagates in the same sense, from the astronomical forcing of the oceans, through the oceans–shelf margin and shelf seas, to the shallow-water dissipation. Figure 7:14 shows a basic scheme for the energy flux: over a sufficiently long period the fluxes, E_1, E_2, E_3, through each interface should be equal (Cartwright, 1977). However, there are some fundamental scientific problems associated with balancing these energy budgets which are discussed in Section 7:9:4. First we consider the nature of the tidal energy processes at each stage, beginning with the processes of local dissipation.

7:9:1 Local dissipation

In equation (7:3) the stress per unit area due to bottom friction is related to the square of the current speed; since the work done is equal to the product of force and the distance moved by the force,

$$\text{Rate of doing work} = \tau_b q = C_D \rho q^2 |q|$$

For a harmonically varying tidal current $U_0 \cos \omega t$,

$$\text{Energy dissipation} = \text{rate of doing work} = C_D \rho U_0^3 \cos^2 \omega t |\cos \omega t|$$

which averages over a tidal cycle to:

$$\frac{4}{3\pi} C_D \rho U_0^3 \text{ per unit area} \tag{7:16}$$

The cubic law of energy dissipation implies that tidal energy loss is a strictly localized phenomenon; as a consequence of this restriction of energy losses to a few areas, the accuracy of any estimates of dissipation, E_3, made by integrating the cubic relationship over an area of shelf will depend critically on the accurate knowledge of currents at a few places where the speeds are greatest. Numerical models are the most effective method of making these estimates, but there is a further difficulty in their application, because the energy lost by combinations of harmonic constituents is substantially greater than the energy which would be lost if they propagated independently.

If we consider the average energy losses through a spring–neap tidal cycle, where the combined M_2 and S_2 currents are given by the appropriate form of equation (4:8):

$$u = \alpha' \cos(2\sigma_1 t - g_{M2} - \beta')$$

replacing the sea-level amplitudes and phases by the corresponding values for the M_2 and S_2 current constituents, assuming rectilinear currents as previously, we seek the average value of the dissipation:

$$(\alpha')^3 = (U_{M2}^2 + 2U_{M2} U_{S2} \cos \theta + U_{S2}^2)^{\frac{3}{2}}$$

over a cycle of θ. Expanding as a binomial in powers of the ratio $U_{S2}/U_{M2} \equiv \gamma$, and neglecting terms in γ^6 and higher powers, this averaged value (Jeffreys, 1976) is:

$$U_{M2}^3 \left(1 + \tfrac{9}{4} \gamma^2 + \tfrac{9}{64} \gamma^4\right)$$

For the Equilibrium Tide the ratio $\gamma = 0.46$, which implies that the average energy dissipated over a spring–neap cycle is 1.48 times that dissipated by M_2 alone. Around the north-west European shelf, where $\gamma = 0.33$ the average dissipation over a spring–neap cycle is 1.25 times the M_2 dissipation. In the Equilibrium case the rate of tidal energy dissipation at springs is twenty times the rate of dissipation at neaps; in the case of the European shelf the spring dissipation exceeds the neap dissipation by a factor of eight.

7:9:2 Regional energy losses

The energy flux across a boundary may be calculated from a knowledge of the tidal changes of sea-level, and of the tidal currents. Details of the development of the formula are given in Appendix 4. This shows that for a unit width of section over a tidal cycle:

$$\text{Average energy flux} = \tfrac{1}{2}\rho g D H_0 U_0 \cos (g_{\zeta 0} - g_{u0}) \qquad \text{(A4:8)}$$

where ρ is the water density, γ is gravitational acceleration, D is the water depth, and $(H_0, g_{\zeta 0})$ and (U_0, g_{u0}) are the amplitudes and phases of the elevations and currents. For a progressive wave the currents and elevations are in phase and the currents are a maximum in the positive direction at the same time as high water. As a result there is maximum energy flux in the direction of wave propagation. For a standing wave there is no net energy flux because the currents and elevations are 90° out of phase, with maximum currents coinciding with the mid-tidal level. In the real world a perfect standing wave cannot develop because there must be some energy lost at the reflection and this energy must be supplied by a progressive wave component.

The energy transmitted across the boundary of a shallow sea, E_2, may be calculated (Taylor, 1919) by integrating the energy fluxes given by equation (A4:8) either using observed currents and elevations or currents and elevations computed by numerical models. Clearly if numerical models represent the currents and elevations over an area correctly, they must include the correct distribution of the areas of tidal energy dissipation. Calculation of the total energy flux due to combinations of all the tidal harmonic constituents is very easy: they may be added arithmetically in the same way as the variance at several frequencies in a time series may be added (see Section 1:5) (Pugh and Vassie, 1976). In general terms the energy flux is proportional to the square of the constituent amplitudes, so that the S_2 energy flux is only about 10 per cent of the M_2 flux, and the energy flux due to other constituents is negligible.

Numerical calculations (Flather, 1976) of energy fluxes onto the north-west European continental shelf show that it is an area of substantial dissipation, as we would expect from the large tidal ranges and the strong currents. The flux of tidal energy from the Atlantic Ocean into the Celtic Sea is more than three times the flux into the North Sea. The major areas of energy dissipation are in the Irish Sea, and in the English Channel. The energy flux through the Dover Straits represents one-quarter of the total energy flux into the North Sea; most energy in this sea is dissipated in the shallow southern area and in the German Bight. Only small energy fluxes are observed through the North Channel and the Skaggerak. Over this whole shelf area the input energy flux E_2 is estimated to be 200 Gw (see also Bokuniewicz and Gordon, 1980).

Energy losses in shallow seas result in a systematic adjustment of the tidal patterns (Pugh, 1981). Consider the case of a standing wave on a rotating earth which produces an amphidromic system (Section 5:2:3) as illustrated in Figure 5:6. If the wave is not totally reflected at the head of a channel, the returning wave will have a smaller amplitude. As a result, in the northern hemisphere the amphidromic point will be displaced from the centre line of the channel towards the left of the direction of the ingoing wave. A very good example of this displacement is shown by the distribution of M_2 amphidromes in the North Sea. Their displacement from the central axis (Figure 5:10(b)) which is due to energy losses, has been discussed in Section 5:4:2. If the incident and reflected Kelvin waves have amplitudes H_0 and αH_0 respectively, then it may be shown that the first amphidrome is displaced from the centre axis by a distance:

$$y = \frac{(gD)^{\frac{1}{2}} \ln \alpha}{2f} \qquad (A4:16)$$

Normally the positions of the amphidromes are plotted for individual harmonic tidal constituents. However, the amphidrome may also be considered as a time-dependent position of zero tidal range, determined on a daily basis for each tidal species, which is the D_2 harmonic discussed in Section 7:3. Conceptually there are advantages in treating dissipation as a time dependent rather than frequency dependent phenomenon because of the strongly non-linear processes involved. When the daily positions of the D_2 amphidrome at the southern entrance to the Irish Sea are plotted (Figure 7:15) they are found to move regularly back and forth within a narrow area: the range of the east–west movement exceeds 70 km, whereas the north–south movement is only 14 km. Figure 7:15 also shows the positions of the degenerate M_2, S_2, and N_2

constituent amphidromes. Within this narrow area the D_2 amphidrome moves systematically, with a period of 14.8 days. The maximum displacement from the centre coincides with spring tidal ranges, whereas the minimum displacement occurs at neaps. For neap tides the D_2 amphidrome is real, with zero tidal range occurring at a point within the Irish Sea. At spring tides the amphidrome is degenerate. The complicated tidal variations plotted in Figure 1:1(a) for Courtown are due to the mobility of this local amphidrome. A similar pattern of amphidrome displacement is observed for the daily movement of the degenerate semidiurnal amphidrome in the English Channel. The reasons for this pattern of amphidrome displacement are related to the cubic law of energy dissipation which means that much more energy is absorbed from the Kelvin wave at spring tides than is absorbed at neap tides. As a result, the reflection coefficient α for the wave is much less at spring tides, and so the amphidrome displacement (equation (A4:16)) from the centre is greater. In the case of the Irish Sea the reflection coefficient at neap tides is 0.65 but this reduces to 0.45 for average spring tides. Similarly complicated tidal patterns are to be expected near other amphidromes, in regions where energy dissipation is high.

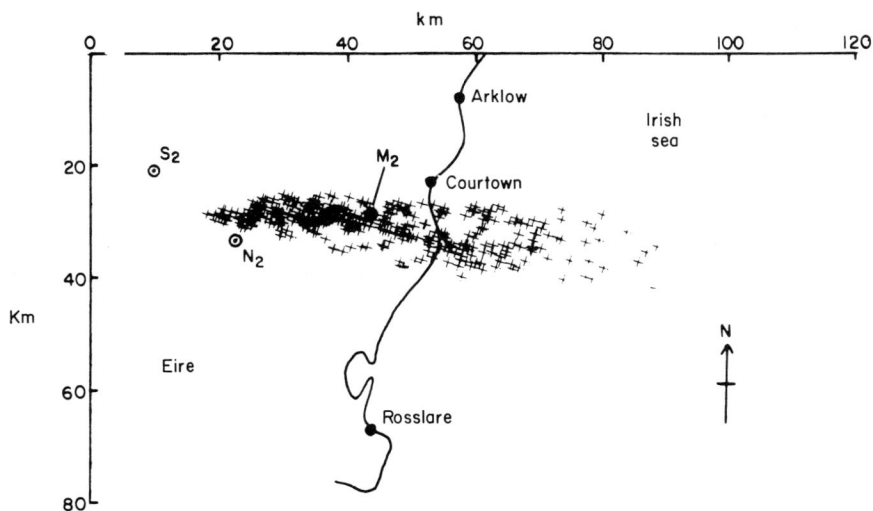

Figure 7:15   Detailed movements of the semidiurnal amphidrome at the entrance to the Irish Sea. The amphidrome is inland at spring tides. The positions of the individual M_2, S_2 and N_2 amphidromes are also shown. (*Reproduced by permission of the Royal Astronomical Society.*)

7:9:3   Global dissipation

Adding together all the calculations of energy flux across all the ocean boundaries of the continental shelf regions suggests a total flux for the M_2 tidal constituent of 1700 Gw (Miller, 1966; Cartwright, 1977). The most important areas of dissipation are the Bering Sea, the Okhotsk Sea, the seas north of Australia, the seas of the north-west European shelf, the Patagonian shelf and the Hudson Bay region of Canada. Very little energy is dissipated on the narrow continental shelves along the west coasts of North and South America, the continental shelves south of Canada, or the shelves around Africa. These total ocean-shelf flux calculations vary in accuracy and in some areas very few measurements of offshore currents and elevations exist, so the results must be treated with some caution.

In Chapter 5 the ocean tides were described in terms of the responses of a series of natural modes of ocean oscillation to the astronomical forcing. The idea of relating the difference in the phase lags for M_2 and S_2 to the energy losses was first proposed in 1823 by Thomas Young who pointed out the similarity between the ocean response to forcing, and the response of a forced linearly damped mechanical harmonic oscillator. As explained in Section 4:2:6 and equation (4:9), the phase difference $(g_{M2} - g_{S2})$ is related to the delay between the time of full or new moon and the time of maximum spring tidal ranges, the *age* of the tide. Globally this lag averages about two days. Newton thought this *age* was due to the inertia of the water, while Bernoulli thought the time taken for gravitation to reach the earth might be important. Inertia alone cannot account for a delay, because forced systems which have only inertial properties are either exactly in phase or exactly out of phase with the driving function. The energy loss affects the response of the global ocean to tidal forcing by allowing a change of phase lag with a change in the frequency of the forcing. Rapid changes of phase lag with frequency are characteristic of linearly damped mechanical oscillators close to resonance. The observed rapid changes of phase lag with frequency in the semidiurnal band are consistent with near-resonant ocean responses. This near-resonant condition is also indicated by the sensitivity of numerical models of global ocean tides to small changes in the imposed boundary conditions.

7:9:4 Astronomical energy sources

The energy losses due to tidal friction have gradually slowed down the rate of rotation of the earth on its own axis, and so increased the length of the day (Lambeck, 1980; Brosche and Sundermann, 1982). Microscopic studies of fossil corals show both yearly and daily cycles in their growth bands and careful analyses of the number of daily cycles in each yearly cycle show that around 400 million years ago there were 400 days in a year, which means that the length of the day at that time was only 22 hours. Modern astronomical observations and analyses of old eclipse data indicate increases of (2.4 ± 0.3) milliseconds per century in the length of the day. One consequence of this reduced rate of rotation is an increase in the mean earth–moon distance of about four centimetres per year, in order to conserve the total angular momentum of the earth–moon system. However, other factors in addition to tidal friction are involved in the changing length of the day. These include the gradual increase in sea-level, discussed in Section 9:5:3, which increases the moment of inertia of the rotating earth, and so decreases the rate of rotation. The rate of rotation also shows unexplained variations over periods of years:

for example, the earth's rotation speeded up during the first decade of this century. These decadal fluctuations are thought to be due to processes in the earth's fluid core.

Returning to Figure 7:14, the loss of energy from the moon–earth system, represented by E_1, may be calculated in several different ways. These include calculations based on the change in the length of the day and analyses of ancient eclipse data, which both give values of 4000 Gw, with an uncertainty of about 10 per cent (Lambeck, 1980). Another independent way of calculating the energy input to the oceans is based on the work done against Equilibrium Tidal forces by movements of the sea surface, making due allowance for the effects of earth tides. Yet another method uses the tidal torque, given by the lag of the observed average tidal bulges in the oceans with respect to the Equilibrium Tide. These direct calculations of the work done by the moon on the tides, E_1, give 3200 Gw for M_2 and 4000 Gw for the sum of all tidal constituents. These concordant estimates of E_1 are considerably in excess of the value of 1700 Gw for E_2 calculated by adding together all the estimated fluxes of energy to the individual shelf seas from the oceans (Miller, 1966; Cartwright, 1978). Estimates of E_3, based on numerical models of individual shelf seas give similar indications of low total dissipation due to tidal friction.

This discrepancy between the energy inputs calculated from the astronomical and global observations, and the much lower rate of identified dissipation in shallow seas has led to a search for other possible mechanisms for tidal energy losses. Frictional losses due to flow around ocean coral reefs, the inelastic flexing of grounded ice floes (Doake, 1978), and inelastic earth tide responses have all been considered in detail. However, none of these could be responsible for losses of sufficient magnitude to explain the difference. It is possible that energy losses due to internal tidal motions have been significantly underestimated. Also, the estimates of fluxes from the oceans onto continental shelves may be greater than estimated, because they are based on only limited data. For example, the energy fluxes onto the Argentine Shelf have been estimated on the basis of only two measurements of currents. Nevertheless, the difference between the astronomical energy inputs and the computed energy losses remains as a challenge to both experimental and theoretical tidal scientists.

The rate of energy loss from the tides may be related to the total tidal energy, both potential and kinetic, possessed at any particular time by the oceans. The total energy for the M_2 constituent has been estimated at 7×10^{17} Joules. Compared with the rate at which the moon does work on the M_2 tide, 3200 Gw, the average ocean tidal energy is replaced in $7 \times 10^{17}/3.2 \times 10^{12}$ seconds = 60 hours. In the next chapter we will discuss the generation of electrical energy by tidal power installations. Compared with such rates of global dissipation, the additional energy losses due to these artificial schemes are negligible.

A further interesting comparison may be made between the rates of tidal dissipation ($E_1 = 4000$ Gw), global energy loss due to earthquakes (300 Gw), geothermal energy losses (30 000 Gw) and average solar energy inputs (1.7×10^8 Gw).

CHAPTER 8

Tidal Engineering

Thus must we toil in other men's extremes
That know not how to remedy our own.

Thomas Kyd, *The Spanish Tragedy*.

8:1   Introduction

Tidal engineering is concerned with the implications of tidal phenomena for the design, construction and operation of maritime engineering works. Examples include coastal defences to resist the ravages of flooding, offshore structures for gas and oil extraction, and schemes for generating power from the energy of the tides themselves. The Harrapans who built the Indian tidal dockyard in Ahmedabad around 2500 BC (Section 1:1) were among the earliest exponents of the discipline, but there must have been people from prehistoric times who built simple dams for protection from the sea.

The engineering design parameters which may affect the design of marine structures include tidal changes of levels, surge effects, currents and changes of mean sea-level. A complete environmental description will also include other factors which are beyond the general scope of this account: for example, waves, winds, earthquakes, marine fouling and ice movement. It is appropriate, however, to give a brief account of the joint effects of waves and tides here.

The environmental parameters and the form in which they are presented to the design engineer will depend on the system being designed. Statistical analyses are most common, and fall into two general classes, those which deal with the normal operational conditions, and those which describe the extreme conditions which the structure must survive. The relative importance of these two aspects will depend on the type of project. Coastal defences must be high enough to protect against the highest expected levels; the normal conditions are less relevant but may have to be considered when adding amenities such as public paths. Conversely, although tidal power schemes and harbours must survive the most extreme conditions, they must also be carefully designed to work efficiently during normal operating conditions. The design of offshore structures and pipelines is concerned with both surviving the extreme conditions and with the normal conditions which control operation and maintenance work. Long-term corrosion and fatigue effects on the lifetime and safety of the structure are also influenced by both normal and extreme conditions.

Other engineering systems which work in a tidal environment include electricity power stations which depend on the sea for their cooling water, and treatment plants which discharge sewage or chemical pollutants into the coastal zone. The local coastal processes of mixing and dispersion by tidal currents are vitally important for their safety and efficient operation. Each individual scheme needs extensive measurements of the environment before installation, and may require subsequent monitoring to make sure that the dispersion is effective.

Authorities which have a statutory responsibility to give approval for new schemes issue general guidelines for the design engineer. The United States Army Coastal Engineering Research Center (1984) publishes a *Shore Protection Manual* for functional and structural design of shore protection works. Both the United Kingdom Department of Energy (Great Britain, 1984) and Det Norsk Veritas (1977) issue *Guidance Notes* and generally accepted rules for the design, inspection and construction of offshore structures. These publications are often phrased in general terms because each proposed construction is slightly different, and because the relative importance of the several environmental parameters depends on the location. It should be realized that it is very unusual to have access to the quantity (or quality) of data which many of the theoretical techniques described in this chapter require. Estimates of the statistics of the relevant tidal and other parameters must be based on only limited observations at the site proposed, so that extrapolation of the available data in both time and space is inevitable. The engineer has to decide how valid it will be to use data from another location, and the best way to make this transfer.

8:2 Normal operating conditions

The patterns of normal tidal behaviour at a site have been considered in Chapter 1. Figure 1:4 shows the frequency distribution of tidal levels at both Newlyn and Karumba, derived from hourly height predictions. Currents may be represented as curves which join the tips of hourly current vectors as shown in Figure 1:2. The common terms (Section 1:4) such as Mean High and Low Water Spring, Highest and Lowest Astronomical Tide and Mean Sea Level have all been developed over many years because of their practical use in approximately defining the tidal environment. More elaborate descriptions in terms of statistical variance are also valid, and are particularly useful for defining the frequency distribution of surges.

The best way to determine the tidal parameters is to analyse a year, or if this is not possible at least a month, of hourly observations of the levels or currents (Section 4:4). Many parameters may then be determined directly from the resulting harmonic constituents. If only a month of data is available, a tidal analysis which makes use of the known tidal characteristics at a nearby site, either as related harmonic constituents or computed through intermediate response functions, may be adequate. The most difficult parameters to estimate from only short periods of data are mean sea level and the probability distribution of the meteorological residuals, because both of these have seasonal modulations. In the case of mean sea level the monthly mean level may be adjusted by comparing the simultaneous mean level at a nearby permanent station with the long-term mean based on several years of data at that station. This difference may then be used to adjust the mean level at the secondary site: such a transfer is acceptable because seasonal changes of sea-level are similar over long distances, as shown for Newlyn and Brest in Figure 9:2. Extrapolating the statistics of meteorological residuals from one site to another is less reliable because local effects can give important regional differences.

Tidal parameters are also summarized in national tide tables. Each of the three volumes of the *Admiralty Tide Tables* contains a summary of the basic constants at each Standard Port, and supplementary tables which enable the parameters to be calculated at many Secondary Ports. The *Tide Tables* published annually by the United States National Oceanic and Atmospheric Administration contain similar details. Information on currents at selected sites is sometimes available on large-scale navigational charts, but these should be used with caution for engineering design because they are based on observations over only a few tidal cycles and because currents are more variable from place to place than are elevations. The United States Army, Corps of Engineers, has also published a useful statistical summary of *Tides and Tidal Datums in the United States* (Harris, 1981; see also Disney, 1955). Tidal characteristics around Britain are summarized on Admiralty Charts 5058 and 5059 and similar charts have been published for other regions. Several examples are given in Chapter 6.

The basic parameters obtained from the published charts and tables are adequate for most straightforward applications, but more elaborate engineering systems may require further specific details. An example might be the engineer who needs to relate elevations and currents along a complicated route for towing a structure from a fabricating yard to an offshore site during a particular month of the year. Other examples will be discussed later. Sometimes these questions may be answered by scrutiny of the harmonic constituents, but in general it is better to make observations and to generate the required information numerically from tidal predictions.

* 8:3 Extreme conditions

Extreme sea levels and currents may be specified in several different ways. For the present our discussions will concentrate on extreme high sea levels. These are sometimes called still water levels to distinguish them from the total levels which include waves (see Section 8:4). They may be parameterized as the probability that a stated extreme will be exceeded during the specified design life of a structure. An alternative is to calculate the level which has a stated probability of being exceeded during the design life. If the probability of a level η being exceeded in a single year is $Q_Y(\eta)$, the level is often said to have a return period $T(\eta)$ of $[Q_Y(\eta)]^{-1}$ years. This makes the implicit assumption that the same statistics are valid for the whole period. Sometimes the level which has a probability of being exceeded once in 100 years, for example, is called the *hundred-year return level*.

The return period may also be related to the *encounter probability* or *design risk* which is the probability that the level will be exceeded at least once during the lifetime of a structure:

$$\text{Risk} = 1 - [1 - Q_Y(\eta)]^{T_L} \tag{8:1}$$

where T_L is the design lifetime. As an example, suppose that the envisaged life of a structure is 100 years, then for a risk factor of 0.1 for exceedence during this period, equation (8:1) shows that the design level should have a probability:

$$Q_Y(\eta) = (1 - 0.9^{1/100})$$

so that
$$T(\eta) = 950 \text{ years.}$$

The appropriate value of $Q_Y(\eta)$ will depend on the value of the property at risk. Nuclear power stations may specify 10^{-5} or 10^{-6}. For the coastal protection of the Netherlands a value of 10^{-4} is adopted (Delta Committee, 1962), but for many British coastal protection schemes values of 10^{-3} or greater are accepted. It should be remembered that a structure has a probability of near 0.635 of encountering a level which has a return period equal to its design life; for acceptable risk factors the design level must have a return period which considerably exceeds the expected lifetime of the structure.

For very expensive structures the most elaborate statistical methods should be used to estimate extremes, but the reliability of even these estimates is limited by the available data. For less expensive schemes approximate methods of estimating have been developed as cheaper alternatives.

8:3:1 Regional Factors

The simplest approach is to compute the ratio between some normal tidal parameter and the level having the return period of N years for a standard port in a region, and to use this factor to scale up the known tidal parameters at the proposed site. The best known of these (Lennon, 1963b) is the *similarity factor*:

$$C_{100} = \frac{100\text{-year high-water level} - \text{Mean High Water Springs}}{\text{spring tidal range}} \tag{8:2}$$

where C_{100} is computed for the Standard Port, and the levels are referred to local mean sea-level. Figure 8:1 shows values of this similarity factor for

268

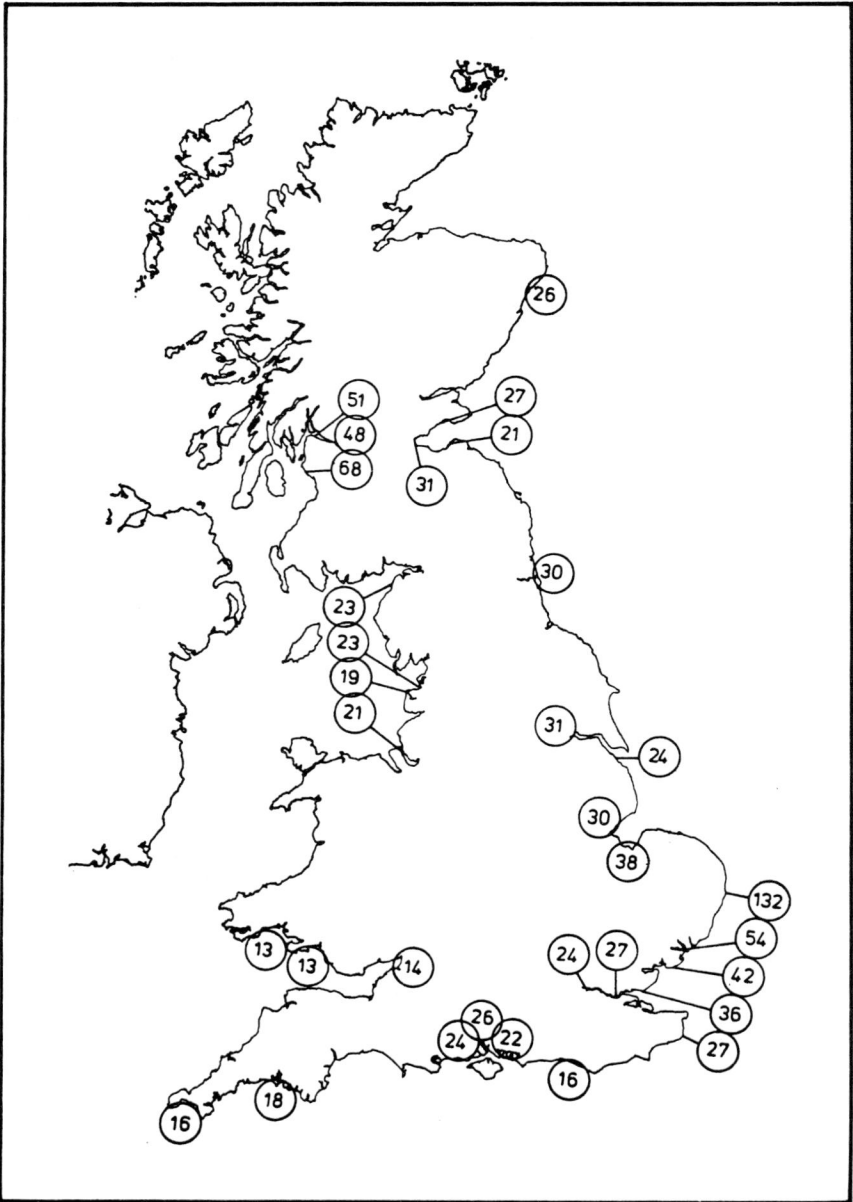

Figure 8:1 Map showing the distribution of the Similarity Factor ($\times 10^{-2}$) for the '100-year flood' level around Britain. (*Reproduced by permission of Academic Press Inc. (London) Ltd.*)

several British ports, where the majority of values lie in the range 0.15 to 0.35 (Graff, 1981). However, near tidal amphidromes in the Southern North Sea and in the area of the River Clyde on the west coast of Scotland, the small tidal range which appears in the denominator of (8:2) produces much larger factors. Larger factors are also found in areas of extensive shallow water where meteorological effects are more severe.

A variety of alternative factors may also be constructed. For example:

$$\alpha_{100} = \frac{\text{100-year high-water level}}{\text{Highest Astronomical Tide} + \text{100-year surge level}} \qquad (8:3)$$

The 100-year surge levels may be estimated from a long series of meteorological residuals with suitable extrapolation, or by analytical or numerical models which relate them to 100-year winds. In the most pessimistic case where the 100-year surge level is assumed to coincide with Highest Astronomical Tide, the value of α_{100} is 1.0. In practice the values are lower than this because extreme surges will probably occur with more normal tidal levels. Table 8:1 gives the values of α_N for five British ports. The values lie in the range 0.80 to 0.90 except for Southend where the tide-surge interaction is reflected in the lower factors because surges avoid tidal high-water levels (Section 7:8). A factor which uses Mean High Water Springs instead of Highest Astronomical Tide may also be used (Great Britain, 1984). These factors are more stable from region to region than the similarity measure, but are initially harder to estimate because the surge levels which have specified return periods are more difficult to determine.

Table 8:1 The adjustment factor α_N relating total levels having a return of N years to the sum of Highest Astronomical Tide and the surge with a return period of N years.

N (years)	20	100	250
Newlyn	0.88	0.90	0.90
Malin Head	0.81	0.82	0.83
Lerwick	0.83	0.84	0.84
Aberdeen	0.81	0.81	0.82
Southend	0.71	0.73	0.73

8:3:2 Analysis of annual extreme values

In order to determine the value of $Q_Y(\eta)$, from which return periods and risk factors may be estimated, it is necessary to tabulate the maximum values reached in as many years as possible (NERC, 1975). Seasonal cycles in extreme levels make the use of sample periods shorter than a year invalid. The annual maxima for 61 years of Newlyn data are plotted as a histogram in Figure 8:2. The level of Highest Astronomical Tide (3.0 m) was exceeded in only 28 of the years. If the annual maxima are plotted in series an 18.6-year cycle due to the nodal changes in the tidal ranges is often detectable. The broken curve in

Figure 8:2 shows the probability of a particular level being exceeded in any single year. For example, the probability of an annual maximum level exceeding 3.10 m at Newlyn is 0.18, because 11 yearly maxima in the set of 61 were higher than this. Expressed in a different way, an annual maximum in excess of 3.10 m has a return period of 5.5 years. Plots like Figure 8:2 are useful for representing the general characteristics of annual maxima, but they cannot be used for the extrapolations necessary when estimating for events which have a very low probability.

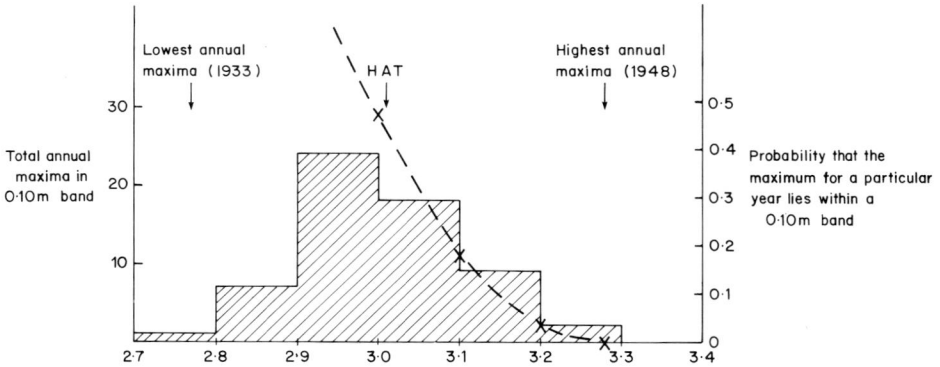

Figure 8:2 Histogram of the distribution of 61 annual maximum levels at Newlyn, 1916–1976. The broken curve gives the probability of a level being exceeded in a particular year; for example, the probability of HAT being exceeded is 0.46.

If the return period of a level is specified as $T(\eta)$ years, this implies that on average one tidal high water out of $705 \times T(\eta)$ will exceed η. This assumes that the extreme total level occurs at high tide, and makes use of the fact that there are 705 semidiurnal high waters in an average year. Hence the probability that a given tide chosen at random will fail to reach the level is $\{1 - 1/[705T(\eta)]\}$ and the probability that all the tides in a year will fail to reach the level is $\{1 - 1/[705T(\eta)]\}$. For such a large exponent the series and hence the probability approximates very accurately to $\exp[-1/T(\eta)]$. This relationship between the probability P that the annual maximum is less than η, and the return period $T(\eta)$ may be conveniently expressed in three ways:

$$P = \exp\left[-1/T(\eta)\right]$$

$$T(\eta) = -1/\ln P$$

$$\ln T(\eta) = -\ln(-\ln P)$$

where ln is the natural logarithm.

The usual procedure is to fit some theoretical curve to values of η plotted against a reduced variate $X = -\ln(-\ln p)$. Plotting the levels against X has the advantages of opening out the two ends of the probability curve ($P = 0$ and $P = 1$) relative to its central position, and of making the transformed curve approximately linear.

Table 8:2   Ranking and probability calculations for $N = 61$ Newlyn annual maximum levels (1916–1976).

η	Frequency	r	P	X
2.75	1	1	0.0082	−1.57
2.80	1	2	0.0246	−1.31
2.85	3	3–5	0.057	−1.05
2.90	9	6–14	0.156	−0.62
2.95	12	15–26	0.328	−0.11
3.00	10	27–36	0.508	0.39
3.05	5	37–41	0.631	0.78
3.10	12	42–53	0.770	1.34
3.15	5	54–58	0.910	2.36
3.20	1	59	0.959	3.17
3.25	1	60	0.975	3.68
3.30	1	61	0.992	4.80

η = level of annual maximum above Ordnance Datum Newlyn in metres.
    frequency is the number of annual maxima at level η.
r = rank of the annual maxima from 1 to 61.
P = mean value of $(2r - 1)/2N$ for each η, which is the observed frequency of annual maxima less than η.
X = reduced variate, $-\ln(-\ln P)$.

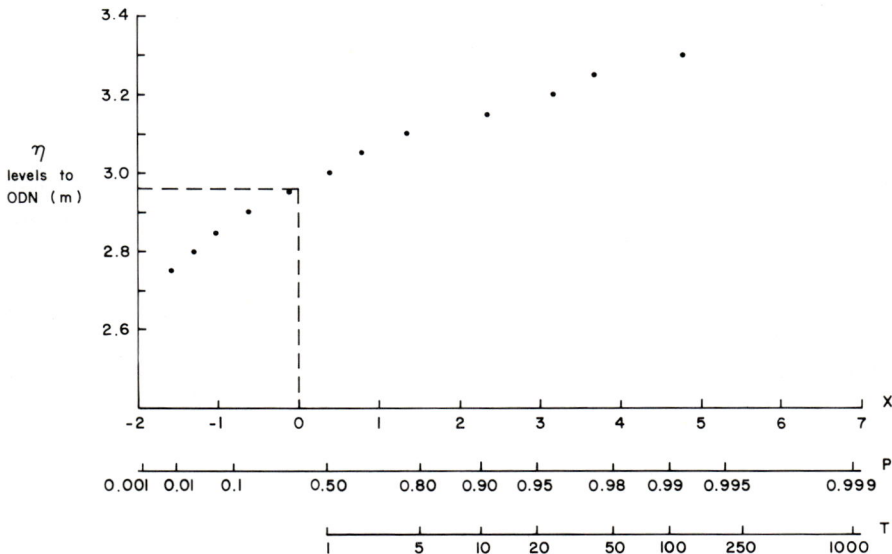

Figure 8:3   Probabilities of annual maximum levels at Newlyn falling below a specified level. T is the return period in years.

In order to plot curves as shown in Figure 8:3, values of P as a function of η are estimated from tabulations of M annual maxima. The M maxima are first arranged in ascending order of magnitude from the smallest, η_1, to the largest, η_M. Data for 61 years of Newlyn annual maxima are ranked in Table 8:2 as an example. Consider a level η_r and η_{r+1}. Of the M observed annual maxima, r are less than η, so the estimated probability of an annual maximum less than η is r/M. The probability P that the annual maximum level is less than η_1 is between $1/M$ and zero whereas the probability that the annual maximum is less than η_m lies between $(M-1)/M$ and unity. Some care is necessary when calculating probabilities of exceeding the level which corresponds to the rth value in the series: by considering values of slightly below and slightly above η_r, the true value for the probability of not *exceeding* η_r in a particular year lies between $(r-1)/M$ and r/M. The usual compromise frequency adopted is the average:

$$P = \frac{(2r - 1)}{2M}$$

The values in Table 8:2 have been calculated in this way; the reduced variate X is then plotted against the levels η in Figure 8:3, which also shows the corresponding probabilities and return periods.

The final stage of the analysis is to plot an appropriate curve through the plotted values. Plotting by hand is feasible, but theoretical curves are preferred. Many such theoretical curves have been proposed, but the most commonly used family of distributions has the form:

$$\eta = \eta_0 + a(1 - e^{-kX}) \qquad (8:4)$$

where η_0 is the value of η when $X = 0$ (2.95 in Figure 8.3) and a and k are parameters determined by the fitting process. This family of curves is known as the Generalized Extreme Value (GEV) distribution. The fit is usually obtained by the method of least-squares.

The group of possible GEV curves may be divided into three classes corresponding to different ranges of k values. In practice the value of k lies in the range -0.6 to 0.6. If k is zero the plot of X against η is a straight line, sometimes called the Gumbel distribution (Figure 8:4). More commonly the distribution is either concave or convex towards the X-axis. The distribution at Newlyn in Figure 8:3 is concave towards the X-axis; for this class the value of k is always positive (0.20 for Newlyn) and a definite upper limit to the extreme levels η is indicated. This class is the one normally found to be applicable for extreme sea levels, but in some areas, notably around the coast of East Anglia and the southern North Sea, the distribution is convex towards the X-axis corresponding to a negative value of the shape parameter k; in this case no upper limit to the extreme levels is indicated. The distributions corresponding to k negative, zero and positive are sometimes referred to as Fisher-Tippett Types 2, 1 and 3 respectively, as shown in Figure 8:4.

The extent to which extrapolation of the fitted curves to levels having very long return periods is justified has been considered by several authorities (Lennon, 1963b; Graff, 1981). As a general rule extrapolation should be limited to return periods not longer than four times the period of annual maximum levels available for analysis, but even within this limit extrapolated values should be interpreted with caution. The form of the extrapolated curve is strongly controlled by the last few points of the plotted values; it is often found that one or two extreme levels observed during the period appear to lie outside the usual distribution pattern and the degree of weight which should be given to them becomes a matter for subjective judgement. The dangers of omitting the most extreme values from an analysis are obvious.

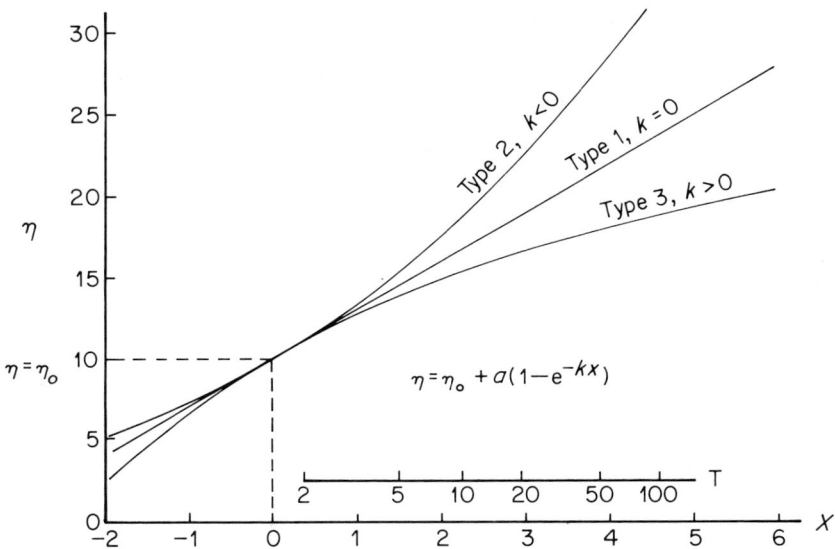

Figure 8:4   The three different forms of the Generalized Extreme Value distribution, where X is the reduced variate.

The method of ranking annual maximum levels should strictly be applied only to data in which no significant trends occur. This means that the probability of a particular annual maximum is the same at the beginning and end of the data set. However, trends, which may be due to changes in mean sea level, tidal ranges or the intensity of meteorological surges, may be removed before the ranking analysis begins. The annual maxima should be trend-adjusted to some common year; the Newlyn annual maxima which we have discussed showed no significant trend. Missing years of data should not affect the validity of the results provided that the gaps are not due to the extreme values themselves, for example because a very severe extreme level damaged the recording instrument.

Although as few as ten annual maxima have been used to compute probability curves, experience suggests that at least 25 values are needed for a satisfactory analysis. The major disadvantage of the method is the waste of data, a complete year of observations being represented by a single statistic. If the largest meteorological surge for the year coincides with a low tidal level, the information is ignored despite its obvious relevance to the problem of estimating extreme level probabilities. One possible development is to establish statistical techniques which take account of the ten highest levels in a year, or some similar group of highest values.

8:3:3 Joint tide–surge probability estimates

An alternative way of estimating probabilities of extreme levels is to make use of the separate distribution of tidal and surge probabilities (Ackers and Ruxton, 1975; Pugh and Vassie, 1980; Walden *et al.*, 1982). The most direct approach is to work with all the observed and predicted high-water levels. In this simple case the assumption is made that any extreme event will occur at a time of the predicted tidal high water; this is valid where the tidal variance greatly exceeds the surge or residual variance. A necessary secondary assumption is that the amplitudes of the residuals at high water are independent of the predicted tidal levels.

The residual and predicted high-water levels over a complete number of years are tabulated to produce normalized frequency distributions. An appropriate tabulating interval is 0.1 m. In the example shown in Table 8:3 the high water and high-water residual distributions have been artificially restricted to five 0.1 m class intervals in each case. Forty per cent of the observed residuals were in the range −0.05 m to 0.05 m, whereas 10 per cent were in the range 0.15 m to 0.25 m. Highest Astronomical Tide lies in the range 3.15 m to 3.25 m above some defined datum.

Table 8:3 Example of tidal high water and high-water residual probabilities. Class intervals are identified by the centre value. For example, a surge of 0.1 m represents values in the range 0.05 to 0.15 m.

| Predicted tidal high-water level (m) | Normalized frequency | High-water residuals | | | | |
| | | −0.2 m | −0.1 m | 0.0 m | 0.1 m | 0.2 m |
		0.1	0.2	0.4	0.2	0.1
3.2	0.1	0.01	0.02	0.04	0.02	0.01
3.1	0.2	0.02	0.04	0.08	0.04	0.02
3.0	0.3	0.03	0.06	0.12	0.06	0.03
2.9	0.3	0.03	0.06	0.12	0.06	0.03
2.8	0.1	0.01	0.02	0.04	0.02	0.01

The frequency distributions of the observations are then assumed to be representative of the probability of future events. The joint probability of a 3.2 m predicted tide and a 0.0 m surge is 0.04, the product of their individual probabilities. Similarly a 3.1 m tide and a 0.1 m residual have a joint probability of 0.04, and a 3.0 m tide with a 0.2 m residual is 0.03. Any of these three joint events will produce a total observed high-water level of 3.2 m, and so the total probability of a 3.2 m level, obtained by scanning along the dashed diagonal, is the sum of the three probabilities, 0.11. Eleven tidal high waters in 100 will lie between 3.15 m and 3.25 m. In this example the highest total level, 3.4 m can only occur when a 3.2 m tide and a 0.02 m residual coincide, which has a joint probability of 0.01 or one tide in 100.

When this method is applied to real data much smaller probabilities of extreme joint events are calculated than in the above example, because the probabilities are distributed over many more class intervals. Suppose that some particular level has a joint probability of 0.0 001 for each high-water level. For a semidiurnal tidal regime with 705 tides in each year, the return period for this level is 10 000 tidal cycles, 14.2 years. This method of estimation from high-water data has the advantage of relatively simple computation provided that high-water observations and predictions are available, but it fails to take account of large positive surges which occur at times other than those of predicted tidal high water.

A more elaborate joint probability technique which takes account of all tidal and surge information is based on the computation of complete frequency distributions for each parameter, as shown in Figures 1:4 and 6:4. Hourly tidal and residual levels are calculated and divided into 0.1 m classes.

Ideally, 18.6 years of tidal predictions should be used and residual distributions should be estimated from several complete years of observations, according to equation (6:1). The probability of joint events is computed by multiplying the separate probabilities as in Table 8:3, and the total probability of a particular observed level is obtained by summing along the appropriate diagonal of the joint probability matrix. This combination may be expressed mathematically as a convolution integral:

$$D_o(\eta) = \int_{-\infty}^{\infty} D_T(\eta - y)D_s(y)dy \tag{8:5}$$

where $D_T(\eta)$ is the probability density function for the tidal levels, where $D_s(y)$ is the probability density function for the surge residual levels, and $D_o(\eta)$ is the probability density function for total, observed levels. For this to be a valid estimate of $D_o(\eta)$ the tide and surge probabilities must be independent. However, an extension of the method which allows the surge probability density functions to vary as the tidal level changes has been applied to the case of Southend (Pugh and Vassie, 1980), where the maximum surge levels have a marked statistical tendency to avoid tidal high waters (Section 7:8).

The probability distributions of both tide and residual levels computed in this way have no units, nor do the joint probability estimates of the extremes which are plotted in Figure 8:5. These probabilities are obtained by summing all the total joint probabilities above the specified level. One advantage of the joint probability method is the automatic generation of probabilities of extremely low levels as part of the computing procedure and these are also plotted. To convert these dimensionless probabilities to return periods some time-scale has to be determined: analysis, supported by comparison with observations, shows that the inverse of the probability may be taken as the return period in hours, at least for British ports. Thus the levels corresponding to $\text{Log}_{10}P = -5.94$ have a 100-year return period.

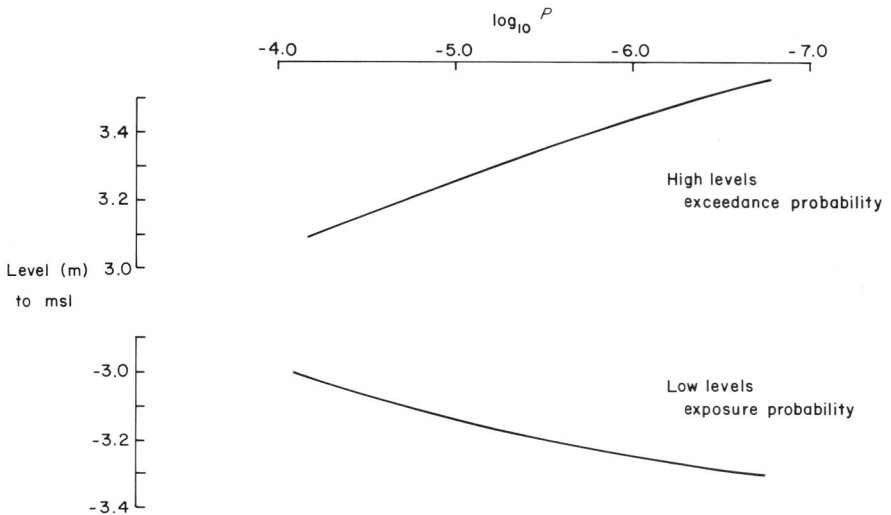

Figure 8:5 Probabilities of observed sea levels exceeding high levels or falling below low levels at Newlyn, based on 18 years of data (1951–1969), from joint-probability distributions.

The principal advantages of the joint tide–surge probability approach may be summarized:

(a) Stable values are obtained from the relatively short periods of data. Even a single year can yield useful results, but four years is a desirable minimum (Pugh and Vassie, 1980). Using the annual maxima method perhaps 25 years of data are normally required.

(b) There is no waste of data.

(c) The probabilities are not based on extrapolation.

(d) Estimates of low-water level probabilities are also produced.

(e) Separate changes in the physical factors which affect levels may be identified and incorporated. For example, projected changes in the tidal regime due to a proposed barrage construction may be incorporated by adjusting the tidal probability distribution. Trends in weather patterns may be identified and incorporated by using variable surge probability statistics.

(f) Projected changes in mean sea-level may be incorporated by simple addition.

Compared with the technique of ranking annual maxima there are some disadvantages:

(a) Data must be of a higher quality, with timing accuracy to better than a few minutes; if not, tidal variations will appear as residuals (see Figure 2:14).

(b)   Extra computational effort is involved.

(c)   Slightly higher values are obtained by the joint probability method, which may be due to weak tide–surge interaction (Section 7:8). Where the surge residual distribution is significantly dependent on the tidal level, more elaborate joint probability computations are necessary.

However elaborate the computations, the results should always be treated with caution because of the limited period of observational data, compared with the equivalent return periods being calculated. The possibility of some rare event cannot be ignored, nor is it easily incorporated into the estimates. Tsunamis are more common in the Pacific than in other oceans, but even for the Atlantic, well documented tsunamis have occurred (Section 6:8). On the west coast of the United States and in Japan, tsunamis are recognized as the most important cause of extreme levels.

On the Atlantic and Gulf Coast of the United States the extreme sea levels are produced by hurricanes, which are too rare at any particular place to permit the calculation of reliable probabilities. Some kind of modelling approach as discussed in the next section is more appropriate. The methods which we have discussed of ranking annual maxima, and of computing joint tide–surge probabilities are most effective for calculating extremes for regions outside the tropics. Other statistical techniques have been developed and applied to the estimates of probabilities of environmental extremes. Smith (1984) describes a method which analyses peaks which exceed some specified threshold. Middleton and Thompson (1986) have developed a rigorous exceedence probability approach which remains effective where surge variations dominate the tidal variations, and avoids the difficulties of relating joint probabilities to return periods.

8.3.4   Modelling from extreme winds

Observations of meteorological variables including atmospheric pressure, wind speed and direction, have been continuous at land stations over many years. Because extreme sea levels are obviously related to extreme weather conditions, one approach to estimating their probability is to apply the extreme winds and pressures estimated from land-based observations, to analytical or ideally to numerical models of the sea area of interest, as discussed in Section 6:5.

The approach is particularly useful for tropical storms, where the distribution of all extreme hurricanes, cyclones or tornados may be estimated from the historical record. The most severe storms may then be used to calculate the response of the water around the port or structure being designed. Both the direction and speed of approach of the storm can be varied in the model until critical maximum level conditions are identified. In the northern hemisphere this will be when the centre of the storm hits the coast a little to the left of the port, giving maximum on-shore winds (see Figure 6:17). In the southern hemisphere the critical incidence is to the right of the site.

Tropical storms usually generate much larger surges than extratropical ones, but their extent is much more limited. For example, the levels near Gulfport, Mississippi, generated by hurricane Camille in August 1969 were raised by more than seven metres. Such narrowly concentrated extreme levels are most unlikely to be monitored by an existing tide gauge. A further refinement is to couple a numerical model which describes a shelf response to specified central pressure, storm size and storm track, to a detailed model for a particular estuary or bay as is done with SLOSH (Section 6:5). The model may be run for several storms and then normalized in some way to estimate return periods or probabilities. The surges may be combined with the known tidal distributions, either by addition to Highest Astronomical Tide, or by computing joint probability distributions.

Similar techniques are available for extratropical storms, which generate smaller surges over wider areas. For example, Figure 8:6 shows the estimated storm surge elevation for the north-west European continental shelf, generated by using the depth-integrated numerical model discussed in Section 6:5. Theoretically, if the continuous operational output of the model over several years were analysed statistically, the extreme level probabilities could be computed for each grid element. In practice the approach was slightly different. Sets of meteorological data corresponding to 16 severe storms in the period were used as input to the model and the maximum level in each grid element, corresponding to each storm, was determined. Some skill in selecting those storms which were supposed to produce the extreme levels was necessary (Flather, 1987). At seven ports where enough data had been collected to estimate the fifty-year surge directly, the ratios between this estimate and both the maximum computed surge and the root-mean-square (r.m.s.) computed surge from the 16 storms were evaluated. These normalization factors were then smoothed from region to region and used to adjust upwards the maximum and r.m.s. computed values from the 16 storms, to estimate the fifty-year surge levels everywhere else. Figure 8:6 is based on the greater value at each site from either of the two methods.

The surge levels may then be added to the Highest Astronomical Tide level to obtain pessimistic estimates of the extreme total levels which may be expected, in the same way as discussed in Section 8:3:1. Downward adjustment of the total level by a reduction factor calculated for a nearby port (see Table 8:1) is a further possible refinement.

8:3:5 Extreme currents

Extreme currents are far more difficult to estimate than extreme levels. The first difficulty is to obtain a sufficiently long series of observations; few series extending over more than a year exist because of the expense and the technical difficulties of making good measurements. Further complications arise because currents are variable with depth at each location, and because they change over short distances, particularly near the shore and around shallow sandbanks.

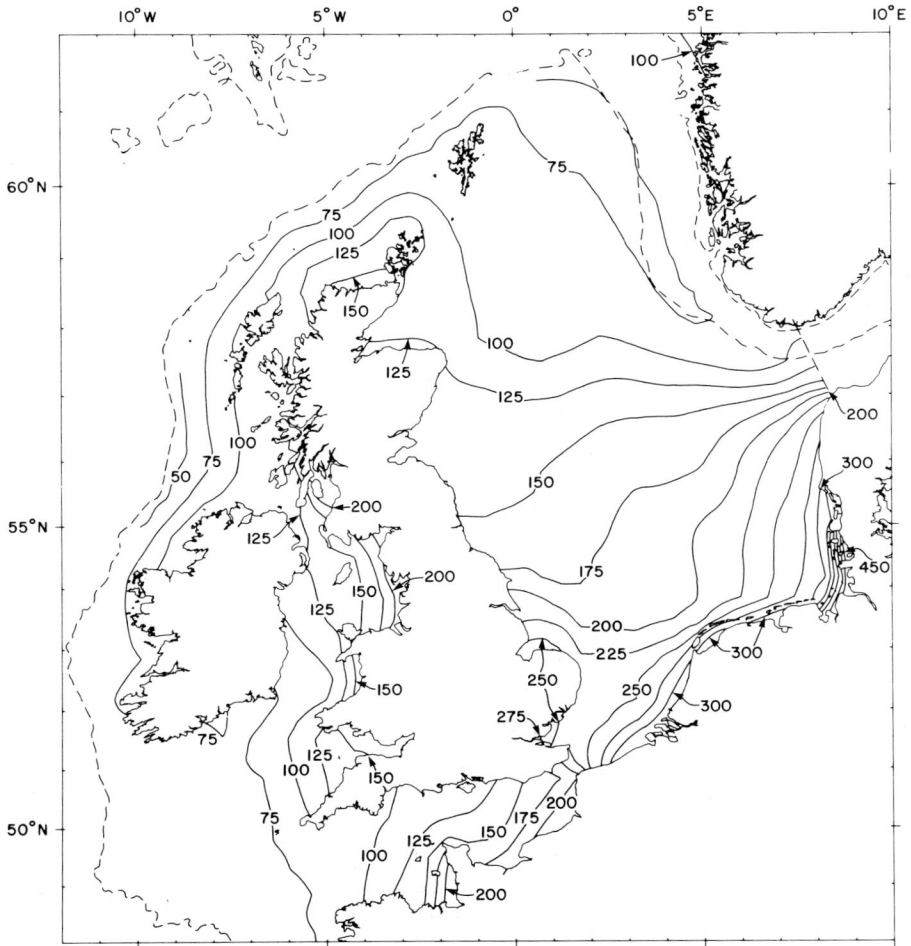

Figure 8:6   Estimated fifty-year storm surge elevations for the north-west European continental shelf, using computer simulations.

Finally, although most of the techniques available for estimating extreme levels are theoretically available for currents, their applications to the speed and direction components of the current vectors is much more complicated.

A very basic approach is to determine the tidal current ellipses from charts, observations or numerical models and then to scale these upwards according to the similarity measures found to be appropriate for levels at a nearby port. It may be sufficient to scale only the major axis of the mean spring current ellipse in this way. The presence of inertial currents can make the use of a similarity measure, determined for levels, invalid for currents because inertial currents do not affect sea levels.

The simplest method of analysing current observations at a particular depth is to produce cumulative frequency distributions of the recorded speeds, averaged over each sampling interval, and to extrapolate according to some fitted distribution. This method may also be applied to produce directional estimated extremes by treating the speeds observed in each directional sector as separate distributions. This method is fairly easy to apply because tide and surge separation is not necessary, but the reliability is invariably limited by the short periods of data available, and the strong seasonal effects.

Crude estimates have also been made by applying techniques similar to those of ranking the annual maximum levels, but in this case the maximum values in each of a series of shorter periods, such as 10 or 7 days, are ranked in order and extrapolated by fitting a suitable distribution.

Table 8:4 Joint frequency distribution of predicted tidal current components at the Inner Dowsing. Total hourly values = 8784. The speeds are for the centre of each 0.1 m sec^{-1} element. (*Reproduced by permission of Pergamon Books Ltd.*)

		West							East		
		−0.4	−0.3	−0.2	−0.1	0.0	0.1	0.2	0.3	0.4	0.5
	1.4										
	1.3										
	1.2										
	1.1										
	1.0			1	2	1	2	2			
North	0.9			7	20	25	20	13	10	1	
	0.8			28	46	74	96	59	29	5	
	0.7		2	44	122	148	142	89	43	3	
	0.6		4	96	143	146	115	88	50	5	
	0.5		2	84	150	125	120	110	41	1	
	0.4		2	100	158	77	112	113	38		
	0.3		1	99	122	38	81	134	40		
	0.2		1	107	113	1	71	158	20		
	0.1			74	110	0	47	185	11		
	0.0			103	87	1	32	182	11		
	−0.1			78	96	4	75	180	3		
	−0.2			79	105	6	92	150	1		
	−0.3			92	101	20	162	130			
	−0.4			64	116	81	216	35			
	−0.5		2	88	121	127	166	13			
	−0.6		3	96	115	146	89	2			
	−0.7		10	85	104	113	41				
	−0.8		20	112	111	77	5				
South	−0.9		25	106	91	31					
	−1.0	1	41	72	34	2					
	−1.1	1	34	39	5						
	−1.2	4	16	7							
	−1.3										
	−1.4										

Extreme currents may also be estimated by separation of the observed current vectors into tidal and surge components, as for levels. Two-dimensional frequency distributions are obtained for each component, but in the simplest case of the currents being rectilinear or if only speeds are considered, the problem may be treated in exactly the same way as for estimating extreme levels. An appropriate class interval is 0.1 m s^{-1}. For rectilinear currents, where the flow is restricted to be backwards and forwards along a particular line, one direction must be arbitrarily defined as positive. The extremes of positive and negative flow correspond to the extremes of maximum and minimum levels. Where the flow is not rectilinear, the flow in two orthogonal directions may be treated separately. North–south and east–west components are usually chosen, but the directions of the major and minor axes of the current ellipses are also suitable. The maximum components in each of the four directions may be then estimated from probability plots produced by combining the probability distributions of the separate tidal and surge components.

It is important to make a clear distinction between the maximum current to be expected in a particular direction, and the maximum expected *component* of current resolved along that direction, as determined above. For example, a truly rectilinear flow, confined along a single axis would have a theoretical zero probability of flow along any other axis, but the component would have finite values in all directions, except the direction at right angles to the flow axis. Any use of a component distribution of current extremes for design purposes must be limited to cases where controlling parameters are linearly dependent on the current component. Where quadratic or cubic laws (for example, of current drag on a structure) are applied, a component distribution is not valid.

Calculations of the full two-dimensional tide and surge frequency distributions and their subsequent recombination to give a total probability distribution begin with tidal analysis of the north–south and east–west current components and their separation into tidal and surge current distributions (Pugh, 1982a). Tables 8:4 and 8:5 show the distributions, in elements of 0.1 m s^{-1}, for nearly a year of observations at the Inner Dowsing light tower.

Table 8:5   Joint frequency distribution of surge current components at the Inner Dowsing. Total hourly values = 8301. (*Reproduced by permission of Pergamon Books Ltd.*)

		West						East				
		−0.5	−0.4	−0.3	−0.2	−0.1	0.0	0.1	0.2	0.3	0.4	0.5
	0.6											
	0.5											
	0.4						3					
North	0.3		1		1	8	28	9		1		
	0.2		1	3	9	38	258	79	4			
	0.1		3	10	38	280	1289	358	36	13	4	
	0.0	2	6	26	95	562	1995	599	104	28	3	1
	−0.1		2	13	47	379	1139	289	28	17	1	
	−0.2	1		3	11	96	248	61	4	2		
South	−0.3				1	12	36	9				
	−0.4							1				
	−0.5											

Table 8:4 shows the tidal distributions based on a year of predictions; zero tidal currents are rare and the most frequent current has components (0.1 m s^{-1} to the east, 0.4 m s^{-1} to the south). The surge distribution of Table 8:5 is more limited, with the most frequent value being zero speed in both components.

The separate tidal and surge distributions are combined by a two-dimensional process similar to that defined in equation (8:5)

$$D_o(u,v) = \int_{-\infty}^{\infty} \int_{-\infty}^{\infty} D_T(u - x, v - y) \, D_s(x, y) \, \mathrm{d}x \, \mathrm{d}y \qquad (8:6)$$

where u and v are the two orthogonal components of total current. The frequency distributions of Table 8:4 are first normalized by the total number of hourly values. The probability of each total current element is then computed as the sum:

$$D_o(u, v) = \sum_i \sum_j D_T(u - ih, v - jh) \, D_s(ih, jh)$$

where i and j are integers and h is the class interval of 0.1 m s^{-1}. In the case of the Inner Dowsing data, summation over the range $i = -5$ to $+5$ and $j = -5$ to $+4$ is sufficient.

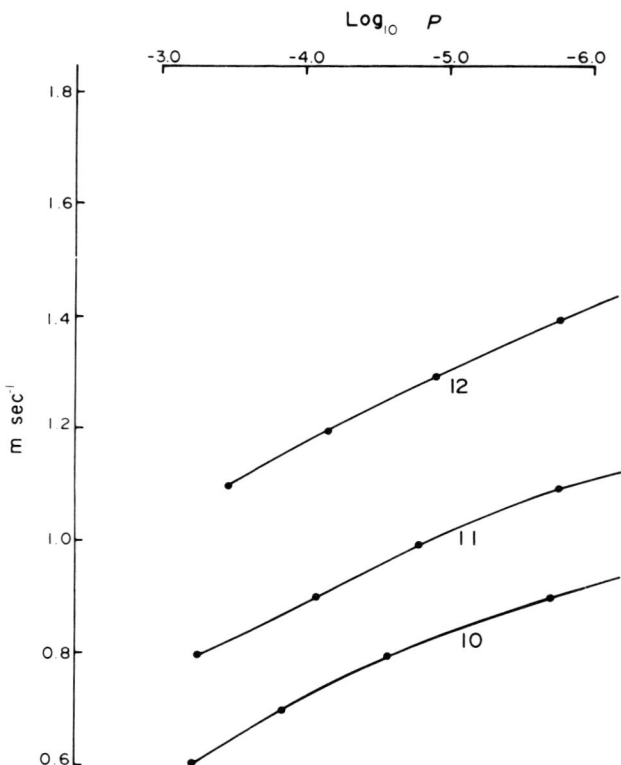

Figure 8:7 Examples of two ways of presenting joint-probability estimates of extreme total currents at the Inner Dowsing light tower: for currents in segments 10, 11 and 12 (see text). (*Reproduced by permission of Pergamon Books Ltd.*)

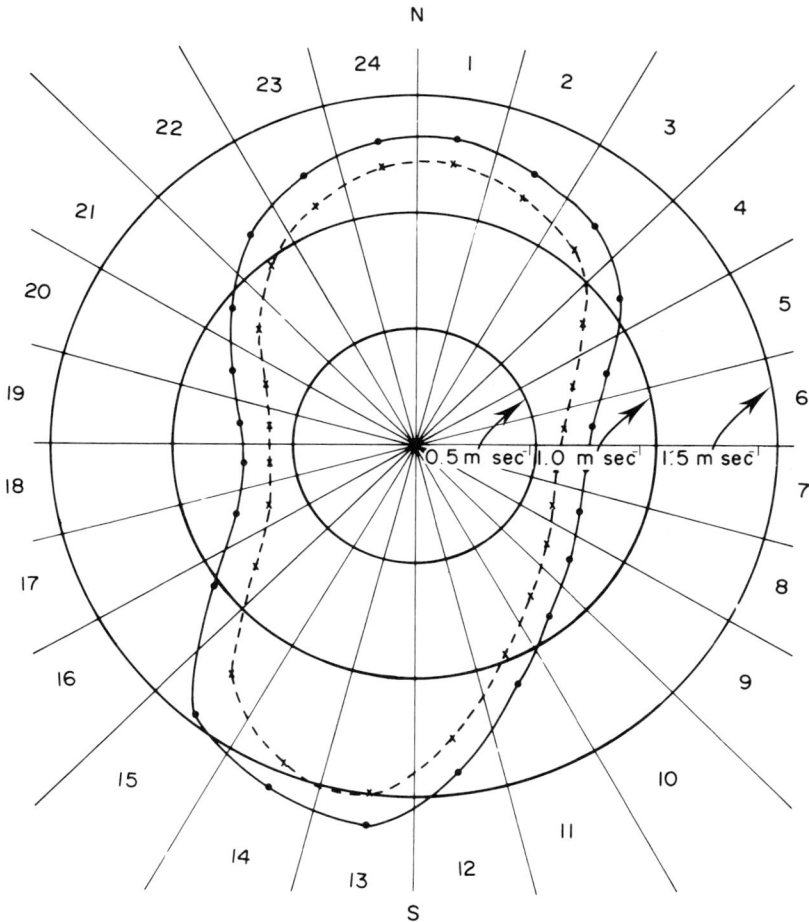

Figure 8:8 As for Figure 8:7: contoured for all segments, with the continuous line representing the 100-year return current and the broken line the five-year return current. (*Reproduced by permission of Pergamon Books Ltd.*)

Although the Cartesian probability distribution matrix obtained in this way contains the basic information on extreme currents, it is not in a form which is easily interpreted. Numerical conversion back to the parameters of speed and direction enables current speeds having specified probabilities to be computed for each 15° sector. Figure 8:7 shows these probabilities for the currents flowing to directions within segments 10, 11 and 12 (numbering clockwise from north: 135–150°, 150–165°, and 165–180°). An alternative presentation, shown in Figure 8:8, involves contouring the speeds which have a particular return period, within the several segments.

The joint probability technique for estimating extreme currents has the same advantages and disadvantages when applied to currents. The main difficulty is always the shortage of good data; a minimum of one year of observations should be available to define the surge distribution. Another approach to estimating surge distribution is to scale the currents according to the local distribution of winds at some permanent coastal meteorological station, but this is only valid when locally generated surge currents are dominant. Usually in extratropical latitudes, surges propagating from elsewhere are at least as important for generating currents as are the local winds.

Where there are insufficient observations to make statistical estimates of extreme currents from the observations, the most powerful approach is to use the results of numerical models in a similar way to that described for levels in Section 8:3:4. Three-dimensional models which give current variations with depth have been developed for research work, but the techniques are not generally applied. An alternative is to use the results of a two-dimensional depth-integrated model, as described earlier, and to adjust for near-surface and near-bottom effects, using assumed distributions and empirically determined constants (Great Britain, 1984).

Tidal currents may be scaled up according to the relationship between spring and highest astronomical tidal levels at a nearby port. Extreme tidal current profiles may be estimated using a formula such as equation (7:9). The extreme surge currents must be added to the tidal currents, but these cannot be estimated reliably on the basis of short-period observations. An alternative approach (Flather, 1987) is to run a numerical model for a series of extreme meteorological conditions, and to scale up the computed currents by comparison with extreme surge levels at a nearby port. Extreme surge current profiles are very difficult to estimate because they depend on the storm-induced eddy viscosity, which is inadequately understood. In shallow water, extreme current profiles vary only slightly over the range of accepted eddy viscosity parameters, but in deep water there is more uncertainty in the computed extreme current profiles (Davies and Flather, 1987). Adding extreme tidal currents to extreme surge currents will give extreme total currents higher than are probable in practice, but present knowledge is not capable of resolving the problem of interaction between them. Eventually, better profiles generated by running three-dimensional numerical models for composite extreme tidal and storm conditions will be generally available. For calculations of extreme levels and extreme currents, it is wise to make conservative estimates and to apply these with caution; however, for currents, because the uncertainties are greater, extra caution is necessary in the interpretation of the estimated extremes.

8:4   Wind, waves, currents and levels

Extreme levels and currents are only two components of the total environment for which systems must be designed and a cautionary discussion is appropriate. The forces due to drifting ice, winds and particularly those due to waves may be equally, or in many cases significantly more important (Vrijling et al., 1983;

Alcock, 1984; Alcock and Carter, 1985). Neither winds nor waves can be considered in detail in this account but there are many books which treat them in depth (for example, Kinsman, 1965). Traditional design practice has been to estimate the probabilities of extreme winds, waves, currents and levels independently, and to add these extreme values, assuming they occur simultaneously, to obtain the extreme environmental design conditions. This assumes that they are statistically totally correlated, which is not the case as we have already discussed for joint probability distributions of tides and surges. Inevitably the traditional assumptions must result in some over-design. However, the assumption of total independence of the parameters is also invalid, as the following chart shows.

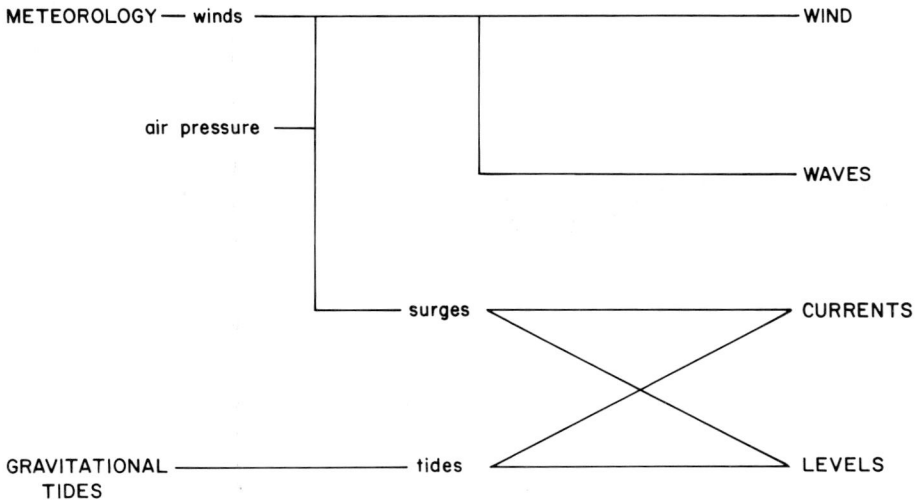

There are two basic factors which govern changes in the marine environment, the weather and the gravitational tides, and these are independent of each other. The wind stresses on the sea surface produce waves, currents and changes of sea-level. The surges combine with the tidal movements to produce the total observed currents and levels. Clearly winds, waves, currents and levels at each site are related, but the statistical correlations, particularly when the directions of the winds, waves and currents are considered, will be different for each site, and can only be determined by long series of simultaneous measurements of each parameter. The problem is further complicated by the strong seasonal variations which occur for winds, waves and surges, and to a lesser extent for tides.

The lack of suitable measurements makes the estimation of extreme wind speeds at sea difficult. Estimates are possible by extrapolating the many years of continuous recording at land stations in conjunction with the irregular observations from ships at sea. These ship measurements are not usually very accurate and will be biased towards low wind speeds because ships seek shelter in extreme weather conditions. The winds are usually defined for a level 10 m above the sea surface where the drag effects are small. For calculating wind forces on structures the three-second gust speed is usually used; values of more than 50 m s^{-1} for the fifty-year return period gust have been estimated for the northern North Sea (Great Britain, 1984). Hourly mean wind speeds are more appropriate for calculations of wave generation. Over the open sea, these may be related to the three-second gust speeds by the empirical reduction factor 0.72 which gives corresponding fifty-year hourly mean wind speeds between 35 and 40 m s^{-1}. Twelve-hour and daily mean maximum wind speeds have been estimated by applying further reduction factors of 0.90 and 0.82 respectively to the hourly mean values. If no directional analysis is available it must be assumed that the extreme winds may blow from any direction. Extreme winds are obviously more probable in winter than in summer. Over the North Sea gales occur for some 20 per cent of the time in January, and may blow from any direction. In July gales blow for only about 2 per cent of the time.

Wind waves are usually the dominant contributor to the environmental forces on offshore structures. Wave heights, periods and directions are important and these may all be related to the winds by a combination of theoretical and empirical rules. The higher waves are associated with longer wave lengths and longer periods. They are also associated with higher wind speeds, longer durations of wind action, and longer fetches—the distance of sea over which the wind has blown before reaching the design site.

Fifty-year waves in the northern North Sea may exceed 30 m in height from peak to trough, whereas in the southern North Sea and the Dover Straits, because of the limited fetch and depth, heights of only 13 m are estimated. The corresponding wave periods are 15 s and 10 s. An alternative method of describing and analysing wave characteristics is to use the wave energy spectrum. The wave spectra for fully and partly developed seas generated by winds blowing for different times and over different lengths have been extensively studied.

The winds which produce extreme waves will also produce extreme surge currents and levels, but the time scales and directions may be different. Wave directions in the open sea may be similar to the wind directions, but near the coast and in other areas where the water depth becomes comparable with (approximately half) the wavelength, the waves will develop a steep leading edge, and they will also be refracted towards the region of shallower water. Changes in the still-water level on which the waves propagate will alter the refraction pattern. These changes of level will also control the height at which the wave forces are imposed on an artificial structure or on a beach.

Currents can also affect the direction and propagation of a wave train, but the most significant interaction between waves and currents occurs when they are opposed to each other (Peregrine, 1976). In this case the wave propagation speed is reduced, but the amplitudes are increased. The physical reason for this is the need to conserve the net energy flux, in the same way as tidal amplitudes increase when travelling over the shallow waters of the continental shelf (see Section 5:4:1). Even more significant for calculating forces on structures, the wave steepness and the associated accelerations are increased.

Levels and currents may be statistically coupled because of dynamically coupling by the laws of long-wave propagation. Where the wave is progressive, maximum forward currents occur at maximum water level, but for standing waves the maximum currents occur near the times of mean sea-level. External surge components propagate as long waves, but locally generated surges will obey different laws. Interactions between tides and surges in shallow water have been discussed in Section 7:8.

The statistical relationships between extreme levels and extreme waves are critical for the design of coastal protection schemes, and can vary over short distances. An example of such a local variation is found in the outer reaches of the Thames Estuary. Extreme still-water levels are generated by the addition of external North Sea surges to high tidal levels. However, the strong winds from the north associated with external North Sea surges produce large waves only on the Kent coast to the south of the estuary, and not on the Essex coast at Maplin to the north of the estuary. Defences designed for simultaneous extreme wave and still-water level conditions would be appropriate for the former case, but would be excessive for the northern coast.

The use made by the design engineer of these various statistics in deciding a suitable protection level requires some compromises, and is not the same in every case. As an example of the balance of factors, consider the design level for the Thames Barrier, which was finally set at 7.11 m (23.33 ft) above Ordnance Datum Newlyn. This figure was taken as a compromise between an inadequate short-term solution to the flooding problem and a very high standard of protection which would be effective until the end of the 21st century (Gilbert and Horner, 1984). Allowing for mean sea-level increases to 2030, on a 1000-year return flood level at Southend, 1.50 m (4.93 ft) was allowed for wind and wave freeboard, and for level increases up the River Thames due to local winds and river discharge.

8:5 Coastal engineering

The two principal fields of coastal engineering involve coastal protection against erosion and flooding, and the development of harbour facilities. Specialized publications deal with many aspects of their design and construction (Dronkers, 1964; Wiegel, 1964; Silvester, 1974; Berkeley Thorn and Roberts, 1981; Muir Wood and Flemming, 1981). Here it will be possible to consider only those aspects which are strongly affected by tidally-related sea conditions. Some further aspects of coastal erosion and deposition are discussed in Chapter 10.

The basic problem of coastal protection from flooding could ideally be solved over periods of several hundreds of years by the construction of suitable continuous sea-walls. In practice of course this would be prohibitively expensive, so only those schemes which protect highly populated areas such as low-lying estuaries (Horner, 1985), vulnerable installations such as nuclear power stations, or those containing valuable property are justified in terms of their cost. Strict cost–benefit analyses may be applied, but other political arguments are also involved in the final decisions. Agricultural land of no special value is unlikely to be given protection. Protection against the sea-level changes which occur over periods of thousands of years is impossible and the correct approach must be one of slow adjustment, for example the gradual migration of residential and industrial areas to higher land. The need for this kind of gradual adjustment was recognized in the United States by establishing the National Flood Insurance Program by Federal legislation in 1968; incentives are given to local communities to prohibit new construction within defined flood zones.

Extensive flooding results when the still-water level exceeds the top of the defence structure, but before this level is reached there may be considerable flooding from overtopping by waves. Several different designs of wall have been built, each intended to minimize wave overtopping. The modern trend is for gradually sloping defences which progressively remove the energy of the incoming waves, rather than for vertical walls which give perfect wave reflection. Unfortunate experiences have shown that these reflected waves can interact with other incoming waves to give enhanced beach scour, lower beach levels, and eventual undercutting of the foundations of the defence walls themselves. Incidentally, the cost of a defence wall grows more rapidly than a simple linear increase with increases of design height: the width of the footings must also be increased in proportion to the height, so that in terms of material alone the increase is more closely proportional to the square of the height.

An interesting application of the joint probability of still-water levels and waves recommends that instead of designing defence walls to prevent overtopping by some specified 'design storm' conditions of still-water level and wave height, engineers should work with the concept of 'design overtopping discharge' (Owen, 1983). The reason for this emphasis on the rate of water discharge over a wall is the recognition that the ultimate test of effectiveness of a sea-wall is the amount of water which overtops, and the frequency with which this occurs. For each given wall design, the factors which govern overtopping at each instant are the still-water level, the significant wave height, the wave period and the wave direction relative to the sea-wall. The overtopping discharge for all possible combinations of these parameters must be determined either by experiment or by calculation, and these define the characteristics of that particular design. The characteristics of the site at which protection is required must be defined in terms of the probability of occurrence of each combination of sea-level and wave conditions. The total probability of overtopping discharges exceeding some specified value is given by summing the probability of all these joint sea-level/wave conditions which individually exceed that value. Curves may be drawn giving the probabilities as a function of the discharge rates. Where the wall design incorporates features such as a wave return wall, which tend to throw water into the air, overtopping discharge rates are much greater when the wind blows onshore. In this case, detailed calculations should also include winds in the total matrix of environmental conditions. Although characteristics of different wall designs can be adequately defined, the major environmental problem with this approach is to obtain a reliable estimate of the probability of occurrence of all possible combinations of water level and wave conditions.

Tidal rivers and inlets give risk of flooding over large areas, and protection against this flooding along the whole length of either bank can be very expensive. For London the problem has been solved at great expense by building the Thames Barrier, which closes only at times of forecast extreme levels in the southern North Sea (see Section 6:10). Along the coast of the Netherlands the solution has been to close off many inlets by completely damming their entrances (Delta Committee, 1962; Leentvaar and Nijboer, 1986). Because these dams take many months or years to build, water exchange between the sea and the inlet continues throughout the construction. Only at the final stages is the gap sufficiently narrow to restrict the exchange and at this time there are special construction problems. As the gap continues to reduce, large differences in level occur between the sea and the inlet, which in turn result in very strong currents of several metres per second through the remaining small connection. Special procedures must be applied during this critical closure stage to avoid severe erosion of the defences already built: Two basically different methods are used: the abrupt closing method using prefabricated caissons which are emplaced during a single short period of slack water, and the gradual closing method whereby the last broad length of dam is built upwards gradually, keeping the crest of the dam nearly horizontal to avoid strong currents and erosion along any particular section (Dronkers, 1964).

Despite the risks of flooding and the costs of protection, several economic benefits are also possible by exploiting the coastal tidal regime. Most of the great ports of the world are situated near the mouths of large rivers and many are a considerable distance inland. London, on the River Thames, and Hamburg, on the River Elbe, are good examples of inland ports. The fact that these ports are only accessible at tidal high water has not been a major disadvantage, and the possibility of marine transport to sites well inland encouraged vigorous industrial development. By travelling inland on a flooding tide and out on an ebbing tide, ships could make considerable savings of fuel and time. The reduced costs, and the feasibility of building harbour walls and docks at sites which are essentially dry for half of the tidal cycle are additional advantages. There are problems associated with the maintenance of navigable channels up rivers and estuaries, but again the vigorous tidal currents serve to keep channels deep. The tidal flows can also prevent harbours freezing during winter, for example in New York, both by their mixing action and by the introduction of salt water which lowers the freezing point. Pollution, inevitably associated with large industrial developments and centres of population, is also more readily diluted and discharged to sea where there are large regular exchanges of tidal water. The design of sewage outfall pipes depends on the water depth and currents at the sea-wall end, and strong tidal flows are advantageous. The conditions for ports where tidal ranges are relatively large, for example around the North Sea, may be contrasted favourably with those around the Mediterranean, for example Marseilles, where tides are small. The pollution problems are much greater in the Mediterranean, and despite their high rates of fresh-water discharge, neither the Rhône nor the Nile has proved navigable for any but the smallest sea-going vessels.

8:6 Offshore engineering

The earliest offshore structures were lighthouses built on rocky reefs, often under severe difficulties, and not always with immediate success because the environmental conditions were poorly understood. The first light on the Eddystone Rocks off Plymouth at the entrance to the English Channel was completed in 1700, but was completely destroyed by a storm in November 1703. The original tower and its replacement, completed in 1709, were both constructed of wood. This second tower, which was destroyed by fire in 1755, was replaced by Smeaton's stone tower in 1759. Smeaton's tower was circular, 26 m high, with each stone dovetailed into the surrounding ones to form a rigid structure. It survived for over 100 years before being replaced by the present 45 m high stone tower.

The vast majority of modern offshore structures are for the production of gas or oil; examples of four design types are given in Figure 8:9. The first such structures for offshore drilling and production were installed in the Gulf of Mexico off the coast of Louisiana in 1947 (Ellers, 1982). A series of steel frames was assembled ashore and towed to the site, where they were pinned to the sea-bed with piles driven through steel sleeves which were part of the structure. The original designs, which formed the prototype for offshore structures built over the next thirty years, became known as steel-template jackets. A more recent alternative design is the tension-leg platform where the topside facilities rest on a buoyant hull designed to yield with the waves. This hull is held down by tubular-steel tethers which are designed never to go slack even in the trough of the maximum expected wave on an extremely low still-water level. The major advantage of the tension-leg platforms is that their cost does not increase rapidly with increasing water depth. Also they may be untethered and anchored at a new site.

Tension-leg platforms were designed for the deeper more hostile waters of the northern North Sea, where the Hutton Field became operational in 1984. Rigs are designed to resist the effects of very high winds and waves, and the platforms must be built to survive the fifty-year or hundred-year storms with the crew remaining aboard. In constrast, rigs in the Gulf of Mexico can be smaller and less robust because the rare extreme hurricane events are monitored sufficiently well for the crews to be evacuated in good time. In the Arctic, icebergs, particularly those small enough to be moved violently by the waves, are a further hazard. For all types of structure the forces due to waves are normally the most important, but currents and sea levels are significant factors. In the southern North Sea where tidal currents are strong and wave amplitudes are depth-limited, currents and waves are of equal importance.

A jacket–type platform

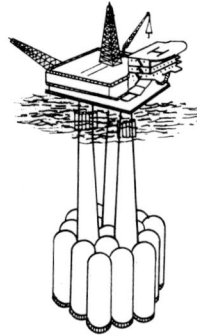

A concrete gravity platform

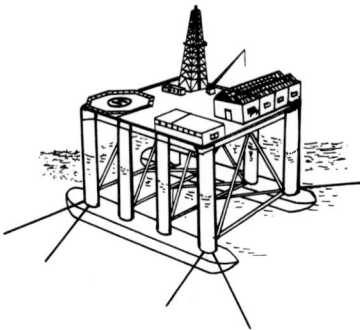

A semisubmersible platform

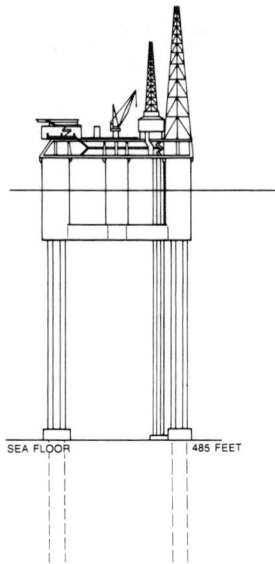

Hutton Tension leg platform

SEA FLOOR   485 FEET

Figure 8:9 Examples of four different designs for offshore platforms, each of which requires different kinds of tidal information. *(From 'Advanced offshore oil platforms' by Fred S. Ellers. Copyright © 1982 by Scientific American, Inc. All rights reserved.)*

The forces on a rig due to water movements may be calculated using an equation developed by Morison and others in 1950, which identifies two separate components, drag and inertia (Sarpakaya and Isaacson, 1981):

$$C_m\rho \, \frac{\pi d^2}{4} \, \frac{\partial u}{\partial t} + \tfrac{1}{2}C_d\rho \, d\,u|u| \qquad (8:7)$$

where d is the pile diameter, ρ is the water density, $\partial u/\partial t$ is the rate of acceleration of the water, u is the water speed and C_m and C_d are empirically determined drag coefficients. The first term, called the inertia coefficient is related to the complementary situation where a body of fluid of volume equal to that of the pile is being accelerated. The second term, the drag coefficient which is usually the more important for extreme conditions, increases, in the same way as the wind stress acting on the sea surface, as the square of the water speed. It is therefore incorrect to enter the velocity components in equation (8:7) to obtain the components of the total force; the correct procedure is to square the total speed and then resolve into components.

The empirical coefficients C_m and C_d have been determined by experiment and found to depend on the Reynolds number:

$$\mathrm{Re} = \frac{U_{max}d}{v}$$

where U_{max} is the maximum flow velocity, and v is the kinematic viscosity of the water (see Section 6:4:4), and on the Keulegan–Carpenter number:

$$\mathrm{K} = \frac{U_{max}\,T}{d}$$

where T is the wave period. The Reynolds number represents the importance of viscous effects, which diminish for rapid flows and larger structures. The Keulegan–Carpenter number, which is also dimensionless, represents the degree of oscillatory flow in wave motions. Small K means small flow excursions with rapid flow reversals whereas large K means large excursions with small accelerations. For K less than about 8 the drag term can be ignored, whereas the inertia term can be ignored for K greater than about 25.

The addition of steady currents to the wave motions affects both the values entered in equation (8:7) and the values of Re and K. The basic procedure, in the absence of a clear understanding of the fluid-mechanical phenomenon involved in the interaction, is to simply add the currents to the waves. If the current is in the direction of wave propagation the wavelength increases but the amplitude decreases. If the current opposes the waves, a less likely event because severe winds tend to drive the surface currents and waves in the same direction, then the waves become shorter and steeper.

If the waves have a steady current q added:

$$\mathrm{Re} = \frac{(U_{max} + q)d}{v} \qquad \mathrm{K} = \frac{(U_{max} + q)T}{d}$$

so that the different values of C_m and C_d apply. Because the drag on a cylinder depends on the square of the design current speed the addition of extreme surge currents to wave currents greatly increases the necessary design strength.

The United Kingdom *Guidance Notes* specify that the safe design of rigs should take into account environmental factors not less severe than the fifty-year maximum wave height and one-minute mean wind speed together with the maximum current. In most cases these will not occur together and so some design redundancy is implied, but it should be remembered (equation (8:1)) that for a structure which has a lifetime of fifty years, designed to resist only the fifty-year extreme environmental conditions, the risk factor is 0.64. For structures with expected lifetimes of twenty years and ten years designed for the fifty-year event, the corresponding risk factors are reduced to 0.33 and 0.18.

8:7　Power generation

The French tidal power plant operated at La Rance near St Malo in Brittany since 1966 (Figure 8:10(a)), was the first modern large-scale scheme to harness tidal energy, which has the advantage of being renewable and pollution-free. Tidal energy manifests itself either as the kinetic energy in tidal currents, or as the potential energy of the water head difference between high and low tide levels. In addition to La Rance, smaller schemes have operated in the White Sea near Murmansk in the Soviet Union, Kiansghsia in China and at Annapolis, near Nova Scotia, Canada. The Annapolis 20 MW scheme may be a pilot for a much more ambitious project to harness the great tidal ranges of the Minas Basin at the head of the Bay of Fundy. Detailed studies have been made for this ambitious scheme (Gray and Gashus, 1972; Stephens and Stapleton, 1981; Charlier, 1982; Garrett, 1984) and for other schemes including those for the Bristol Channel and the Mersey Estuary in Great Britain, for Garolim Bay in South Korea and for other sites in India, Brazil and Argentina. Previous to this considerable modern interest, the energy in tides has been exploited on a smaller scale for many centuries. An early tide mill, mentioned in the Parish Records of 1170 for Woodbridge, Suffolk on the English coast of the North Sea, has recently been restored to working order (Figure 8:10(b)). Mills were used for grinding corn and for cutting wood for the local community. Similarly, tidal mills were common in the USA north of Cape Cod (Redfield, 1980) where the tidal ranges exceed 3 m. The smaller ranges to the south of Cape Cod were less favourable, but Brooklyn, New York had a mill in 1636.

The potential energy contained in a basin of area S, filled at high tide, and discharging into the open sea at low tide is:

$$S \int_0^{2H} \rho g z \, dz$$

where H is the tidal amplitude, ρ is the water density and g is gravitational acceleration, which gives:

$$2 S \rho g H^2$$

Navigation
lock

Power plant

Sluices

(a)

Mill race

(b)

Figure 8:10 (a) La Rance tidal power station, France. (*Reproduced by permission of Electricité de France.*) (b) The old English mill at Woodbridge, Suffolk. (*Reproduced by permission of the Woodbridge Tide Mill Trust.*)

The process could be repeated by filling the basin again from the low-water level, at the time of high water in the open sea. The total energy theoretically available in each tidal cycle is:

$$4S\rho g H^2$$

and the mean rate of power generation is:

$$4S\rho g H^2/(\text{tidal period})$$

The energy actually extracted by a scheme depends on several other design factors.

The ideal site for tidal power generation is a basin of large area, where tidal ranges are large. The large tidal range is most important because of the square-law dependency. The power available from semidiurnal tidal regimes is twice that available from diurnal tides. In addition to these power requirements there are obvious advantages if the basin has a relatively narrow and shallow entrance so that the cost of building a tidal barrage is less. Other factors which can make a prospective site more or less suitable are the distance to the area where power is needed, the existing uses of the basin for navigation or recreation, and the local price of competing power sources such as coal, gas and hydroelectric stations. Strict cost comparisons are difficult because the future costs of alternative power sources and the operational life of a tidal power scheme are both uncertain.

Table 8:6   Characteristics of some areas suitable for tidal power generation. All have semidiurnal tidal regimes (tidal amplitudes and basin areas from Banal, 1982).

	Mean tidal amplitude (H)	Basin area (S)	Maximum theoretical mean power output
La Rance	4.0 m	17 km^2	250 Mw
Bay of Fundy (Minas Basin)	5.5	240	6500
Annapolis, Nova Scotia	3.2	6	55
Severn Estuary (outer scheme)	4.0	420	6100
South Korea (Garolim Bay)	2.5	85	480

Table 8:6 summarizes the mean tidal ranges and basic basin surface areas for four schemes, including La Rance. Potentially the schemes for the Minas Basin and the Severn Estuary could produce many times the energy of La Rance. Table 8:7 compares tidal energy with other sources and sinks of energy. Some of the proposed schemes are shown in Figure 8:11. The Minas Basin scheme has the disadvantage of being distant from major areas of power demand. In the Severn Estuary, although the major cities of Cardiff and Bristol are close by, they are also major ports whose commercial interests would be affected if access for shipping were restricted.

Table 8:7   A comparison of estimated tidal energy fluxes with other sources and sinks of energy.

Sources	
Solar radiation incident on the earth	1.7×10^{11} Mw
Large coal-fired power station	1 000 Mw
Early tide mills	0.1 Mw
Annapolis tidal power scheme	20 Mw
La Rance scheme	240 Mw
Theoretical optimum Minas Basin scheme	6 500 Mw
Total United Kingdom generation capacity (1981–2)	55 000 Mw
Sinks	
Peak load, United Kingdom (1981–2)	42 600 Mw
Peak load, New England (1980)	22 000 Mw
Dissipation due to M_2, north-west European shelf	200 000 Mw
Dissipation to M_2, worldwide	3 200 000 Mw

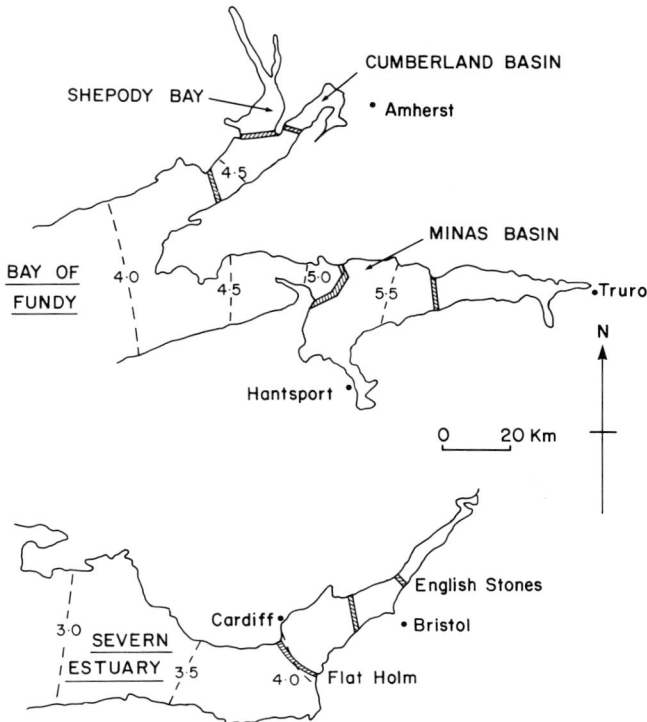

Figure 8:11   Locations of various tidal power schemes proposed for the Bay of Fundy and the Severn Estuary; the M_2 amplitudes are shown by the broken lines.

The demands for power are closely linked to the 24-hour diurnal cycle, with weekly and strong annual cycles superimposed. The availability of tidal power from simple schemes is regulated by the phase of the M_2 tide, and is at its maximum at a time which advances by 52 minutes each day. Unlike available wave power, which reaches a maximum during the winter months, coinciding with the maximum seasonal demand, tidal power has its greatest variation over the spring–neap cycle. Because the available power varies as the square of the tidal range, the variation of available power between spring and neap tides may be a factor of ten. The mismatch between supply and demand has led to the development of several schemes which sacrifice optimum total power generation in favour of a more controlled supply. Power which is available to meet demand in this way is called 'firm power'.

Many of the original tidal mills used two or more ponds to store water and to control the water-level difference being used to generate the power. Even for the simple single-pool ebb-flow system it is impossible to make use of the total theoretical energy capacity. Straight-flow turbines which have been designed to operate most efficiently with water heads of several metres are less efficient at very low water heads; for these lower water speeds, larger rates of discharge are necessary to give the same power. The simplest way to achieve more direct control over the available power is to operate with two basins, one filled when high-water conditions allow, and the other drained at low tide. Turbines, which are sited between the two basins only need operate for one direction of flow, which makes them cheaper, but less than half of the theoretical energy is then available. These schemes are usually expensive because of the need for two sets of sluices to control the flow of water between the basins and the sea, and because of the additional barrage construction necessary for separate basins. Some further control is possible by using off-peak power from the sources to adjust the levels in either the high or low basin.

Despite several ingenious designs for tidal power schemes, the timing of the tide makes them basically inferior to other schemes for firm power production. However, if they can be used principally as energy producers in conjunction with a separate energy storage scheme, their capacity can be maximized. The normal method for retiming the energy supply is hydroelectric storage, and such storage systems are part of most national electricity networks; careful programming is necessary to fit the periodic tidal generation of energy into the overall system which includes other sources, such as nuclear and coal-fired stations that have different requirements for efficient operation. Energy from La Rance is now integrated into the French network by manipulation of hydro-storage on the River Rhône. A more exotic proposal is for energy storage in the form of compressed air in airtight underground caverns, with power generation by gas turbines at times of peak demand. For tidal schemes in the Bay of Fundy it has been suggested that these caverns could be excavated within extensive thick local underground salt deposits by solution-mining.

Any scheme which removes energy from the tides must also have an effect on their behaviour. Since the most favourable sites are those where large tidal amplitudes are generated by local dynamic resonances, they are particularly vulnerable to imposed changes. The only satisfactory way to predict the effects of a proposed scheme on the local tides is to apply a numerical model with its outer boundaries well removed from the area of the scheme. The effects of several different schemes proposed for the Bristol Channel have been predicted by models which take the western edge of the continental shelf as their boundary. These schemes were all shown to have a tendency to reduce the tidal ranges by 10 per cent or more in the vicinity of the barrage. However, at the entrance to the Bristol Channel the effects on the tides were limited to an amplitude reduction of 1 per cent or less, and an advance in the tidal phase by a few minutes.

The addition of a tidal power scheme to a natural tidal system oscillating near resonance need not automatically reduce the tidal range. Tidal resonance usually occurs when a basin has a length which is a quarter of a semidiurnal tidal wavelength (Section 5:2:2). If the natural length is slightly longer than this critical length, then it is quite probable that the introduction of a barrier near the head may tune the estuary closer to the resonant condition. It has been suggested that both Cook Inlet in Alaska and the Gulf of Maine, Bay of Fundy systems could be tuned this way.

Despite the long and detailed studies which have been made for several schemes at many potential sites favoured by large tidal ranges, only a few schemes, including those at La Rance and Annapolis, Nova Scotia, have become operational. The reasons for this reluctance for a more enthusiastic commitment by governments and power-generating authorities are not difficult to identify. Firstly, even the most ambitious schemes for an area such as the Bristol Channel could supply only perhaps 10 per cent of the power requirements of the United Kingdom, so the contribution can only be marginal to that of other sources such as nuclear or conventional coal stations. The price for this marginal contribution is a substantial capital investment coupled with major changes to the environment, the consequences of which cannot be predicted with certainty. Practical lessons learned in the development of the smaller schemes may eventually lead to a major project such as those in the Bristol Channel or the Minas Basin.

Opponents of schemes to extract power from the tides have used many arguments, both economical and environmental to support their case. Perhaps the least valid of these arguments is the suggestion that by extracting power from the tides, we would fundamentally change the dynamics of the earth–moon system and of the earth's rotation on its axis. Table 8:7 shows that even the most ambitious scheme, in the Minas Basin, could never extract more than a few tenths of a percent of the energy that is dissipated naturally by the lunar semidiurnal tides through bottom friction and other natural processes.

CHAPTER 9

Mean Sea-level

But who would say that a sea surface is inclined?

Strabo, *Geography*

9:1 Introduction

In Chapter 1, equation (1:2), sea-level at any instant of time was defined as the mean sea-level plus the tidal and the surge components. In this chapter we consider the mean sea-level, written as $Z_0(t)$. This term was deliberately written as a time-dependent function because changes in the mean level are evident in the long series of measurements now available. There are many regional analyses of trends in annual mean sea levels (for example, Arur and Basir (1981), Aubrey and Emery (1983), Woodworth (1987)). However, these changes are very small when compared with the daily tidal and surge changes of level. Long-term changes are typically one or two decimetres per century. Nevertheless, the geological record shows that over periods of millions of years enormous vertical movements of the land relative to the sea have occurred. Even during the Quaternary Period (the most recent 2 million years of the geological record) there is evidence from erosion platforms, gravel terraces, and shallow marine deposits, of relative sea levels having been 100 m lower than at present. Since the last ice-age 10 000 years ago sea levels have increased in many places by more than 40 m.

Conversely, these differences might be considered as land levels 100 m higher and 40 m lower than at present. The change of mean sea-level relative to a fixed point on land is a measure only of the difference between the vertical movements of the mean sea-level and of the land itself. During periods of glaciation the sea-level falls because water is locked into the polar ice-caps; as the glaciers recede the global sea-level increases, but this general increase in level may not be apparent along coasts which have only recently been relieved of their ice burden. Along these coasts there is an isostatic land uplift, which is measured as a decrease in local sea-level. This is the case at present around the Baltic Sea, where vertical land movements of up to 1.0 m per century dominate the more gradual global sea-level increase of around 0.10–0.15 m per century.

The idea of measuring mean ocean levels relative to fixed bench-marks at a given epoch was apparently first suggested by Baron von Humboldt (see Ross, 1847, p. 23).

Long-term changes of mean sea-level are called *secular* changes. Global changes in the mean sea-level are called *eustatic* changes. Vertical land movements of regional extent are called *eperiogenic* movements. One of the major problems of mean sea-level interpretation is the identification of separate eustatic and eperiogenic changes when only secular changes are directly measurable at a particular location.

Figure 9:1   The chapel at Strava, near Corinth, Greece, shortly after a sequence of three earthquakes in 1981. This part of the coast subsided by 1.5 m, but other parts were uplifted by up to 0.2 m (Supplied by G. C. P. King.)

It is sometimes supposed that the mean sea-level determined by averaging the effects of tides and surges over a year or even several years, is the local level of the geoid (see Section 3:5). Although mean sea-level is a good first approximation to the geoid, there are other oceanographic effects such as water density variations, permanent ocean circulation patterns, and atmospheric effects such as mean air pressures and winds, which sustain permanent displacements of the mean sea-level from the geoid. These differences may exceed 1.0 m but are usually much less than this; for example, the mean sea-level at Balboa on the

Pacific Coast of the Panama Canal is about 0.20 m higher than the mean sea-level on the Atlantic Coast, due to a lower average density of Pacific water (Reid, 1961), and perhaps also influenced by the restriction of the east-going currents around Antarctica by the Drake Passage connection between the Pacific and Atlantic Oceans.

Scientific and public interest in the possibility of accelerated sea-level increases due to global warming has made this an area of active research into both the processes of sea-level change and their social impact. In this account we focus on the basic oceanography. Several references are given to recent publications; valuable earlier accounts are given by Rossiter (1962) and Lisitzin (1974). We first consider the methods available for determining mean sea-level and sea-level trends from tide gauge records. We then outline the oceanographic reasons for the observed changes in time and in space. Chapter 10 discusses the implications of the present upward trends for the stability, erosion or development of existing beaches, and some of the geological evidence for sea-level changes over longer periods. The importance of sea-level increases is dramatically emphasized in Figure 9:1, which shows the chapel at Strava, Greece, where the coastal level subsided by 1.5 m after three earthquakes in 1981.

9:2 Determination of mean sea-level

It is now normal practice for national authorities to calculate monthly and annual mean sea levels from hourly values of observed sea-level. There are several ways of calculating the average value in order to eliminate the short-term changes of relatively large amplitude, due to tides and surges. Some care is necessary to avoid the high-frequency variations at periods of a few days or less being introduced into the monthly mean values. This phenomenon is called aliasing (see also the discussion of instrument design and data reduction in Section 2:4). One instrumental technique for performing this averaging is to restrict the entrance hole of a stilling-well so severely that not only are waves eliminated (Section 2:2:2), but so too are the tidal and surge level changes. For a well with a diameter of 0.30 m, a suitable tide elimination with a well time-constant of 2 days would theoretically be achieved with an orifice diameter of 0.5 mm. The practical difficulties of ensuring that such a small hole is not blocked by small suspended particles make this method of averaging generally unsuitable.

Arithmetic mean values

The most direct way of calculating monthly mean levels is to add together all the hourly values observed in the month, and then divide the total by the number of hours in the month. The annual mean level can be calculated from the sum of the monthly mean levels, weighted for the number of days in each month. Any days for which hourly values are missing should be excluded, but only these days are lost in the analysis. Small errors are introduced by the incomplete tidal cycle included at the end of the month. This method is used by many authorities because it requires little mathematical insight, yet produces values close to those obtained by more elaborate tide-eliminating techniques (Rossiter, 1958). The maximum contribution, due to aliasing of tidal changes, to a 30-day monthly mean sea-level is 0.055 per cent of the M_2 amplitude, 0.267 per cent of the K_1 amplitude and 0.401 per cent of the O_1 amplitude. Over a 365-day year the maximum M_2 error is 0.035 per cent. The S_2 component will of course average to zero over any period of complete days. If the averaging is done manually, the hourly levels can be tabulated in rows for hours and in columns for days; the monthly total may then be checked by summing each row and each column, and comparing the total for the rows with the total for the columns, although such laborious manual procedures are now seldom necessary.

Low-pass filtered mean values

The tidal aliasing of monthly mean levels is best removed by applying a low-pass numerical filter to the hourly values to get a smoothed daily noon value, before calculating the average of these values. Any of the filters detailed in Appendix 1 may be used for removing diurnal and semidiurnal tides. The Doodson X_0 filter, which requires only 39 hourly values, is suitable for data containing occasional gaps, because not too much data is lost on either side of the gap. The monthly means differ insignificantly from the means calculated after applying the longer 72-hour and 168-hour filters. The standard deviation of the differences between the 168-hour filtered monthly means at Newlyn over a year, and the corresponding 72-hour filtered means, the X_0 means and the arithmetic means, were 0.2 mm, 1.5 mm and 2.0 mm respectively. Even if the data is free of gaps, there is little point in applying more elaborate filters for mean sea-level studies alone, because the oceanographic variability is much greater than the errors in filtering.

3-hourly values

In some cases, where the labour of producing hourly levels is considered too great, a slightly less accurate value of the monthly mean may be obtained by averaging 3-hourly values (Rossiter, 1961). This method should not be used where shallow-water components at periods shorter than 6 hours are present.

If values are read from a chart at 0, 3, 6, 9, 12, 15, 18 and 21 hours for each complete day of data in the month and then added, a simple method is available to adjust the total for the incomplete tidal cycle at the end. Calculate the terms δ:

$$\delta = 0.5 \text{ (first 3 values following last day } - \text{ first 3 values on first day)}$$

of each block of daily values. For a complete month of data only one value of δ is calculated. The values of δ are added to the monthly total, which is then divided by $8n$, where n is the number of days of complete 3-hourly readings. This method of averaging, called the Z_0 filter, is only appropriate where hourly techniques cannot be used. However, the mean sea-level values obtained are usually reasonably close to the more accurate values.

Mean tide level

The mean tide level is the average of all the high and low water levels in a specified period. To give an equal number of both high and low waters the last maximum or minimum in a month may have to be omitted. Mean tide level is not the same as mean sea-level because of the influence of the shallow-water tidal harmonics, although variations in the two values are highly correlated. To illustrate the systematic difference for a semidiurnal tide, consider a sea-level with an M_2 and an M_4 component:

$$\zeta = Z_0(t) + H_{M2} \cos (2\omega_1 t - g_{M2}) + H_{M4} \cos (4\omega_1 t - g_{M4}) \qquad (9:1)$$

If H_{M2} is much greater than H_{M4}, ζ has a maximum value of:

$$Z_0(t) + H_{M2} + H_{M4} \cos (2g_{M2} - g_{M4})$$

and a minimum value of

$$Z_0(t) - H_{M2} + H_{M4} \cos (2g_{M2} - g_{M4})$$

The average of these two extremes gives a difference:

$$\text{Mean tide level} - Z_0(t) = H_{M4} \cos (2g_{M2} - g_{M4})$$

At Newlyn the systematic difference is $0.10 \cos (101°) = -0.02$ m, but in general it may be as large as $\pm H_{M4}$ if:

$$2g_{M2} - g_{M4} = 0, \pi$$

This systematic error is only introduced by even harmonics of the dominant tidal constituent because these shift both high and low waters in the same direction. Odd harmonics shift the high and low levels in opposite directions, thereby producing effects which cancel. There may be a similar distortion of mean tide level in a diurnal regime by interaction between the K_1 and the K_2 constituents. Many of the very old estimates of average sea levels were computed as mean tide level because there were no continuously recording instruments available.

9:3   Datum determination and stability

Although the accuracy of a single hourly reading of sea-level may be accurate at best to only 0.01 m, the mean of several hourly values is potentially much more accurate, provided that the error in each reading has a random probability distribution. If this distribution is statistically normal the mean of n independent readings is more accurate than a single reading by a factor of $n^{-\frac{1}{2}}$. On this basis, one hundred readings would reduce the error from 0.01 m to 1 mm. It would be unrealistic to claim any greater accuracy than this for mean monthly and annual sea levels, because of unidentifiable systematic errors.

The main reason why errors are unlikely to be less than 1 mm is that the original hourly readings will also contain systematic errors due, for example, to incorrect chart fitting, bias in reading the charts, or perhaps incorrect gauge calibration. These errors can only be reduced by careful probe checks of well levels against chart levels, and by sound observational techniques.

The major limitation to the study of sea-level changes over periods of many decades is the stability of the Tide Gauge Bench-mark datum (Section 2:2:2). Unfortunately, these marks are often destroyed in the process of harbour development. To enable the recovery of the original datum, and also to check against very local subsidence in the vicinity of the Tide Gauge Bench-mark, it is now customary to connect to at least three auxiliary marks. The levelling between Tide Gauge Zero and the Tide Gauge Bench-mark, and onward to the auxiliary marks should be checked annually to an accuracy of 1 mm for compatibility with the accuracy of the averaged tide gauge levels. The absolute reference level at a site is defined relative to the Tide Gauge Bench-mark.

The connection between a national geodetic levelling datum such as Ordnance Datum Newlyn or the United States National Geodetic Vertical Datum of 1929, and the local Tide Gauge Bench-mark is not an essential part of the mean sea-level definition. Formerly many authorities defined mean sea-level relative to their geodetic datum, with the inevitable result that each relevelling of a country, if it caused the local values of the national datum to be redefined, introduced discontinuities in the sea-level record which were not oceanographic in origin. Defining sea levels relative to the Tide Gauge Bench-mark has the advantage of clearly separating the oceanographic and the geodetic aspects of the problem of sea-level trends.

Monthly and annual mean sea-level series for a global network of stations are collected and published by the Permanent Service for Mean Sea Level, together with details of gauge location, data gaps and definitions of the datums to which the measurements are referred. If the datum history can be clearly established so that a homogeneous series of levels can be prepared, these levels are adjusted to a Revised Local Reference datum (RLR) which is *defined* relative to the Tide Gauge Bench-mark so that the mean level in a specified year is approximately 7.0 m. This arbitrary value of 7.0 m has been chosen to avoid troublesome negative quantities and to emphasize its difference from any other system of datum definition. The Permanent Service for Mean Sea Level holds bench-mark related data from over 1000 stations; of these stations, 112 have recorded data from before 1900. The longest record held, which is from Brest, France, begins in 1806, when recording gauges were not available. The record for Sheerness, Britain begins with Palmer's 1832 gauge, but is unfortunately not continuous. One of the longest existing records is for San Francisco, which the United States National Ocean Service has assembled by careful studies of bench-mark histories of three successive tide gauges on three slightly different sites since 1854 (Hicks, 1978; Smith, 1980).

* 9:4 Analysis of monthly mean sea levels

The monthly mean sea-level data must be analysed in a systematic and uniform way. At first glance the values plotted in Figure 9:2 appear very noisy. As with the analysis of the tidal components of sea-level change, the accepted analytical approach is to fit a mathematical function to the time series of $Z_0(t)$ by a least-squares technique. The function should contain a number of factors which are determined by the fitting process. The form of the fitted function should relate to the factors expected to influence sea-level in a way which allows these factors to be interpreted physically.

Careful examination of Figure 9:2 shows that there is an annual cycle present. Extending our knowledge of meteorological influences at periods of hours and days indicates that local air pressure and winds will be significant influences. The winds are often most conveniently parameterized as the gradients of the air pressure in space, the isobar separation, rather than as the measured speeds and directions; this approach is based on the geostrophic balance which exists between atmospheric pressure gradients and winds. The separation of mean sea-level changes and surge changes of sea-level is rather arbitrary but a division may be conveniently made at around 2 months, the Nyquist frequency (Section 4:2:5) associated with monthly samples of mean sea-level.

It is also necessary to include a long-term trend of sea levels in the analysis. At present there is insufficient evidence to justify the inclusion of any trend functions more elaborate than a linear upward or downward trend. However, there is now considerable interest, associated with studies of climate change, in the possible detection of significant increases or decreases in these trends.

A suitable expression incorporating these effects is:

$$
\begin{aligned}
Z_0(t) = \overline{Z}_0 + at & \qquad \text{long-term mean and trend} \\
+ \mathbf{N} + \mathbf{S}_a + \mathbf{S}_{sa} & \qquad \text{periodic tidal terms} \\
+ b_0 P_A + b_1 \frac{\partial P_A}{\partial x} + b_2 \frac{\partial P_A}{\partial y} & \qquad \text{meteorological effects} \\
+ e(t) & \qquad \text{residuals} \qquad (9:2)
\end{aligned}
$$

where $\overline{Z}_0$ is the long-term mean level, a is the linear trend in Z_0, $\mathbf{N}$ is the nodal tide of period 18.6 years, $\mathbf{S}_a$, $\mathbf{S}_{sa}$ are the seasonal variation and its first harmonic, P_A, $\partial P_A/\partial x$, $\partial P_A/\partial y$ are the local air pressure and its gradients, b_0, b_1, b_2 are coefficients relating the sea levels to the meteorology, and $e(t)$ are the residuals not included in the model; they will include errors in the measure-

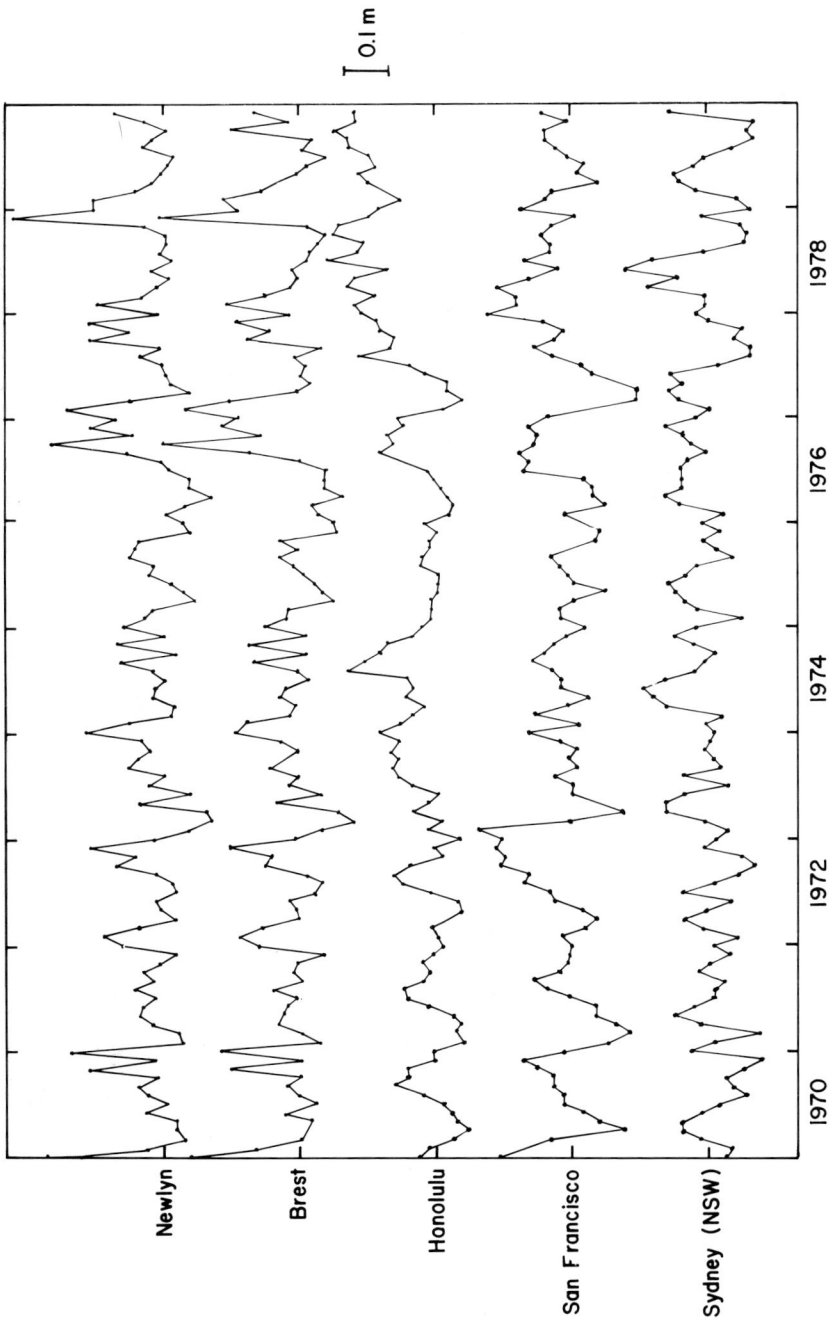

Figure 9:2   Ten years of monthly mean sea-level values from five ports.

ments as well as factors not included in the basic model. They may contain significant information about changes in the behaviour of the oceans.

The normal procedure would be to fit as long a series of $Z_0(t)$ as possible to the function (9:2), using a matrix inversion technique, similar to those used for tidal analyses (Chapter 4). Details of these techniques are beyond the scope of this book (Thompson, 1980). Here we are interested in the factors determined by the fit, Z_0, a, the amplitudes and phases of N, S_a and S_{sa}, and the meteorological terms b_0, b_1, b_2.

Table 9:1 gives the values for $\overline{Z}_0$, a, S_a and S_{sa} for the five stations and 10 years of Figure 9:2 and compares the initial and final standard deviations. The value of $\overline{Z}_0$ is for epoch 1970, and is close to 7000 mm in all cases because of the choice of a Revised Local Reference datum, relative to the specified benchmarks. The proportion of the total sea-level variance removed by the model which did not include the b_0, b_1 and b_2 terms is less than 50 per cent in all these analyses. For Newlyn and Brest, in more stormy northern latitudes, because of the omission from the model of all meteorological effects except those included in the S_a and S_{sa} terms, the variances removed are only 26 per cent and 19 per cent respectively. For both Newlyn and Brest, the proximity of a continental shelf has enhanced the importance of these weather effects. The best model fit is obtained at Honolulu which is an oceanic site located in low latitudes, where the weather effects are small and regular. The small 18.6-year nodal tide was not included in the fitted model because it could not be isolated from only ten years of data.

Fitting a model to observed data is only the first part of a proper scientific description; the next stage is to interpret the model parameters in terms of the different physical factors responsible for the sea-level changes.

9:5 Changes of mean sea-level with time

Changes of mean sea-level are due to the same basic factors which are responsible for changes at shorter periods: gravitational tides and the weather. For mean sea-level the direct and indirect effects of weather, particularly the seasonal and interannual changes in winds and solar heating, are the more important. Figure 9:3, which shows a spectrum of the monthly sea-level variations at Newlyn from 1915 to 1981, is an expanded low-frequency version of Figure 6:6. Even with such a long record of high quality data the statistical confidence limits are sufficiently broad for only the annual changes to be clearly defined. However, several other harmonics can be seen against the general background activity.

9:5:1 Seasonal and other periodic changes

For the five stations listed in Table 9:1 the annual harmonic S_a has amplitudes between 40 mm and 50 mm. This uniformity is misleading because more generally, the amplitudes observed from the global network of stations actually show considerable regional variations (Pattullo et al., 1955; Pattullo, 1963;

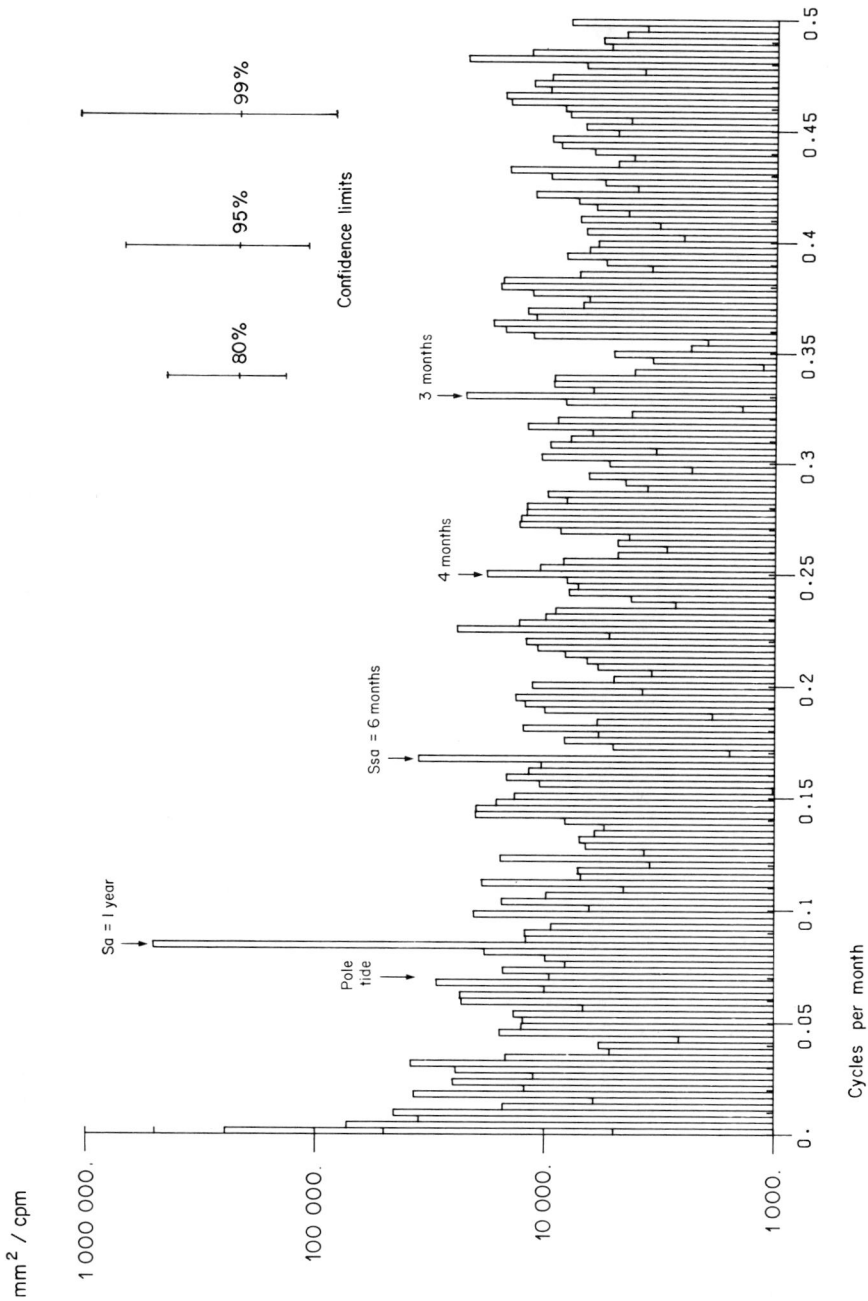

Figure 9:3   Low-frequency spectrum of 65 years of Newlyn sea levels. This is an expanded analysis of part of Figure 6:6.

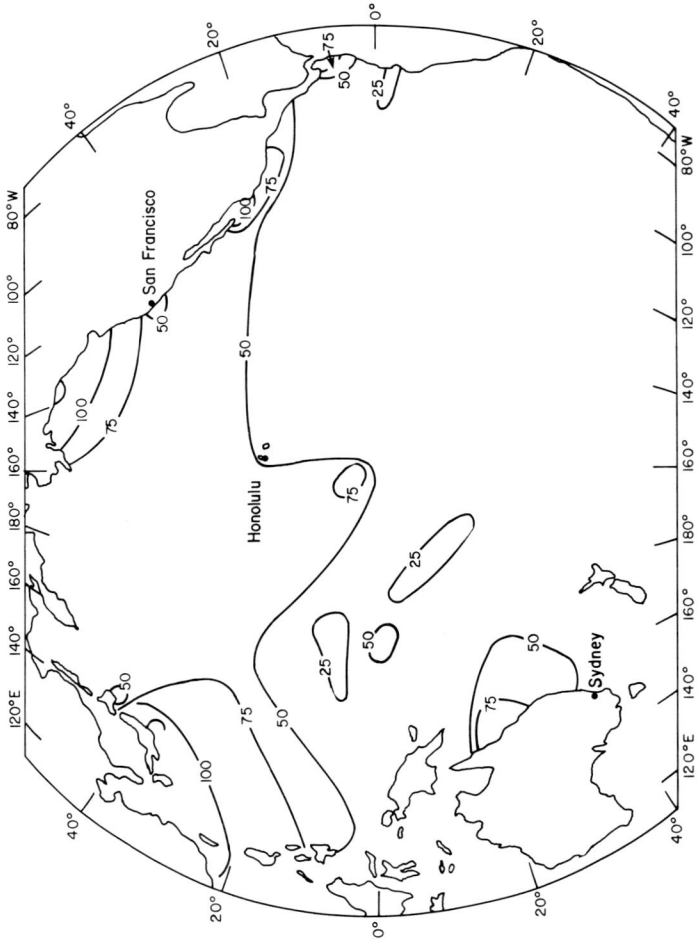

Amplitude of the annual variation H_t in millimetres

(a)

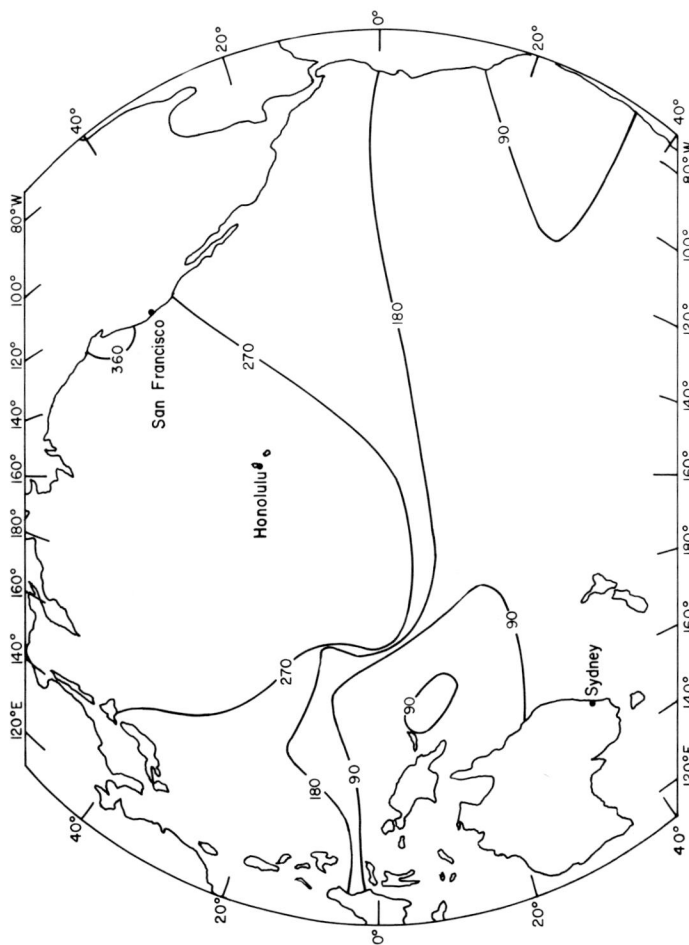

Phase of the annual variation g_{Sa} in degrees

(b)

Figure 9.4 Amplitudes and phases of the annual variations of sea-level in the Pacific Ocean. (From Wyrtki and Leslie, 1980.)

Woodworth, 1984). Analysis of Pacific sea levels has shown (Figure 9:4) that amplitudes in excess of 100 mm occur in the East China Sea and along the coast of southern Japan, whereas amplitudes of less than 25 mm are found at some ocean island groups, including the Marshall Islands and Samoa, and between the Galapagos Islands and Ecuador (Wyrtki and Leslie, 1980). The analyses in Table 9:1 show a maximum in the annual cycle during the northern hemisphere autumn at Honolulu and San Francisco and during the southern hemisphere autumn at Sydney. This difference between the northern and southern hemisphere is confirmed in Figure 9:4. Conveniently, the phases plotted are almost the same as the day in the calendar year for which S_a is a maximum. Along the coast of North America, from Mexico to Alaska, the month of this maximum increases systematically from September in the south to December or even January in the north. The limited data available from the South Pacific shows a maximum in March along the South American coast, but generally the maximum values in Australia and New Zealand occur slightly later, from April to June. Similar patterns are found in the Atlantic Ocean, but they have not been studied in the same detail.

The semi-annual term S_{sa} is almost as large as the annual term S_a (30–40 mm) for San Francisco and Sydney, but is only 6 mm at Honolulu. For most of the Pacific Ocean the semi-annual terms are of little significance, but at Truk Island and other stations in the western equatorial Pacific S_{sa} is more important than S_a. Even here, although the analyses give an amplitude of 35 mm, the oscillations are irregular, and the signal is often not obvious in the record from individual years.

The amplitudes and phases of the observed S_a and S_{sa} are more variable from year to year than those of the diurnal and semidiurnal harmonic constituents because of the year-to-year changes in the seasonal weather cycles. It is appropriate to describe them in terms of the average, coherent component, as determined in Table 9:1 and an irregular or anomalous component which has the same period. The close agreement between the seasonal variations at Newlyn and Brest, and the ocean contours plotted in Figure 9:4 both show that these variations are coherent over large areas of ocean.

Both S_a and S_{sa} include a small gravitational tide which arises from the long-period term (equation (3:12)):

$$\bar{\zeta} = a \frac{m_s}{m_e} \left(\frac{a}{R_s} \right)^3 \left(-\tfrac{3}{2} \sin^2 d_s - \tfrac{1}{2} \right) \left(-\tfrac{3}{2} \sin^2 \varphi - \tfrac{1}{2} \right) \tag{9:3}$$

where m_e and m_s are the mass of the earth and sun, a is the mean earth radius, R_s is the solar distance, d_s is the solar declination and φ is the station latitude.

Only the sun produces annual gravitational tides. The maximum Equilibrium amplitudes occur when $\varphi = 90°$, that is, at the poles where the latitude term is unity. The varying distance to the sun over a year, R_s, influences the annual harmonic S_a. The semi-annual effects are generated by the $\sin^2 d_s$ term as d_s cycles between 23° 27′ N and 23° 27′ S. The term in $\sin^2 \varphi$ which describes the latitude variation of the amplitudes, shows that at 35° 16′ north and south of

Table 9:1 Results of fitting a model: trend + annual cycle + semi-annual cycle to the monthly mean sea-level data plotted in Figure 9:2.

	Data standard deviation	Residual standard deviation	Model fit to original variance (%)	Trend (mm/year)	Mean level to 1970 (mm)	Datum	Sa Amplitude (mm)	Sa Maximum	Ssa Amplitude (mm)	Ssa Maximum
Newlyn	79	68	26	4.8	6996	11.7 m below BM SW 4677 2856	45	Dec.	16	Jan./July
Brest	82	74	19	5.0	6992	12.5 m below BM NGF1	48	Jan.	15	Feb./Aug.
Honolulu	56	40	49	−0.4	7012	12.0 m below BM no. 2	49	Sept.	6	Mar./Sept.
San Francisco	71	58	33	3.3	6997	10.1 m below BM no. 180	39	Nov.	33	Feb./Aug.
Sydney	63	48	42	−1.1	7000	12.7 m below BM 101	42	May	32	May/Nov.

the equator, the amplitudes of S_a and S_{sa} in the Equilibrium Tide are zero. During the annual cycle, the value of R_s varies by 1.7 per cent above and below its mean value. At present the nearest approach of the Earth to the sun, called the perihelion, occurs early in January. Entering the known values in equation (9:3) gives the Equilibrium Tidal constituents in mm:

$$H_{S_a} = -3(\tfrac{3}{2}\sin^2 \varphi - \tfrac{1}{2})\cos(h - p')$$

$$H_{S_{sa}} = -20(\tfrac{3}{2}\sin^2 \varphi - \tfrac{1}{2})\cos 2h$$

where h is the mean longitude of the sun, which increases by $0.0\,411°$ per mean solar hour and which is zero at the equinoxes on 21 March and $\pi/2$ on 21 June; p' is mean longitude of solar perigee (perihelion), which changes over a cycle of 21 000 years, the present value (year 2000) is 283°. Because the solar radiation effects at a frequency ω_3 dominate, modern practice is to dispense with the p' of the gravitational tide. The frequency change is irrelevant, but the reference phases are changed (see Section 4:2:1).

The negative sign implies a 180° phase change, but is not relevant for our discussions. Even these very small amplitudes must be reduced to allow for the elastic response of the solid earth to the tidal forces, and for the gravitational attraction of the tidal bulge on itself. This is done by multiplying by a factor of 0.69 which is derived from the geophysical 'Love numbers' (Section 3:2:3). The final amplitudes of 1.0 mm for the gravitational S_a component, and 6.8 mm for the gravitational S_{sa} component, are very small compared with the values for the constituents derived by analyses of observations, as shown in Table 9:1. Theoretical considerations of these long-period tides suggest that the ocean responses should give observed tides close in both amplitude and phase to those of the Equilibrium Tide; the substantial difference between the observed and the Equilibrium Tide makes it clear that the meteorology is the dominant factor responsible for the annual and semi-annual variations of sea-level.

For surges it is often supposed that an increase in atmospheric pressure will produce a compensating 'inverted barometer' decrease in sea-level with a 10-mm sea-level rise for a 1-mb fall in atmospheric pressure. In Section 6:3 we discussed why this is not strictly true for daily variations of air pressure; there is an additional reason why the relationship is not strictly true for the seasonal variations over the oceans. To understand why, consider an enclosed lake over which the atmospheric pressure changes slowly. There can be no compensating adjustment of the water level because the whole lake is similarly affected. In the same way, there is an annual cycle in the mean atmospheric pressure averaged over all the oceans, which reaches a minimum of 1012 mb in December and has a maximum of 1014 mb in July; the lower winter pressures over the oceans are mainly due to a shift of the air mass towards Siberia. This seasonal cycle in the mean ocean air pressures must be removed from the observed cycle in the atmospheric pressure at a particular site, before proceeding to calculate the effective seasonal 'inverted barometer' factor. Calculations of this corrected static S_a effect due to seasonal air-pressure variations have shown amplitudes of less than 30 mm for most of the oceans, with larger amplitudes of 60 mm to 80 mm in high northern latitudes. Generally the atmospheric pressure effects are small, but not negligible. It is not possible to isolate the seasonal changes due to direct atmospheric pressure changes from those due to effects of wind stresses on the sea surface, because of the correlations between winds and atmospheric pressure. For stations on continental shelves the seasonal effects of winds on sea-level may be more significant than the direct atmospheric pressure effect.

Changes of *steric levels* result from changes in the density of a column of sea water without a change of its total mass, due either to salinity or to temperature changes. Ocean gauges in lower latitudes and particularly in subtropical latitudes show good agreement between annual sea-level changes and those computed from observed density changes in the water. In the tropical Pacific Ocean sea-level fluctuations are a reliable measure of thermocline depth fluctuations between about 15° N and 15° S. They are also a good indication of the total heat content of the water (Rebert *et al.*, 1985). In the eastern tropical Atlantic Ocean, sea levels are clearly related to the total heat content of the upper 500 m of water (Verstraete, 1985). Lower sea levels in tropical regions is an indicator of potential upwelling of nutrient-rich deep colder water.

In the Bay of Bengal very large seasonal amplitudes in excess of 1 metre occur, due to seasonal monsoon effects and to changes in steric levels because of fresh-water flow. Even here, however, the conditions are close to isostatic. Other estuary regions will also be significantly affected by fresh-water flow in their annual cycle of sea-level changes.

There is a periodic component of the Equilibrium Tide which has a much longer period and a smaller amplitude than the seasonal variations (Pattullo *et al.*, 1955; Rossiter, 1967). This is the *nodal tide*, **N**, so called because it is due to the regression of the moon's node, described in Chapter 3, which has a period of 18.6 years. The Equilibrium form with the amplitude in mm is

$$H_N = 18 \; (\tfrac{3}{2} \sin^2\varphi - \tfrac{1}{2}) \cos N$$

where N is the mean longitude of the moon's ascending node. This is zero, corresponding to maximum amplitudes of the nodal tide in March 1969, November 1987, June 2006, and at 18.6-year intervals thereafter (Tables 4:2 and 4:3). These are also the times of minimum M_2 amplitudes because of nodal modulations, as explained in Section 4:2:2. In the same way as for the gravitational part of the annual and semi-annual tides, the Equilibrium amplitude should be reduced by the Love number factor to allow for the response of the earth to the tidal forcing (see Section 3:2:3). The value of 0.69 used for the diurnal and semidiurnal tides, which assumes an elastic Earth, may not be appropriate for the slower 18.6-year changes. Indeed, if the correct factor could be determined from analyses of sea-level records it would help to define the geophysical transition between the elastic and inelastic behaviour of the solid earth.

Unfortunately there are two reasons why this distinction may not be possible. Firstly, the amplitude of the nodal tide is very small, which makes it difficult to determine against the high level of background noise at low frequencies. At Newlyn, for example, the noise level in the spectral band containing the nodal tide has a variance more than three times the variance of the Equilibrium nodal tide of 7 mm amplitude.

An analysis of more than 6900 total years of European sea-level observations from several locations has been averaged (Rossiter, 1967) to give a mean amplitude and phase of 4.4 mm and 354°. This phase is sufficiently close to 0°/360° to support theories of an Equilibrium response, but the observed amplitude of 4.4 mm was less than expected. The other factor which could have affected this result is the generation of non-linear tides of low-frequency by the 18.6-year modulation in the amplitude of the diurnal and semidiurnal tides. The semidiurnal tides have maximum amplitudes when the nodal tide is a minimum. Such a cycle has been observed in sea temperatures and attributed to non-linear tidal mixing of surface and colder deeper waters being more effective when the currents are strongest (Loder and Garrett, 1978).

Another variation found in long sea-level records is the *pole tide*, generated by the Chandler Wobble of the axis of rotation of the earth with a period close to 436 days. Because the Chandler Wobble is not excited in a regular way, the phase is not constant, and so the phase of the resulting pole tide must also vary. One method of analysis is to compute the sea-level response to the displacements of the polar axis observed by astronomical methods. Even with very careful analysis the theoretical amplitude, which is typically 5 mm, is so small that very broad confidence limits are inevitable. The Newlyn spectrum in Figure 9:3 shows a peak near to 0.07 cycles per month which is probably due to the pole tide, but although labelled as such, it cannot be said to be significant within the calculated confidence limits. There are indications of a pole tide at other ports with long records, including both Honolulu and San Francisco, but the noise levels are again very high. However, in the North Sea and the Baltic Sea areas the amplitudes determined are larger. The nodal tide of more than 30 mm observed in the Gulf of Bothnia is several times the expected value. It is not known why this local amplification of the pole tide occurs (Miller and Wunsch, 1973).

There is a continuing search for other peaks and cycles in the long-period sea-level records, either in attempts to verify the predictions of a particular geophysical model, or as an application of some new statistical technique. For example, maximum entropy spectral analysis has shown peaks near to the nodal tide of 18.6 years and close to the solar sunspot cycle of 11 years, each with amplitudes of about 9 mm (Currie, 1981). The 11-year cycle may be generated indirectly by meteorological variations, but again the mechanism is not clear. At much shorter periods, Enfield (1986) has found 50-day oscillations in Pacific sea levels, which may be related to 7-week fluctuations in the length of the day (Hide, 1984).

9:5:2 Meteorological effects

We have already discussed the influence of atmospheric pressure and seasonal winds on the annual cycle S_a, and the semi-annual cycle S_{sa}. Figure 9:3 shows that these meteorological effects are also present at all other frequencies. Apart from their direct forcing, winds and air pressure may also modify the circulation patterns and surface gradients established on the oceans in their adjustment to geostrophic balance. In general, the direct meteorological effects on monthly and longer period changes of mean sea level are greatest in higher latitudes where storms are more common, and in the vicinity of wide continental shelves, as is also the case for the surges discussed in Chapter 6. This explains why the models summarized in Table 9:1 were least effective for Newlyn and for Brest.

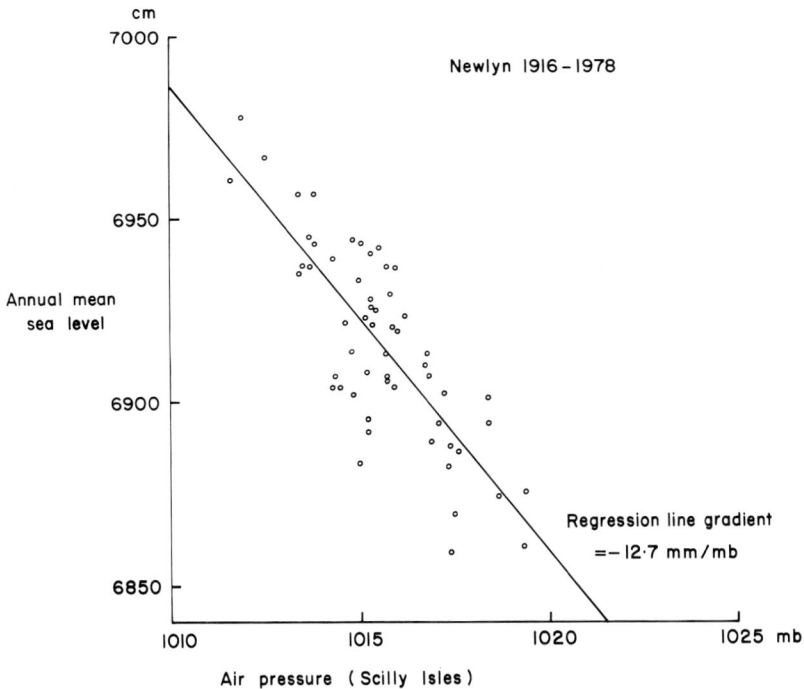

Figure 9:5 The relationship between annual mean levels and annual mean air pressures at Newlyn.

Figure 9:5 shows a graph of the annual mean atmospheric pressures plotted against the annual mean sea levels for Newlyn, for the period 1916–1975. Clearly higher annual air pressures are associated with lower mean sea levels, but the correspondence is not exactly the 'inverted barometer' response of − 10 mm for each millibar of pressure increase. The fitted regression line gives a coefficient of − 12.7 mm per mb. The systematic difference between the two is due to the additional effects of winds which are correlated with the air pressures. The more random variations may also be due to interannual oceanographic changes. At Southend, at the shallower southern end of the North Sea, a similar regression analysis gave -4.4 mm per mb, an even more dramatic departure from the 'inverted barometer' response.

In the model of equation (9:2) the winds are incorporated by atmospheric pressure gradients and by the statistical coefficients b_1 and b_2 rather than by direct observations. The usual way of computing the gradients is to use a set of at least three atmospheric pressure series, preferably with the stations located at the corners of an approximately equilateral triangle.

More elaborate multiple regression models can include several air-pressure stations. A model analysis (Thompson, 1980) which also incorporated a seasonal cycle and a trend factor and in which a network of nine stations was used to represent the meteorological forcing of British sea levels was able to account for more than 80 per cent of the sea-level variance. Theoretically the wind stress equation (6:12) requires a regression against the square of the wind speeds, but in practice this is found to make little improvement to the statistical fits (Garrett and Toulany, 1982; Pugh and Thompson, 1986).

The statistical removal of meteorological effects from mean sea-level changes makes it possible to reduce the background noise to an extent which allows other cycles and secular changes to be extracted with greater confidence (Rossiter, 1962; Pugh and Faull, 1983). However, the disadvantage of statistical methods is that the empirical constants may not be easy to interpret in any physical sense.

It should be remembered that the dynamic responses of the ocean as a whole to meteorological forcing on a global scale may result in sea-level gradients and current variations which are uncorrelated with the local weather. Some examples of these global adjustments are discussed in Section 9:6 which deals with sea-level gradients in space.

9:5:3 Secular trends and eustatic changes

The changes of mean sea-level over periods of decades or longer are of great importance for coastal development and for the design of coastal defences against flooding. Over periods of centuries or longer, the changes may be too great for any economically viable artificial barriers to give protection against flooding. However, it is sensible to allow in the design of coastal defence systems for predicted changes a century or so ahead. For the oceans as a whole, the level has risen tens of metres in the 10 000 years since the last glacial recession, due to the release of the water which was previously frozen in the polar ice-caps. The recent rate of rise is now much slower than previously, being only about 1.5 mm per year (Barnett, 1984); however, because there are considerable variations from this rough average, the value of a global mean, eustatic, increase has proved difficult to determine, and it is now recognized that this simple concept is not valid.

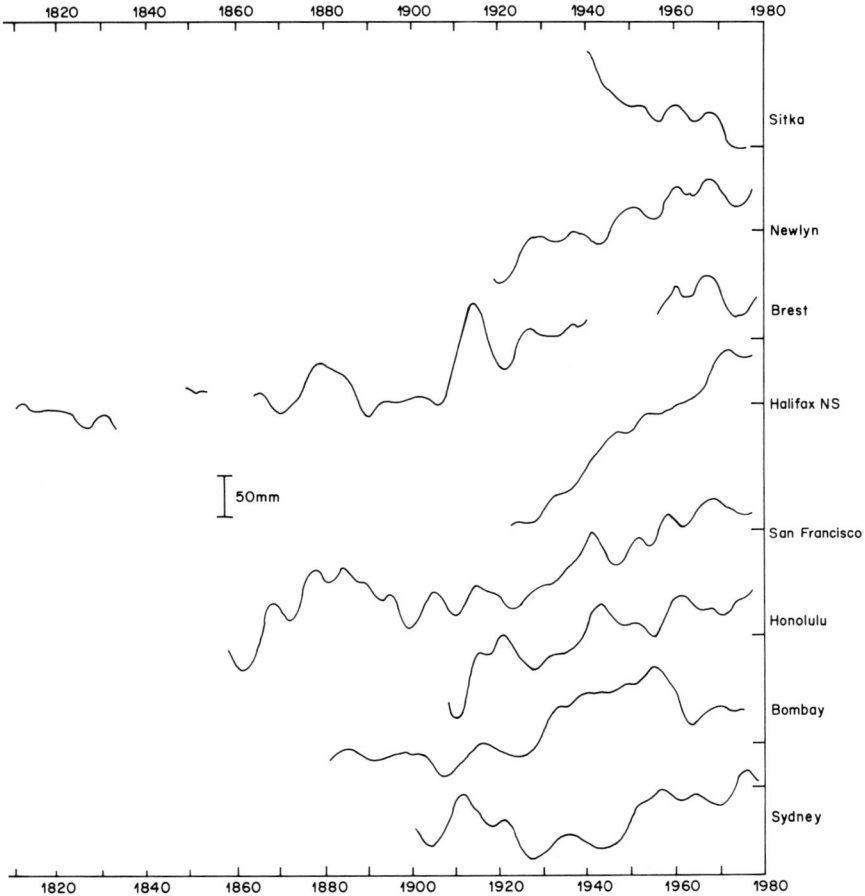

Figure 9:6   Low-frequency changes in annual mean sea levels at several stations showing a general upward trend and the interannual variability.

There have been several recent reviews of sea-level increases and the consequences of climate change. Barnett (1983a) gives a full discussion of the possible causes. Other valuable accounts are given by Revelle (1983), Barnett (1984), United States Department of Energy (1985) and Robin (1986). In this section we discuss the observed changes of sea-level and some of the factors responsible. The impact of sea-level change is discussed in more detail in Section 11:7.

Figure 9:6 shows the changes of sea-level at eight ports calculated from annual mean values by applying a seven-year filter with weights $(1, 2, 3, 4, 3, 2, 1)/16$. The trends shown are all upwards except for Sitka, Alaska, but these upwards movements are not constant. Superimposed on them is a background of interannual variability. This interannual variability is clearly

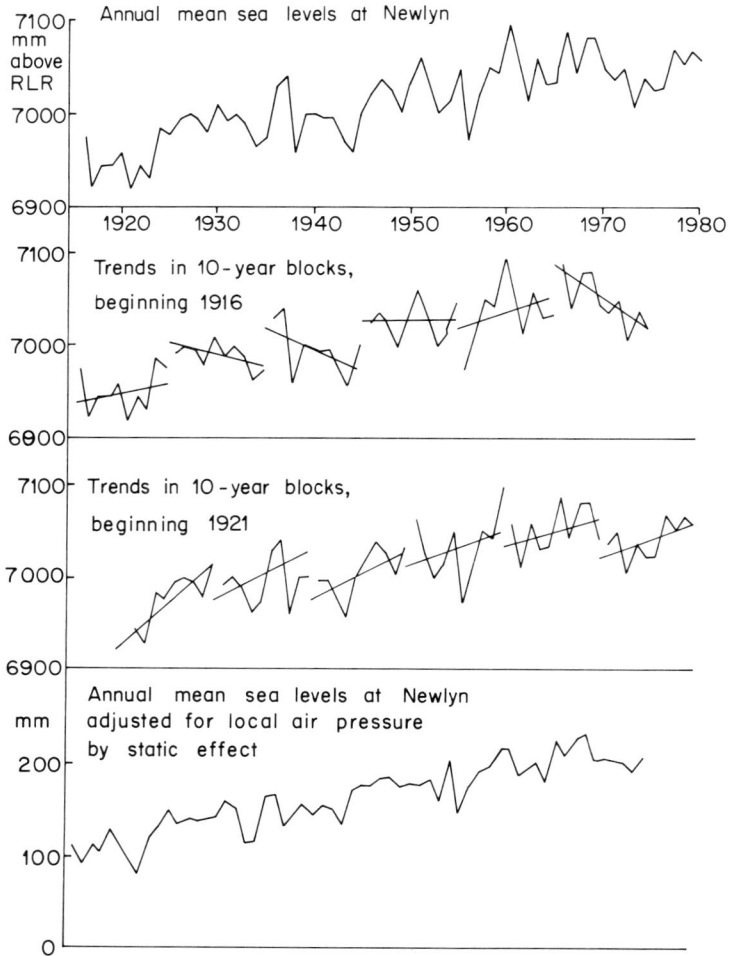

Figure 9:7 Comparison of trends in annual mean sea-level at Newlyn, taking successive ten-year blocks from 1915, and from 1921. The lower figure shows the reduction in the noise which results after correcting for the mean annual air pressure. (Copyright © 1983 The Institution of Civil Engineers.)

coherent between Newlyn and Brest, and similarities exist between the Honolulu and San Francisco records. The variations are not coherent, however, on a global scale which means that large-scale horizontal pressure gradients are sustained in the oceans over periods of years. The changes in weather patterns and in ocean circulation which are responsible for the interannual changes makes their study particularly relevant to an understanding of the climate changes which take place over several years. Their influence makes analysis of trends which are based on only a few years, or even a decade or so of annual mean levels, potentially misleading. Nevertheless, from time to time the popular press seize on these short-period trends, extrapolate them for a century or longer, and predict devastating consequences.

The problem of extrapolating short-period trends may be illustrated by an example. Figure 9:7 shows the annual uncorrected sea levels at Newlyn with trends fitted through successive 10-year blocks from 1916 onwards, and from 1921 onwards (Pugh and Faull, 1983). For three of the blocks sequenced from 1916 the trends are downwards, while the overall trend is still clearly a gradual increase in sea-level. Better estimates of trends can be made if the meteorological effects are first removed by statistically fitting a model. The lower trace shows the more regular result obtained by making just a simple direct 'inverted barometer' correction for the annual mean atmospheric pressure. Thompson (1986) has shown a further marked reduction in the Newlyn interannual sea-level variations by separating and removing the combined influence of local winds and air pressure. The trends fitted to the total periods of data available for the sites plotted in Figure 9:6 are given in Table 9:2. They confirm that the trend varied from a maximum upward rate at Halifax to a downward rate of change at Sitka. The 1940–1980 trends which are also given for comparison show that there has been no recent general increase or decrease in these local trends, a conclusion which is confirmed by the statistical fitting of more elaborate curves such as polynomials to the data. In almost every case the improved fits to the trends obtained by curves which allow rates to change with time, are not statistically significant.

Returning to Table 9:1, the 10-year trends over the 1970–1979 period are seen to be sometimes very different from the long-term trends given in Table 9:2 for the same site. Sydney and Honolulu have slight negative trends for the 1970–1979 period. If the values of upward trend at Newlyn and Brest were extrapolated over a century the increase would be around 0.5 m, whereas on the basis of previous evidence, the lower long-term trend is the more likely to prevail (but see Section 11:7).

Table 9:2 Fits of linear trends to long series of annual mean sea levels (Figure 9:6); the estimated rates of sea-level rise are given in mm per year. Standard errors are given in parentheses.

| | Latitude | Period | Total years | Trend (s.e.) | |
				Complete series	1940–end
Sitka (Alaska)	57° 03′ N	1939–1979	41	−2.5(0.4)	−2.6(0.3)
Newlyn	50° 06′ N	1916–1980	65	1.7(0.2)	1.5(0.4)
Brest	48° 23′ N	1807–1981	141	0.9(0.1)	0.0(0.5)
Halifax, Nova Scotia	44° 40′ N	1897–1980	62	3.7(0.1)	3.2(0.3)
San Francisco	37° 48′ N	1854–1979	127	3.8(1.5)	1.5(0.4)
Honolulu	21° 19′ N	1905–1980	76	1.6(0.2)	0.8(0.4)
Bombay	18° 55′ N	1878–1978	101	1.0(0.1)	−1.0(0.4)
Sydney	33° 51′ S	1897–1981	85	0.7(0.1)	2.0(0.3)

In Alaska and Northern Scandinavia there are large downward trends of sea-level. Juneau, Alaska, has a downward trend of more than 1.0 m per century although it is barely 150 km north of Sitka, where the downward rate is only 0.25 m per century. Large downward trends of sea-level are also observed in the northern Baltic Sea, particularly in the Gulf of Bothnia along the coast of both Sweden and Finland. Of course, these apparent downward trends of sea-level which are often observed in high latitudes are actually due to vertical upward land movements. One of the major difficulties in the study of longer-term changes of level is the separation of sea-level changes from the eperiogenic changes of land level. Because mean sea-level changes are measured relative to a fixed bench-mark on land, the observed secular changes may be due to the movement of the fixed bench-mark relative to the geoid, or to a change of mean sea-level. In most cases both effects are present. Eperiogenic vertical land movements are discussed in section 10.6.

Sea-level changes may be due to a change in the volume of water in the oceans, a change in the shape of the ocean basins or long-term changes in the pattern of circulation and atmospheric pressure which control the shape of the sea surface relative to the geoid. The shape of the sea surface is considered in Section 9:6. In this section we are concerned with the two effects, changes in water volume and the shape of the ocean basins. These will raise or lower sea-level over the global oceans uniformly, producing what are defined as the eustatic changes. Their effect is slightly to raise or lower the level of the geoid. However, it is important to realize that the idea of a simple eustatic change of sea-level is incorrect, as the redistribution of water mass leads to a new geoid shape and there are also isostatic adjustments in the earth's interior (see Section 10:7).

Perhaps the most obvious reason for changes in the volume of water in the oceans is the melting of glaciers and polar ice. These changes are called glacial eustacy. At present, knowledge of the changing volumes of the ice-caps is not good enough to say whether all of the recent changes could be due to melting ice. Melting of glaciers other than those in Antarctica and Greenland may also be a significant factor (Meier, 1984). The West Antarctic ice-sheet and the Greenland ice-sheet are important potential sources of melting ice. It has been calculated that a reduction of 2 m in the thickness of the West Antarctic ice-sheet would produce a 10 mm increase in global sea-level. If the melting were greatly accelerated because of the warming of the earth by a 'greenhouse effect' due to increases in atmospheric carbon dioxide, the consequent flooding could be serious. Meteorological measurements show that during the period between 1890 and 1940 the mean surface air temperature of the northern hemisphere rose between 0.3 °C and 0.6 °C. The southern hemisphere shows a slightly smaller increase but the data are sparse and unrepresentative. Between 1940 and the mid-1970s there were relatively steady conditions, followed by a rapid warming (Jones et al., 1986). The sea-level record fails to show similar trends, but it is not inevitable that the link between air temperatures and sea-level should be so direct.

Statistical correlations between sea-level and the global temperature trend, allowing for a time difference, have suggested an 18-year lag of sea-level on temperatures, with an increase of 0.16 m for each 1 °C (Gornitz *et al.*, 1982). The results are only indicative of possible relationship because there is at present insufficient data to have much confidence in the correlations. If significant melting of the polar ice is taking place this will have an initial tendency to slow down the global temperature increase because the latent heat required to melt the ice will not be available to contribute to the global warming. Global warming or cooling must also change the temperature of the upper layers of the ocean. The resulting thermal expansion would cause increases in sea-level which would be included in the 0.16 m per °C coefficient, and which would account for a time delay due to the thermal inertia of the water. Estimates show that if the top 70 m of the oceans were warmed by 1 °C and the lower levels by an amount which decreases exponentially with a scale height of 367 m, there would be an 80 mm increase in surface levels. Analysis of the surface water density structure in selected regions has shown no significant trends (Barnett, 1983b), but the early measurements were relatively crude, and it is difficult to allow for the evolution and application of different measuring techniques through the period. The deeper layers of the ocean would probably cool slightly during the initial stages of enhanced glacial melting due to the increased rate of formation of Antarctic and Arctic bottom water. However, at the low temperatures (0 °C to 4 °C) of the deep water, the coefficient of thermal expansion is much less than for the warmer near-surface waters, so that changes in the temperature of deep waters would have a smaller effect on sea levels.

Sea-level measurements alone, even if they could uniquely define a eustatic trend, could not distinguish between volume increases which were due to melting of polar ice-caps, and those due to thermal expansion. Direct measurements of changes in the surface levels and volumes of the ice-caps may help separate the two contributing factors. Other geophysical measurements are also relevant to the problem (Barnett, 1983a). Melting of polar ice-caps, and the subsequent distribution of the water mass over the surface of the oceans, particularly in equatorial latitudes, will increase the moment of inertia of the earth; conservation of angular momentum requires a compensating small decrease in the rate of rotation which can be measured as a slight increase in the length of the day. A rise of 10 mm in sea-level would increase the length of the day by 0.06 ms. Increases of perhaps 5 ms in the length of the day have been observed since 1850, but other factors which include momentum losses due to tidal friction are also involved (see section 7:9). Even if all the estimated eustatic rise of sea-level were due to the melting of polar ice, it would only account for 1 ms of the observed change. Changes in volume due to thermal expansion of the water would produce a negligible change in the earth's moment of inertia and in the length of day, because the redistribution of mass is not significant. In theory, melting of polar ice may also be distinguished from thermal expansion of the water because the position of the earth's pole of rotation is very sensitive to small changes in the mass of the polar ice and in its position. It should be possible to distinguish between the melting of Antarctic ice and the melting of Greenland ice because the Greenland ice is less symmetrically positioned about the mean axis of rotation. For example a 150 mm increase due to melting of Greenland ice would displace the pole 8.3 m towards 37° W. Astronomical observations of the axis of rotation are consistent with the total eustatic rise being due to the melting of equal quantities of Antarctic and Greenland ice, but even after averaging out short period effects such as the Chandler Wobble, the interannual polar movements are rather irregular.

Changes in the shape of ocean basins are unlikely to contribute significantly to the observed sea-level increases. Rates of sedimentation in the oceans are of the order of 1 mm per century, and the increases of ocean dimensions due to constructive plate tectonics at the ocean ridges would be equally negligible even if the compensating subduction of crustal material elsewhere is ignored. Evidence for sea-level changes over geological times are discussed in Section 10:6.

The problem of separating eustatic and eperiogenic contributions to the secular trend at a particular site remains. If absolute gravity could be measured at the tide gauge site at regular intervals, vertical movement of the land would be observed. However, a change of gravity of 6 μ gals, equivalent to 30 mm of vertical land movement, is the present limit to measurements of absolute gravity (6 parts in 10^9). Gravity differences between two sites can be measured to 3 μ gals, equivalent to level differences of 15 mm. Modern geodetic methods for position fixing, including Very Long Baseline Interferometry (VLBI), the satellite-based Global Positioning System, and Laser Ranging to satellites are all capable of measuring the relative positions of two sites to within a few centimetres (Carter et al., 1986). Of course, even if a survey network of existing tide gauge stations were fixed now to these accuracies, it would be several decades before significant trends would be detectable. Measurement of movement by space techniques is more satisfactory than measurement by gravity changes because the latter may also include mass adjustment within the Earth. With both gravity measurements and geodetic ranging, the short-term movements of the coastal site due to tidal loading and other local effects have to be eliminated by averaging over a suitably long period.

To summarize, the best recent analyses of eustatic sea-level trends (Barnett, 1984. See also Etkins and Epstein, 1982; Gornitz et al., 1982; Barnett, 1983a) based on a global average, and assuming positive and negative eperiogenic movements are randomly distributed, give increases from 120 to 150 mm per century. There are not enough observations to determine the source of the present sea-level increase, but observations of astronomical and climate change are compatible with this upward trend being due to the melting of ice from Antarctica and Greenland. However, only the average trends over a century in the several related observations are mutually consistent. Uncorrelated variations over several decades are found in all the series. In addition to ice melting, some contribution to the observed increase of levels from warming of the surface layers of the ocean is also likely.

9:6 Changes of mean sea-level in space

If the waters of the ocean were all of the same density, if they remained motionless relative to the earth, and if the atmospheric pressure acting on them were the same everywhere, then the water surface would coincide with the geoid. This would be the total geoid including not only the potential due to the mass distribution within the earth and the effects of the earth's rotation on its axis, but also the permanent part of the tidal potential due to the moon and sun. Because these ideal conditions do not apply for the real oceans, the sea surface is displaced from the geoid, usually by a few decimetres but sometimes by more than a metre. These excursions of mean sea-level from the geoid are roughly two orders of magnitude less than the geoid's deviations from an ellipsoid of revolution.

Considering first the long-term mean distributions of air pressure, these may be assumed, to a first approximation, to influence sea-level according to the inverted barometer principle. However, as previously discussed for the seasonal changes, the additional effect of the mean winds associated with the mean pressure distributions is very difficult to determine. One approach is to use a detailed numerical model of the area of interest and to compute the sea's response to the observed winds. Another method involves computing the statistical relationships between the low-frequency changes of sea levels and the winds, and then to assume that these responses can be extrapolated to and used at zero frequency, to give the permanent effects on sea-level. However, both methods involve assumptions which are not completely satisfactory. As we have previously noted, the effects of winds on sea-surface gradients are roughly inversely proportional to the water depth, so that they become more important for mean sea levels on continental shelf regions.

The other factor which causes sea-level gradients is the variation of water density. Classical oceanography assumed that at some depth in the ocean there are no currents, and therefore no horizontal pressure gradients. Above this level, a knowledge of the temperature and salinity profiles enabled the water density profile to be calculated. The horizontal differences of density could then be computed at several levels, extending from above the 'level of no motion' up to the sea surface. In the absence of any other forces and provided that a state of equilibrium has been reached, these pressure gradients must be balanced geostrophically by a current flowing at right angles to the gradient. From equation (3:23) we have:

$$fv = \frac{1}{\rho} \frac{\partial P}{\partial x} \qquad (9:4)$$

where $f = 2\omega_s \sin \varphi$ is the Coriolis parameter, φ is the latitude, ω_s is the rotation rate of the earth, ρ is the water density, $\partial P/\partial x$ is the horizontal pressure gradient; v is then the current speed perpendicular to the X direction.

The pressure gradient just below the sea surface gives not only the difference in levels, but also the average speed of the surface currents across the section between the stations. For any pair of deep sea hydrographic stations these sea

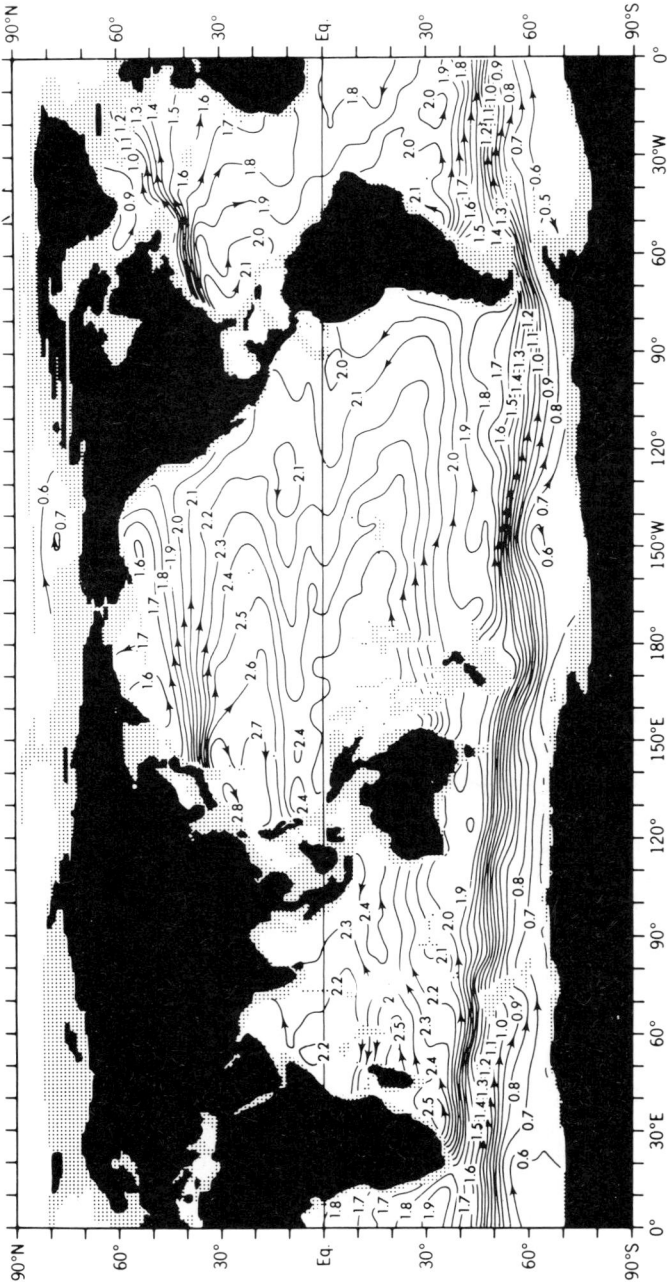

Figure 9.8 The global dynamic sea-surface topography relative to the 2000-dbar level, determined from historic hydrographic data. Contours are in geopotential metres (see Section 3:2:1). (From S. Levitus, 1982, *Climatological Atlas of the World's Oceans* Fig. 67, United States Department of Commerce, National Oceanic and Atmospheric Administration, Professional Paper, No. 13.)

surface gradients are indirectly measured in terms of the water temperature and salinity. Hydrographic surveys have the fundamental disadvantage of being slow and expensive to operate.

Figure 9:8 shows the distribution of steric levels, the different heights of mean sea-level, assuming no motion at a depth of 2000 m. The lowest sea levels are found off Antarctica, whereas the highest levels are found in the tropics and particularly in the tropical Indian and Pacific Oceans, where the average densities in the vertical water column are lowest. The difference of these steric levels across the Panama Canal explains why Pacific levels are some 0.20 m higher than the Atlantic levels. In the northern hemisphere the flow is in the direction which has the upward slope to the right of the direction of flow, but in the southern hemisphere the slope is up to the left. In Figure 9:8 the surface gradients in the Southern Ocean, being upward to the north, are consistent with the strong west-to-east circulation.

If the sea surface were mapped to an accuracy of a few centimetres relative to the geoid by satellites which covered the whole surface in a few days as described in Section 2:2:6, the assumptions of a level of no motion would be unnecessary, because the geostrophic currents at the surface would be directly determined and the interior currents could be computed relative to this from the water density distributions. Measurement of permanent currents from satellite altimetry data is limited by our knowledge of the geoid, as explained in Sections 2:2:6 and 4:2:7. Nevertheless, variations in surface levels and circulation patterns can be determined. Figure 9:8 shows the average long-term deformation of the sea surface, but there are also surface level changes which persist for a few years or so. These interannual variations which are clearly visible in Figure 9:6 form a background against which secular trends must be distinguished. For ocean stations these changes of level must be associated with interannual changes of circulation patterns, but in shallower waters a variation in the local winds may be more directly responsible.

Some examples from the Pacific Ocean show how monitoring interannual changes of sea levels can be used to describe ocean dynamics. The North Equatorial Current which flows from east to west, roughly between 10° N and 20° N, is associated with a level increase from south to north. This gradient, which can be monitored by measuring the sea-level difference between Honolulu (21° N, 158° W) and Kwajalein (9° N, 168° E), produces level differences of perhaps 0.30 m. However, there are large variations in the difference between 0.15 m and 0.45 m, implying weaker or stronger currents (Wyrtki, 1979). Sometimes a clear annual signal is present, such as that in Figure 9:2 for 1970 and 1971 at Honolulu, but at other times the fluctuations are less regular. When the monthly mean sea levels predicted by a model such as that represented by equation (9:2) (but excluding the air-pressure gradient terms), are subtracted from the observed levels, the residuals $e(t)$ may be plotted and contoured. Figure 9:9 shows that these residuals may have coherence on an ocean scale. The occurrence of an accumulation of warm water and higher sea levels in the Western Pacific shown clearly in this example has been related (Wyrtki, 1979;

1985) to the onset of El Niño, the irregular appearance of unusually warm water off the coast of Peru on the eastern side of the ocean. These high surface temperatures have an effect on the winds, which in turn inhibits ocean upwelling. The consequent loss of nutrients which would normally be available for the development of the biomass reduces the fishing catches to levels which may be economically disastrous.

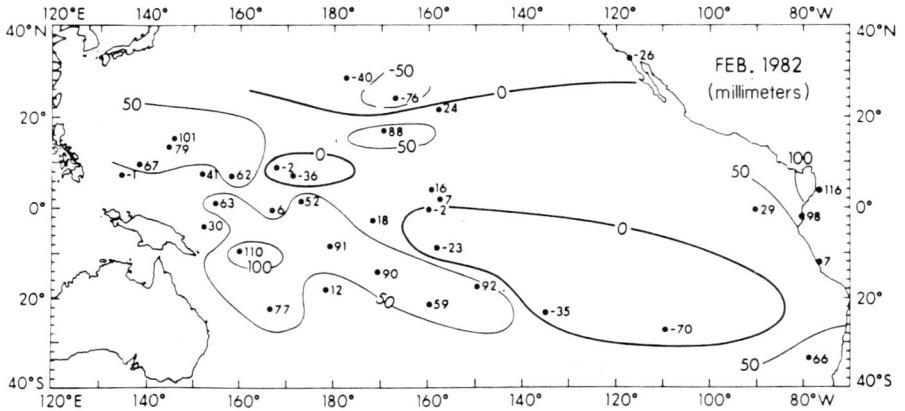

Figure 9:9   The deviations of the mean sea-level from the normal level in the tropical Pacific Ocean during February 1982, showing their coherence on an ocean scale. Values in millimetres show high sea levels in the Western Pacific and lower sea levels to the east, prior to the 1982/1983 El Niño (Wyrtki, 1985).

The mechanism which makes what appears to be an improbable connection between anomalously high mean sea levels in the Western Pacific and the subsequent collapse of a major factor in the Peruvian economy involves a chain of related events. El Niño is known to be preceded by a prolonged period of southeast trade winds lasting more than 18 months. This wind increases sea levels by about 0.10 m above the average levels in the western equatorial Pacific Ocean as shown in Figure 9:9, at the expense of levels in the east Pacific which are lowered. When the winds eventually relax their stress on the sea surface, the accumulated water begins to adjust to the now unbalanced pressure gradient along the equator, by transmitting an internal equatorial Kelvin wave eastwards along the thermocline interface. The depth of the thermocline may be depressed by tens of metres off Peru, which in turn prevents the normal upwelling of the nutrient-rich waters below the thermocline. There is a related rise of mean sea-level off Peru and a gradual fall in the Western Pacific. These apparently remote effects are really part of a complicated ocean–atmosphere interaction which, if fully understood, would enable forecasts of climate changes, perhaps over time scales of years. Even without an understanding of the full dynamic system, monitoring the sea-level anomalies and the relaxations of the winds enables valuable forecasts of El Niño, several months ahead.

SEA LEVEL

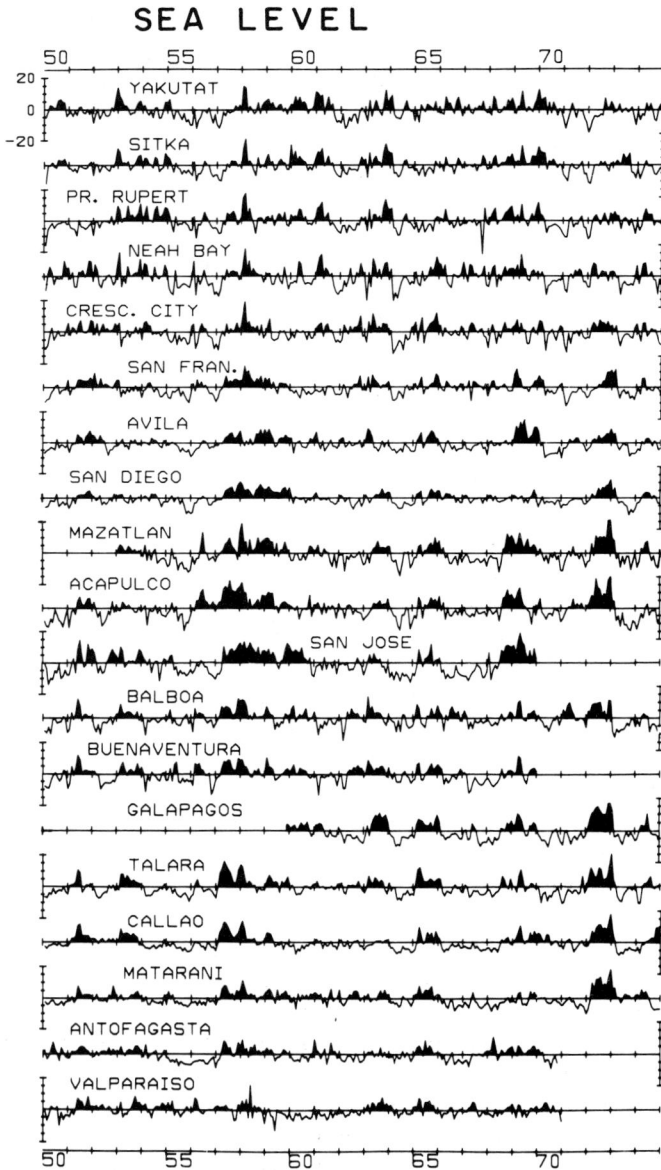

Figure 9:10   Time series of monthly anomalies of sea-level along the east Pacific coast of America from 1950 to 1974. Positive anomalies are shaded black (Enfield and Allen, 1980).

The coastal sea-level variations from the long-term mean, associated with El Niño are not restricted to the coast of Peru. Figure 9:10 shows the monthly mean sea-level anomalies along the whole of the eastern Pacific coast from Valparaiso (33° S) in Chile to Yakutat (60° N) Alaska, over the period from

1950–75 (Enfield and Allen, 1980). The large positive anomalies which are clearly visible between Matarani and Galapagos are associated with the El Niño events of 1957–8, 1965–6 and 1972–3. Statistically these anomalies can be traced as far south as Valparaiso and as far north as San Francisco (the 1972–3 event can also be seen in Figure 9:2). The anomalies probably propagate north and south from the El Niño region as coastally trapped internal Kelvin waves. North of San Francisco the anomalies are highly correlated with the local wind stresses, showing the increased importance of the energetic winter storms at higher latitudes, where they have generated local shelf responses which mask the disturbances freely propagating from near the equator. The levels from Crescent City to Sitka all have a maximum correlation with the local alongshore wind stress (see Section 6:4:5). These local effects on coastal sea-level anomalies at higher latitudes means that they cannot be interpreted directly in terms of ocean dynamics. If necessary, sea surface profiles measured by satellite altimetry could enable coastal measurements to be adjusted for shelf effects and projected outwards across the shelf edge for computing ocean circulation.

The most direct application of equation (9:4) is to the determination of flow through straits. Wyrtki (1961) studied the annual variation of the circulation in Indonesian Waters. Flow through the Drake Passage has been monitored over the period 1976–9 by bottom pressure measurements on either side (Wearn and Baker, 1980). Pressures at the bottom in 500 m water depth showed fluctuations of about 25 mbar on both the north and south sides, over periods from weeks to a year or longer. In these latitudes a 10-mbar pressure difference across the Passage corresponds to a mean velocity of $0.011 \, \mathrm{m \, s^{-1}}$. The observed pressure fluctuations were equivalent to current variations of $0.033 \, \mathrm{m \, s^{-1}}$. In the Florida Current, water transport has been related to coastal sea levels and bottom pressures (Maul et al., 1985; Maul, 1986); sea levels on the west side of the Strait of Florida, corrected for weather effects, were strongly correlated with the transport, but the strongest correlation was with bottom pressure measurements in 50 m depth off Jupiter, Florida. Near-shore effects on sea levels limit the use of coastal gauges for computing flow. Pugh and Thompson (1986) found that coastal levels around the Celtic Sea have a residual noise level, which is not coherent with the regional dynamics, of 0.02 to 0.03 m, whereas the bottom pressure measurements have a residual noise level of about 0.5 mb, equivalent to 0.005 m. This makes them potentially more useful for defining regional pressure gradients, and hence regional dynamics. Cartwright and Crease (1963) used currents through the Dover Straits, estimated by other means, to compute the sea-surface gradient, and hence to transfer datum levels, as discussed in the following section.

9:7 Mean sea-level, surveying and the geoid

Surveyors have traditionally attempted to measure heights relative to the geoid because it was conceived to be the true figure of the earth. Sea-level was thought to be a close approximation to this abstract shape, and so it became standard practice for each country or group of countries to evaluate mean sea-level for a selected tide gauge and to relate the datum of all future levelling to this particular plane. We now know that permanent gradients on the sea surface due to currents, density changes, atmospheric pressure and winds cause the mean level to vary from the undisturbed geoid level by more than a metre (Figure 9:8). Nevertheless, the original concept of mean sea-level being used to define the geoid was very farsighted and has stood for a long time as a close approximation to the reality of the original concept.

Mean sea-level can only be measured and related to land levels along the coast, but the geoid is conceived as a continuous completely closed surface. Under the continents the geoid can be thought of as equivalent to the water surface which would exist on narrow sea-level canals cut through the land masses.

To determine a true mean sea-level, measurements must be made over long periods of time in order to average tidal variations. However, no matter how long a period of data is averaged, as Figures 9:2 and 9:6 show, the ideal true mean sea-level is unattainable because changes are taking place over short and long time scales. In Britain the first primary network of levelling was made between 1840 and 1860, to a datum which was supposed to approximate mean sea-level at Liverpool (Doodson and Warburg, 1941). The sea levels were measured from 7 to 16 March 1844, so the approximation must have been very crude. For the national relevelling which began in 1912, three gauges were established, one at Dunbar in Scotland, one at Felixstowe on the east coast, and one at Newlyn in the extreme southwest. Separate sea-level determinations at these three gauges were supposed to give some control over the adjustments of the levelling network. When the survey was complete, the differences between the land levelling and the mean sea levels were found to be larger than the supposed errors within the levelling itself (Close, 1922). After careful consideration, the British Ordnance Survey decided to relate all levels to the mean sea-level at Newlyn, where distortions due to local shelf oceanographic effects should be smallest, and asserted that the 0.25 m higher sea-level apparent at Dunbar was a genuine sea-level difference. Hourly levels from 1 May 1915 to 30 April 1921 were averaged and Ordnance Datum defined as 9.412 m above Observatory Zero which in turn was defined as 25 feet below a specially constructed Fundamental Bench-mark. The Newlyn records which have been maintained continuously since 1915 now constitute an invaluable data set for studies of sea-level and tidal trends. The recent mean levels are some 0.10 m above the original mean level, and Figure 9:6 shows that the Newlyn gauge was established during a period when sea levels at Brest and at Newlyn were recovering from levels which were below the long-term upward trend.

Surveys of the United States showed consistently higher mean sea levels on the Pacific coast than on the Atlantic coast, and also a gradual apparent rise of sea-level to the north. The difference between Pacific and Atlantic levels was due at least in part to real oceanographic differences (Reid, 1961; and Figure 9:8). The survey authority took a different view to the British Ordnance Survey:

it seemed desirable for the zero of the geodetic levelling network to coincide with local mean sea-level, where both quantities were known. A general adjustment was made in 1929, in which it was assumed that mean sea-level at 26 selected tide gauge sites in the United States and Canada coincided with a 'true' geoid. This datum is known as the 'Sea Level Datum of 1929'. The various mean sea levels were determined at different times between 1853 and 1928 over periods sometimes as short as two months, although for most sites the recording time extended over at least a year. Unfortunately, the oceanographic variability such as that shown in Figures 9:2, 9:6 and 9:8 will have been embedded into this adjusted levelling network. Before adjustment, relative to the mean sea-level at Galveston, Texas, the maximum apparent mean sea-level excursions were $+0.59$ m at Fort Stevens, Oregon and -0.28 m at Old Point Comfort, Virginia. The 1929 Datum is to be replaced shortly (Zilkoski, 1985) by a new adjustment termed NAVD 88. Similarly, the Great Trigonometrical Survey of India in the late nineteenth century assumed all mean sea-level differences were due to errors in the levelling and adjusted the heights of the intermediate points accordingly (Eccles, 1901).

The levelling networks of several European countries from Spain and Portugal to Norway have been related to a common datum, Normal Amsterdam Peil (NAP), under the auspices of the International Association for Geodesy. The possibility of providing closure checks on the combined network by comparing mean sea-levels at several coastal stations has occasionally stimulated new work to estimate the gradients of the sea-level surface on the north-west European continental shelf.

The idea of correcting the real sea surface for oceanographic and meteorological effects, to obtain a 'true' geoid level is called oceanographic or hydrodynamic levelling. Applying these techniques to the Atlantic Ocean and to the shelf seas of northwest Europe shows that there is a decrease of mean air pressure to the north of approximately 0.4 mb per 100 km, equivalent to a 0.04 m increase in adjusted sea between Newlyn and the north of Scotland. However, the main effect is a raising of levels from west to east due to the winds associated with these mean atmospheric pressure gradients. Average wind and air pressure conditions were shown to lower Newlyn levels by 0.04 m while levels in the northern Baltic Sea were increased by up to 0.20 m. Hydrodynamic levelling has also been applied to the problem of datum transfer between Europe and Britain by careful analysis of the mean sea levels across that Strait of Dover between Ramsgate and Dunkerque (Cartwright and Crease, 1963). The computed higher level at Dunkerque, 0.079 m above the mean level at Ramsgate was largely due (0.055 m) to the geostrophic slope associated with a mean northward current through the Dover Strait, from the Channel into the North Sea. The remaining small adjustments accounted for the effects of the mean wind, for air pressure and for the local rotation of the mean currents. Adjusting the mean levels on the two coasts for this slope places Ordnance Datum Newlyn at 0.196 m below the 'Zero officiel du Nivellement General de la France' (NGF). This is consistent with an expected downward slope on the mean ocean surface relative to the geoid between Marseille, where NGF was determined, and Newlyn. (NGF was calculated as the mean sea-level between February 1885 and January 1897.) It should be realized, however, that transfer of levels over such long distances with an accuracy of a few centimetres makes very severe demands on the techniques of geodetic land levelling.

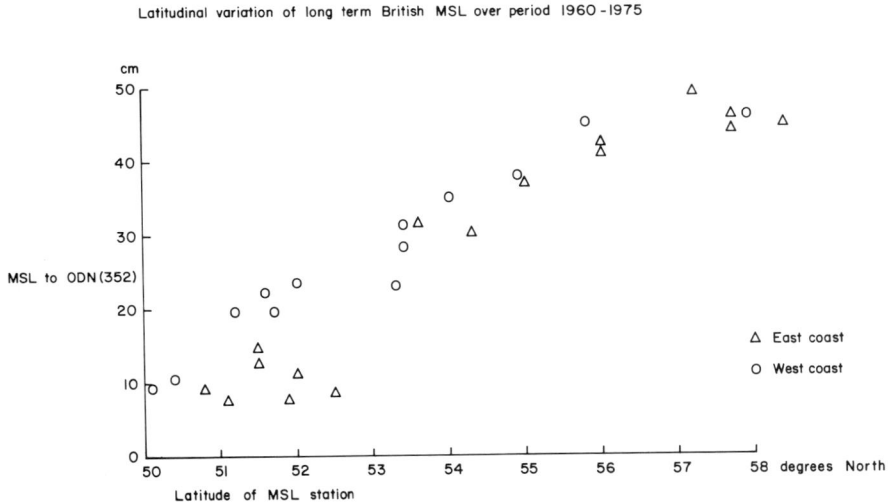

Figure 9:11 The observed and corrected latitudinal slope of British mean sea level, relative to Ordnance Datum Newlyn, for the Epoch 1960–1975. Corrections are made for the atmospheric pressures and pressure gradients (winds) (Thompson, 1980). (*Reproduced by permission of the Royal Astronomical Society.*)

In general it has not proved possible to reconcile the results of geodetic and hydrodynamic levelling in a way which satisfies the exponents of either discipline. Oceanographers cannot use differences in mean levels measured along a coast to compute residual water movements, and modern surveyors seldom adjust their networks for mean sea-level anomalies. One particularly persistent pattern in these anomalies is an apparent upward slope on the mean sea-level surface from south to north as noticed from Newlyn to Dunbar, and in the USA surveys; this south to north tendency is also observed in the Australian network in the southern hemisphere and in Japan. Probably the first geodetic survey to produce these anomalies was the 1839–42 survey of Ireland (Dixon, 1979). The apparent upward slope of British mean sea levels, shown in Figure 9:11, is consistent along both the west and east coasts, and results in a level at Aberdeen (57° N) which is higher by 0.3 m than the level at Newlyn (50° N), and this slope persists even after adjustments have been made for mean meteorological conditions. Oceanographically, if such a difference is real, the pressure gradients must result in very strong residual currents, but such currents are not observed in practice. The apparent gradients along the Australian and the western United States coasts have recently been reduced by new computations. The western United States coast has been levelled repeatedly to see whether the anomaly persisted. The 1968–9 and 1968–71 levellings placed San Diego mean sea levels at 0.6 m below San Francisco mean levels, an upward northern slope over a horizontal separation of 1000 km. Subsequent levellings reduced this slope and the levelling of 1977–8 actually placed San Diego mean sea levels at 0.15 m above San Francisco levels. During these repeated revisions of the levels, neither the tide gauge at San Diego nor that at San Francisco showed any sign of dramatic eperiogenic movements, which suggests that each separate geodetic levelling must contain individual systematic errors (Balazs and Douglas, 1979).

Possible sources of errors in geodetic levelling are beyond the scope of this discussion: the oceanographer's experimental expertise extends no further inland than the Tide Gauge Bench-mark. However, there is one levelling adjustment which relates directly to oceanography, the vertical land movements due to tidal forcing. There are two contributing factors. The earth's body tide response to direct gravitational forcing of the moon and sun may be calculated directly by adjusting the Equilibrium Tide for the diurnal and semidiurnal elastic yielding of the Earth using the appropriate Love numbers. It has maximum amplitudes of a few tens of centimetres but the errors introduced into the levelling are much less. For spring tides in British latitudes, the maximum error is less than 0.02 mm per km of levelling. The second effect is the tilting of the land under direct loading of the ocean tides. The effect is largest near the coast where the geodetic leveller approaches the Tide Gauge Bench-mark, but even without the advantage of random cancelling over the time of the survey, the maximum error is unlikely to exceed 5 mm in the 5 km nearest to the coast, and is totally negligible elsewhere. For the most precise geodetic connections to Tide Gauge Bench-marks it is possible either to correct for the loading or to programme the forward and backward traverses of each section so that the tilts due to tidal loading cancel.

The dichotomy between oceanographic and geodetic land levelling will eventually be resolved by applying other independent techniques such as level measurements by satellite altimetry and the Global Positioning System. To do this, the accuracy must be better than a few centimetres over several thousand kilometres. If the systematic errors are found to be an inherent part of the geodetic land levelling, the measured sea levels, corrected for oceanographic effects by applying suitable numerical models which can allow for local details, may be the best available representation of the geoid.

At present geodetic and oceanographic levellings agree to better than a metre, but it has not always been so. When Napoleon Bonaparte planned to build a canal to join the Red Sea to the Mediterranean in 1798 he made a personal tour to inspect the isthmus of Suez before appointing an engineer, Le Peré, to carry out detailed surveys, but Le Peré reported that the construction of a Suez Canal could lead to disaster because the levels in the Red Sea were some 10 m higher than in the Mediterranean. We now know from modern surveys, that annual mean levels at Port Thewfik at the southern entrance to the modern Suez Canal are only 0.15 m higher than those at Port Said in the north.

CHAPTER 10

Geological Processes

Go on to the foreshore if it's fighting you want, where the rising tide will wash all traces from the memory of man.

J. M. Synge, *The Playboy of The Western World*

10:1 Introduction

Tides, surges and changes of mean sea-level are all important agents for effecting geological changes. In this short account it is possible to touch on only a few aspects of their relevance to geology. Selected references are given for those who require more detail. Here we consider two particular areas which are of major interest: the relationships between tidal currents and sediment dynamics, and the geological implications of long-term mean sea level changes. This largely descriptive account is intended to provide sedimentary geologists, geographers, geomorphologists and coastal engineers with an insight into tidal dynamics and its implications for their disciplines, and to encourage physical oceanographers to consider their work in ways which will increase its relevance to the geological arguments. We begin with a brief summary of sediment characteristics and the basic laws of sediment movement in water.

10:2 Laws of sediment movement

The sediments affected by tidal currents consist of clays, silts, sands and gravels. Sediments are generally classified in terms of particle sizes, according to the Wentworth scale which is summarized in Table 10:1. Sands usually consist of quartz minerals which are chemically and physically very stable; over long periods of reworking they become progressively more rounded by erosion. Clay particles are usually produced by chemical and physical weathering of igneous rocks, and when examined under a microscope are seen to be shaped as flakes, a consequence of their crystal structure. Muddy sediments consist of clays with some variable silt content. Large particles such as sands settle much more rapidly out of suspension in water than fine sediments such as clays.

Table 10:1 The Wentworth scale of sediment grain size.

Particle	Size range (mm)	
Boulder	>256	
Gravel, pebbles, cobbles	2–256	
Sands:		
Very coarse	1–2	
Coarse	$\frac{1}{2}-1$	
Medium	$\frac{1}{4}-\frac{1}{2}$	
Fine	$\frac{1}{8}-\frac{1}{4}$	
Very fine	$\frac{1}{16}-\frac{1}{8}$	62–125 μm
Silt	$\frac{1}{512}-\frac{1}{16}$	2–62 μm
Clays	$<\frac{1}{512}$ – down to dissolved material	<2 μm

However, some small particles may aggregate together to form larger particles with faster settling rates. These aggregates are called *cohesive* sediments, and usually consist of clays and silts. Coarser quartz sands which do not stick together are termed *non-cohesive* sediments. The behaviour of cohesive sediments in the sea depend on their physical and chemical properties, whereas the behaviour of non-cohesive sediments is largely controlled by their physical characteristics.

Coastal and shelf sediments derive from a number of sources; the relative importance of these depends on the particle size and availability. Because of their size, deposits of boulders and gravels are not very mobile except during severe storms, and so they tend to be found near the shore where they are released as a result of coastal erosion by waves and currents; offshore deposits are more probably residuals left after glacial recession. Sand, which is finer-grained and more mobile, is supplied by coastal erosion and by river discharges during storms or after heavy rain. Silts and clays are generally supplied continuously by rivers, but there are strong seasonal variations. The extent to which the sediment particles of different sizes are separated or sorted depends on the extent to which they have been reworked by waves and currents, and is said to be a measure of the sediment maturity: thus beach sand, which is usually well sorted, is mature, whereas glacial deposits are not mature.

Sediment dynamics research has looked for methods of estimating the mass transport of sediment from a knowledge of the current at a fixed level (usually 1.0 m) above the bed, and some knowledge of the sediment sizes. Conceptually the current is the driving input forcing the sediments which constitute the sea-bed; the response is the net sediment transported. The problem of developing formulae which accurately represent the relationship between the input and the system response is complicated. Many transport formulae attempt to describe the sediments only in terms of grain size. The *total* sediment transport may be estimated by empirical formulae, but it is more usual to separate the transport into two components, *suspended load* which consists of particles in suspension several particle diameters above the bottom, and *bed load* which is the material partly supported by the bed as it rolls or bounces along. Suspended-load transport is more important for the finer sands ($< 200\mu m$), silts and clays which are more easily maintained in suspension by turbulence owing to their slower settling rate. For the coarser sediments bed load is also an important component of the total sediment transport.

Measurements in laboratory flumes have shown that as the current speed is gradually increased from zero, there is a speed at which the sediment begins to move. This is called the *threshold speed* q_{CR}; its value depends on the roughness length factor z_0 as defined in equation (7:8) and the sediment particle diameter. Table 10:2 shows experimentally determined values of q_{CR} for a level 1.0 m above the sea-bed. Note that the z_0 dependence allows for different types of bed-form such as ripples or other small features.

Table 10:2 Threshold speeds (q_{CR}) m s^{-1}, for different grain sizes and roughness lengths, where q_{CR} is measured 1.0 m above the bed. From Heathershaw (1981).

	d (cm)			
z_0 (cm)	0.01	0.02	0.05	0.10
0.01	0.30	0.30	0.36	0.50
0.05	0.25	0.25	0.30	0.41
0.10	0.23	0.23	0.27	0.37
0.50	0.17	0.17	0.21	0.29

Starting from physical principles, several formulae have been developed to estimate the bed-load transport (Heathershaw, 1981; Pattiaratchi and Collins, 1985; Dyer, 1986). Bagnold related the sediment bed-load transport to the work done against bottom friction by the fluid moving over the bed. As discussed in Section 7:9:1, this increases as the cube of the current speed. However, to overcome the difficulty of applying the formula at speeds below the threshold speed, where it gives finite transport where none exists in practice, Bagnold's concepts have been developed to express bed-load transport as:

$$Q_{SB} = \beta \, (q_{100} - q_{CR})^3 \qquad q_{100} > q_{CR}$$

$$(10:1)$$

$$Q_{SB} = 0 \qquad q_{100} \leqslant q_{CR}$$

where q_{100} is the current 1 m above the sea-bed. The factor β includes the density of the particles and of the water, as well as the gravitational acceleration and an efficiency factor which depends on the grain size and the amount by which the shear stress on the bed exceeds the threshold level. An alternative formula (Yalin, 1972) considers the average lift forces exerted on a sediment particle: increased transport at higher speeds is due to increased particle path length in suspension, rather than due to an increase in the number of mobile particles. Again, the transport increases as the cube of the excess current speed above a threshold level.

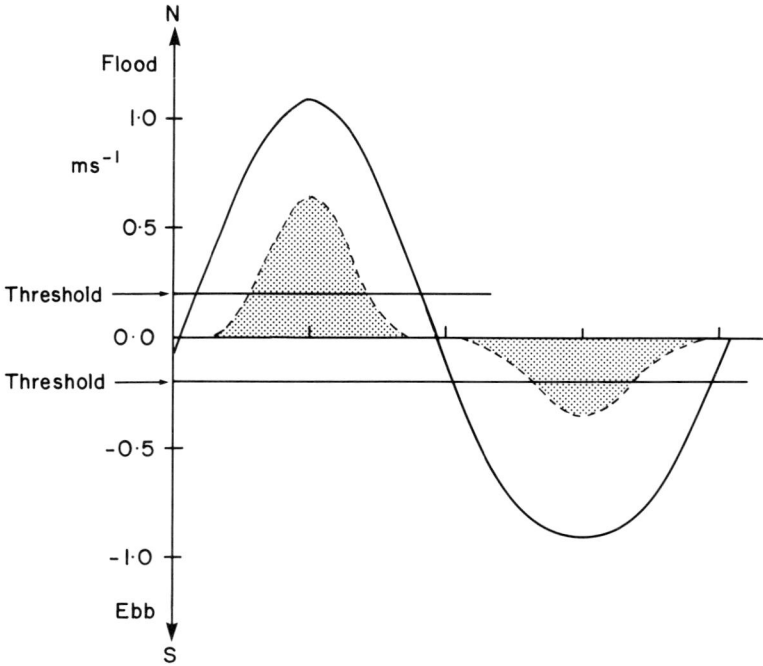

Figure 10:1   Showing the relationship between net bed-load transport, and the asymmetry in the strength of flood and ebb flows, assuming a threshold speed of $0.2\,\mathrm{m\,s}^{-1}$ and a cubic law of transport.

The relationship between the bed-load transport and the cube of the current speed makes any asymmetry between flood and ebb tidal current speeds very significant for net sediment transport. This effect is illustrated in Figure 10:1 where the flood flow is stronger and of shorter duration than the ebb flow, a condition often found in estuaries and shallow water (Section 7:3:3), and one which corresponds with the distortions of a progressive tidal wave shown in Figure 7:5. The broken line shows $(q_{100} - q_{CR})^3$ which reaches a maximum value of $0.66\,\mathrm{m^3\,s^{-3}}$ on the flood tide, but reaches only $-0.36\,\mathrm{m^3\,s^{-3}}$ on the ebb tide. The shaded areas represent the sediment transported in each half of the tidal cycle for material which has a threshold speed of $0.2\,\mathrm{m\,s^{-1}}$. In this simple example, approximately twice the volume of material is transported in the direction of the flood tide as is transported back by the ebb tidal currents, even though there is no net residual water movement. The significance of this net transport for estuarine sediment dynamics will be discussed in Section 10:2:2.

The relative importance of a net residual water flow and of higher harmonics which distort the symmetry of the basic tidal harmonic on the bed-load transport may be investigated (Howarth, 1982) by writing the current as:

$$u = \bar{u} + U_{M2} \cos (2\omega_1 - g_{uM2}) + U_{M4} \cos (4\omega_1 - g_{uM4})$$

If $\bar{u}$ and the U_{M4} are substantially less than the amplitude of the main constituent of the semidiurnal current variation U_{M2}, then over a complete tidal cycle the average value of u^3 (ignoring the threshold factor for simplicity) is:

$$U^2{}_{M2} \left(\tfrac{3}{2} \bar{u} + \tfrac{3}{4} U_{M4} \cos 2g_{uM2} - g_{uM4} \right) \qquad (10:2)$$

The amplitude of U_{M2} is the dominant factor for controlling the bed-load transport. For the same magnitudes of $\bar{u}$ and U_{M4}, the effects are comparable, but residual flow is more important than the higher harmonic. The effect of the higher harmonic depends on the phase difference between the $\mathbf{M_2}$ and $\mathbf{M_4}$ currents; the relationships are the same as those which define the shapes of the elevation curves in Figure 7:2, except that the axes now relate to rectilinear flow, and the cases of strong asymmetry in the maximum flow speeds (and hence bed-load transport) in the positive and negative X directions are (d) and (b) respectively. It is a matter of simple trigonometry to show that odd harmonics such as $\mathbf{M_6}$ will not affect the asymmetry of the tidal currents and hence have no influence on bed-load transport; the higher-order even harmonics, for example $\mathbf{M_8}$, are usually too small to be significant.

The suspended load increases very rapidly with current speed once the critical threshold value is exceeded. For the finer sediments suspended-load transport may be dominant, as is the case for the medium and fine sands found on the European shelf in the southern North Sea, the Irish Sea and the Bristol Channel. For sediments which remain in suspension for times which are comparable with tidal periods, the direction of net sediment transport is likely to be in the direction of mean water circulation. Measurement of this net transport is not easy because it requires integration of the product of current speed and sediment concentration through the benthic boundary layer over several tidal cycles. Expressing the net transport as a function of current speed requires cubic or even higher power law relationships:

$$Q_{ss} \propto u_*{}^{\alpha + 2}$$

where the power α generally lies between 1 and 5. Detailed studies of suspended sediment transport in the Sizewell region of the southern North Sea, where there are fine sands, have shown that in this area $\alpha = 0.8$, close to the cubic law for bed-load transport (Lees, 1983). The problem of estimating net sediment transport is further complicated because the suspended sediment load can affect the dynamics of the benthic layer so that the assumptions of a logarithmic near-bottom profile may not be valid in conditions where turbulence and hence bottom stress are reduced.

Detailed studies (Heathershaw, 1981) have also compared sediment transport laws in the Swansea Bay area of the Bristol Channel with those at Sizewell in the North Sea. In Swansea Bay Bagnold's formula gave the best estimates of bed-load transport, but at Sizewell the same formula overestimated the actual transport by two orders of magnitude and the Yalin formula gave the best estimates. Sediment studies in the New York Bight found Bagnold's equation gave transport rates an order of magnitude greater than that proposed by

Yalin. In Swansea Bay the dominant mode is bed-load transport, whereas at Sizewell, suspended transport was two orders of magnitude more important. The practical approach to estimating sediment transport due to currents must be to evaluate and calibrate possible transport equations by field measurements to find a relationship which is appropriate for a particular region.

Formulae developed for steady conditions may require modification when applied to oscillating tidal flows. For example, sediments raised into suspension at times of maximum currents will be transported by the flow at later times, and the bottom boundary dynamics which they then experience may be those of a location different from that at which they were eroded. Also, turbulence and the associated bottom shear are greater on the decelerating phase of a tidal cycle compared with the turbulence and bottom stress when the currents are increasing (Gordon, 1975). As already discussed, there is an increasing phase lag (Section 7:4) for current reversals as the level increases above the sea-bed, and this too will have important implications for suspended sediment transport. This lag can significantly alter the directions and volumes of sediment transport, and may prevent the currents and bottom deposits reaching theoretical stable equilibrium conditions of transport.

In general, the cubic and higher laws relating sediment transport to current speed imply that transport will be most significant in restricted locations and at times of storms, when extreme currents are generated by tide and surge currents acting together, and when simultaneous wave activity helps to initiate erosion (Dyer, 1986).

Cohesive sediments have variable particle size and geometry, which makes their behaviour even more complicated than that of non-cohesive sediments. Particles of clay carried in suspension by rivers into estuaries become aggregated into particles which have more rapid settling rates. These particles have negative surface charges due to broken intermolecular bonds and other variations in the structure. In river water the particles are surrounded by a layer of positive ions which prevent their combination by the electrical repulsion between the charges. In sea water the ion concentration tends to make the layer of positive ions thinner and less stable so that the particles have a higher probability of colliding and forming aggregates, a process called flocculation. Physically the particles may collide because of turbulence, Brownian motion and differential rates of settling. For example, the larger particles tend to grow more rapidly because they come into contact with other particles which they overtake during their more rapid settling. Particle dynamics also favour combinations between those of similar sizes. The result is a high rate of sediment deposition in estuaries, where river waters mix with sea water, an area known as the *turbidity maximum*. In some cases layers of suspended mud may form, especially at slack water (Kirby and Parker, 1982). Particle aggregates may also be destroyed as salinity continues to change, or if turbulence increases sufficiently to break them physically. Both these processes have a strong dependence on tidal conditions.

10:3 Coastal processes

In the coastal and near-shore environment tides and waves work together to produce very complicated geomorphology (Komar, 1976; Holman and Bowen, 1982; Allen, 1985a; Dyer, 1986). Characteristic features include estuarine deposits, lagoons, barrier islands and coastal sand banks. Any simple analytical model of the processes involved can only roughly represent the real situation; conversely, complete geographical descriptions in terms of morphological processes inevitably lack numerical precision. The role played by waves in the development of coastal features is not part of this account, but cannot be overemphasized: breaking waves are very effective agents of erosion and for driving sediments along the coast. For any particular length of coast there is usually a direction of net long-shore sediment transport, the magnitude of which depends on the supply of sedimentary material and the direction of wave attack along the shore. If local tidal vorticity (Section 7:6) produces an asymmetry between ebb and flood currents, the direction of the resulting residual tidal flow will also influence the direction and extent of the long-shore transport. However, in terms of coastal development, the major influence of the tides is due to the changes of water level which raise and lower the level at which wave attack occurs. In this section we consider first a two-dimensional approach to beach profiles and erosion. We then describe the influences of tides on the three-dimensional development of estuaries and lagoons, and some consequences of engineering works for coastal sediment movements.

10:3:1 Two-dimensional beach profiles

Beaches are areas of continuous interaction between sediments, waves and tides. The shape of a beach is controlled by several factors: the most important are the wave energy and the grain sizes of the sediment particles. In general, high energy waves result in gently sloping beaches because of greater erosional power, whereas small amplitude waves tend to move sand gradually up the beach; however, this relationship is invariably complicated by other factors (Komar, 1976). During the summer, low amplitude waves may gradually build up beach levels which are then reduced by the high waves incident during winter storms. For any particular combination of sediment, wave and tide conditions, equilibrium beach profiles are never reached because these conditions themselves change before adjustment is complete. The interaction between beach profiles and wave conditions introduces further complications. If the beach slopes are steep, the waves are little attenuated and so can be very destructive, removing sediment down the beach, which in turn reduces beach gradients. However, if this process advances too far, the amplitudes of the incoming waves are attenuated over the shallower offshore beach and they become constructive again: there is then a net sand transport up the beach and a gradual return to steeper slopes. Coarser grains such as shingles are associated with steeper beaches because the down-beach erosion effected by the back-swash from waves is weakened by some percolation of the water through the sediments themselves. Many low flat sandy beaches have steep shingle storm ridges at the top.

The influence of tides is most apparent as a factor controlling the level at which activity is concentrated, and the range over which beach profiles are developed (Komar, 1976; Carr and Graff, 1982). These levels are described by their probability density functions as shown in Figure 1:4. In the usual case where the tide is dominated by semidiurnal variations, as in Figure 1:4(a), there are two levels of frequent wave attack. In the case of mixed tides there is more likely to be a single level of most frequent attack, located near to the mean sea-level. Beach profiles have been roughly related to sea-level frequency distributions, but there is considerable scope for more precise analysis (Heathershaw et al., 1981; Webb et al., 1982; Wright et al., 1982; Wright and Short, 1984).

Coastal geomorphologists often define three types of tidal regime. Where the tidal range is less than 2 m, and is comparable with storm wave heights, the wave energy is effectively concentrated at one level and the tidal effects are negligible (Davies, 1964). In these circumstances beaches are often narrow and steep and such tidal regimes are termed *microtidal*. Where the tidal ranges are greater than 4 m, tidal effects are important and the term *macrotidal* is applied. On macrotidal beaches there are often well defined zones related to the tidal levels. Beaches which experience intermediate ranges are termed *mesotidal*. The type of tidal regime may also be significant: for example, semidiurnal regimes have stronger tidal currents than diurnal regimes with the same amplitude variations, and semidiurnal tides work over beaches twice as often.

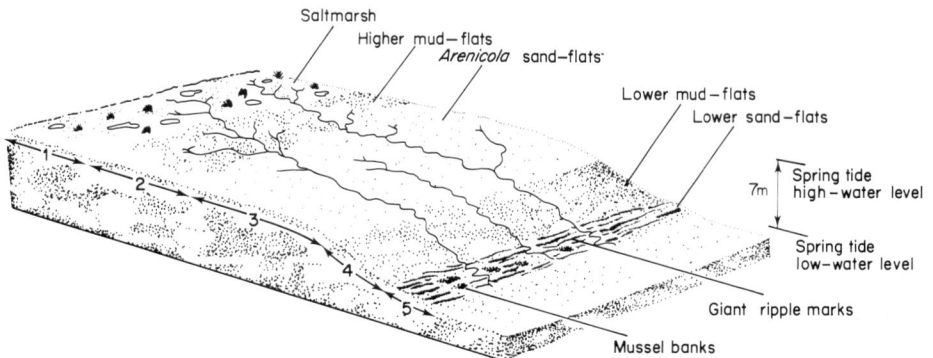

Figure 10:2 Sediment zonations in a macrotidal environment, for a typical estuary in temperate latitudes. (From Evans, 1965.) (*Reproduced by permission of the Geological Society.*)

Figure 10:2 shows the different sedimentary zones characteristic of a well developed beach profile in an estuary with a large semidiurnal tidal range (> 7 m) which is partly protected from waves, the Wash bay on the North Sea coast of England (Evans, 1965). In this case five zones are identified.

(1) The *salt-marsh*, the highest level influenced by marine processes, is only covered at high water on very high spring tides. The landward limit is sometimes a low cliff, but elsewhere the transition is more gradual. Silt and clay are trapped by the vegetation at high water, and this gradually increases the marsh level until inundations become less and less frequent.

(2) The *higher mud-flats* are located around the mean high-water level: most of this zone remains exposed at high water on neap tides. The vegetation is more sparse and there are shallow drainage channels separated by low raised areas. Intertidal drainage in the Severn Estuary at these higher levels has been described by Allen (1985b). The waves which reach here are attenuated by passage over the lower flats, and have insufficient energy to raise the muds into suspension.

(3) The *sand-flats* consist of rippled and well-sorted sands, normally fine and very fine grained, with abundant worm casts (see Section 11:3:2). They are covered during each tidal cycle and are subjected to regular wave activity. The sands settle back out of suspension quickly, but the finer silts and clays which constitute the muddy sediments are carried further inshore where they are deposited at higher levels at high-water slack, or offshore as the tide ebbs. This flat extensive area is also called the *low-tide terrace*. It is characterized by *ridge and runnel* formations, normally 100 m apart with an amplitude rarely exceeding 1 m. These ridges, which are aligned parallel or at a slight angle to the shoreline, are broken in places by gaps through which the rising tide penetrates to create temporary lagoons. These lagoons persist until a further tidal rise of sea-level overtops the ridge itself. Wave action advances in steps from ridge to ridge, with little activity within the intermediate runnels.

(4) The *lower mud-flats* consist of muddy sediments eroded from the sand-flats. They coincide roughly with the more steeply sloping part of the intertidal zone and the mean low-water level.

(5) The *lower sand-flats* are only exposed at low water on spring tides. They consist of well-rippled sands and have a lower gradient than the more steeply sloping mud-flats. When the lower sand-flats and the lower mud-flats are covered by water at high tide, long-shore tidal currents may be significant factors in controlling sediment transport

These divisions of beaches into different levels are not absolute, nor can they be precisely related to tidal levels. Cable Beach, a macrotidal beach on an open coast with full wave exposure in north-western Australia, has been described in terms of three zones (Wright *et al.*, 1982). Maximum gradients were found in the upper zone, between the level of high-water neaps and the maximum tidal levels. The lowest gradients in this case were found in the lowest zone, between mean low-water neap levels and the lowest tidal levels.

The response of a beach profile to a change in the mean sea-level, assuming a closed sediment system, may be estimated using the Brunn Rule of erosion as illustrated in Figure 10:3. A closed system means that there is no loss or additional supply of sediment except from the beach and nearshore, over the active beach width, L. By equating the sediment volumes in the erosion and deposition zones it may be shown that the shore erosion resulting from an increase of mean sea-level ΔZ_0 is given by:

$$\text{Shoreline recession} = \frac{L\Delta Z_0}{D}$$

$$(10:3)$$

irrespective of the detailed shape of the profile (see Section 11:7). Despite the limiting two-dimensional assumptions made in developing this formula, it has been applied with some success for beaches along the east coast of the United States, particularly when allowances were made for vegetative matter in the eroded material, and for known long-shore transport (Brunn, 1983).

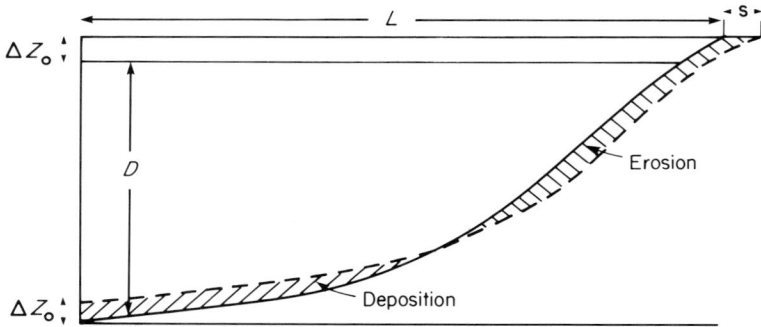

Figure 10:3   The basis for the Brunn rule of coastal erosion, relating coastal erosion to sea-level changes.

10:3:2   Three-dimensional dynamics

The ridge and runnel features found on the foreshore where there are large tidal ranges are an example of the role played by tides in shaping characteristic coastal features. So too are the marshlands and mangrove swamps which develop where tidal ranges are large and the exposure to waves is low. These marshlands and mangrove swamps are drained and flooded through complicated systems of channels which are often rich in wildlife (Section 11:3:2). In this section we discuss in general terms the role of tides in the development of barrier islands and lagoons, and their influence on the physical character of estuaries.

10:3:2:1   Barrier islands and lagoons

Apart from the supply of sediment itself, the dominant factor which controls the rate of transport of sediment along a beach, the long-shore drift, is the direction of the wave attack. Tides and tidal currents may play a secondary role in the net transport where, for example, sediments are lifted into suspension at a particular part of a cycle, and transported by the currents which prevail in a later part of the cycle. McCave (1978) reports significant changes in grain size along an East Anglian beach which he attributes to the offshore dispersal of the finer material by tidal currents. The net transport of sediment along the coast may result in the development of spits such as Spurn Head at the entrance to the Humber Estuary in the North Sea. In many parts of the world barrier islands are developed as a result of long-shore sediment drift; these islands run parallel to the shore, and enclose lagoons or tidal flats.

There are two mechanisms which are important for the development of barrier islands. The first of these is the separation from the coast of a long spit as a result of its being breached by waves and tidal currents. The second factor is the change in mean sea-level over geological time, which converts features such as inland sand dunes into offshore barrier islands. Barriers are well developed in many places, including the south-east coast of the United States, the southern North Sea where the Frisian Islands extend along the coasts of Holland and Germany, and along the coast of Victoria, Australia.

The lagoons and marshes behind the barriers are related to the tidal range on the seaward side. Along the Atlantic coast of the United States (Hayden and Dolan, 1979) wide complex lagoons are typical of areas where the range exceeds 2 m and the offshore beaches are broad and shallow. Narrow simple lagoons are found where the tidal ranges are small and the offshore beach profiles steep. However, there are significant variations from this basic pattern.

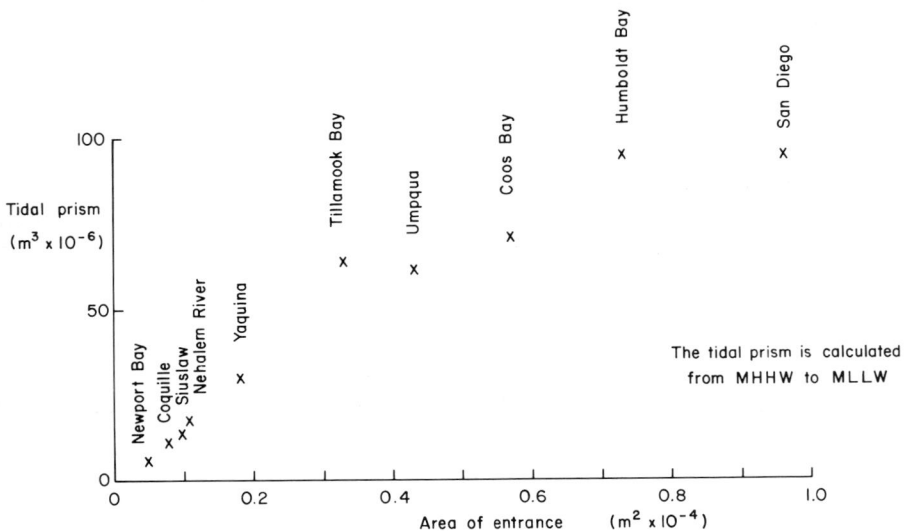

Figure 10:4 The relationship between the tidal prism, the volume of water exchanged in a tidal cycle, and the area entrance (based on O'Brien, 1931).

If the barrier island has a gap which connects the lagoon with the open sea, a subtle balance is struck between the long-shore transport of sediment which tends to close the gap, and the flow of water through the gap as the tide floods and ebbs which keeps it open. The volume of water exchanged between a lagoon or estuary and the open sea in the course of a complete tidal cycle is sometimes called the *tidal prism*. Figure 10:4 shows the relationship between the tidal prism and the area of the entrance system at its narrowest point for the Pacific coast of the United States. In this case the tides are mixed, and so the volume of the tidal prism is calculated for the range between mean higher high water and mean lower low water. Gray's Harbour, Willapa, Columbia River and San Francisco Bay are not plotted but fall on extrapolations of the same trend (O'Brien, 1931). If land within a lagoon is reclaimed by engineering works, this relationship implies that there will be a subsequent reduction of the entrance area. In general this shrinking is a natural progression because lagoons are very effective traps for sediment which is transported in on the rising tide and remain after settling out at high water.

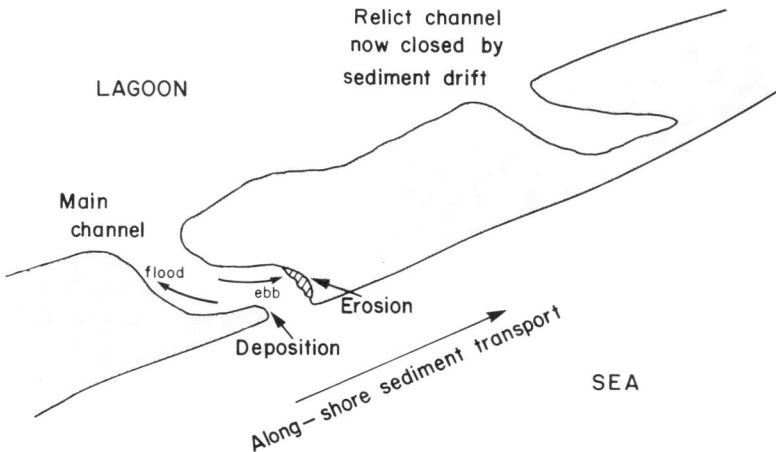

Figure 10:5 Showing how channel entrances to lagoons tend to migrate in the direction of alongshore sediment transport, partly because of the erosion caused by the tidal currents.

Lagoon entrances and river mouths are often deflected by the spits which develop in the direction of along-shore sediment transport (Brunn, 1978; van de Kreeke, 1985). Also, the transverse ebb and flood of the exchange tidal currents causes erosion which enhances the development of channels parallel to the coast. Lagoon entrances may migrate in the direction of the transport and become sealed at times of low tidal ranges and currents, after which new entrances are developed elsewhere. The process is illustrated in Figure 10:5. The development of a new entrance can be due to wave erosion or to freshwater flows from the land forcing a breach which is subsequently enhanced by tidal flows. A good example of this is the Batticaloa lagoon on the east coast of Sri Lanka, where access to the sea from the lagoon and fish-processing facilities is periodically interrupted by the closure of the lagoon entrance; when this closure occurs, boats can go fishing only after the fishermen have reopened the access channel.

Under some circumstances, separate tidal flood and ebb channels may develop. Flood currents use one channel, or a set of channels which are interleaved with channels used by the ebb currents. The flood channel decreases in depth as it leads inshore, whereas the ebb channels are deepest inshore and shallow shoals are found at the seaward end. Water continues to flow in along the flood channel for some time after the return flows have been established in the ebb channels, with the result that there is a net circulation along the channels. This net circulation of the water affects the movement and distribution of coastal sediments. Studies along the east coast of the United States have shown that where the tidal range is small (less than 2 m) ebb-tidal deltas are also less well developed, but flood tidal deltas are enhanced (Hayes, 1980). Larger tidal ranges have larger ebb-tidal deltas associated with them. Large waves also tend to accentuate the growth of flood-tidal deltas.

10:3:2:2 Estuaries

The pattern of ebb-tidal and flood-tidal channels is also characteristic of estuaries such as the Wash bay, where complicated patterns of residual circulation are developed. The protection from extreme waves afforded by the estuarine environment can allow full development of the zonation patterns of sediment and vegetation discussed in Section 10:3:1, but if the wave activity is too low or the supply of sand is insufficient, sand-flats may fail to develop. In areas where the wave activity is greater, the mud-flats may be absent, replaced by a sandy beach backed by shingle or cliffs.

Estuaries are continually being supplied with river sediments which in some cases may be completely retained within the estuary whilst in others most will pass out to coastal waters. The fate of sediments brought down by the river is dependent upon the vertical circulation patterns and on the variations of salinity along the length of the estuary (Allen et al., 1980).

Figure 10:6 shows the residual flow in a typical partially mixed estuary. This mixing of the fresh water discharged by the river with the incoming sea water is enhanced by the turbulence induced by the tidal currents (Dyer, 1973; McDowell and O'Connor, 1977). As a result there is an entrainment of salt water into the surface seaward flow of river water, and the sea water carried out to sea in this way is replaced by a net residual flow at the bottom, directed up the estuary. The place when up-stream residual bottom flow and down-stream river flow converge is called the *null point*. In most estuaries the null point is associated with a turbidity maximum, where the suspended sediment load may be as high as 10 kg m^{-3}. This area of maximum turbidity is due to the transport of sediment by the converging bottom currents, and in many cases to the local flocculation of river clays because of the sudden increase of salinity. The actual salinity value in the region of the null point is less than 1 ppt in partially mixed estuaries, but higher salinities are found where the mixing is vigorous. In the Thames Estuary the salinity at the null point is nearer to 10 ppt.

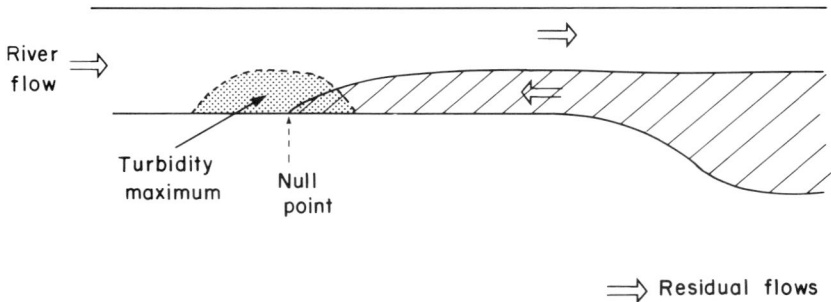

Figure 10:6 Circulation patterns in a partially mixed estuary, showing the formation of a null point turbidity maximum, and on-shore bottom flow.

The null point and the turbidity maximum are unsteady features which change on semidiurnal, fortnightly and seasonal time scales. Within a tidal cycle, in the Columbia River Estuary on the Pacific Coast of the USA (mean range 2 m), the turbidity maximum may move up and down river by 20 km (Gelfenbaum, 1983). During neap tides the turbidity maximum decreases in concentration, whereas at the time of spring tides it regenerates. Seasonal increases of freshwater discharge from the river cause higher levels of turbidity and the turbidity maximum is pushed downstream. Similar patterns are observed in other estuaries which have large tidal ranges. The spring–neap variations are even more important in the estuaries of the northern European shelf, where semidiurnal tidal ranges are higher. As an example, the Gironde estuary changes from a well-mixed state during spring tides to a partially mixed or even well-stratified state during neap tides. An additional factor which influences sediment discharge to the sea is the fortnightly residual circulation due to a build-up of levels at spring tides and the subsequent discharge as the range decreases (see Section 7:7). In the River Amazon the river discharge is so strong that the mud deposition takes place outside the estuary in the coastal zone, from where it moves westward along the coast of Venezuela.

Within the Severn Estuary, strong cross-channel gradients of sediment concentration are observed, showing that the process of sediment movement is essentially three-dimensional (Kirby and Parker, 1982). Strongest gradients are observed on the ebb currents associated with spring tides, when the suspended load is mixed throughout the water column. These fronts between waters of different suspended sediment concentrations persist throughout the spring–neap cycle in approximately the same geographical positions although they are less sharply defined during the neap tides. The mechanism for their development is not fully understood, and it is surprising that they are not associated with salinity gradients, as is the case where river sediments are normally discharged into the sea. In the Severn Estuary the salinity concentrations imply effective mixing of river and sea water, whereas the sediment concentrations imply separate water masses. In some estuaries (including the Gironde and the Thames), spring tidal currents may generate pools of water containing very high concentrations (as much as 100 kg m^{-3}) of suspended mud, which have their own dynamics and identity because of their very high density. They tend to settle out at neap tides, but are regenerated when the currents return to the higher speeds at spring tides.

10:3:3 Sediment control; engineering

Most of the world's population lives near to the coast, where good transport facilities encourage the development of industries, and where the agricultural land is usually flat and fertile. The opportunities for recreation are also appreciated by yachtsmen, fishermen and holiday-makers.

As a result there are conflicting pressures from different community interests, which also include the common need for effective discharge of sewage and other effluents, and the problem of controlling the continual process of coastal erosion and deposition. In Section 8:5 we discussed aspects of coast protection against flooding, closure practices for the final stages in the construction of protection barrages, and the estimation of extreme levels to assist in the design of defences. Here we expand the discussion to consider the interaction of tidal flows and sediment movements in response to coastal engineering projects (McDowell and O'Connor, 1977).

Any such project must be carefully designed not only to achieve its direct purpose, but also to avoid any secondary consequences which can easily result from interference in the dynamic equilibrium between coastal geomorphology, sediment fluxes, and tides and waves. Tides have always played an important role in river navigation, allowing access to inland ports, and maintaining minimum channel depths by scouring. However, the large modern ships require deeper channels and bigger harbours than those afforded by ports up rivers. New harbours have been developed, often on the open coast to save time and navigation difficulties, and to allow access at all states of the tide.

Several barriers have been constructed across tidal estuaries and rivers, either for the generation of tidal power, or to provide protection against flooding. River sediments may be trapped behind these barriers and prevented from reaching the open coast. Further, the tidal prism is reduced, and so too is the speed of the tidal currents in the downstream channels. As a result sediment deposited below the barrier is no longer eroded on the ebb tide, and there is a gradual build-up of bed levels, a further reduction in the tidal prism and more deposition until a new equilibrium is established.

Like lagoons, harbours are excellent sediment traps because the sediment brought in on the flood tide is able to settle out at slack water; unfortunately, the vicinity of the harbour walls, where the ships berth, is most vulnerable to

350

siltation. Alongside the walls the currents are weak, and there is no wave
activity to encourage erosion, so that frequent dredging is necessary. Secondary
flows associated with eddies near harbour entrances form unstable mud-banks
which are particularly hazardous for navigation. Even if a fully protected
harbour is not necessary, the drag of the pile structures alongside loading and
unloading jetties also reduces bottom turbulence, and siltation develops.

The normal pattern of sediment flux in an extensive estuary is an inward
movement along the flood channels and an outward movement along the ebb
channels. At low water a system of ebb channels is seen traversing the foreshore
and over several years the locations of these channels oscillate about an average
position: if an ebb channel moves too far from the average position, a new
channel will form and the old one will be filled in gradually.

Coastal engineers in the last century introduced the practice of stabilizing
these channels by building training walls on either side, so that the ebb flow was
concentrated between them. These training walls are flooded as the tide rises
(but clearly marked for navigation!), and better port access results. However,
there are secondary effects, not all of which are beneficial. As silt and fine sand
enter the estuary on the flood tide, they accumulate permanently behind the
walls because meandering channel erosion has stopped. The tidal prism is
reduced, and eventually the higher reaches of the estuary will develop as salt
marshes. An example of this is found in the Mersey Estuary and the approaches

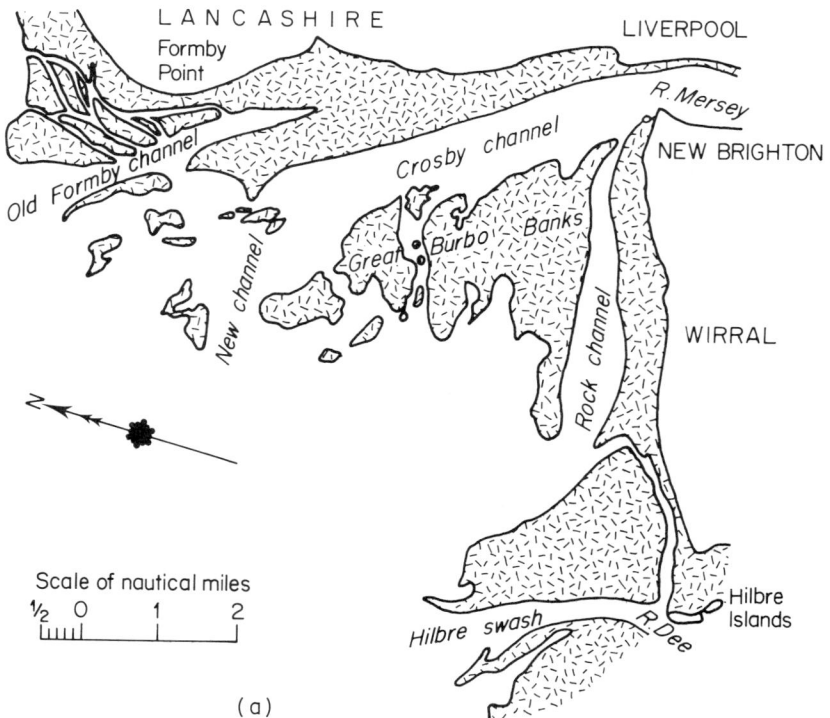

(a)

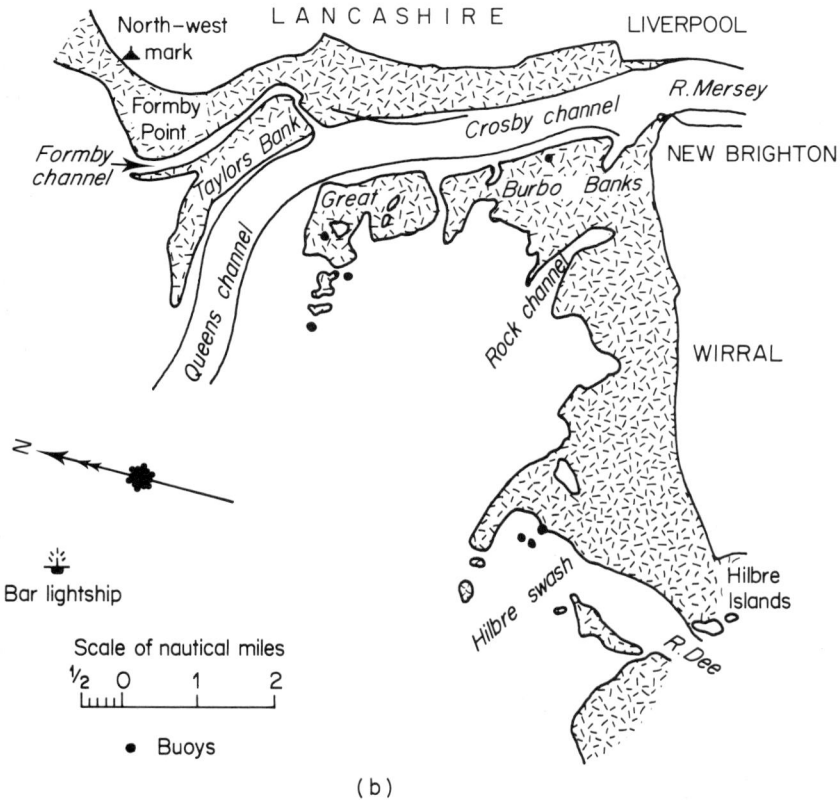

(b)

Figure 10:7   Approaches to the Port of Liverpool (a) 1833, before the training walls were built, when three main channels existed and (b) 1955, several decades after the building of training walls, showing their influence on the pattern of sedimentation. (*Reproduced by permission of the Institution of Civil Engineers.*)

to the Liverpool port facilities from the Irish Sea (Price and Kendrick, 1963; McDowell and O'Connor, 1977). Figure 10:7 compares the estuary approaches before the building of the training walls, and a century later. Originally there were three channels, a central channel and two coastal channels, the Formby Channel, and the Rock Channel parallel to the Wirral foreshore. After the main channel was stabilized in the period 1909–1910, the Formby Channel and the Rock Channel gradually silted up. Also, as a greater volume of water and sediment entered on each flood tide because the deeper dredged channel was more efficient at emptying the estuary, estuary siltation was enhanced.

Another nineteenth-century structure which has changed the local tides and beach structure is the causeway which connects Mont-St-Michel in northern France with the mainland. This has prevented flow around the island and enhanced the level of the sediment deposits so that fewer and fewer tides reach this popular tourist attraction.

Removal of beach or coastal-dredged sand or gravel for building material is occasionally practised on a small scale (Section 10:4:3), but extensive exploitation is rare because of the dangers of erosion which result. More common is deliberate reclamation of land by dykes or polders for agricultural and other purposes. Polders are usually built in deltas or other low-lying coastal areas and embayments (Volker, 1982). As a result they are flat level lands, fertile and easily worked. Examples are found along the Dutch coast, in Japan, the Indian State of Kerala, Bangladesh and Sri Lanka. The most suitable areas for reclamation are those where the land is at a level equal to or slightly above normal high-water levels, as is often the case in the United Kingdom and European countries. In other more densely populated countries where there is a more urgent need for arable land, reclamation of lower land has often been undertaken. In northern Honshu, Japan, a lagoon whose depth was almost 5 m below mean sea-level was reclaimed in 1966. In Kerala a coastal lagoon has been impoldered with a level 2.5 m below mean sea-level.

Where the level of the reclaimed land is below mean sea-level problems of irrigation must be overcome by pumping, but care is necessary to avoid seepage of salt water through the ground. One advantage of the level character of the reclaimed land is the possibility of controlling the water table, through the growing season, depending on the specific requirements of the crops. Artificial drainage systems must be able to cope with extreme rainfall. In the case of reclaimed land which is at or above mean sea-level, such as the shallow polders, drainage can be economically effected by sluice gates under gravity at low tide, but irrigation of deep polder requires machinery and is more expensive.

Poldering is an expensive, long-term investment, which often needs political commitment to the sustained national advantage. At present, more cost-effective ways of increasing productivity are often found by improved drainage of existing polders. In Bangladesh the polders developed along the coast are usually formed by circular banks around the islands. The risks of flooding in these low-lying areas has been described in Chapter 6. As a precaution, refuge mounds called *kilas* have been constructed several metres above sea-level.

10:4   Offshore tidal sediments

Away from the coast, as the influence of surface waves on the sea-bed is progressively diminished in deeper water, tides and storm surge currents become more important in the development of sediment structures and their distribution. Sediments offshore under the influence of tides have been discussed in detail elsewhere (see, for example, Stride (1982) for a series of review papers). The characteristics of bottom sediments are far from uniform. Indeed, an intimate knowledge of their distribution was a traditional method of ship navigation for many centuries, where a plug of tallow wax was inserted at the bottom of the sounding lead to obtain a bottom sample: details of bottom sediments are still given on many nautical charts. Offshore deposits of sand and gravel are mined for building purposes, so their distribution and movement remain of great practical importance. Many offshore deposits remain from times of lower sea levels and different tidal conditions, and are no longer mobile (see also Section 10:6); if the bottom currents fail to exceed the threshold velocity for movement, then these so-called *relict* sediments are now stable. However, many shelf deposits are still being reworked, particularly as a result of storm currents, and there is a form of dynamic equilibrium involving erosion, transport under tidal currents, and deposition elsewhere. The variety of bed-form configurations depends on the grain sizes as well as the water movements.

10:4:1 Sediment deposits and fluxes

There are two sources of mobile marine sediments. In addition to the reworking of offshore deposits, the other main sediment source is the coastal zone, from which particles of all sizes have been eroded by wave action. Some of these particles are carried away from the coast into deeper water by the tidal currents. Off shore the effects of the waves are reduced, and so, once deposited, these particles can only be reactivated if the local current exceeds the threshold velocity. As a result there is a general tendency for the grain size to diminish as the water depth increases. However, the simple notion of decreasing grain size with increasing water depth fails to take account of relict deposits or the more complicated effects of tidal currents.

Table 10:3 Ratio between progressive wind wave currents and progressive tidal wave currents, each of unit amplitude, at the sea-bed in water of different depths; the effects of bottom friction are neglected for this first-order comparison.

Water depth (m)	Wave period (s)		
	5	10	15
10	0.48	0.87	0.94
20	0.14	0.72	0.88
50	0.00	0.36	0.70
100	0.00	0.07	0.42
200	0.00	0.00	0.10

Table 10:3 compares the currents associated with a progressive tidal wave and with surface gravity waves of the same amplitude at the sea-bed, for a series of different water depths: for this first-order comparison the effects of bottom friction have been neglected. It is important to realize that the combined effects of waves and tides are much more important than their individual effects. For example, waves and storm surge currents may be critical even in areas where

tidal currents are dominant, because of the rapid increase of transport rates with current speed. However, it is worth noting that because of the cubic law of transport, if the waves and surge currents have a random directional distribution, they will tend to enhance the directional patterns of tidally-driven sediment transport; when they are acting together with tidal currents they enhance the transport whereas if they are acting against the tides there is no transport. As a result the *patterns of tidal transport* are still dominant, but the long-term *rates of transport* may exceed those expected on the basis of unaided tidal currents.

If tidal currents are important for controlling sea-bed characteristics, then the areas of strongest tidal currents should correspond to the areas where coarse surface sands and gravels are found. In many cases this is confirmed by observations. For example, in the English Channel, gravels are found roughly in the areas of strong currents shown in Figure 5:16. Gravel deposits have also been related to areas of maximum tidal currents on Georges Bank off the north-east coast of the USA. However, the correspondence is not exact because of storm effects and the presence of relict deposits. Gravel is also found in other places where it has been deposited at times of lower sea levels, or by glacial action as the glaciers receded.

The sediments of the North Sea may be compared with those of the Yellow Sea which has similar tidal conditions (Section 5:4:2 and Figures 5:10(b) and 5:12). In the North Sea little new sediment is supplied and there is extreme reworking of existing sands in the south, but only intermittent movement, during extreme storms, of the sediments in the northern part. In general, surface sand and gravels are found in the shallower areas, with mud in the deeper parts. The Yellow Sea has an abundant supply of sediments from the rivers, particularly from the Yangtze and Hwang Ho (Sternberg *et al.*, 1985). Sands dominate on the eastern side where the tidal currents are strongest but on the western side fine sand and mud are dominant. They remain here because the weaker currents cannot move them after deposition. Muds can be carried in suspension over many tidal cycles, and so their movement is strongly influenced by the tidally generated basin-wide anticlockwise circulation, which brings the sediments from the more energetic tidal areas.

Tidal currents have a strong influence on the direction of bed-load transport. Figure 10:8 shows the direction of bed-load transport for sand on the continental shelf around the British Isles based on direct evidence such as tracer movements, and indirect evidence such as the bed-form developments described in Section 10:4:2. The heavy broken lines show areas of bed-load parting, where bed-load transports diverge. In the English Channel, the Irish Sea, the southern North Sea and the North Channel, these partings correspond with the locations of semidiurnal tidal amphidromes (Pingree and Griffiths, 1979). There is also a tendency for the transport to be directed into bays and estuaries such as the Wash and the Thames Estuary, Liverpool Bay in the Irish Sea, and the Gulf of St Malo in the Channel. Both of these sediment transport features can be related to the phase differences between the M_2 and M_4 tidal currents, because

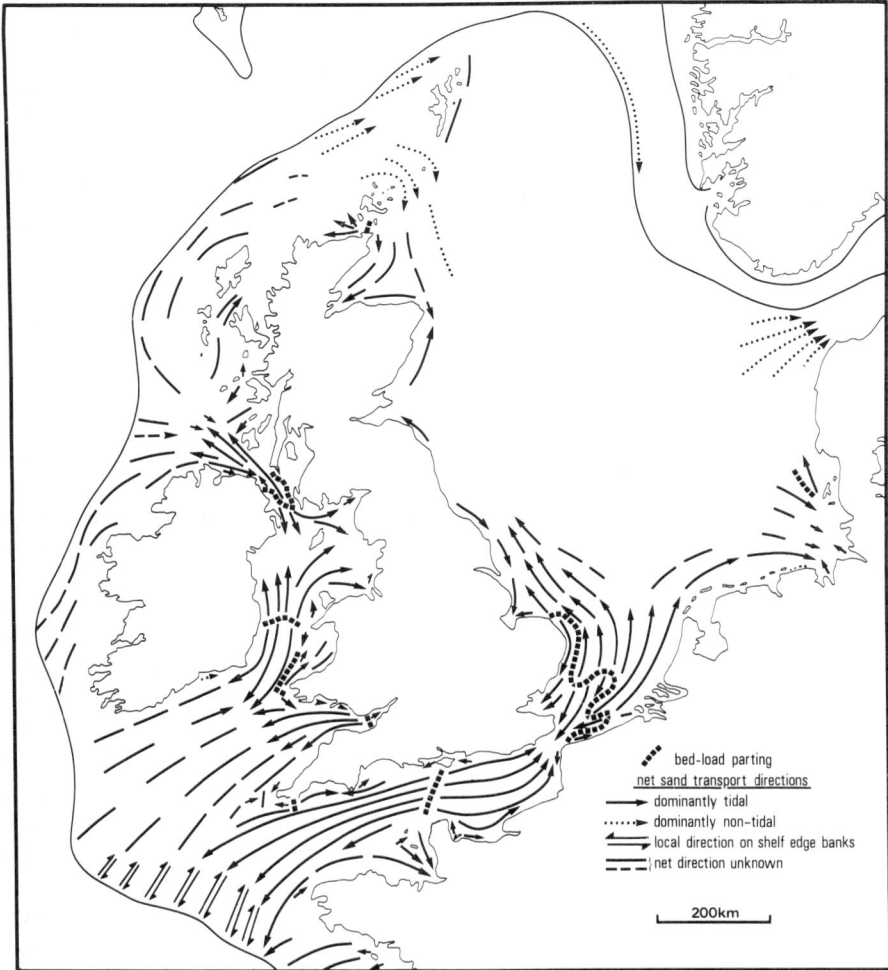

Figure 10:8 The net sand transport directions on the continental shelf around the British Isles. Note the relationships between the location of tidal amphidromes and bed-load parting. (From Stride, 1982.) (*Reproduced by permission of Chapman & Hall.*)

of the significance of this phase difference for the difference between maximum currents on the flood and ebb phases of the tidal cycle. For tidal flows into estuaries and bays the stronger flood currents discussed in Section 7:3:3 are responsible for the gradual filling process. In the case of divergence away from amphidromes, the phase of the M_4 currents are critical. Figure 10:9 shows the relationship for levels and currents between a semidiurnal and a quarter-diurnal standing wave configuration in the same channel where the waves are in phase at the head of the channel. The M_2 and M_4 currents are in phase between the head of the channel and the first semidiurnal amphidrome, giving maximum

Barrier

Levels

2   1   2

1   Semidiurnal amphidrome

2   Fourth-diurnal amphidrome

Currents

$g_{u_{M_2}} = 0°$

$g_{u_{M_4}} = -180°$ $g_{u_{M_4}} = 0°$

$2g_{u_{M_2}} - g_{u_{M_4}} = 180°$ $2g_{u_{M_2}} - g_{u_{M_4}} = 0°$

| ← Down channel sediment flux | Up channel sediment flux → |

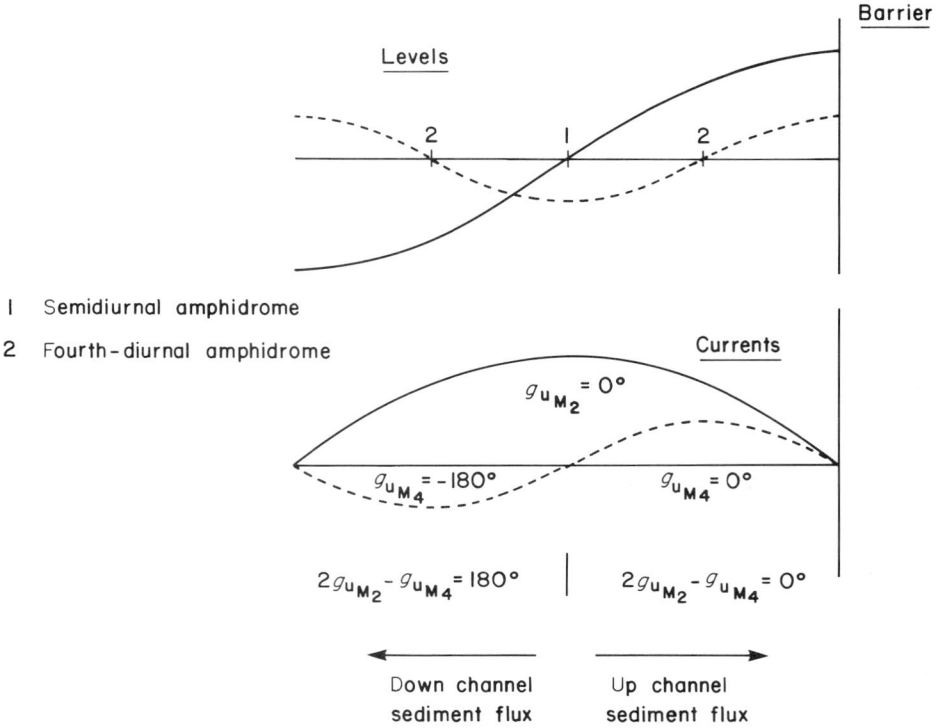

Figure 10:9   Showing the relationship between phases of M_2 and M_4 for a standing wave pattern, and the development of a bed-load parting near the amphidrome. Conditions in the English Channel are close to this theoretical model.

tidal currents towards the head of the channel as shown in Figure 7:2(d). However, beyond the first amphidrome, the M_2 and M_4 currents are out of phase and the net sand transport under a cubic flux law is down the channel away from the amphidrome as shown in Figure 7:2(b). Any sediment mobilized by the strong tidal currents near the semidiurnal amphidrome will be carried either up or down the channel so that the amphidrome area is a region of bed-load divergence. The M_2 and M_4 tidal conditions in the English Channel shown in Figures 5:10(b) and 7:3 correspond very closely to this idealized model. The asymmetry of the actual tidal current speeds can account for the central bed-load parting shown in Figure 10:8.

Figure 10:8 also shows a region of banks near the shelf edge. Most sediments along the shelf edge are relict sediments from times of lower sea levels and there is a noticeable absence of muds and fine sands (see Belderson *et al.*, 1986). The finer particles have probably been removed from the shelf and deposited down the continental slope by the strong bottom currents generated as the tidal waves move onto the shelf (Section 5:4:1). Internal waves, including internal tides, which can be focused at the shelf break by the ocean's density structure may also be important factors (Heathershaw *et al.*, 1987). So too are internal tides in the canyons of the continental slope, but these are little understood and difficult to measure.

10:4:2 Bed-forms

The shape of the sedimentary features on the sea-bed depends on a number of local conditions (Stride, 1982; Dyer, 1986). Of these, the most important is the strength of the tidal currents, often represented in terms of the maximum on mean spring tides. In addition, sediment supply and grain size, water depth and wave exposure are also significant. The observed bed-forms are the integrated response of the sediments to all water movements. Some bottom features are permanent while others gradually migrate. Of the permanent features, some relate to past conditions of tides and mean sea-level while others are in a state of dynamic equilibrium. The characteristic distribution of bed-forms made by tidal currents of different speeds under conditions of average sediment abundance is summarized in Figure 10:10. Nevertheless, because of the many variables involved in their formation, and the general non-linear behaviour of sediment responses to water movements, a precise mathematical and physical description of bed-form formation is still only a distant prospect.

Several basic types of bed-form have been described and related to different development processes of formation. In general, the stronger the current the larger the scale of the resulting bed-forms. There is a strong tendency for the bed-forms to align either transverse to or parallel with the main axis of the tidal flows. Transverse alignment is characteristic of lower current speeds (less than $0.5 \, \mathrm{m \, s^{-1}}$), whereas parallel alignment occurs when the tidal currents are stronger than about $1 \, \mathrm{m \, s^{-1}}$. Any net flow of current in one direction, or an asymmetry in the strength of the flood and ebb tidal flows may result in a net flux of sediment as discussed previously. As a result there is a corresponding asymmetry in the cross-section of the bed-forms with gentler slopes on the upstream side as defined by the direction of net sediment transport, and steeper slopes on the downstream side.

Sand patches are associated with areas of limited sand supply where the peak tidal currents do not exceed $0.5 \, \mathrm{m \, s^{-1}}$. The sand gathers in patches between beds of relict gravels which are themselves immobile. The main axis of these patches is aligned transverse to the main direction of tidal flow. They are typically 2 to 3 m wide, and have sharply defined boundaries. In plan view they are usually ragged, or take the shape of a crescent. Their persistence is thought to be due to the effects of the different gravel and sand roughnesses on the bottom turbulence: sand stirred into suspension by greater bed roughness and turbulence above the gravel will settle out and remain in the quieter sand patches. Where the sediment supply is more abundant these patches become extensive and form sheets.

Rippled sheets are also found in regions of low tidal flows; ripples only form if the median grain diameter is less than 0.7 mm, and they are often transient features which may be removed by stronger spring currents or by storms. The wavelengths between ripples are typically 0.5 m and their heights are typically 0.05 m. In terms of the sediment grain size, the wavelength spacings are typically 1000 grain diameters and the heights are of the order of 100 grain diameters.

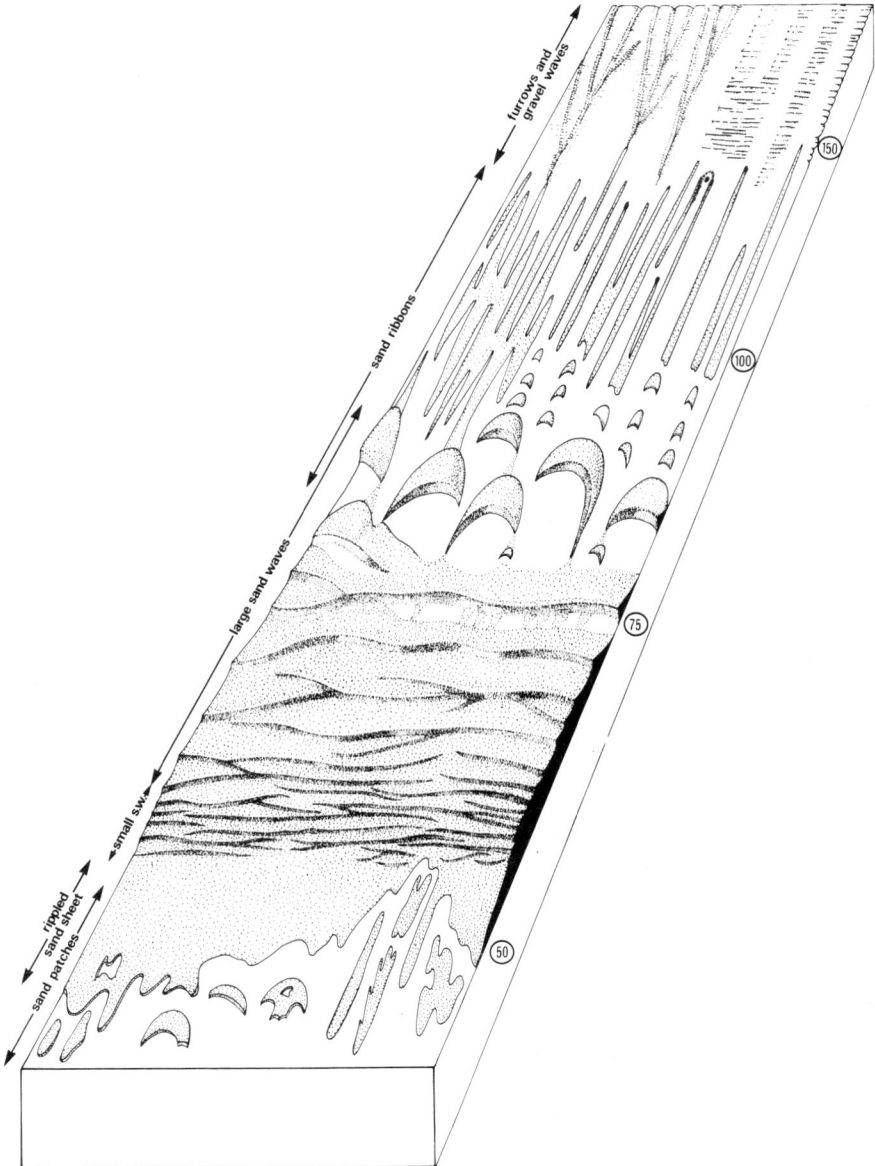

Figure 10:10 Block diagram of the bed-forms made by tidal currents on the continental shelf related to the mean spring peak currents in cm s^{-1}. (From Stride, 1982.) (*Reproduced by permission of Chapman & Hall.*)

Sand ripples are often found superimposed on other larger scale features such as sand waves. *Sand waves* themselves (often called dunes by hydraulic engineers) are distinguished from ripples by their much longer wavelengths (tens to hundreds of metres) and heights (1.5–25 m). Like sand ripples, their main axis is transverse to the currents. Spring tidal currents in excess of 0.65 m s^{-1} and a moderate abundance of sediments are minimum conditions for their formation. Over several tidal cycles they may be seen to migrate slowly in the direction of net sand transport, which is controlled by the asymmetry between the speeds of the flood and the ebb tidal currents. There is gradual erosion of material from the gentler sloping upstream face of the sand wave, and deposition on the steeper downstream face. Where the ebb and flood current speeds are approximately equal there may be erosion and deposition during both phases and a complicated internal sand-wave structure will develop. However, where the asymmetry between the flows is marked, erosion and deposition will be in only one direction and a clear sequence of sand and mud drapes may be identified with their spacing related to the spring–neap tidal cycle (see section 10:5 for a discussion of the use of these drapes to study ancient tidal conditions). Sand waves are sometimes found in conditions where the observed mean spring tidal currents are too weak for their formation; for these sand waves, residual currents and storm conditions must also be important factors in their formation and development. As a general rule these sand waves are smaller than the average. Active sand waves tend to be larger where the tidal currents are stronger.

Sand ribbons develop where tidal currents are strong (typically 1 m s^{-1}). They may be several kilometres long but only a hundred or so metres wide. The length is usually more than 40 times the breadth. The sharp boundaries found for sand patches are also typical of sand ribbons and a similar mechanism related to greater turbulence above the surrounding gravels is thought to be responsible, probably related to secondary circulation patterns which are generated (McLean, 1981). The characteristic bottom fauna associated with these bed-forms has been described by Holme and Wilson (1985).

Furrows and *gravel waves* are found where the currents are very strong (typically 1.5 m s^{-1} for gravel waves). Furrows may be about 1.5 km long and 15 m wide. At speeds greater than 1.5 m s^{-1} *scour hollows* (long elongated troughs perhaps tens of metres deep) may be formed by erosion of the sea-bed. Once formed a scour channel experiences strong currents because it provides a path of low resistance for tidal flows. The strong currents in the vicinity may cause strong turbulence and 'boiling' of the sea surface during maximum spring flows. Many present-day scour hollows, located where currents are now not very strong, were formed at times of lower sea levels and stronger local tidal currents. Examples include the Inner Siver Pit in the southern North Sea, and Hurd Deep in the English Channel; Hurd Deep is an exceptional feature, 150 km long, 5 km wide, and 150 m deeper than the surrounding sea-bed.

For current speeds around 1 m s^{-1} *sandbanks* may develop instead of sand ribbons, if there is an abundant supply of sediment. Nearshore sandbanks tend to be parabolic in form, and appear to evolve into linear sandbanks further off

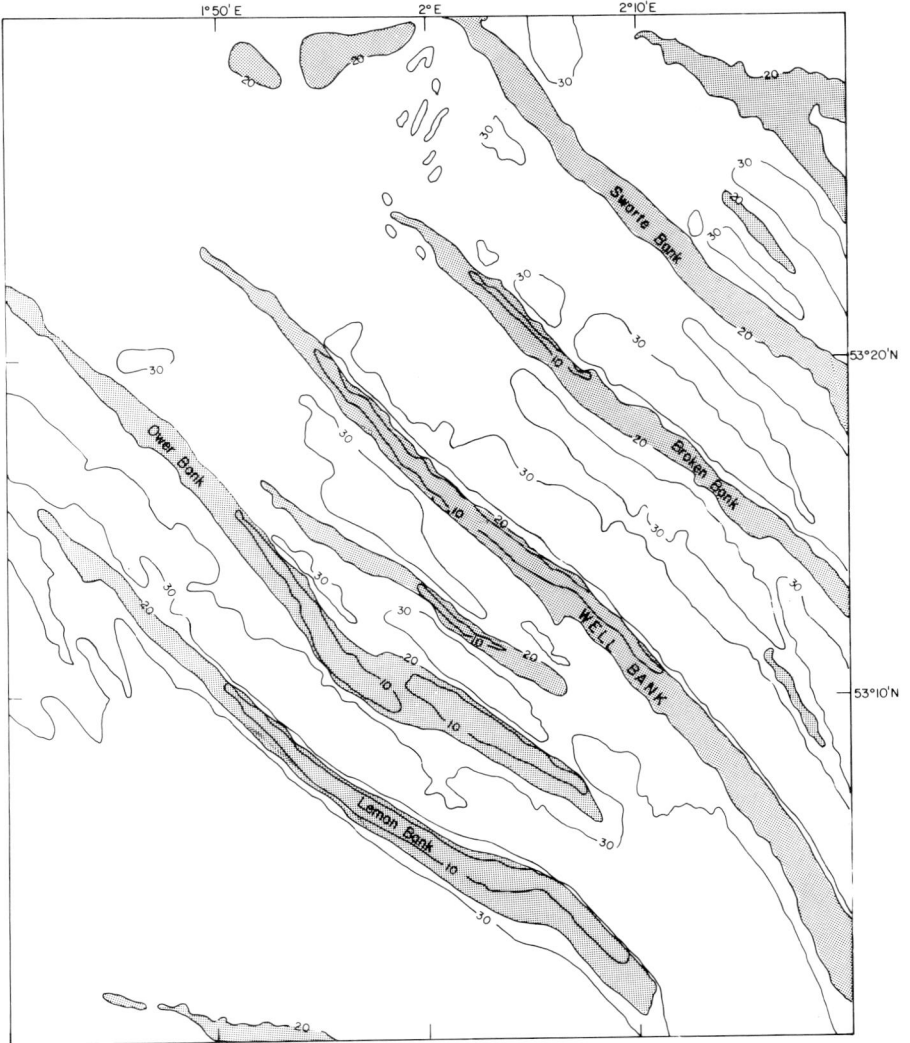

Figure 10:11 Part of a sandbank field in the North Sea. (*Reproduced by permission of Academic Press Inc. (London) Ltd.*)

shore. They are found in many areas of the southern North Sea as illustrated in Figure 10:11. They are also found in many other parts of the world, including the east coast of the United States and the Yellow Sea (Zimmerman, 1981). As described in Section 7:6, the expected residual circulation around banks in the northern hemisphere is clockwise, due to the combined effects of depth variations and bottom friction on the tidal vorticity. These effects reinforce each other if the axis of the sandbanks is inclined to the left of the axis of the tidal streams, and this inclination of sandbanks is often observed. If the axis is at an angle, then one side of the bank, the side facing away from the dominant current is steeper, as in the case of sand waves. The complex relationship between the currents and the sandbanks and the way in which they modify each other is still only partly understood, although there have been significant recent theoretical advances (Huthnance, 1982).

Offshore, parallel sandbanks are often found in groups. Near-shore sandbanks may be intimately connected with coastal processes and may be supplied with sand from the beach system, as has been proposed for East Anglian sandbanks by McCave (1978). Near headlands single or pairs of sandbanks are sometimes formed; for example, sandbanks are found to the east and west of Portland Bill in the English Channel (Pingree, 1978; Pingree and Maddock, 1979) (Figure 7:8). Coastally located sandbanks may be related to the tidal eddies generated by headland vorticity (section 7:6). To understand how a surrounding residual circulation can reinforce a central bank by driving convergent sediment flows at the bottom, it is necessary to consider the gradients on the sea surface required to produce the curved flow. These forces must be directed downwards towards the centre of the eddy to give the centrifugal acceleration appropriate for the speed of circulation. At the bed the circulating current is much reduced by bottom friction, but the pressure gradient towards the centre remains. This pressure gradient drives a secondary convergent component of the flow towards the centre. Similar dynamics can be observed driving sediment at the bottom of a stirred cup of coffee. The formation and location of coastal sandbanks can be controlled by other bottom features such as rock outcrops. Two pairs of sandbanks may be formed near islands, which are equivalent in this case to a pair of headlands (Figure 7:9).

Sandbanks in water 120 m deep in the Celtic Sea, and others in the North Sea, must have been formed when sea levels were much lower than at present. Careful analysis of sandbanks and other relict bed-forms can give indications of tidal conditions in the past. For example, Belderson et al. (1986) have shown that tidal currents in the Celtic Sea when sea levels were 100 m lower than at present would have been sufficiently strong to generate these sandbanks, which are now moribund.

10:4:3 Mining, renewal, licensing

Shortage of building material on land has encouraged the exploitation of offshore sand and gravel. It has been estimated that some 11 per cent of the total sand and gravel production in the United Kingdom is from the sea, and certain procedures are necessary to control the practice, and to avoid damage to adjacent coastlines and fisheries. In the United Kingdom the Crown Estate Commissioners grant licences for extraction, but only after wide consultation. Hydraulics Research Limited, who are consulted about coastal effects, have developed four criteria which should be satisfied (Price et al., 1978).

(1) Is the dredging far enough off shore that beach drawdown into the hole cannot take place? On the basis that the approximate limit for onshore/offshore movement is about 10 m below low water, this is usually taken as the minimum depth to ensure no beach drawdown.

(2) Is dredging to be carried out in deep enough water to ensure that the hole will not interfere with the onshore movement of shingle? An 18 m depth is considered sufficient for no shingle movement, but this may be increased in regions of strong tidal currents.

(3) Does the dredging area include banks which if removed would increase wave activity at the shoreline?

(4) Is the area sufficiently distant from the shore and in deep enough water so that changes in wave refraction over the dredged area do not lead to changes of littoral transport at the shoreline?

If these conditions are satisfied, further consultations are held with Fishery Authorities and other interested parties, before a licence to dredge is issued.

10:5 Tides past

Tidal amplitudes and phases, although stable from year to year, are known to have changed over longer periods. Comparisons of historical and recent tidal observations have shown only small changes, except where extensive developments such as river dredging or harbour building have occurred (Cartwright, 1972). Between 1761 and 1969 the oceanic semidiurnal tides at St Helena in the South Atlantic were constant in amplitude and phase to within 2 per cent and $2°$. Similarly, between 1842 and 1979 the amplitudes of the tides at two sites on the east and south coast of Ireland had increased by less than 2 per cent and the phases were identical to within $1.0°$, equivalent to 2 minutes (Pugh, 1982b).

The role played by tides in the development of recent shorelines and continental shelf sediments was certainly also significant in the geological past. Over geological time-scales the responses of the oceans to tidal forcing must have changed, as must the character of the Equilibrium Tidal forcing itself. In Section 7:9:4 we discussed how the energy lost by tidal friction came from the dynamics of the earth–moon system, and principally from the reduction in the rate of rotation of the earth. In order to conserve the total angular momentum of the two-body system, this loss of rotational energy is compensated for by an increase in the earth–moon separation. Although this increase in the separation is estimated at only 0.04 m per year, over geological time major changes in the separation and hence the tidal forcing have accrued. For example, extrapolating the present rate backwards in time, the separation was one earth radius less 150 million years ago than today. In turn this implies, on the basis of the cubic relationship between the distance and the magnitude of the tidal forces (equations (3.5) and (3.12)), that the tidal amplitudes should have been 5 per cent higher then than they are today.

If these increased tidal amplitudes are taken to imply corresponding increases in the energy losses, then further extrapolation suggests that the moon was within a few earth radii of the earth between 1000 and 2000 million years ago. The close proximity indicated by these gross extrapolations is called the Gerstenkorn event. The geological record, however, gives no indication of such an extreme event at any time in the past 4600 million years. Clearly the assumptions made about past tidal ranges and energy dissipation need further examination. It may be that the near-resonant semidiurnal responses of the present ocean are not typical of previous ocean and continent configurations. Similarly the different lengths of day discussed in Section 7:9:4 would have excited different ocean modes. The more rapid rotations of the earth would have increased the tidal frequencies so that different responses would be excited, matching the shorter tidal wavelengths. Models of the average energy dissipation in an ideal hemispherical ocean show that on this basis the dissipation rate falls rapidly if the tidal frequency is increased, and that the Gerstenkorn event was much earlier than previously estimated (Webb, 1982a). Numerical models which take into account variations in the actual ocean configurations and their response to tidal forcing over the past 450 million years show that ocean tides were not dramatically different then from the present day (Stride, 1982).

Modelling the shapes of the ancient ocean basins fails to include the very important influence of the continental shelf regions, which is where the tidal energy is dissipated. Some information about past tidal conditions in shallow seas can be gleaned by careful analysis of sedimentary rocks which were deposited at that time. Geological indicators of the tidal regime under which sediments were established include sand waves, longitudinal sandbanks, sand and mud sheets, and scoured surfaces. These features have been tentatively identified by geologists in sedimentary rocks from as early as 1500 million years ago. In some circumstances sand and mud layers are interleaved, indicating periodic changes in the different conditions of transport and deposition (Section 10:4:2). Advancing sand waves may have such interleaved layers of sandy foresets and mud drapes, corresponding to periods of vigorous flow and slackwater respectively. One example from Oosterschelde (Netherlands) is shown in Figure 10:12 (a). These sediments are thought to have accumulated 200–300 years ago at a depth 10–15 m below sea-level (Visser, 1980; Allen, 1981; Allen, 1985a). The spring–neap modulations are clearly visible, with the mud-drape separation an order of magnitude greater during periods of vigorous spring tidal currents than during the intermediate period of neap tides. Assuming a cubic law relating current speeds to the resulting sand transport it was possible to estimate the ratio of spring to neap current speeds. Similar mud drapes from Cretaceous sand deposits (about 135 million years ago) from southeast England show periodic variations in the drape separation which are much less regular than the recent Oosterschelde mud-drape spacings. Figure 10:12 (b) also shows the spacing in the Folkestone beds, where the shorter lengths of the cycles of spacings have been attributed (Allen, 1981) to a strong diurnal tidal influence. On the basis of simple transport models, it was

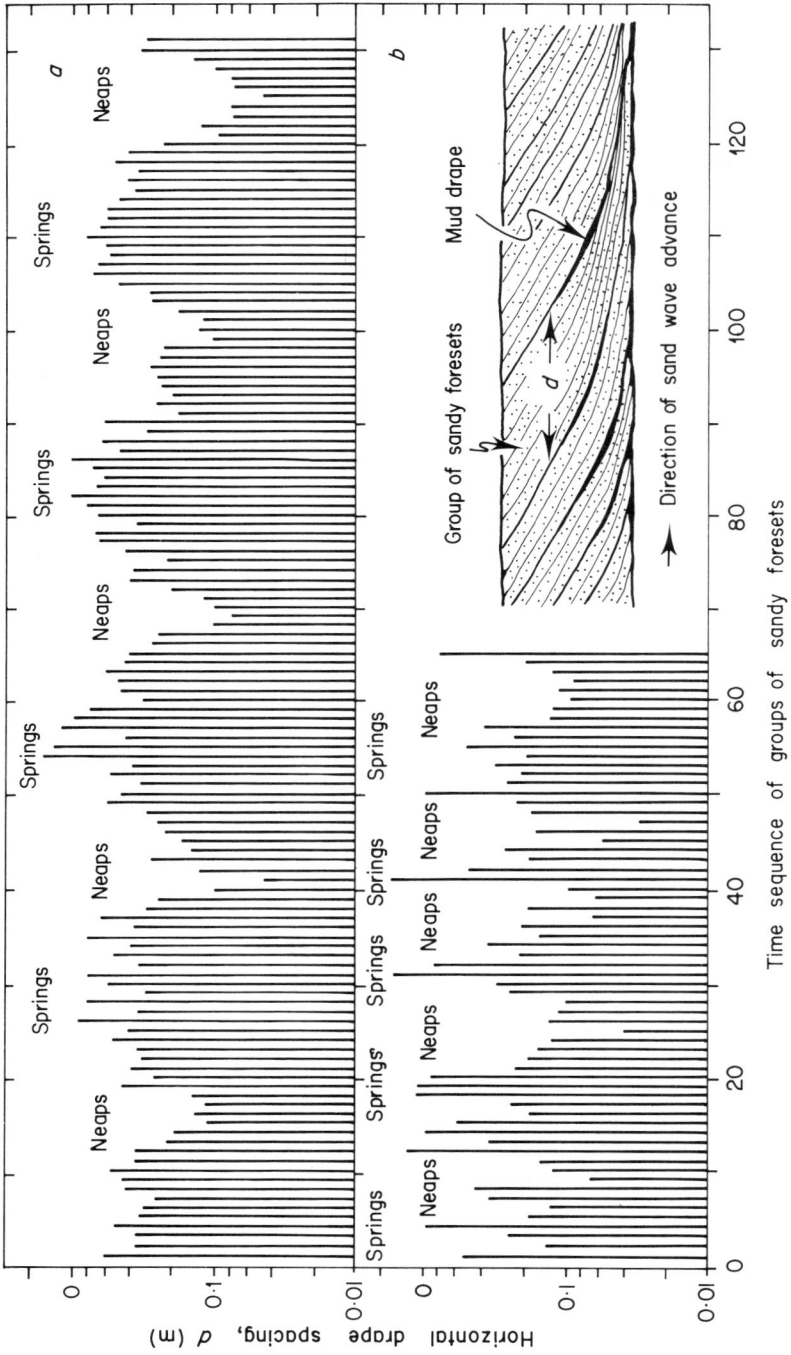

Figure 10:12 Relationships between drape separation and spring–neap tidal currents in sedimentary deposits. Separations are much greater during spring tides. (From Allen, 1981. Copyright © 1981, Macmillan Journals Limited.)

estimated that the current speeds and spring–neap ratios were similar to those observed at present around the northwest European continental shelf, but these estimates are limited by a poor understanding of the quantitative aspects of sediment transport.

Organic evidence for tidal rhythms is found in fossils. Stromatolites, which are the fossilized remains of structures produced by blue-green algae and bacteria and which today form largely in the intertidal zone, show rhythms of tidal growth in Precambrian times, but again, quantitative interpretation is controversial.

Growth rates of animals, particularly bivalves, have been studied in relation to tidal cycles, a process further discussed in Section 11:3:2. Spring–neap cycles have been observed in the modern growth of *Clinocardium nuttalli* on the Oregon coast of the USA and the common cockle *Cetastoderma edule* in European waters. Similar periodic variations in the growth of fossil bivalves, notably *Limopsis striatus-punctatus* in the Upper Cretaceous, may also be due to spring–neap tidal variations. Some caution is necessary in this interpretation because growth rates may also be influenced by diurnal variations in solar radiation, and so the interpretation of spring–neap patterns solely in terms of variations in tidal range and exposure must be questionable. For example, the advance in time through a solar day of a normal lunar semidiurnal tide of constant amplitude could also give rise to a spring–neap modulation due to the changing relationship between maximum water depth and maximum solar radiation as discussed in Section 11:2. Finally, growth may cease during the winter in harsh conditions (Section 11:3:2 and Richardson *et al.* (1980b).

Tidal conditions in ancient continental seas have been speculatively constructed using estimated water depth and basin geometry together with ideas of amphidrome development and energy losses. Such models may eventually prove useful tools for the coherent analysis of the relationship between geologically simultaneous sediment deposits over extensive regions, but this kind of interpretation is still at a very early stage of development.

One interesting area of interaction between geological processes and tidal dynamics is the relationship between basin development and tidal resonance. The adjustment of continental shelf basins towards or away from resonance as a result of erosion or deposition is a little studied aspect of tidal dynamics: certainly at times of much lower sea levels, such as those discussed in the next section, the shelf responses would have been different because of the smaller basin dimensions and the shallower water depths, and so too would have been the rates of energy loss and the energy available for coastal and bottom sediment erosion. Large tidal ranges, corresponding to near-resonant conditions would tend to enhance coastal erosion, and so increase basin dimensions. The increase in the natural period of the basin would then lead to a less-resonant response, and smaller tidal ranges. When tidal ranges become smaller, different deposition processes and coastal changes will change basin periods over geological time.

10:6 Mean sea-level: the geological record

The practical importance of describing and understanding how sea-level has changed over historical and geological times is accentuated by the large areas of land which would be flooded by relatively small increases in level, and conversely the large areas of shelf seas which would be exposed by a small fall. Some 20 per cent of the earth's land surface lies within one hundred metres above and below the present mean sea-level.

For oceanographers the identification of global changes of sea level due to an increased volume of water, or those related to changes in ocean dynamics would be easier if it were possible to identify areas of geological stability, where there have been no vertical movements over the past few thousand years. For geologists one of the long-term goals has been the identification of a global curve of eustatic sea-level changes with time (Gutenberg, 1941; Morner, 1980); however, recent work has shown that it is an over-simplification to represent the global response to changes in the ocean volume in terms of a single curve. As we discuss later (Section 10:7), redistribution of the water mass has secondary effects.

The first detailed records of mean sea-level were started in Brest in 1807, but even before that, measurements of high and low water levels were being made at Amsterdam and Stockholm, from 1682 and from 1774 respectively. Despite some uncertainty about their datum stability and the problems associated with using mean tide level as a substitute for mean sea-level (particularly in regions of shallow-water tides), trends consistent with the observed climate changes are present (Morner, 1973). To see these trends clearly it was first necessary to subtract estimated land movements of $0.4 \, \text{mm yr}^{-1}$ subsidence for Amsterdam and an uplift of $4.9 \, \text{mm yr}^{-1}$ for Stockholm. After a period of little sea-level change from 1682 to 1740, there was perhaps a fall of 0.25 m over the next 100 years. The present sea-level rise began in the middle of the nineteenth century. The Brest record shown in detail in Figure 9:6 confirms this increase after 1850. The apparent fall of 0.25 m coincided with a worldwide climatic cooling and a corresponding glacial advance in many areas. The last major glaciation was at its maximum 20 000 years ago. At this time most of Britain, Northern Europe and much of North America particularly North-West Canada were covered by extensions of the polar ice-caps and the sea levels were perhaps 100 m lower than today (West, 1968).

Evidence for lower sea levels than at present includes submerged erosion notches (Figure 10:13(a)), submerged shore-lines and deltas. The extension of river valley systems well out across the continental shelf, sometimes as far as a submarine canyon at a shelf break, such as the Hudson Canyon off New York, are particularly striking. Submerged forests, peat beds and shells which contain organic material may be dated by radiocarbon methods to determine the time when they were at mean sea-level. A proportion of the carbon dioxide absorbed by plants and animals during their lifetime (1 in 10^{12}) contains radioactive C_{14}. Decay of this radioactive proportion to nitrogen by electron emission with a

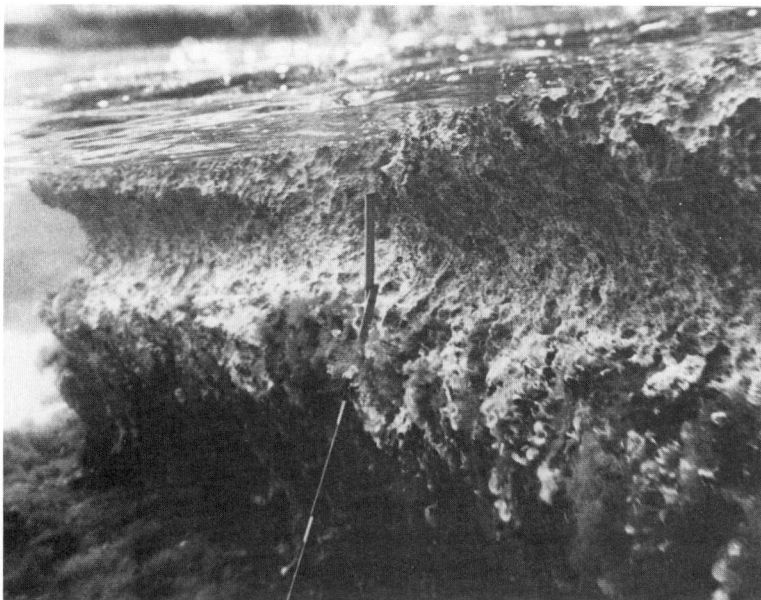

(a)

(b)

Figure 10:13 Photographs of coastal features showing (a) lower former sea levels than at present from Kalpathos, Greece, showing submerged erosion notches, and (b) higher sea levels indicated by notches at Kametoku, Tokuno-Shima, Japan. (Supplied by P. A. Pirrazoli, *Reproduced by permission of the Institut Océanographique, Paris.*)

half-life of 5730 years continues after the death of the organism. Measurement of the residual radioactivity enables the date of the death to be determined with an accuracy of a few hundred years throughout the 20 000 year period since the last major glaciation (Greensmith and Tooley, 1982). Evidence for relative upward land movement is seen in raised beaches, tidal flats and salt-marshes as well as in wave-cut terraces and elevated sea cave systems in cliffs as shown in Figure 10:13(b). These may also be dated by radiocarbon techniques if suitable biological material can be linked to their formation. Where tidal ranges are small, such as in the Mediterranean, relative changes of sea-level since 2000 BC are indicated by underwater archaeological surveys of ancient harbours which are now submerged (Flemming, 1978).

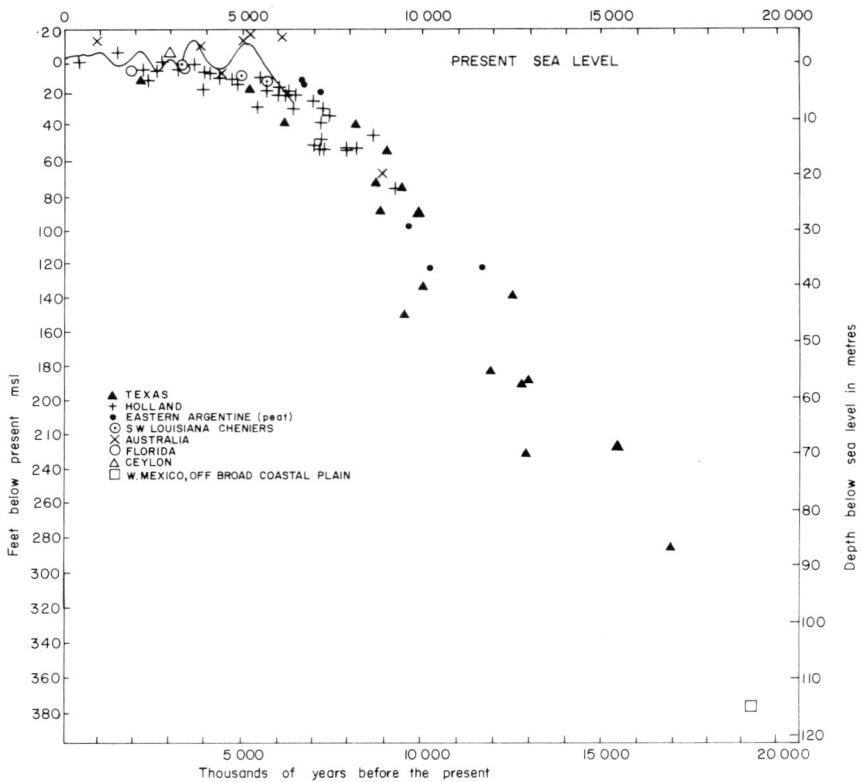

Figure 10:14   Apparent sea-level rises since 10 000 years BP, from geologically relatively stable areas of the earth's crust, based on carbon-14 dates on peats and near-shore shells. *(Figure 126 (p. 267) from SUBMARINE GEOLOGY, Second Edition, by Francis P. Shepard. Copyright © 1948, 1963 by Francis P. Shepard. Reprinted by permission of Harper & Row, Publishers, Inc.)*

The direct interpretation of sea-level changes measured by these several techniques in areas of relative stability (such as those where previous exposed shore lines are level over long distances and which have not been subjected to glacial loading) shows a relatively rapid rise following the glacial recession from

20 000 years ago, which gradually slowed down 8000 years ago when levels were some 15 m below those of today. Figure 10:14 shows that the rise then proceeded more gradually until present levels were reached some 4000 years ago. Since that time the mean sea-level changes have consisted of oscillations of small amplitude. On this evidence, the present slow rates of change are not typical, and in the geological past changes of 1 m or more per century were common.

The recent ice age was only the latest in a succession of advances and retreats which have repeated throughout the past two million years. Throughout this Quaternary period 17 glacial–interglacial cycles have been identified. However, over geological time these periods of large glacial oscillations are unusual and have only been seen on four other occasions in the past 900 million years.

Not all marine transgressions and regressions are caused by glacial cycles. Changes in the shapes and volumes of the ocean basins have also greatly affected sea levels over periods of millions rather than tens of thousands of years. For example, high sea levels can also be correlated with times of very active formation at mid-ocean ridges. This is because the hotter less-dense crust forms a broader higher ocean ridge, which reduces the mean ocean depth and so increases the sea-level. Desiccation of isolated basins such as the Mediterranean Sea and the Red Sea would also cause higher sea levels in the oceans (Donovan and Jones, 1979).

10:7 Isostatic adjustment and geoid changes

Tide gauge mean sea-level trends show the largest variations from an 'ideal' eustatic increase or decrease in sea-level in regions which have been subjected to loading by the glacial advances and retreats. In Figure 9:6 the sea-level trend for Sitka is strongly downward, whereas sea levels globally are increasing by 0.10 to 0.15 m per century. Even more extreme examples where vertical land movements are rapid enough to keep ahead of the eustatic water-level increases can be found at Juneau 150 km north of Sitka where rates of more than 1.0 m per century are measured, and in Scandinavia. This relatively rapid uplift has caused approaches to ports such as Oulu and Vaasa in Finland and Sundsvall and Gavle in Sweden to become progressively shallower and more difficult for navigation. Figure 10:15 shows the falls of sea-level relative to the land in mm per year for north-west Europe (Rossiter, 1967; see also Emery and Aubrey, 1985). There is a maximum value of nearly 10 mm per year in the Gulf of Bothnia but the trend reverts to the global average eustatic rise around the coast of Denmark. Further south, the apparent rates of sea-level rise actually exceed the eustatic rate, which implies a sinking of the land. This sinking also occurs around other areas of post-glacial uplift and is responsible for the very rapid rise of 3.7 mm per year in apparent sea-level at Halifax, Nova Scotia, plotted in Figure 9:6.

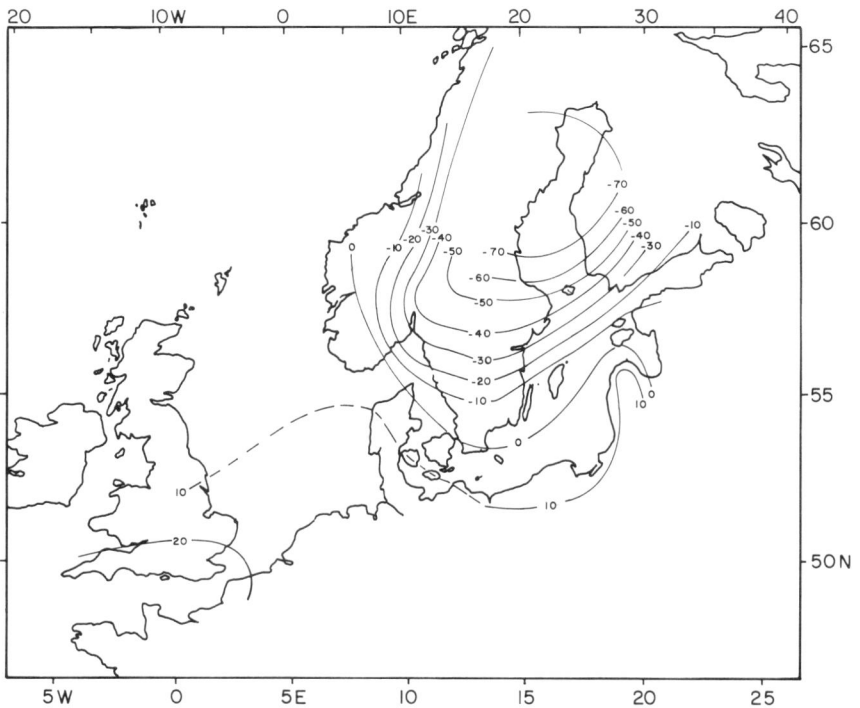

Figure 10:15 Relative rate of sea-level changes, from tide gauge measurements, in cm per century; positive values denote sea level rise. The isostatic uplift of Scandinavia is shown as a negative sea level rise. (*Reproduced by permission of the Royal Astronomical Society.*)

The vertical upward movement of areas which have recently been relieved of their glacial load was first interpreted in terms of glacial isostasy by the Scottish geologist Thomas Jamieson in the second part of the nineteenth century as a result of his studies of the raised coastlines in Scotland. The basic concept of isostasy assumes that at some depth in the earth's mantle the horizontal pressure gradients disappear. If a particular area of the crust is loaded with sediment then that area will be depressed, whereas the area from which the sediment was eroded will be uplifted. The adjustment of levels is by small horizontal flows of material in the mantle. Similarly, when the crust is loaded by an ice-sheet there is a compensating depression of the surface. For example, some parts of the base of the Greenland ice-cap lie below present sea levels. If the flowing material has a density four times greater than that of the ice, assuming full compensation, the depression of the underlying crust will be a quarter of the thickness of the ice burden. Around the ice-cap a band of relative uplift will be caused by the displaced mantle material. This surrounding area of forebulge was also recognized by Jamieson in his Scottish studies. As the ice-caps recede there is an increase in the eustatic sea-level and an uplift of the unburdened area. Around the uplifting area there will be a compensating zone of gradual sinking as the mantle material adjusts. The recent sinking of Halifax and of the southern North Sea are associated with this compensating zone, but in the latter area there is also a long-term subsidence which has led to the accumulation of 1000 m of sediment in the past million years. If the mantle were perfectly fluid the adjustments would be immediate and widespread: the rate at which uplift and downwarping are still continuing today, several thousand years after the glacial recession, can be used to help define the mantle's viscoelastic properties. At this stage it is necessary to augment the simple local isostatic model by several additional factors, including the adjustment of the whole global ocean under the changing water burden, and the modification of the geoid itself by the worldwide redistribution of the masses of the water and the solid earth. The importance of these factors in turn places restrictions on the use of the concept of a single globally-uniform eustatic curve of sea-level rise.

If the concept of a 'black box' which has an input, a response, and an output, is applied to the glacial loading process, in the same sense as it is applied to the analysis of tidal variations, the input or driving force is the varying load of the ice-caps. The 'black box' is the physics of the earth's response to this forcing, part of which involves the rheology of the mantle, which is imperfectly known, and part the laws of the gravitational potential for a given mass distribution, which are well established. The model output is the observed relative sea-level movement around the world's coastlines, as determined from geological observations and from tide gauge measurements. The two major uncertain factors, the glacial loading history and the viscoelastic properties of the mantle, are constrained within mutually compatible limits in computations (Peltier *et al.*, 1978; Clark, 1980; Peltier *et al.*, 1986).

When such a computer model is applied to the recession of the northern polar ice sheets, assuming a uniform mantle viscosity, six distinct zones of relative sea-level changes are identified; these are shown in Figure 10:16 for a eustatic sea-level rise which stopped 5000 years ago. The first zone covers areas glaciated during the last ice age. The elastic and viscous uplift of the ocean floor combined with the fall of the geoid because of the mass migration, causes the land to rise rapidly relative to the sea and to continue to give apparent sea-level falls to the present time at a diminishing rate. Thus Sitka is at a later stage of isostatic recovery than Juneau. Between the first and second zones is a transition zone where immediate land uplift is followed by land subsidence as the forebulge moves through in the train of the receding ice-cap. Present day sea levels will appear to be rising relatively rapidly in this intermediate zone, which includes the Halifax region. The characteristic of the second zone is a gradual upward movement of relative sea-level through to the present day. Zones three and four both reached sea levels close to the present levels some thousands of years ago, after which time the sea levels in zone three started to fall slightly whereas the levels in zone four continued with a gradual rise. Honolulu is one of the tide gauge stations in the fourth zone, where a gradual rise is observed. In zone five, which covers most of the southern oceans, a rapid

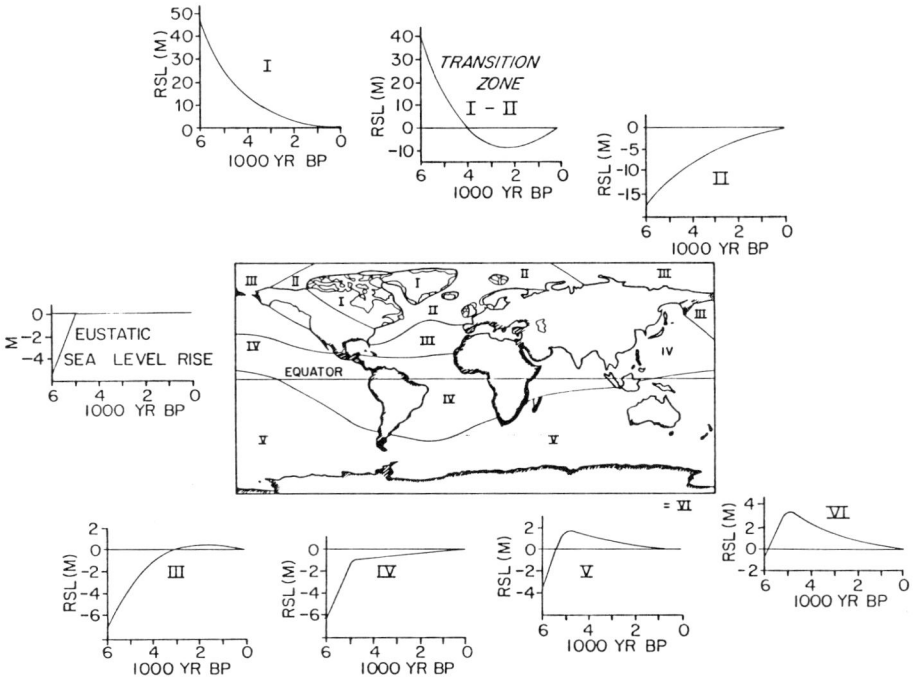

Figure 10:16 The six predicted zones of sea-level change following the melting of ice sheets 20 000–7000 years ago. The surrounding curves show typical predicted changes of mean sea-level relative to the present value, at particular coastal sites in each zone. (From Clark, 1980.) (Copyright by the American Geophysical Union.)

initial response in the form of a rising relative sea-level is followed by a slow fall of sea-level and relative land uplift to the present time. A similar behaviour is predicted for zone six, which consists of all continental margins except those in zone two; along these margins the initial rise of relative sea-level which followed the rapid ice-melting is followed by a gradual land uplift as the adjacent ocean crust adjusts isostatically to the increased weight of water.

All the different responses show a residual recent sea-level movement relative to the land, despite the fact that the glacial melting and the increases in the total volume of sea water stopped 5000 years ago. Although we may continue to define a eustatic rise of sea-level as the change in sea-water volume divided by the ocean's surface area, it is now accepted that eustatic rises cannot be observed free of residual post-glacial distortions, even in the tropics. The network of tide gauges around continental margins and the concentration in the northern latitudes in zone two is not an unbiased distribution. Estimates of the recent eustatic changes using tide gauge trends must eliminate data from the areas of rapid land movements, particularly from zone one and from the transition area to zone two. The relative sea-level trends observed in the other areas may be averaged to give an indication of the increase in the volume of ocean water, but the results even from glacially remote sites such as India must be interpreted with the total global behaviour in mind.

10.8   Earthquakes and local crustal movements

The mean sea-level movements associated with the advance and retreat of ice-caps have a global character, but there are more local land movements caused by other geological processes (Kasahara, 1981). The most spectacular of these are movements due to earthquakes. Charles Darwin reported the movements following the Concepcion, Chile earthquake of 1835: putrid mussel-shells were found ten feet above high-water mark, for which the inhabitants had previously dived at low-water spring-tides. Darwin proceeded to argue that the discovery of sea shells several hundreds of feet above sea-level was further evidence that the whole Andes chain was formed by intermittent systematic uplift.

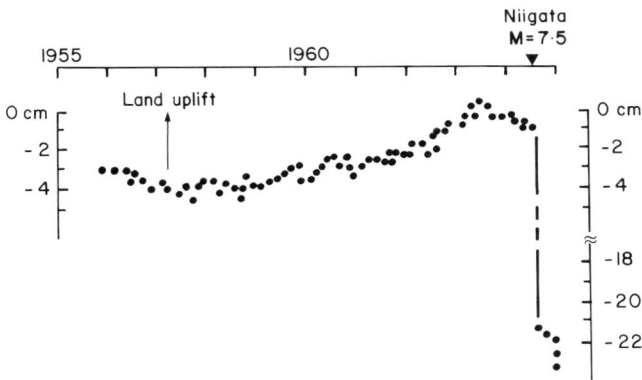

Figure 10:17   Vertical land movement associated with the 1964 Niigata earthquake in Japan (magnitude 7.5) from 1955–1963 and the drop one year before the earthquake itself. (Reproduced by permission of Elsevier.)

However, the major Chile earthquake (of magnitude 8.6 on the Richter scale) in 1960, failed to show any change in mean sea-level at a gauge 100 km north of the epicentre, which shows that the movements in individual earthquakes are only of local extent. The existing network of tide gauges cannot monitor such local earth movements, except along the coast of Japan, where a dense concentration of gauges is installed as part of an earthquake monitoring system. Following the Niigata earthquake (magnitude 7.5) in 1964, a fall of the land relative to the sea of more than 0.20 m was recorded on the tide gauge. The sea levels shown in Figure 10:17 reveal another interesting feature: the steady land uplift from 1955 is seen to reverse a year before the major earthquake movement. Similar systematic rises and falls were observed by repeated careful levelling along more than 100 km of coast. The possibility of identifying the onset of this reversal of vertical land movement as a basis for forecasting future earthquakes has encouraged further measurements and studies. It appears that the larger the earthquake magnitude, the longer is the time of anomalous vertical crustal movement.

Movements along the destructive plate margin of the South American coast are more likely to result in vertical land movement than those along the conservative plate margin between North America and the Pacific Ocean. The San Francisco earthquake of 1906 (magnitude 8.3) shows no discontinuity in the sea levels nearby, plotted in Figure 9:6.

Similarly, the sea levels at Sitka, also plotted in Figure 9:6, show no effects after the strike-slip motion earthquake (magnitude 7.1) of July 1972, which was centred 40 km offshore. The problem of measuring land movements is made very difficult because they are usually much smaller than the natural oceanographic variability of the monthly and annual sea levels as shown in Figures 9:2 and 9:6. One way of overcoming this difficulty is to monitor the difference between mean sea levels at a pair of nearby stations, making the assumption that the extent of the monthly sea-level changes is greater than that of the tectonic movements. However, applying this differencing technique to the Sitka data (Juneau was not used as a comparison gauge because of its very rapid postglacial uplift), showed only a possible 0.03 m uplift during the period 1966 to 1968. It is unlikely that tide gauge measurements alone could give useful indications of impending earthquakes in areas vulnerable to this type of slip movement (Wyss, 1975a, 1975b, 1976).

In some cases relative movement of the land/sea levels may be due to human activities. The Japanese city of Osaka has suffered subsidence due to ground water extraction, as has Bangkok, Thailand (Volker, 1982). Increasingly regular flooding of Venice is in part due to the pumping of water from city wells and the accelerated subsidence which resulted. The relative increase of sea-level for Venice is reported to be 3 to 5 mm per year, more than twice the rate along the more stable nearby coasts. It remains to be seen whether the recent provision of an alternative supply of fresh water will reduce the relative rise of sea-level in Venice to the more normal regional values of around 1.5 mm per year.

CHAPTER 11

Biology: some tidal influences

It is advisable to look from the tide pool to the stars and then back to the tide pool again.

John Steinbeck, *The Log from the Sea of Cortez*

11:1   Introduction

Plants and animals have certain essential requirements if they are to survive and prosper. Each organism has developed special characteristics to enable it to compete successfully in its particular environment: in particular, conditions on land and in water and the species which live there are very different. At the coast, certain species and ecosystems have developed to thrive in an environment which changes between these two extremes in a pattern defined by the rise and fall of the tide (Southward, 1965; Nybakken, 1982; Barnes and Hughes, 1982). For survival in this highly variable region species must be able to cope not only with the relatively uniform conditions of submersion (uniform temperatures, plentiful supply of dissolved oxygen, abundant nutrients and organic debris and micro-organisms which can be extracted relatively easily as food), but also with varying periods of exposure to air (extremes of temperature, salinity and solar desiccation). These periods of emersion, which may last for several hours, or at higher levels, days, must be survived until the next submergence.

Over time there has been a general tendency for marine plants and animals to evolve upwards from the sea to colonize the land. Those which have developed mechanisms to survive the longest exposures are better able to compete effectively at the higher levels, whereas others, such as corals, are unable to survive more than a brief period of exposure, and so their upward extension is restricted to the lowest tidal levels. As a result of these varying degrees of adaptation, a complicated pattern of species zonation can be observed from the bottom to the top of the tidal range. This zonation is most apparent on rocky shores where it has been extensively studied. In estuaries, salt marshes and mangrove swamps, the development of ecosystems is also strongly influenced by the changes of tidal levels and the rhythms which they impose, not only in terms of submersion and exposure to air but also in terms of the resulting temperature and salinity changes, sediment movements and nutrient fluxes.

Coastal ecosystems, particularly those in estuaries, are very vulnerable to human influences, either directly because of exploitation, or indirectly because of local or even global pollution. Permanent damage may result from disturbing the fine balance between living organisms, sediments and tides. But the coast also has great potential for beneficial uses: careful management of these uses can allow multiple complementary activities including general recreation, food production, and transport. One example of a valuable coastal activity is mariculture, the rearing of selected plants and animals under controlled semi-natural conditions for enhanced food production. Mariculture has a long history but has now established itself as a major industry in many countries. Modern cultivation includes shrimps, prawns, oysters and mussels. Careful management of the coastal environment is essential to avoid health hazards, for example through inadequate arrangements for sewage disposal. Marshes and low-lying vegetation, which are only occasionally flooded by tides, may also provide a highly suitable habitat for malaria-carrying mosquitoes and the hosts of other diseases. A proper understanding of the relationships between all the aspects of coastal systems is an essential first step towards their effective management, and the role played by tides must be an integral part of this understanding (Nature Conservancy Council and Natural Environment Research Council, 1979). In this chapter we can outline only a few basic principles which relate tidal rhythms to the functioning of coastal ecosystems. Accounts of the biology of the intertidal zone are given in several textbooks.

The influence of tides on living organisms is not limited to the shore. For example, the different communities which develop over different types of sea-bed, are indirectly controlled by the effects of tides on sediment movement as described in Chapter 10. More directly, some types of fish have adapted to changing tidal currents to assist in their migration. Further, the biologically productive fronts found in many shelf areas, between stratified and unstratified water, are located in positions controlled by the water depth and the strength of the tidal streams. These offshore tidal influences in living systems are discussed in the second part of the chapter. Although the biological responses to long-term sea-level changes are outside the scope of this book, some potential implications of interannual sea-level changes for fisheries development, and of possible correlations between fish catches and sea-level are indicated. Human intervention and reaction to tides has already been considered, particularly in Chapters 8 and 9. The final section of this chapter discusses the environmental and possible social impact of enhanced rates of sea-level rise.

11:2   Tidal rhythms and statistics

Marine biologists have often looked for relationships between the local tidal regime and the zonation of coastal species. In attempting to relate biological distributions to tidal patterns and to possible Critical Tidal Levels (CTLs) it is necessary first to present the tidal patterns in an appropriate form. There are many different ways of doing this. The most direct is to plot sea-level or current variations against time, as shown for example in Figure 1:1(a). However, the frequency distribution of tidal levels at Newlyn and at Karumba as plotted in Figure 1:4 are in a form which allows them to be related more easily to particular levels of vertical zonation. The astronomical basis for other well known patterns such as the diurnal inequality of high and low water levels, spring–neap amplitude modulations and the recurrence of extreme ranges are discussed in Section 3:4. Tidal patterns can be found and presented in a number of other ways, each of which may be relevant to the adaptive mechanisms of some particular plant or animal. In the discussion which follows, the tidal characteristics at Newlyn (strongly semidiurnal) and at San Francisco (mixed) are analysed in detail as examples of different methods of presentation.

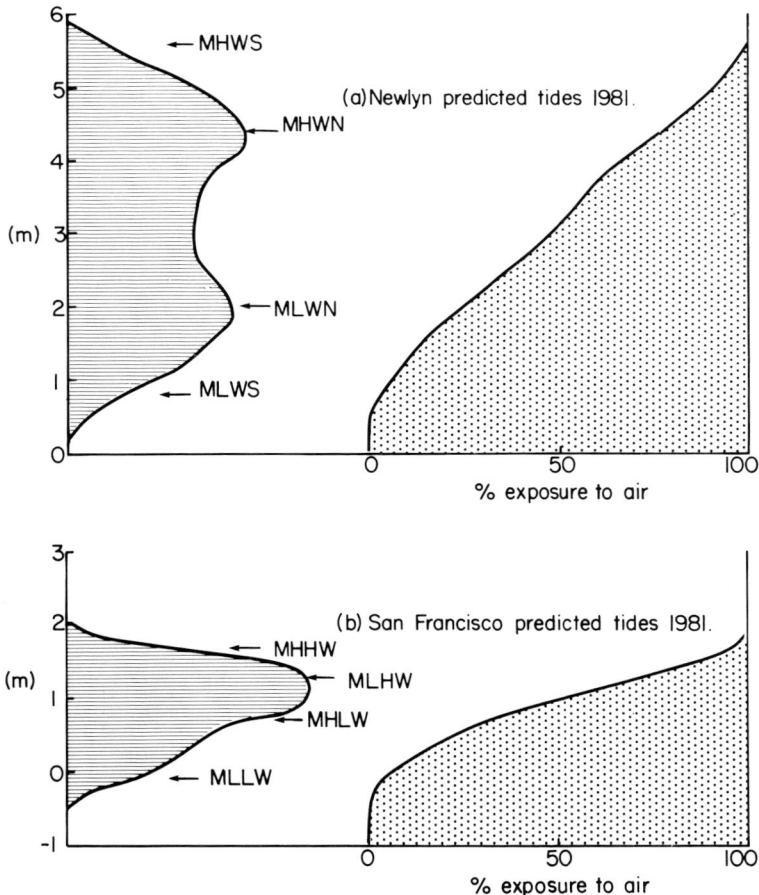

Figure 11:1 Frequency distribution of tidal levels and curves showing percentage exposure to air of each level for (a) Newlyn, which has a strong semidiurnal tide, and (b) San Francisco, which has mixed tides. Statistics are for 1981 predictions. Levels are to Chart Datum in both cases.

(a) Emersion/submersion curves

The extreme difference experienced by a species between emersion (exposure to the air) and submersion by the sea may be presented statistically as in Figure 11:1. This shows the *frequency distribution* of tidal levels for a year at Newlyn and at San Francisco and the overall percentage of time for which each level is exposed to the air. The Newlyn frequency distribution is also shown in more detail in Figure 1:4, together with a similar curve for the diurnal tides at Karumba. The San Francisco frequency distribution is intermediate between the two peak frequencies found in the dominant semidiurnal regime and the single peak of the diurnal regime. These frequency distributions define the levels at which disturbances and stress due to wave activity are most likely to be concentrated. The percentage *exposure curves* are termed *emersion curves*; it is easy to present the same statistics in the form of *submersion* or *immersion curves*. Lewis (1964) has discussed the use of the term 'exposure' which may have two meanings in this context: 'exposure to air' meaning not submerged, and 'exposure to wave action'. We follow his recommended latter use for 'exposure', and use 'emersion' for exposure to air. Confusion between emersion and immersion is avoided by using 'submersion' as the opposite of emersion. Causal relationships between the boundaries of intertidal zones and Critical Tidal Levels (CTLs) defined by changes in the gradients of emersion curves were originally proposed by Colman (1933), but as the Newlyn and San Francisco curves show, the annual percentage exposure changes quite smoothly as the level increases and there are no sharp changes of gradient (see also Doty (1946), Underwood (1978) and Swinbanks (1982)).

(b) Diurnal and semidiurnal emersion patterns

One of the critical factors for any particular intertidal plant or animal is likely to be the length of the periods of exposure to air which it must survive. The lower curve of Figure 11:2, which shows the tides for a particular day at San Francisco, may be used to illustrate the existence of five different zones. These are separated by Higher High Water, Lower High Water, Higher Low Water and Lower Low Water. For the first zone, above HHW there is continuous emersion (exposure to air). Between HHW and LHW flooding occurs only once during the day and there is a long period of emersion. Between LHW and HLW there are two submersion and two emersion events in the day whereas between HLW and LLW there is a single relatively brief exposure period. Below LLW there is continuous submersion. For San Francisco the fourth zone (between HLW and LLW) is generally significantly broader than the second zone (between HHW and LHW) because of the local character of the diurnal tidal inequality. Around the Hawaiian Islands and the Philippines the inequality is greater in the high waters, while at Hong Kong the inequality is approximately equal for both high and low waters. For semidiurnal tidal regimes such as Newlyn the tidal range is dominated by zone three and the second and fourth zones are only a small part of the whole tidal range.

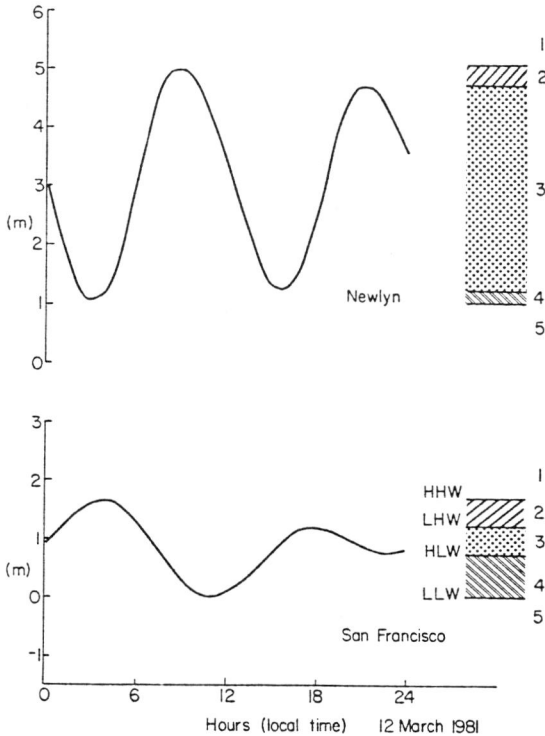

Figure 11:2 Tidal levels for Newlyn and San Francisco, 12 March 1981. The right-hand column distinguishes the five separate zones. (1). Not flooded during the day. (2) Flooded once for a short period. (3) Flooded and exposed to air twice per day. (4) Exposed to air once per day. (5) Continuously flooded throughout the day.

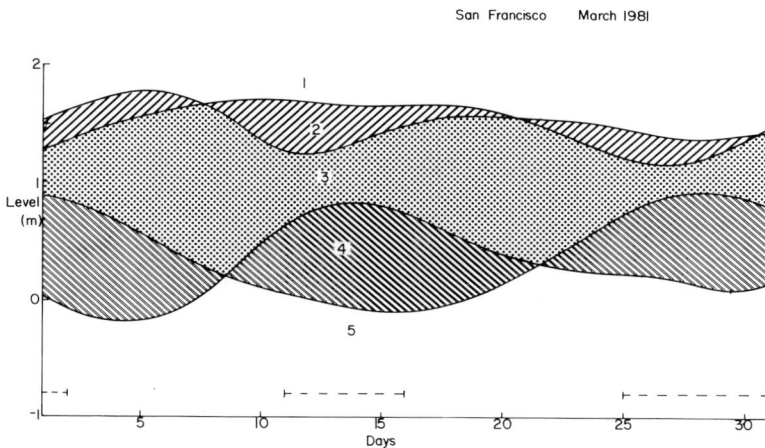

Figure 11:3 The variation through March 1981 of the extent of the five zones defined in Figure 11:2, for San Francisco.

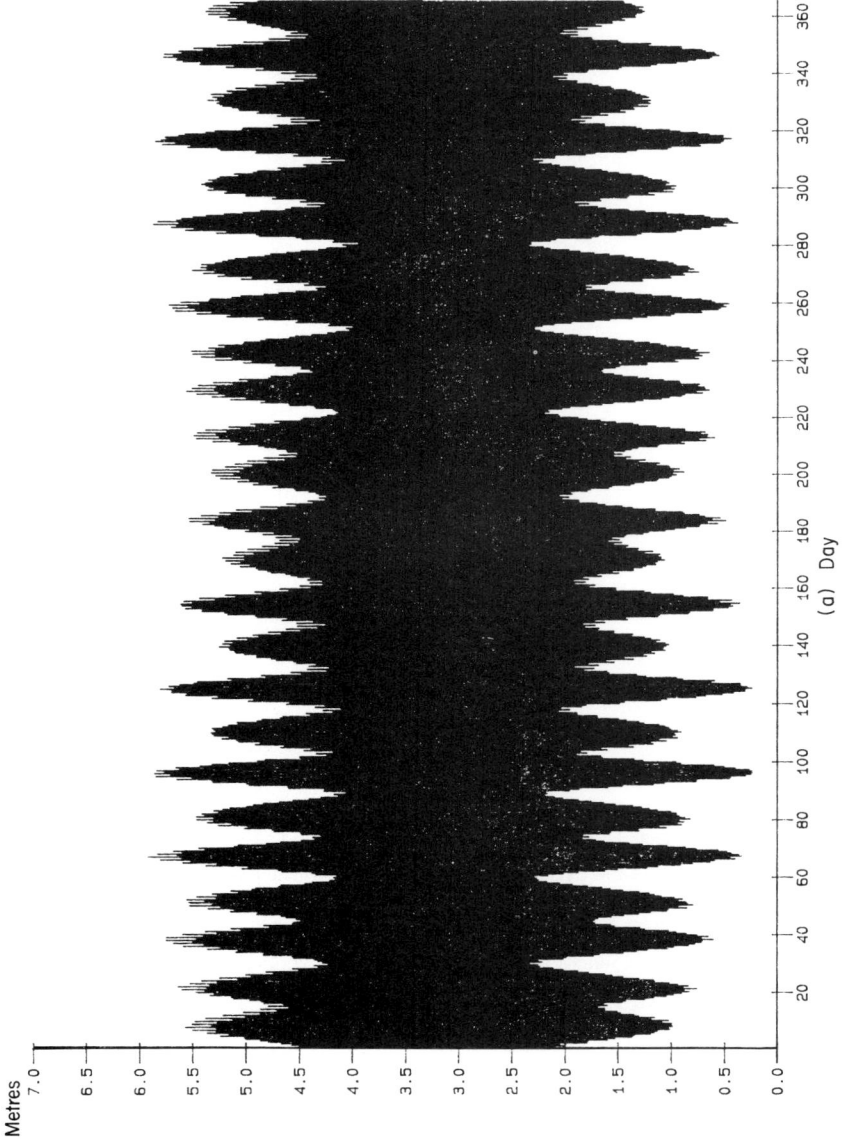

(a) Newlyn
Predictions 1981

(a) Day

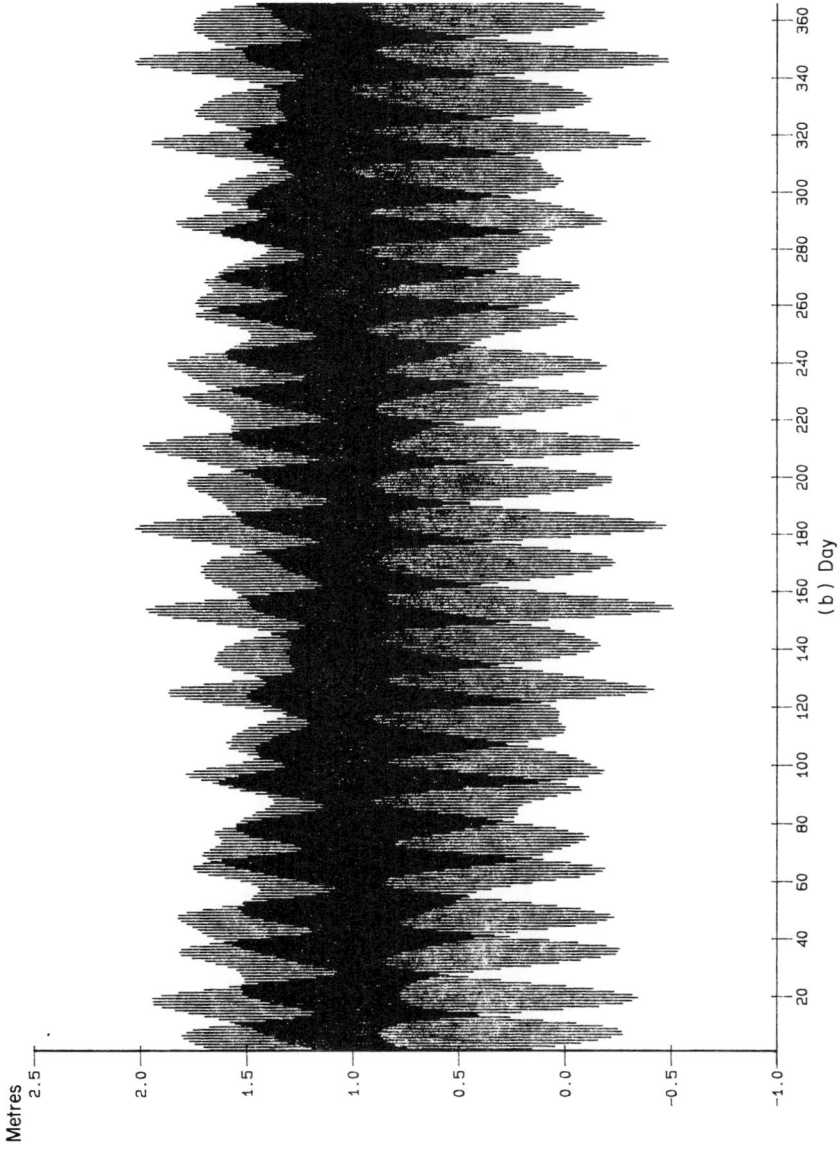

(b) San Francisco
Predictions 1981

Metres

(b) Day

Figure 11:4 Plots of the variations of tidal range through 1981 at (a) Newlyn and (b) San Francisco.

Table 11:1(a) The periods of exposure to air of different levels at Newlyn (1981 predicted tides).

Level (m)	Period of exposure to air (h)																Maximum emersion
	0–2	2–4	4–6	6–8	8–10	10–12	12–14	14–16	16–18	18–20	20–22	22–24	24–26	26–28	28–30	>30	
6.0																1	Continuous
5.6						33	1					6	2			9	2472 h
5.2					24	166	1					23	3			24	334 h
4.8					260	156	3					9				24	186 h
4.4					486	100	1					12		6		13	112 h
4.0				312	365	24						2					
3.6				671	34												
3.2				643													
2.8		1	62														
2.4	10	161	704														
2.0	49	461	534														
1.6	118	387	136														
1.2	172	142															
0.8	106	17															
0.4	17																
0.0	Continuously submerged																

MHWS 5.6 m
MHWN 4.4 m
MSL 3.2 m
MLWN 2.0 m
MLWS 0.8 m

Table 11:1(b)   The periods of exposure to air of different levels at San Francisco (1981 predicted tides).

Level (m)	Period of exposure to air (h)																Maximum emersion
	0–2	2–4	4–6	6–8	8–10	10–12	12–14	14–16	16–18	18–20	20–22	22–24	24–26	26–28	28–30	>30	
2.1																1	Continuous
1.9												17	1			5	3154 h
1.7											29	79	4			21	596 h
1.5										9	145	69	2			12	223 h
1.3							14			43	55	5					
1.1		30	14	4	52	62	12			4							
0.9	45	114	180	159	164	149											
0.7	35	82	132	168	292	26											
0.5	34	89	148	234	136												
0.3	36	128	233	253													
0.1	48	155	174	70													
−0.1	53	59	32														
−0.3	28	10															
−0.5	2																
−0.7	Continuously submerged																

Datum levels:

MHHW	1.7 m
MLHW	1.3 m
MSL	0.9 m
MHLW	0.7 m
MLLW	−0.1 m

Table 11:2 The periods of submergence by sea water of different levels at Newlyn (1981 predicted tides).

Level (m)	Period of submersion by sea water (h)																Maximum submersion
	0–2	2–4	4–6	6–8	8–10	10–12	12–14	14–16	16–18	18–20	20–22	22–24	24–26	26–28	28–30	>30	
6.0	4																
5.6	65	5															
5.2	139	94															
4.8	144	287	3														
4.4	90	429	100														
4.0	18	239	445	1													
3.6		9	671	24													
3.2			248	456													
2.8				659	45												
2.4				369	310	11	1					3	1			1	35 h
2.0				25	504	104						6	2			7	124 h
1.6					317	146						6	1			22	186 h
1.2					79	191	2					2	2			24	261 h
0.8					1	77	2					5	3			16	645 h
0.4						5										2	4620 h
0.0	Continuously submerged																

Table 11:3   The periods of exposure to air of different levels at Newlyn (1981 observed levels).

Level (m)	Period of exposure to air (h)																Maximum emersion
	0–2	2–4	4–6	6–8	8–10	10–12	12–14	14–16	16–18	18–20	20–22	22–24	24–26	26–28	28–30	>30	
6.4																	Continuous
6.0																	5290 h
5.6																2	1367 h
5.2					1	1	2					8	3			15	583 h
4.8				1	47	40						14	2			24	210 h
4.4				26	234	145	1					17	2			26	136 h
4.0				284	452	157	1					11	6			11	
3.6			8	642	381	111						3					
3.2			167	538	55	31											
2.8		12	687	6													
2.4	12	180	507														
2.0	68	452	129														
1.6	120	373															
1.2	166	140															
0.8	95	10															
0.4	8																
0.0	Continuously submerged																

MHWS 5.6 m
MHWN 4.4 m
MSL 3.2 m
MLWN 2.0 m
MLWS 0.8 m

The levels of HHW, LHW, HLW and LLW may be considered to define the Critical Tidal Levels for a particular day, but these levels are themselves changing continuously. Figure 11:3 shows the variations in the extent of the five zones through March 1981 at San Francisco, and Figure 11:4(a) and (b) show considerable annual variations in the tidal range, which defines the limits of the upper and lower zones, over a complete year. Figure 11:4(b) also indicates the variations in the extent of the second and fourth zones, where the plotting density is reduced at the upper and lower margins.

(c) Emersion period frequency distribution

A more complete picture of the emersion pattern at any particular level is given statistically in Table 11:1 in terms of levels and duration. For example, a plant or animal at a 4.4 m level at Newlyn is normally exposed for periods between 8 and 12 hours, but in addition on 18 occasions in the year it is exposed for periods between 22 and 26 hours, and on 13 occasions it is exposed for periods greater than 30 hours. The longest period of exposure is 112 h. A similar pattern of exposure is seen for San Francisco, but there are many more periods of exposure between 18 and 26 hours because of the greater diurnal inequality. At any level, the broader the spread of the periods of exposure to air, the greater is the stress on any species at that level which prefers water. Not unexpectedly, the most regular cycle of emersion and submersion is found for mean sea-level where, at Newlyn, there is exposure for each tide of the year, and the average period of exposure is 6 h 16 min with a standard deviation of only 12 min. At San Francisco the conditions at mean sea-level are more varied: periods of exposure vary up to 9 h 3 min about an average value of 5 h 51 min with a standard deviation of 2 h 14 min. Table 11:2 shows the complementary statistics for the periods of submersion or immersion for predicted tides at Newlyn: in this case the splitting of the distribution into semidiurnal, diurnal and longer periods occurs for the low levels. The statistics in Tables 11:1 and 11:2 are for predicted tidal levels, and in practice are modified by the weather. The actual observed sea-level changes at Newlyn given in Table 11:3 for the same year show that the basic pattern in the predicted tides remains, but slightly higher levels (6.0 m) are reached and the emersion times at particular levels are more variable: for example, the mean level has an average exposure time of 6 h 11 min with a standard deviation of 23 min.

(d) Time of day at emersion

The dangers of desiccation for a marine plant or animal exposed to the air are obviously much greater if the emersion takes place during the heat of the day. The broken line along the bottom of Figure 11:3 shows the days at San Francisco on which LLW falls between 0900 and 1500 local time, and it is significant that the LLWs on these days are among the lowest of the month. Night-time exposures are not without hazard, as winter frosts can also inflict

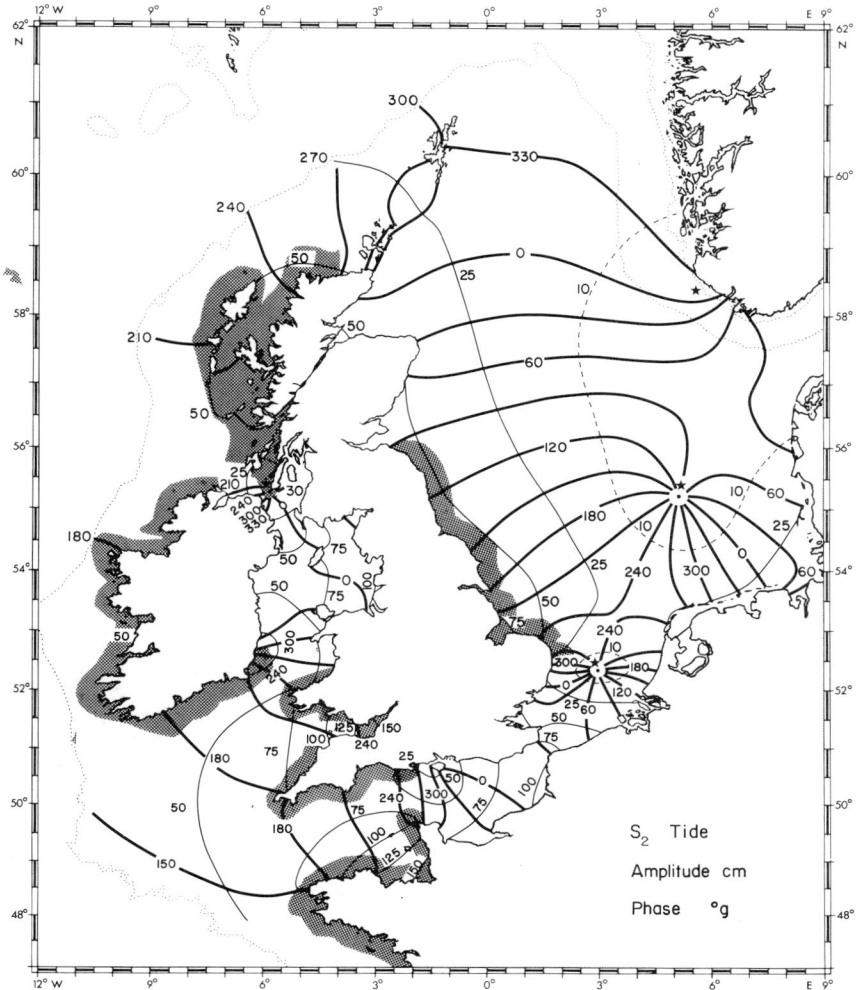

Figure 11:5 The S_2 cotidal chart for the north-west European continental shelf. Shaded areas where $90° < g_{S_2} < 270°$, corresponding to spring tidal low water falling between 0900 and 1500, local time (after Pingree and Griffiths, 1981). (*Reproduced by permission of Cambridge University Press.*)

serious stress on plants and animals whose normal preferred habitat is within the more stable temperatures of coastal waters. Coral reefs exposed to cold or rainy weather may suffer extensive damage and subsequent decay of cell tissue. For a semidiurnal tide, the time of day at which extreme low-water levels occur is related to the phase of the S_2 constituent, expressed in local time (Lewis, 1965; Pugh and Rayner, 1981). If the phase (g_{S_2}) is close to 0° (or 360°), then the maximum values of S_2 occur near the time of solar transit, noon and midnight. In this case, maximum spring tidal levels will occur near noon and midnight at times when the S_2 and M_2 constituents are in phase. However, if the S_2 phase is near $180°$, then S_2 will have a minimum value near noon and midnight, and the extreme low-water spring tide levels will occur at these times. Species which live at low levels will be exposed to the air at noon when the sun is at its zenith, and will require strong protective mechanisms to avoid desiccation. Examination of a cotidal map of S_2 will show those areas where low-level species are at greatest risk. Figure 11:5 shows those areas of the north-west European continental shelf where low water on spring tides falls between 0900 and 1500 ($90° < g_{S_2} < 270°$) (Pingree and Griffiths, 1981); these include the coasts of the west and south-west British Isles which is convenient for littoral biologists but not necessarily so for less mobile species which need to avoid high temperatures and evaporation. Lawson (1966) has shown a relationship between seasonal tidal changes and seasonal variations in zonation on the coast of west Africa, where the tides are mixed. During the months of northern summer the lower of the two daily low waters occurs at night, whereas in the northern winter it occurs during the day. As a result several species which exist near low water invade higher levels in the northern summer.

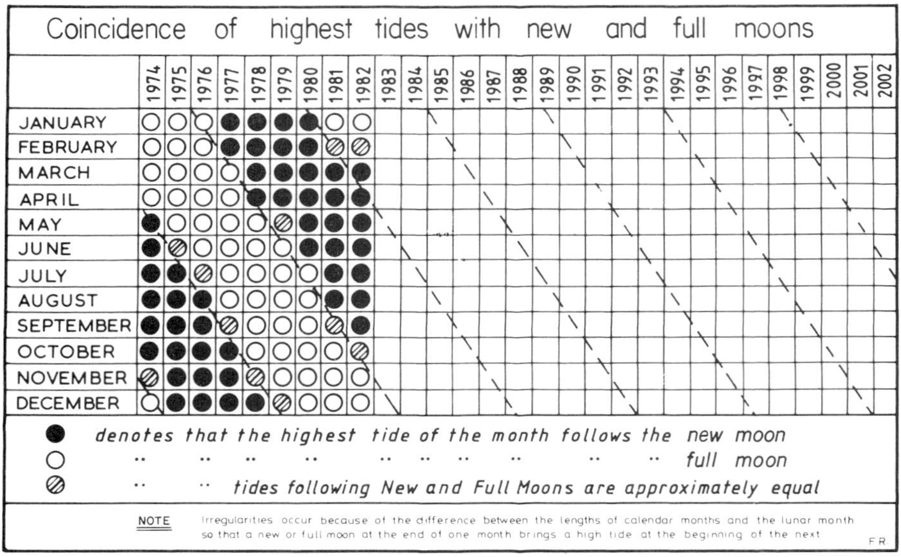

Coincidence of highest tides with new and full moons

	1974	1975	1976	1977	1978	1979	1980	1981	1982	1983	1984	1985	1986	1987	1988	1989	1990	1991	1992	1993	1994	1995	1996	1997	1998	1999	2000	2001	2002
JANUARY																													
FEBRUARY																													
MARCH																													
APRIL																													
MAY																													
JUNE																													
JULY																													
AUGUST																													
SEPTEMBER																													
OCTOBER																													
NOVEMBER																													
DECEMBER																													

● denotes that the highest tide of the month follows the new moon
○   "   "   "   "   "   "   "   "   "   " full moon
⊘   "   " tides following New and Full Moons are approximately equal

NOTE   Irregularities occur because of the difference between the lengths of calendar months and the lunar month so that a new or full moon at the end of one month brings a high tide at the beginning of the next

F.R

Figure 11:6 The relationships between high spring tides and the phases of the moon. (From Rowbotham, 1983. Copyright © F. W. Rowbotham.)

The influence of the weather, through surges and waves, on the emersion and submersion of different levels cannot be neglected, but is generally secondary to that of the tidal changes of level. Surges raise or lower the observed levels relative to the predicted tidal levels, and their importance will depend on the amplitude of the surge variations compared with those of the tidal variations. One obvious consequence of surges is the extension of the observed range of sea-level variations above and below Highest Astronomical Tide and Lowest Astronomical Tide respectively (see Figures 6:4 and 8:2). During storms coasts exposed to high waves are submerged to higher levels than those reached by the tides alone.

Both waves and surges blur the boundaries of Critical Tidal Levels where these are indentified, and this shading of boundaries makes it unproductive to speculate on the influences of a fine structure of several Critical Tidal Levels (Swinbanks, 1982). While certain tidal zones and boundaries can be identified by statistical analysis of the tides, these are transition levels rather than sharp limits. The sharp boundaries found in the vertical zonation of species must be accentuated by mechanisms within the ecosystems themselves.

Tidal patterns become apparent in many other ways. Figure 11:6, which shows the relationships between the highest spring tidal ranges in a month and the phases of the moon, emphasizes the approximate 9-year cycle in the coincidence of a full moon and maximum spring tides. This particular diagram was prepared to plan when people might observe the Severn Bore (Section 7:7) by moonlight during the high equinoctial spring tides of March and April! Less romantic species have adapted to patterns in the tides for the more basic reason of survival.

11:3   Shore processes

The intertidal area between highest and lowest water levels generally shows high biological productivity and contains a very rich and diverse range of species. The many reasons for the high level of productivity include the regular availability of nutrients during each tidal cycle in water which is shallow enough for photosynthesis to proceed. But, as already emphasized, it is also a region of great environmental stress: plants and animals must be able to withstand the temperature and salinity extremes generated by exposure to wind and sun, the physical forces exerted by waves and the predation of both marine and terrestrial animals including birds. The potential for high productivity can only be realized by those species which have highly adapted survival mechanisms.

The three non-biological factors which determine the distribution of shore species are the patterns of tidal emersion, the degree of exposure to waves, and the nature of the bottom material or substratum. Rocky, sandy and muddy shores each have their particular plants and animals. Shingle and gravel beaches, especially those exposed to waves, support little life because of the mobility of the material.

11:3:1   Zonation on rocky shores

The more usual approach to describing the vertical divisions of the shore is to subdivide them in terms of the plants and animals which are found there (Yonge, 1949; Brusca, 1980; Lobban et al., 1985; Moore and Seed, 1985).

Ricketts and Calvin (1968) proposed a four-zone classification appropriate for the Pacific Coast of North America. Extensive surveys of sea-shores in many parts of the world led Stephenson and Stephenson (1949, 1972) to propose their 'universal' system of zonation whose terminology is summarized in Figure 11:7. In the following discussions most of the species named are those found on the coasts of north-west Europe.

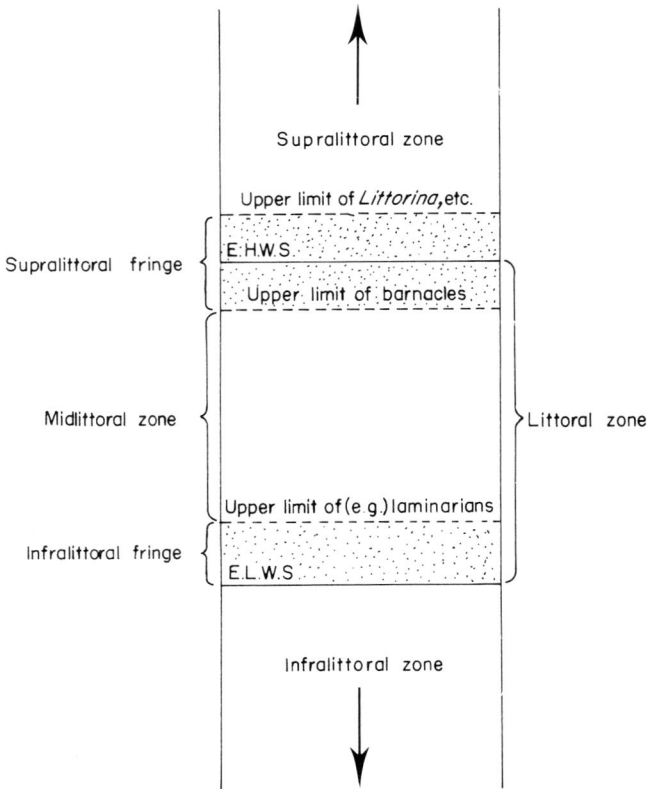

Figure 11:7   The general scheme for intertidal zonation, proposed by T. A. and A. Stephenson.

The *littoral zone* extends between Extreme Low Water Springs and Extreme High Water Springs, which approximate to the Lowest and Highest Astronomical Tidal Levels. Around EHWS, for coasts with limited wave exposure, there is a relatively arid zone subject to the harsh conditions transitional between terrestrial and marine. This zone, which is termed the *supralittoral fringe* is inhabited by few species; in the United Kingdom these include the periwinkle *Littorina neritoides* and yellow and orange lichens. The lower limit of this zone is frequently marked by a zone of black lichen.

The *midlittoral zone* is the principal zone of the sea-shore and contains a greater number of species than the zone above. Barnacles of different kinds are a general feature, and the upper limit for barnacle cover may be taken as the upper limit of this zone. Near this upper limit is found *Pelvetia canaliculata*, the channel wrack, a brown seaweed which can survive long periods of exposure: in very hot weather it may turn black and shrivel up, returning to normal when submerged by the next high sea-level. Limpets and winkles are prevented from drying up and protected from predators such as birds by their stout shell. The mussels, *Mytilus*, close their shell valves when the tide is out and hold sea water to avoid exposure to low salinities.

Intertidal anemones appear to have special characteristics which allow them to retain a relatively large volume of water (Stotz, 1979). All organisms in this zone must have strong attachment mechanisms to be able to resist wave action. Seaweeds are usually abundant; for example, in the Firth of Clyde on the west coast of Scotland, there is a downward transition from *channel wrack* through *spiral wrack* and *knotted wrack* to *bladder wrack* and *serrated wrack* (Schonbeck and Norton, 1978; Norton, 1985). These seaweeds can provide shelter for animals when the tide is out, and attenuate the strongest wave currents (Druehl and Green, 1982; Lobban *et al.*, 1985).

The *infralittoral* or *sublittoral fringe* is characterized by rich but extremely variable populations. In temperate and cold regions it consists of forests of large brown algae, particularly the laminarian seaweeds. Animal groups include crustacea (for example, crabs, lobster, shrimps and prawns), molluscs (for example, bivalves, limpets and sea-slugs), echinoderms (for example, starfish and sea urchins), sea anemones and sponges. On coral reefs this zone corresponds to the upper limit of growth, with highest rates of photosynthesis for those corals which can withstand brief periods of exposure. Massive coral forms such as brain corals subjected to emersion during low tides lose their upper tissue so that the colony increases in width but not in height.

Below the *infralittoral fringe* is the true *infralittoral* or *sublittoral zone*, which is never exposed to the air, but in which there are high nutrient levels and potential for benthic photosynthesis especially at low tide. The upper parts of this zone have a rich population of plants and animals which gradually merges into the true marine benthic community. Studies of transitions in the *sublittoral zone* generally require SCUBA equipment.

The Stephensons' scheme is sufficiently flexible to allow it to be applied to many varying tidal climatic and wave exposure conditions although it cannot fully explain the causes of the zonation so described. Lewis (1964) has refined this general zonation scheme: the *midlittoral zone* is divided into an upper *littoral fringe* and a lower *eulittoral zone*. The *supralittoral zone* is renamed the *maritime zone*. Figures 11:8 and 11:9 show zonations characteristic of the coasts of West Britain and Ireland. Lewis showed that these different biological zones are determined by the actual exposure time during which desiccation develops so that there is an upward extension as the degree of coastal exposure to waves is increased, as illustrated in Figure 11:10 (see also, Burrows *et al.*, 1954).

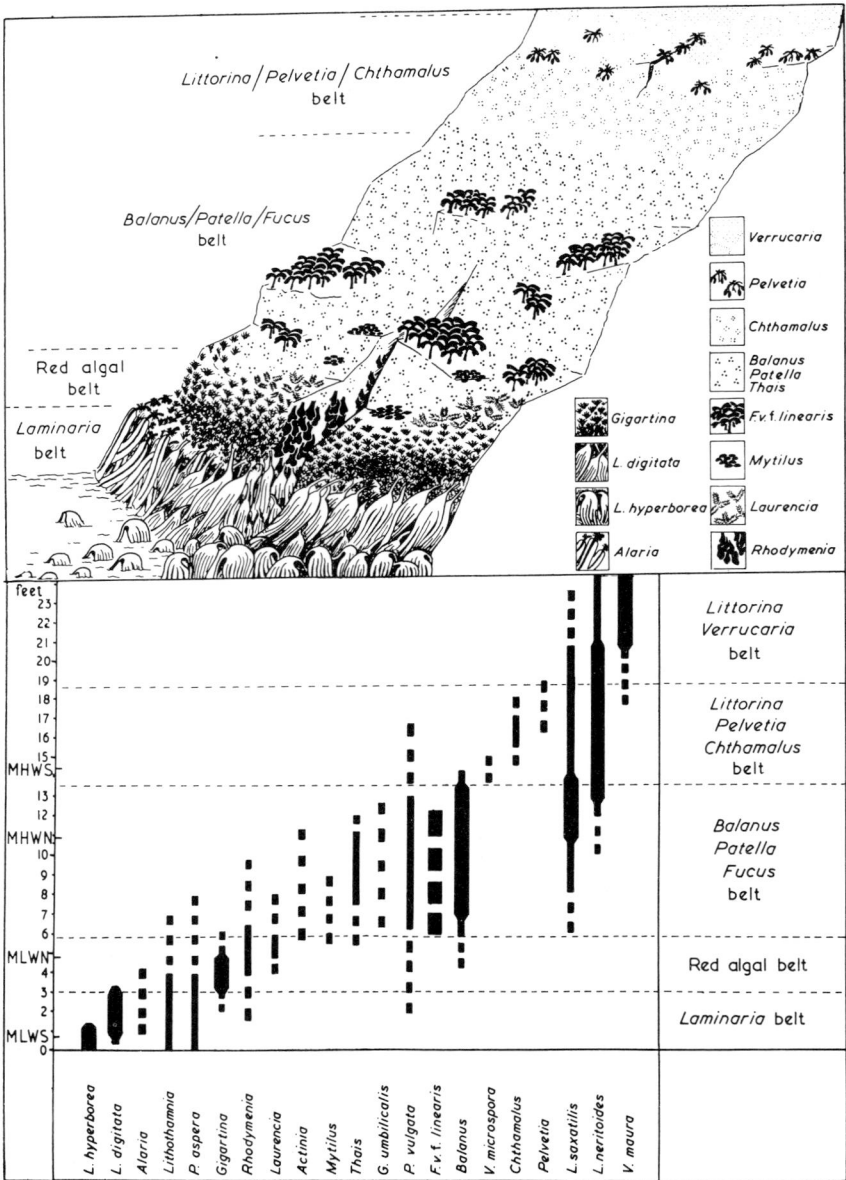

Figure 11:8 Zonation on a barnacle-dominated slope, commonly found on moderately exposed shores of north-western Scotland and north-west Ireland (from Lewis, 1964). (*Reproduced by permission of Hodder & Stoughton Limited.*)

Figure 11:9   Zonation on (a) a moderately exposed shore on the South of Jura, western Scotland, and (b) a severely exposed face near Kerry, south-west Ireland (from Lewis, 1964). (*Reproduced by permission of Hodder & Stoughton Limited.*)

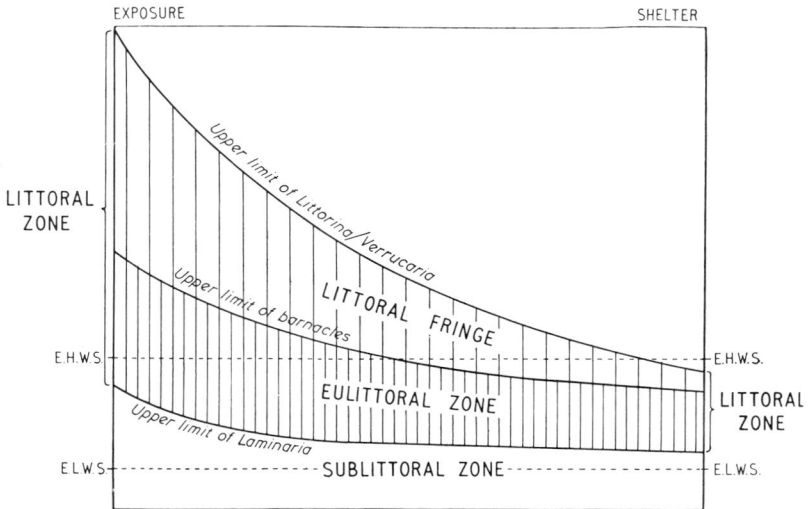

Figure 11:10   The change in the vertical extent of biological zones with the degree of wave exposure (Lewis, 1964). (*Reproduced by permission of Hodder & Stoughton Limited.*)

An interesting enhancement of growth in the infralittoral fringe and infralittoral zone occurs in rocky channels through which tidal currents ebb and flow. This enhancement is due to the continuous stream of nutrients and essential foods, and because the agitation of sediments prevents clogging of the filter-feeding mechanisms of organisms; the water is usually clear and free of the turbidity often generated by waves, and which may reduce photosynthesis. Rock pools in the intertidal regions also provide special biological niches. Individual species have different physical and physiological mechanisms for coping with changes of exposure, temperature, higher light intensity, salinity and other variables such as pH and the partial pressures of oxygen and carbon dioxide. These chemical parameters can vary considerably in small shallow intertidal rock pools where many species shelter during periods of low tide, preferring the possible extremes of salinity, temperature and of oxygen depletion to total exposure to the air.

To summarize, there is usually an increase in the variety and number of species as one descends from the higher dryer parts of the shore to the lower wetter parts. The biological basis of zonation is the most generally accepted method of dividing a shore, with loose correlation to any chosen system of Critical Tidal Levels. However, no universal scheme can be developed based on biological indicators because community composition varies geographically. Modern research has tended to concentrate on understanding the response of individual species to environment changes, regarding this understanding as an essential precursor of integrated studies of the ecology and zonation through the whole vertical extent of the shore.

11:3:2   Zonation on sedimentary shores

Coasts formed by gravels, sand and mud deposits are in a state of continuous erosion and deposition, and their structure is strongly influenced by tides and waves. In many cases the biological flora and fauna are important agents in the dynamics of coastal stability and change (Section 10:3:1). Sedimentary shores which support life vary from very sheltered embayments where fine muds accrete, to exposed beaches which are reworked by waves during each storm. Shingle beaches, consisting of gravels and pebbles, are generally too mobile to allow organisms to establish themselves.

Zonation is less obvious on sedimentary shores than on rocky shores because most of the species there can survive by burying themselves from extremes of emersion, temperature and salinity (Evans, 1965; Eltringham, 1971). Within the sediments large populations of individual species can thrive. When the tide is out the sediments retain a high proportion of water (more than 50 per cent by weight for some muds, falling to 20 per cent for coarser sands) so that burrowing animals in the trapped water avoid exposure to low salinities. In addition, the sands filter the water during each tidal cycle and concentrate the particulate matter to the advantage of burrowing animals.

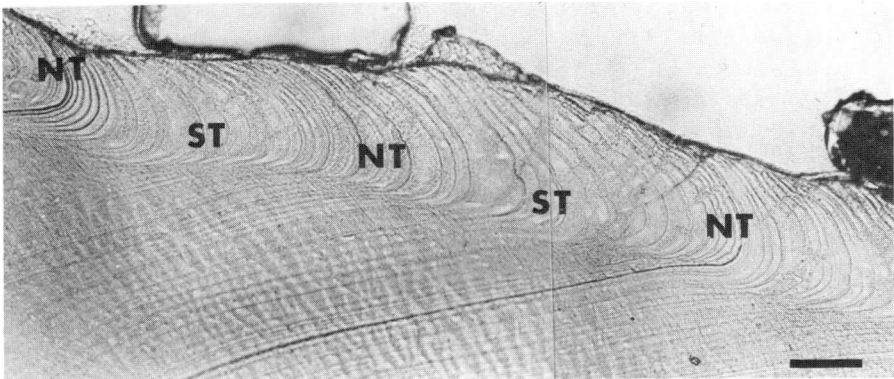

Figure 11:11   Acetate peel replica of a polished and etched shell section of the common European Cockle, *Cerastoderma edule* collected from a position of high-water neap tides in the Menai Straits, North Wales. Growth from left to right. The growth bands are thin dark lines deposited during low tides when the cockle becomes emersed, and the wider areas between the bands are the growth increments, laid down when the animal is actively feeding and growing. There is a clear lunar periodicity of formation with rapid growth during spring tides (ST), and slow growth at neaps (NT). Scale bar, 100 μm. (Supplied by J. H. Simpson.)

The most productive areas and the greatest number of species are found at the lowest levels, where they experience the longest periods of submersion. The common cockle (*Cerastoderma edule*) reaches its greatest size at low water levels but is more abundant at mid-tide levels. Studies of the cockle (Richardson *et al.*, 1980a, 1980b) have shown significant differences (5–10 per cent) in growth rates during the spring–neap cycle. At mid- and high-tide levels the animals grow fastest at springs and slowest at neaps. Figure 11:11 shows the narrowing of growth increments during neap tides (NT) and fast growth during springs (ST). Near low water the least growth occurred at springs and the most at neaps. This anomaly was ascribed to the reduced periods of submersion during spring tides and continuous submersion during neaps at this level. Shells from the Irish Sea showed a reduced growth rate in winter, but Norwegian shells showed no growth in winter because of the severe conditions. The annual number of growth bands in shells from severe climates is much lower than the number of tidal intervals, and so care is necessary when interpreting numbers of growth bands in terms of past tidal and geophysical conditions (Section 10:5).

The cockle is a suspension feeder which gets its food by filtering water through a short tube which sticks above the sand, and so the longer it is covered by the sea, the more food it can gather. Conversely, the 'thin tellin' shell (*Tellina tenuis*) is most abundant at low levels, but increases in size higher up the beach; this trend has been attributed to its long flexible syphon which can actively seek food, for which there is less competition at the higher levels. These suspension feeders take advantage of the bottom sediments which are disturbed by the tidal currents. Other molluscs found near low-water levels include the razor-shell (*Ensis*) which if disturbed, can avoid capture by retreating very quickly to deeper layers in the sand. Few macroalgae can survive at this low level on sedimentary shores, though there may be a seasonal growth of brown seaweed attached to gravels.

The middle region of sandy beaches between high and low water on neap tides is generally dominated by the lugworm *Arenicola* (see Figure 10:2). The lugworm constructs a characteristic U-shaped burrow which is marked at the surface by a depression at the top of one end of the shaft and a worm cast at the other. The worm itself is up to 20 cm long and 15 mm wide. It feeds by swallowing sandy material; nourishment is derived from organic matter within the sand, and the worm is able to spend its whole life in one location because the constant replacement of the surface deposits by tidal cycling ensures a continually renewed food supply. The worm casts are formed when the worm pushes its tail to the surface through the other shaft of its burrow to excrete the sand. Lugworms are preyed upon by fishes when submerged, and by birds when their level of beach is exposed to the air. As a general rule the biggest lugworms are found at lower levels of the beach.

The highest levels on exposed beaches are not as productive as the middle levels, but there may be a series of tidal high-water lines consisting of dried cast-up seaweed and other debris, each line supporting its own characteristic fauna. At these higher levels, the influence of vegetation in shore development is most significant, and may result in the establishment of salt-marshes or mangrove swamps. In temperate regions salt-marshes, which are close swards of herbaceous plants, can form above MHWN levels if there is only limited exposure to waves. In the tropics, forests of salt-tolerant shrubs and trees, for example mangrove swamps, can occupy the upper half of the littoral zone from slightly above mean sea-level to MHWS.

Although salt-marshes and mangrove swamps appear very different, they both consist of salt-tolerant plants, and both have been established by these plants slowing the water movements so that sands and muds settle out and are not subsequently eroded. As a result, there is a progressive increase in the levels of both salt-marshes and mangrove swamps over a period of years (see Section 10:3:1). A sandy or muddy intertidal area is first colonized by the species that are most tolerant of salt and submersion, which then begin the process of trapping sediments. On many sandy beaches *Spartina* grasses are the first to become established at the upper tidal levels. As the sediments accumulate, other less tolerant but more competitive species begin to dominate as the shore levels continue to rise. A network of creeks develops for tidal inflow and outflow (Allen, 1985b).

Mangrove swamps support a rich ecosystem which contains terrestrial species in the branches of the mangroves and marine species within the underlying roots and sediments. Mobile animals such as crabs are particularly well adapted: the fiddler crab harvests the scum of nutrients left by the retreating tide, so helping to recycle minerals and organic matter; as the tide rises they are able to seal their burrows against flooding. Their reproductive cycle is linked to the tides as they release their young at the highest tide of the month. At low tide tree-living snakes and monkeys are able to feed on the crabs. In addition to mobile organisms, there is also a rich epifauna attached to tree trunks and branches, including many species such as barnacles which are characteristic of rocky coasts, and there is similar strong vertical zonation of species.

Birds are the most mobile of all coastal fauna and estuaries are particularly noted for the variety and activity of their bird life (Evans, 1979; Feare and Summers, 1985). Geese and ducks feed mostly on salt-marsh plants and on the sea-grass *Zostera* which is found at lower tidal levels. The majority of estuarine birds, including large populations of waders, take advantage of the times of low tidal levels to feed on the worms and molluscs of the mud-flats. Ducks also eat worms and shrimps. The long bills of the oyster-catchers, dunlin, knots and curlew allow them to penetrate into deep worm burrows such as those of the lugworm. The rich availability of food is illustrated by the fact that the Severn Estuary has been estimated to support about 50 000 dunlin each winter. The lives of these shore birds are dominated by the tides: they feed whenever the mud is exposed, by day and in many cases by night. Efficiency of feeding is lower at night and some species including the sand hopper, *Talitrus saltator*, emerge at night to minimize predation risks. At high tide the birds are forced back to the edge of the shore or onto mid-estuary islands where they congregate in large numbers. Migrant waders in transit during the very high equinoctial tides of early spring and autumn may be concentrated into very limited areas at high water, making these excellent seasons for bird watchers.

Oil pollution is a persistent hazard for shore species, and the physical smothering which coatings of heavy oil produce on the intertidal floras can be lethal if not removed quickly. Exposed rocky cliffs are usually cleaned rapidly by waves, but salt-marshes, mangrove swamps and coral reefs are the most vulnerable to permanent damage.

11:4   Offshore processes

For biological systems offshore, changes in tidal levels are not significant, but tidal currents can have strong direct or indirect influences on the distribution and behaviour of organisms. As an example of indirect influences, the different bed-forms described in Chapter 10 associated with different types of tidal current regimes have different associated flora and fauna. For example, diverse stable fauna, including varieties of sponges are found in a restricted zone immediately adjacent to narrow longitudinal furrows and it seems likely that one of the characteristics of this limited region is the steady lateral movement of small amounts of sand and other particles.

Many near-shore temperate regions experience the advection by tidal currents of extensive patches of red or brownish-coloured water. These events, called 'red tides', are caused by massive blooms of dinoflagellates, in which the concentrations of these tiny plants are so high that the water colour is changed. In some areas the red tides are a regular seasonal occurrence, for example the summer dinoflagellate blooms in coastal British Columbia, California, Florida and eastern Canada. In other areas the bloom is less regular. In all cases their intensity is very variable from event to event and there are indications that the intensity is increasing (LoCicero, 1975; Anderson et al., 1985). During red tides toxins may be secreted, and absorbed by organisms such as clams, oysters, mussels and other shellfish which feed on the dinoflagellates. These can affect coastal fisheries and fish farms and may in turn be harmful to humans who eat the seafood. A particularly severe red tide outbreak along the New England coast in 1972 resulted in complete closure of the Massachusetts coastline for harvesting shellfish and the declaration of a public health emergency. Conditions for spectacular coastal blooms are related to high concentrations of nutrients and chemicals in coastal waters, and hence may be indirectly linked to land drainage. Once a bloom is established its continuation depends on physical factors such as winds, tidal and other currents, temperature and salinity. Balch (1986) examined the occurrence of 226 red tide blooms for possible relationship to the spring–neap tidal cycle, but was unable to find any significant correlation.

Further discussions of indirect tidal influences on offshore biological processes are beyond the scope of this account, but three examples of more direct tidal influences will be considered in more detail.

11:4:1   Selective tidal stream transport

The movements of plaice which spawn in the Southern Bight of the North Sea have been studied in sufficient detail to show that migration patterns are related

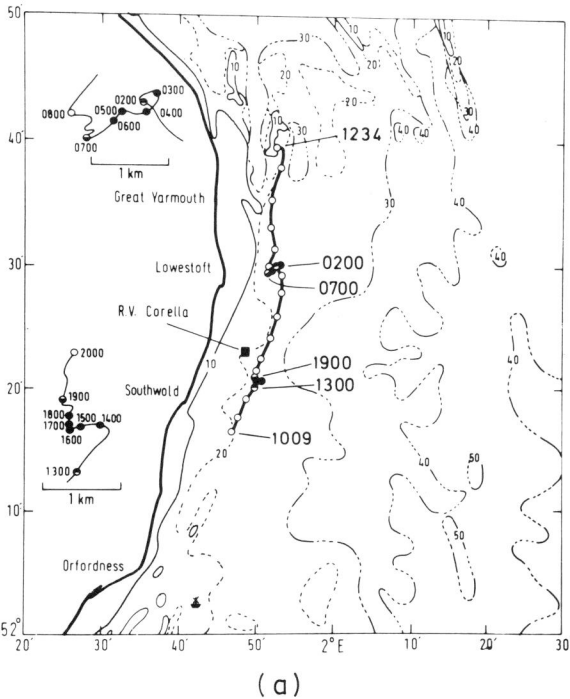

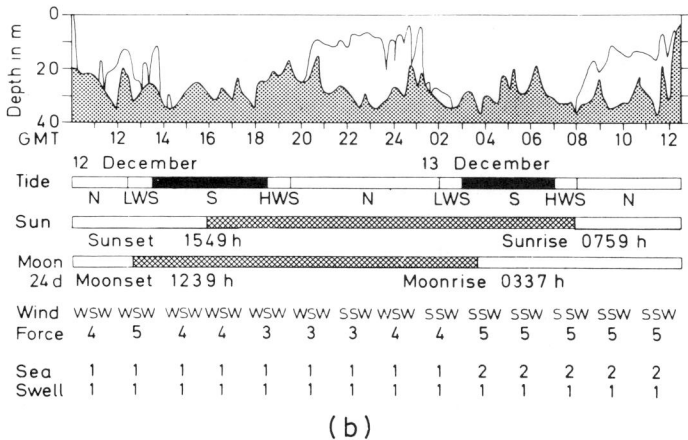

Figure 11:12 Movement of an acoustically tagged plaice in the southern North Sea. (a) Track chart: release was at 1009 h GMT, 12 December 1971. Hourly positions of the fish are indicated and the times of slack water are given. ○ north-going tide; ● south-going tide and (b) depth variations in relation to the direction of the tide. (*Reproduced by permission of the International Council for the Exploration of the Sea.*)

to tidal streams. In the late autumn maturing fish migrating from more northerly feeding grounds into the Southern Bight are caught most frequently by midwater trawls on the south-going tidal currents. In winter spent females returning to the north are caught more often in midwater on the north-going tidal currents than on the south-going tidal currents. Clearly travelling in midwater when the tidal currents are in the direction of intended migration allows a fish to make the maximum movement over the ground for the minimum expenditure of energy.

Movements of 12 acoustically tagged plaice released and followed off the East Anglian coast have been reported by Greer-Walker *et al.* (1978). Eight of the fish were tracked over distances of more than 15 km; all of these fish showed a pattern of regular vertical movements, coming off the bottom near slack water and remaining in midwater for the following 5–6 hours before returning to the bottom. When on the bottom individual fish activity was greatly reduced and in some cases the fish did not move for 2–3 hours. When swimming at midwater depths fish moved rapidly over the ground in the direction of migration. Figure 11:12(a) shows the movement of a plaice which showed this behaviour, and Figure 11:12(b) shows the depth of this fish in relation to the direction of the tidal current. It was noticed that the ascents were more closely related to the time of slack water than the descents; usually settling on the bottom was preceded by a series of exploratory excursions, probably to check for reducing water movements relative to the fixed bottom reference.

Selective tidal stream transport has also been found to occur for sole, cod, dogfish and silver eels. Although other factors are involved, the mechanism is thought to be significant to the movement and distribution of many stocks of fish on the continental shelves where tidal currents are strong. Selective tidal stream transport will be more effective in regions, such as the coast of East Anglia (Figure 11:12), where the tidal current ellipses are highly rectilinear because in these areas the currents define an axis along which the fish move, and because there are clear signals in the reduced currents to switch movement away from and back down to the bottom as the tidal currents change direction. More circular ellipses will give weaker directional signals and triggers for vertical movement. There would be clear advantages for fish which travelled using selective tidal stream transport at times of spring tidal currents but the relationships between the times of fish migration and the phase in the spring–neap cycle have not been studied in detail.

11:4:2 Tidal-mixing fronts

The loss of tidal energy due to the frictional drag of the sea-bed resisting the currents discussed in Section 7:9, has important effects on the tidal dynamics, but also some biological significance. Some of the lost energy takes the form of turbulent kinetic energy which is available for vertical mixing of the water column. During the summer at mid-latitudes, there is a net input of heat to the surface waters with a resulting tendency for stratification to develop with a warm surface layer separated by a thermocline from deeper cooler water. However, if there is sufficient turbulent kinetic energy, this tendency to stratify is overcome by the vertical mixing, and there is a smaller temperature increase, distributed uniformly through the whole water depth. Areas of mixing are often those where biomass enhancement occurs. Tidal turbulent energy may result in mixing at fronts, in the wake of islands and vertically where water masses tend to stratify (Bowman *et al.*, 1986). In these areas of effective vertical mixing the surface temperatures are lower than in the stratified areas (Simpson, 1981; Simpson and James, 1986). A satellite infra-red image taken of the north-west European shelf seas in the summer shows a complicated pattern of temperature variations (Figure 11:13) due to different intensities of vertical mixing. The boundary between stratified and unstratified water is often quite sharply defined, with temperature gradients of $1\,°C\,km^{-1}$ or greater. These transition zones between different water masses are called *fronts*.

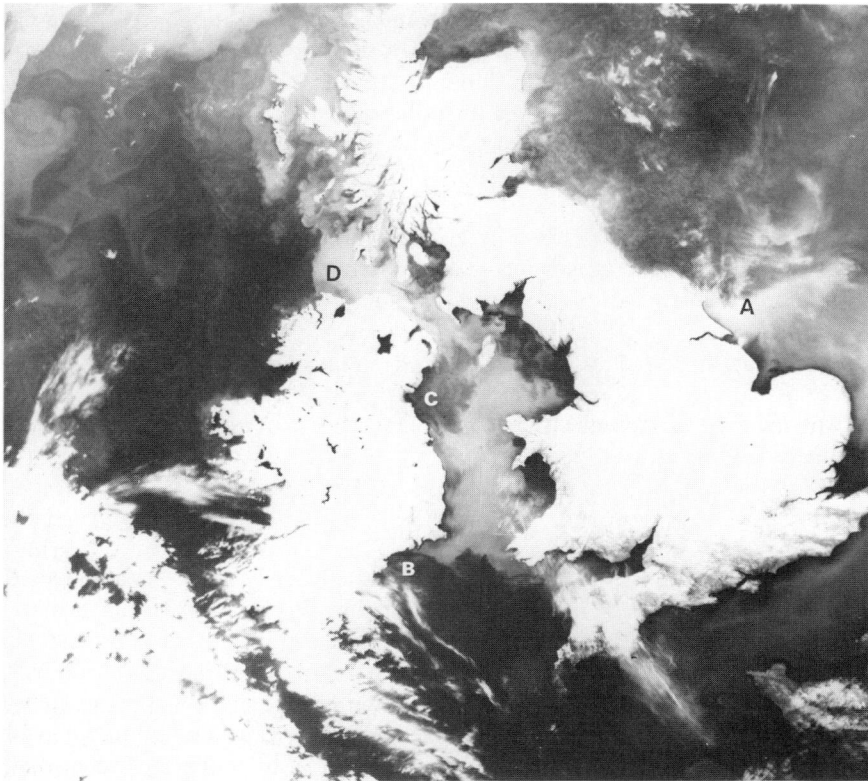

Figure 11:13   Satellite infra-red image of the north-west European shelf are from NOAA-9, 2 June 1985, 0332 GMT. The fronts shown are : (A) Flamborough Head front, (B) Celtic Sea front, (C) western Irish Sea front, (D) Islay front.

For the north-west European shelf region shown in Figure 11:13 the role of tidal mixing is normally much more important than wind mixing, with the result that the fronts are found regularly from summer to summer in the same positions. For example, a frontal zone surrounds the area of low tidal currents where stratification develops in the north-west Irish Sea; south of St George's Channel there is a clearly defined front between the stratified waters of the Celtic Sea, and the tidally mixed waters of the southern Irish Sea; another front forms west of Scotland, the so-called Islay front; and yet another forms between the mixed waters of the southern North Sea and the stratified waters further north, the Flamborough Head front. Similar regular fronts occur in other areas of strong tidal currents such as the Bay of Fundy, the Bering Sea and Cook Strait, New Zealand.

From equation (7:16), the rate of energy removal per unit area from the tidal wave averaged over a tidal cycle is:

$$\frac{4}{3\pi} C_D \rho U_0^3 \text{ per unit area} \tag{7:16}$$

where U_0 is the amplitude of the harmonic current variations, C_D is the drag coefficient and ρ is the water density. Only a small fraction ε of this is converted into turbulent kinetic energy available to mix the water column. For a rate of surface heat input Q, the demand for potential energy to destroy stratification and maintain the mixing in water of depth is:

$$\frac{\alpha Q g D}{2C} \tag{11:1}$$

where α is the linear expansion coefficient and C is the specific heat. For complete mixing:

$$\frac{4}{3\pi} \varepsilon C_D \rho U_0^3 > \frac{\alpha Q g D}{2C} \tag{11:2}$$

and at the front between stratified and unstratified regions, these terms are balanced:

$$\frac{Q D}{\varepsilon U_0^3} = \frac{8 C_D \rho C}{3\pi \alpha g} \tag{11:3}$$

Provided that the heat input and the mixing efficiency are constant over the limited time and area of interest, the parameter D/U_0^3 controls the formation and location of the front, because the terms on the right-hand side are essentially constant. If the surface tidal stream amplitudes are taken as representative of U_0, the critical value of the parameter D/U_0^3 (D in metres, current speed in m s^{-1}) is found by observation to lie in the region 50 to 100. This range corresponds to an efficiency factor of only about 1 per cent of the lost tidal energy being used for vertical mixing (Simpson and Hunter, 1974).

The increase in the strength of the currents from neap to spring tides increases the factor U_0^3 around the north-west European shelf by a factor of 6,

Figure 11:14   Coastal Zone Colour Scanner image (5 March 1983) showing tidal fringe patterns southwest of the island of Jersey. (*Reproduced by permission of Academic Press Inc. (London) Ltd.*)

which should cause the front to move into deeper water. However, the extent of the movement is much less than theoretically predicted because the efficiency of mixing appears to reduce once stratification is established, so that the stratification and the position of the front survive the stronger spring currents (Simpson and Bowers, 1979). At high values of D/U_0^3 the vertical density profile of the water column generally consists of two layers separated by a sharp thermocline, whereas at low values of the parameter the water column is well mixed. Surface temperature differences give horizontal gradients across a front which should be balanced by geostrophic flows parallel to the front. Although this flow is difficult to observe directly with moored current meters because the tidal currents periodically advect the front to and fro over several kilometres, experiments with drifting drogues have shown approximately geostrophic flows of $0.20\,m\,s^{-1}$ parallel to the Islay front (Simpson *et al.*, 1979). Other fronts show a more complex picture with strong eddy development which may play an important role in cross-front mixing.

Even in regions where vertical mixing is sufficiently strong to prevent stratification, differences in the total water depth through which the heat is distributed leads to horizontal temperature gradients and water masses with different densities. So too do differences of residence time, water circulation rates and freshwater river run off. Figure 11:14 shows a satellite infra-red image of a series of tidal fringes which extend to the southwest of Jersey, one of the Channel Islands. The large semidiurnal tidal ranges and the associated currents are sufficiently strong for the water column to be mixed throughout the year, but the different depths cause the development of two distinct water masses, separated by a strong tidal front between Jersey and Guernsey (Pingree *et al.*, 1985). The front is particularly conspicuous in summer and late winter. Water from the Jersey side of the front is entrained by the tidal currents and progressively advected to the southwest by the residual circulation to form a sequence of fringes which are best developed at times of spring tides. Although these fringes become thinner and weaker as they are diffused by horizontal tidal mixing, they persist for several days.

Frontal regions are known to have high levels of biological productivity. The Channel Islands front is an area of convergent currents, and can be clearly identified by surface rafts of floating sea-weed, debris and oil lumps. Fishing vessels often concentrate their activities in the vicinity of fronts, and high fish concentrations are often seen on echo sounder records as patches of dense, layered, scattering, particularly on the stratified side of tidal-mixing fronts. In the case of the front in the western Irish Sea the numbers of birds on the stratified side of the front are typically an order of magnitude greater than on the well mixed side. Among the birds likely to be found are puffins, shearwaters and terns, all presumably feeding on the enhanced biomass in the vicinity.

The detailed relationships between these top predators and lower trophic levels in frontal ecosystems are poorly understood, nor is it clear which processes link the biology and physics, and are common to different types of frontal systems. Detailed studies of the Irish Sea front (Fogg *et al.*, 1985) have shown that the three main water types that are formed seasonally (surface stratified, bottom stratified and mixed water) can be regarded as separate ecosystems, their biotas showing distinct differences in composition and activity. The surface stratified water builds up its phytoplankton population rapidly in the early spring following the development of the thermocline, after which the levels stabilize and remain about the same throughout the summer; average chlorophyll concentrations are twice those in the mixed water. Carbon levels in this water are fixed by nutrient limitation with populations of phytoplankton, bacteria and zooplankton in balance. This balanced community breaks down when the water column becomes destabilized in the autumn. The bottom stratified water which is too deep for light penetration and photosynthesis to be significant, shows low zooplankton stocks, and the lowest level of community organization. In the mixed water phytoplankton levels are limited by light rather than nutrients because the vertical mixing limits the time spent near the illuminated surface. The interdependence of the trophic levels in the mixed water was intermediate between those in the other two water bodies.

Studies of the tidal fronts in the approaches to the English Channel (Pingree et al., 1975) have shown very high concentrations of phytoplankton blooms which may develop into red tides. It is proposed that the frontal region is an area of recently stabilized previously mixed water in which conditions are created which encourage the rapid growth of phytoplankton. However, it is not certain which physical and biological processes are most important in determining the flux of nutrients and the extent of the subsequent bloom.

11:4:3 Internal tides

The conditions of high nutrient concentrations in the near-surface zone which favour primary production and high biological productivity may also be brought about by internal waves. Several such examples have been observed associated with the shelf break (Pingree et al., 1983, 1986). The pycnocline between the lighter surface water and the denser, nutrient-rich deeper water is disturbed vertically by the tidal flows (Section 5:6) and such displacements in pycnocline depth may travel onto and across the shelf. Peak-to-trough pycnocline displacements in excess of 50 m have been observed at the shelf break of the Celtic Sea. The internal tide is highly distorted by the shelf tidal currents: when the off-shelf currents oppose the on-shelf propagating internal tide, the latter behaves like a bore (Section 7:7) and a series of short high-frequency internal waves called solitons are formed. The turbulence associated with the breaking of these solitons mixes as pulses of nutrients into the near surface layers. The turbulence interacts with the surface wind waves to produce variations of sea-surface roughness which can be seen in satellite images of the Celtic Sea break and the eastern shelf break of North America. These patches are obvious in images taken by Synthetic Aperture Radar which shows them to be widespread, especially at spring tides.

Internal tides also propagate away from the shelf break into the deep ocean, where similar high levels of nutrient concentrations associated with large internal tides and surface turbulence patches have been observed 240 km away from the shelf break, a travel time for the internal waves of more than two days (Holligan et al., 1985). Similar processes of internal wave mixing contribute to high productivity along other shelf break regions, but the total energy losses which can be attributed to these internal wave dissipation processes are unlikely to exceed 1 per cent of the total global dissipation of tidal energy, as discussed in Section 7:9:3.

11:5   Long-term changes

Do the long-period tidal cycles have any corresponding cycles in biological systems? The nodal cycles discussed in Section 4:2:2 and the nodal tide discussed in Section 9:5:1 are possible long-term influences. However, the nodal tidal amplitude in mean sea-level of a few mm over 18.6 years is unlikely to be significant compared with the general trends in mean sea-level. Nodal changes in the M_2 amplitude of 3.7 per cent (Table 4:3) are likely to cause cycles in the development level of mangrove swamps and salt-marshes, particularly where surge distortions of the tides are small. The large modulations of O_1 (18.7 per cent) and K_1 (11.5 per cent) must make nodal changes significant for coastal development where diurnal tides are dominant. Changes in tidal range are most significant where there are extensive flat intertidal areas.

In addition to the changes of tidal range, the corresponding changes of the tidal speeds will cause changes in the turbulent kinetic energy available for mixing stratified water. Sea surface temperatures from several North American ports have been analysed (Loder and Garrett, 1978) for temperature cycles having an 18.6-year period; small amplitude temperature variations (less than 0.5 °C) were found which were in phase with the nodal tide. Similarly, British Columbia sea temperatures showed small changes corresponding to the O_1 and K_1 nodal modulations.

British Columbia sea temperature, salinity and sea levels have also been analysed for long-term changes (Mysak et al., 1982): a 5–6 year oscillation was statistically related to the annual catches of sockeye salmon and herring recruitment. Periodic temperature and salinity fluctuations which could affect fish populations were associated with long-period baroclinic waves. This kind of statistical relationship, although possibly correctly linked to a physical process, is seldom of use in a predictive sense because the fish may behave differently if some other factor changes. With so many interactive factors it is seldom possible to identify a single environment parameter which relates directly to the behaviour of a biological system, in marked contrast to the responses of physical systems.

11:6   Tidal and lunar rhythms in plants and animals

Biological systems have developed complicated mechanisms for adjusting to natural rhythms, and adjustments to tidal cycles by plants and animals which live on the sea-shore have been widely studied (Palmer, 1974; Brady, 1982; Neumann, 1981; Naylor, 1982 and 1985).

One of the best known and most spectacular cases of tidal adaptation is the small (15 cm) California coastal fish *Leuresthes tenuis*, better known as the grunion (Ricketts and Calvin, 1968). On the spring tides of the second, third and fourth nights after full moon, in the months from March to June, pairs of male and female fish swim up the beach on breaking waves just after the tide has turned. The female digs and deposits her eggs into the sand where they are simultaneously fertilized by the male, before both swim back out to sea on the next high wave. After an interval of 10–14 days, on the slightly higher spring tides of the subsequent new moon, immediately on being submerged, the eggs hatch and the young fish swim out to sea.

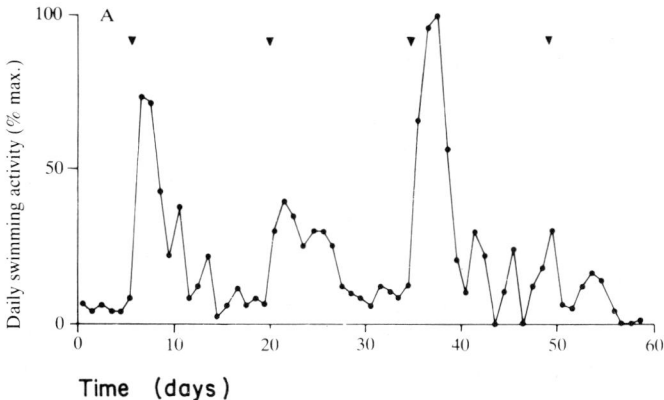

Figure 11:15 The total swimming activity of 20 freshly collected specimens of the sand-beach isopod *Eurydice pulchra* recorded in constant conditions for 50–60 days in the laboratory, showing an enhanced activity at times of spring tides. (From Naylor, 1985.) (*Reproduced by permission of Cambridge University Press.*)

An approximate fourteen-day cycle, called a semi-lunar cycle, has been observed in the activity of several intertidal crustaceans in laboratory conditions of constant darkness. Figure 11:15 shows the activity of the sand-beach isopod *Eurydice pulchra* over 60 days in the laboratory; maximum activity corresponds to the times of spring tidal ranges at the beach where they were collected. *Eurydice* avoids being carried away with the falling tide by ceasing to swim, burrowing in the sand before the tide has fully ebbed. The lower activity at times of neap tides appears to provide a mechanism whereby the isopods avoid stranding in sand above the high-water level (Naylor, 1985). Similarly the sandhopper *Talitrus saltator* shows maximum activity at spring tides. It has been suggested that this activity helps the sandhopper to avoid being stranded above the level of high water. Moulting and hatching in this species are also correlated with the semi-lunar activity rhythm, so that freshly-emerged animals appear when the beaches are fully wetted by the sea, and desiccation risk is reduced (Naylor, 1982).

Observations have also shown that many marine algae, including the brown seaweed *Dictyota dichotoma*, release their sperm and eggs every 14–15 days within a few hours of each other and this behaviour continues when they are removed to the laboratory (Palmer, 1974). One difficulty in identifying a genuine semi-lunar endogenous clock mechanism through the laboratory studies of organisms, is eliminating the possibility that daily (circadiurnal) and lunar (circatidal) rhythms are beating together to give activity modulations at spring–neap tidal periods.

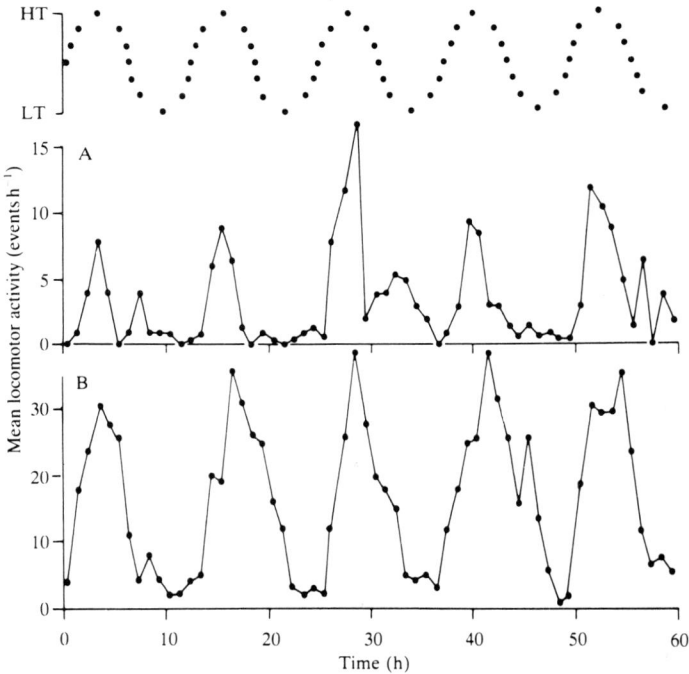

Figure 11:16 Endogenous free-running circatidal rhythms of two marine crustaceans in constant conditions in the laboratory. (A) *Carcinius maenas*, walking rhythm of three crabs in tilting box actographs. (B) *Eurydice pulchra* swimming rhythm recorded as numbers (of five individuals) swimming past an infra-red light source. The top trace is the tidal regime at the collection site. HT—high tide, LT—low tide. (*Reproduced by permission of Cambridge University Press.*)

Endogenous rhythms of circatidal period (12.4 hours) have been observed to persist in the physiology and behaviour of many coastal marine organisms. Figure 11:16 shows the activity patterns for groups of shore crabs, *Carcinus maenas*, in the laboratory under constant light conditions. Temperature appeared to have no effect on the time-keeping ability of the different groups. On natural beaches these crabs secrete themselves under stones long before low tide in order to avoid desiccation or attacks by birds. *Eurydice* also shows enhanced activity at times of high water. This clock-controlled anticipatory behaviour contrasts with that of other organisms such as sessile barnacles which cease feeding and close their shells as a direct response to reducing tide levels.

Circatidal activity patterns have been observed in constant laboratory conditions in many organisms over periods from a few cycles to several months depending on the species and the individuals involved. Those from beaches with different tidal phases maintain these phase differences in the laboratory. Some species also show signs of having adapted to persistent diurnal modulations of tidal range.

There are many ways in which tidal rhythms can be reinstated or shifted in phase. A single short cold treatment can initiate new tidal rhythms in green crabs which have become tidally inactive in the laboratory. This stimulation process is an example of biological entrainment, and the mechanisms which can induce it are called zeitgebers, literally, time-givers. Laboratory zeitgebers include temperature, pressure, salinity and mechanical disturbance, each of which would be encountered with tidal periodicity on beaches. Species taken from one beach to another with a different tidal phase adapt quickly to the new conditions. The strength of the endogenous rhythmic behaviour appears to depend on many factors including the species itself, the season, as well as the tidal amplitude and pattern.

An unusual rhythm related to tidal conditions has been reported from the island of Fiji (Price and Karim, 1978). The term *matiruku* in Fijian means literally 'low tide in the morning', but it also means someone who is periodically insane. Local descriptions of the symptoms of *matiruku* mention elation, violence, increased speed and strange statements, but why these have been related to local tidal conditions is not known. Local ideas suggest direct influences of the moon, which in Fiji is in its first or third quarter when low tide occurs in the morning.

11:7   Responses to increasing sea levels

General popular and scientific concern about the 'greenhouse effect', increasing global temperatures due to increased carbon dioxide and other gases in the atmosphere, has identified rising global sea levels as a potentially major problem (Barth and Titus, 1984). Various models of responses to an expected doubling of present atmospheric concentrations of carbon dioxide suggest a global warming of between 1.5 °C and 4.5 °C. The total effect on sea levels of this warming is uncertain and there are many factors involved. A recent report by the United States Department of Energy (1985) estimated that partial melting of the Antarctic Ice Sheet will add between 0 and 0.3 m to present sea levels by the year 2100; the Greenland Ice Sheet will contribute between 0.1 and 0.3 m, and melting of smaller ice-caps and glaciers could also add between 0.1 and 0.3 m. Sea-level may also increase by expansion of water in the upper layers of the ocean due to warmer temperatures. Each of these estimated contributions is subject to major uncertainty: for example, the estimate for the contribution by the Antarctic Ice Sheet to total sea-level change by the year 2100 ranges from −0.1 m to 1.0 m. Thomas (1986) has examined these four factors and the range of their possible contributions are summarized in Figure 11:17, where the dark shading indicates the most probable responses based on present knowledge. The total of these four factors indicates an increase of global 'eustatic' sea levels of 1.0 m by the year 2100. These 'eustatic' increases must be added to local coastal sinking, and to any changes in the ocean surface dynamic topography (Figure 9:8). Changes in the geoid (Figure 3:14) due to redistribution of water mass must also be considered.

The most obvious consequence of increasing sea-level would be coastal erosion and the flooding of low-lying land. The extent of the erosion may be estimated by the Brunn rule (equation (10:3)), which, although basically two-dimensional, is often applied three-dimensionally. Along the low-lying southeast coast of the United States shore-line retreat of the order of 1 m for each 0.01 m of sea-level rise has been estimated (Titus, 1986). Delta regions are very vulnerable, and up to 20 per cent of the land in Bangladesh could be flooded with a 2 m rise of sea-level. Similarly, over 20 per cent of the Nile Delta would

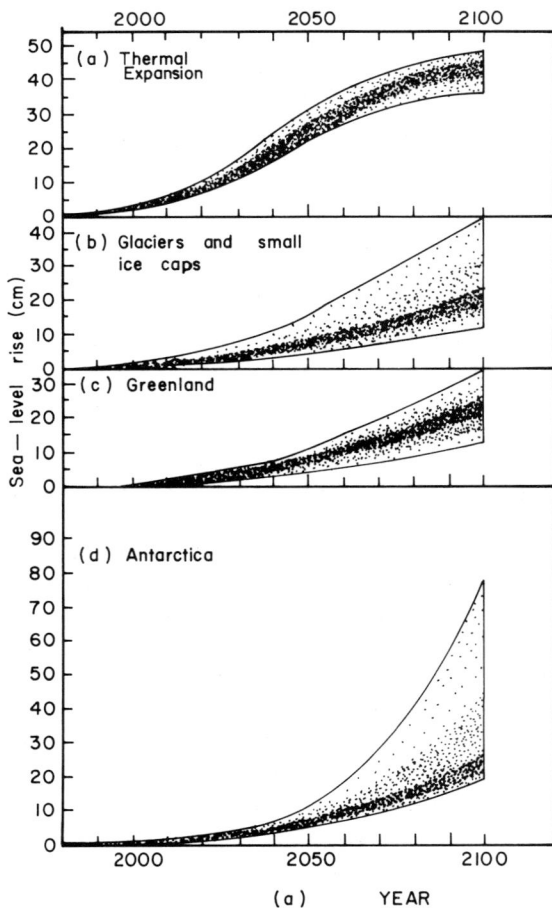

(a)   YEAR

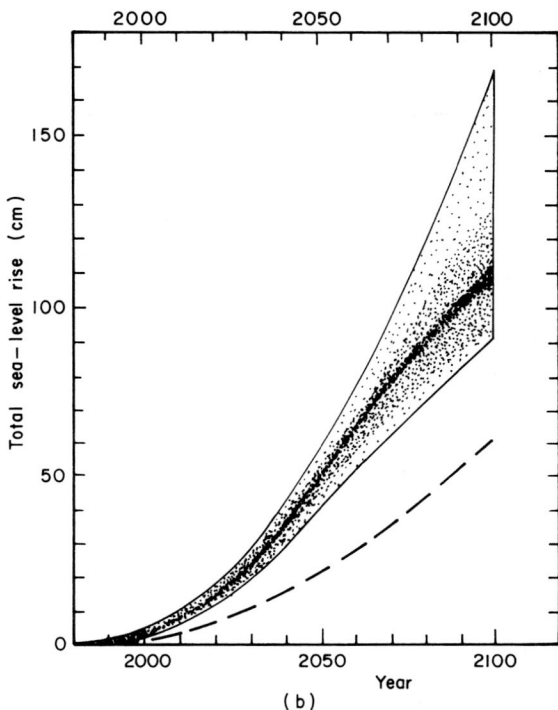

Figure 11:17 (a) Estimates of eustatic sea-level rise during the next century. Surface air tempera-tures are assumed to increase linearly until 2050, to an average value 3 °C higher than at present, and then to remain constant. The dark shading indicates the 'most probable' response, based on our current understanding of these processes. (b) Total sea-level rise during the next century. The broken line depicts the response to a warming trend delayed by 100 years by thermal inertia of the ocean. (From Thomas 1986).

be threatened. Typically a 0.3 m rise of sea-level would cause recreational beaches to erode by 20 to 60 m. Coastal barrier islands (Section 10:3:2:1) may maintain themselves despite slowly rising sea levels by gradual landward transport of sand, which washes over the island upward and landward. Coastal wetlands, which develop at levels near present high-water levels, account for vast areas of land which would be particularly vulnerable. Over the past several thousand years coastal wetlands and marshes have usually built up through sedimentation and peat formation at a rate comparable with the sea-level increases. However, more rapid rates of sea-level increases may result in substantial wetland losses. Inland extension of wetlands is possible if the adjacent land is low-lying and available, but not if the existing limits are protected by the construction of flood barriers. The total significance of the loss of vast areas of wetlands is not easy to estimate, but salt marshes serve as nursery grounds for half the species of commercially significant fishes in the south-eastern United States, and are the favoured habitat of many species of plants and animals. In addition, wetlands can provide coastal protection from storms.

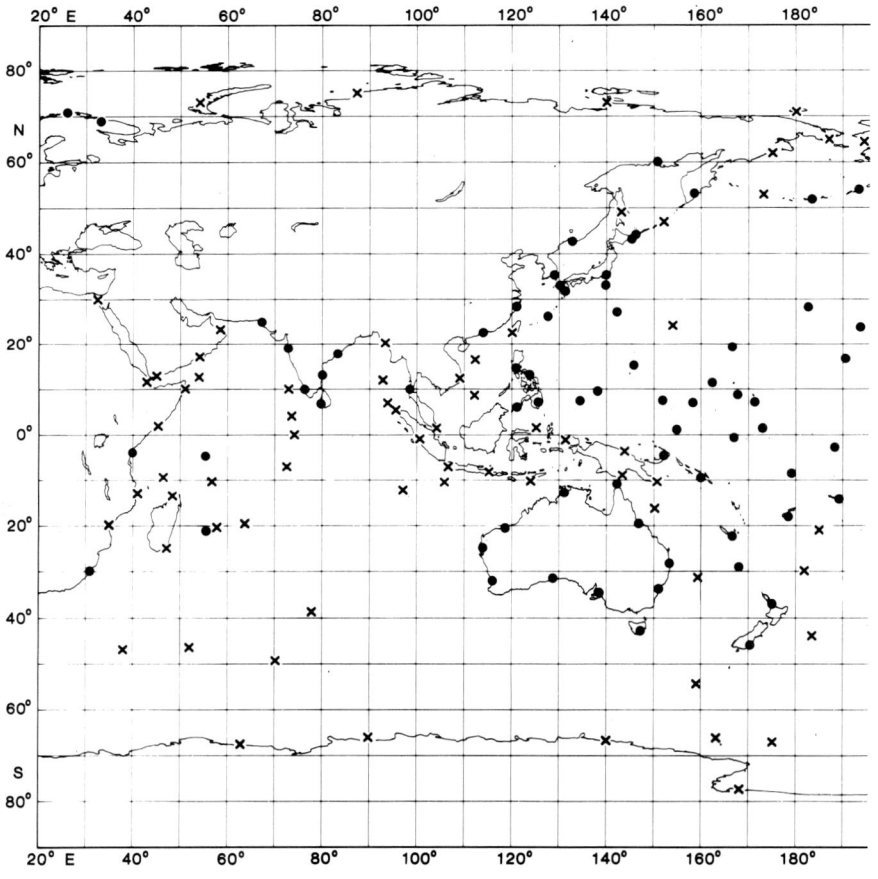

Figure 11:18   Positions of proposed sea-level stations for the Global Sea Level Observing System

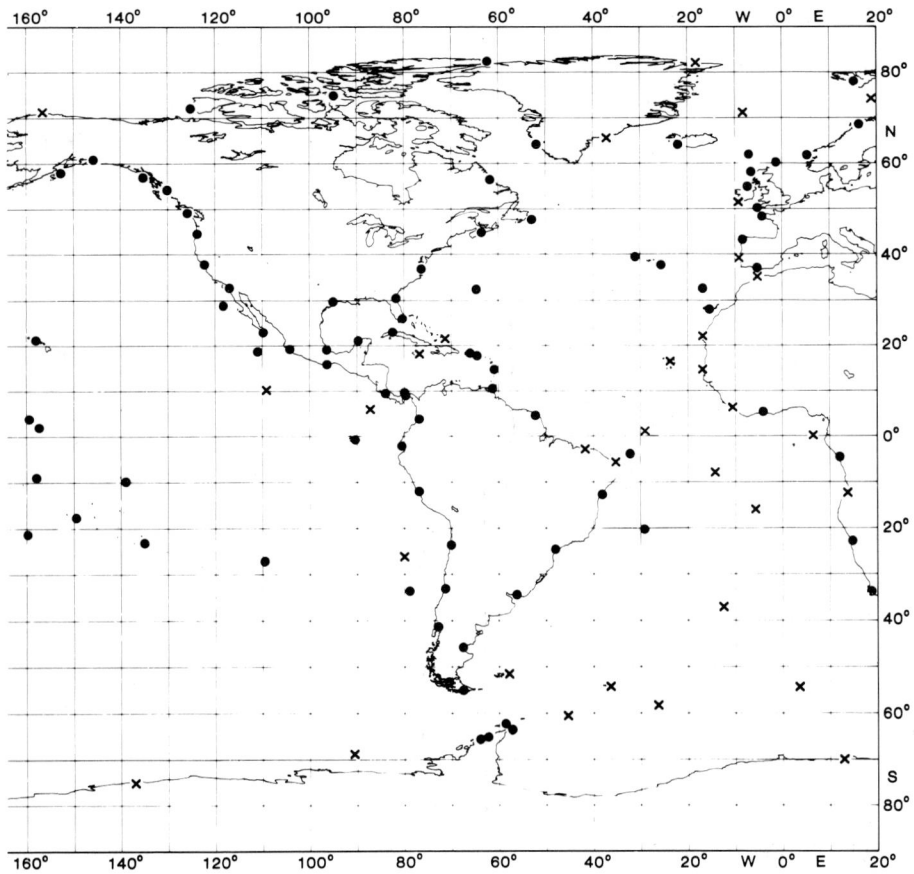

(GLOSS). The dots indicate the existing stations and the X's those additional stations proposed.

Coastal erosion is a continual process, and the adjustment to higher sea levels would take the form of increasingly frequent attacks by storms at higher levels, and consequent erosion. The return periods discussed in Section 8:3 would need to be recalculated as the sea levels increased. For example, a 1.0 m rise in mean sea-level would impel a reappraisal of the efficiency and operation of the Thames Barrier as the design is based on present rates of sea-level rise. Simple adjustment of existing statistics by addition of mean sea-level increases may be incorrect because of changes in the tidal dynamics and the surge generation in the resulting deeper water offshore and in tidal rivers.

Existing estuaries, where there is a delicate balance between salt and freshwater biology would become more saline. The effect can be compared with the salinity increases at times of droughts and low river flow. Gradual salinity increases can be accommodated by upstream migration of brackish and freshwater species, provided that space and water quality conditions are favourable. Intrusion of salt water into groundwater aquifers may require engineering countermeasures such as injecting freshwater to reverse the flow locally, or even the development of alternative sources of freshwater for consumption.

The problem of sea-level rise may be addressed in a number of different ways. At one extreme vast resources might be expended on maintaining the present coasts and coastal activities; this solution would be very difficult to apply for areas such as the Ganges and Nile Deltas, but quite feasible and cost-effective for areas such as London, at least for the first metre or so of sea-level rise. As described in Section 10:3:2:3, reclamation of low-lying land has proved worthwhile in many cases, even for the value of the agricultural land. At the other extreme a minimum response would be to plan an ordered withdrawal to higher land. The gradual adjustment of low-lying estate prices to reflect their finite availability will be a financial response over many decades, similar to the present purchase of leasehold property, provided sea-level increases are predictable and sustained. Some legislation in the United States, notably the National Flood Insurance Program of 1968, already encourages communities to avoid risky construction in flood-prone areas (Section 8:5).

The question remains: what should we be doing now in anticipation of sea-level rise? It would be premature to develop enhanced and expensive sea defences. Global increases in atmospheric carbon dioxide are well established. There is also evidence of increased global temperatures (Jones *et al.*, 1986). However, there is no evidence yet of global sea-level increases over and above the 10–15 cm per century which has been proceeding since tide gauge records began in the early nineteenth century (Section 9:5:3). It is supposed that an increase in sea-level should follow from an increase of air temperature due to the melting of ice and the warming and expansion of ocean water. This is a likely hypothesis based on careful but limited scientific analysis. A first priority must be to improve the scientific understanding on which these predictions are based, in order to reduce the uncertainties indicated in Figure 11:17. As discussed in Section 9:5, present knowledge cannot determine even the relative importance of the various factors contributing to the observed global sea-level increases of 10–15 cm per century. This research will involve many separate scientific disciplines, including glaciologists, oceanographers, geodesists, geologists and atmospheric scientists. The danger is that although there may be only a small rise of sea-level initially, the real effects may be concealed for several decades by the thermal inertia of the ocean; meanwhile an irreversible process may be established. Another priority is the establishment of a global sea-level monitoring system, with gauges measuring to common standards as part of a well distributed network. The network shown in Figure 11:18 is being developed by the Intergovernmental Oceanographic Commission, based on proposals made by Professor Klaus Wyrtki and the Permanent Service for Mean Sea Level, with the enthusiastic endorsement of member states. Selected gauges (Section 9:7) will also be connected into a geometric global coordinate system to distinguish between changes of sea-level and vertical land movement. These measurements are a long-term commitment to identifying and understanding sea-level changes for future generations.

APPENDIX 1

Filters for Tidal Time Series

This appendix contains two sets of useful filters for the analysis of data which is dominated by tidal phenomena. The first set allows the computation of hourly values from observations taken or tabulated at intervals of 5, 10, 15 and 30 minutes. The second set contains multipliers for calculating low-pass values from hourly values, by eliminating the tides of diurnal and higher species. These are useful for mean sea-level computations and for studies of sub-tidal meteorological effects. Godin (1972) has an extensive discussion of tidal filters. Other accounts are found in Karunaratne (1980), Thompson (1983) and Dijkzeul (1984).

All the filters described here are symmetrical so no phase shifts are introduced. The filtered value $X_F(t)$ at time t is computed:

$$X_F(t) = F_0 \cdot X(t) + \sum_{m=1}^{m} F_m [X(t + m) + X(t - m)]$$

For the Doodson filter, which was designed for use before digital computers were available, the computed value of $X_F(t)$ must be normalized by dividing by 30. For all the other filters the weights have already been normalized to unity.

Figure A1:1 shows the response curves for the conversions to hourly values. The 5, 10, and 15-minute filters have been designed to have the same characteristics. The 30-minute filter has a slightly sharper cut-off. Note that for M_8 and higher harmonics there is significant attenuation. If tidal records from areas of severe shallow-water distortion are being analysed it may be necessary to work with the data at the original sampling interval.

Figure A1:2 shows the characteristics of the three low-pass filters. If the data contains many gaps which cannot be interpolated, the Doodson X_0 filter, which has a half-length of only 19 hours, should be used, as less data is lost in filtering. For special studies of long periods of unbroken data the 168-hour filter is appropriate. The 72-hour filter is a good compromise between minimizing the loss of data at the beginning and end of each record, and maximizing the sharpness of the cut-off at frequencies just below the diurnal tidal band.

The Doodson X_0 filter, unlike the others listed, has finite values between tidal species at frequencies higher than those plotted. At a frequency of 0.39 cycles per hour the characteristic has an amplitude of -0.57, which means that the X_0 filter should be used with caution where energy at these frequencies may be

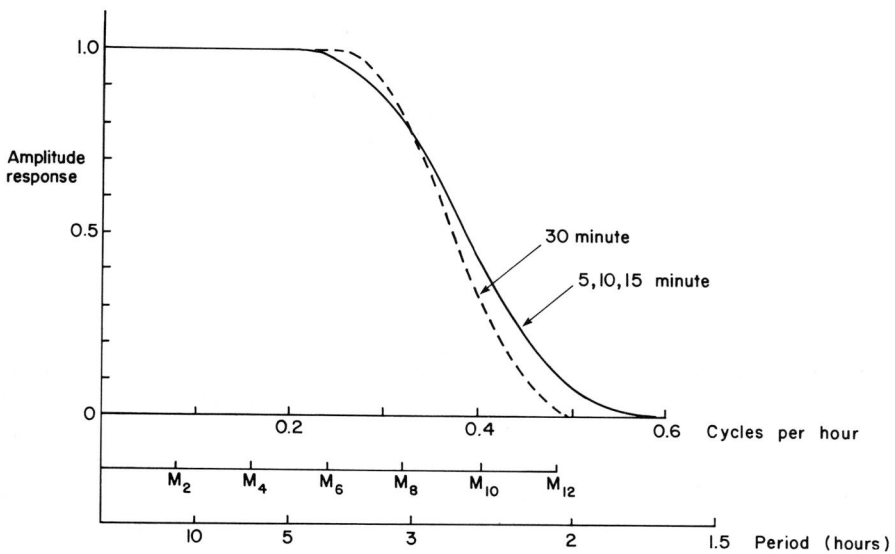

Figure A1:1   Characteristics of some filters used to compute hourly values.

present, if this energy has not been eliminated by earlier filtering. The Demerliac (1974) filter is longer than Doodson's, extending from $(t - 35\,\text{h})$ to $(t + 35\,\text{h})$, but it has a sharper cut-off and no unexpected characteristics at higher frequencies.

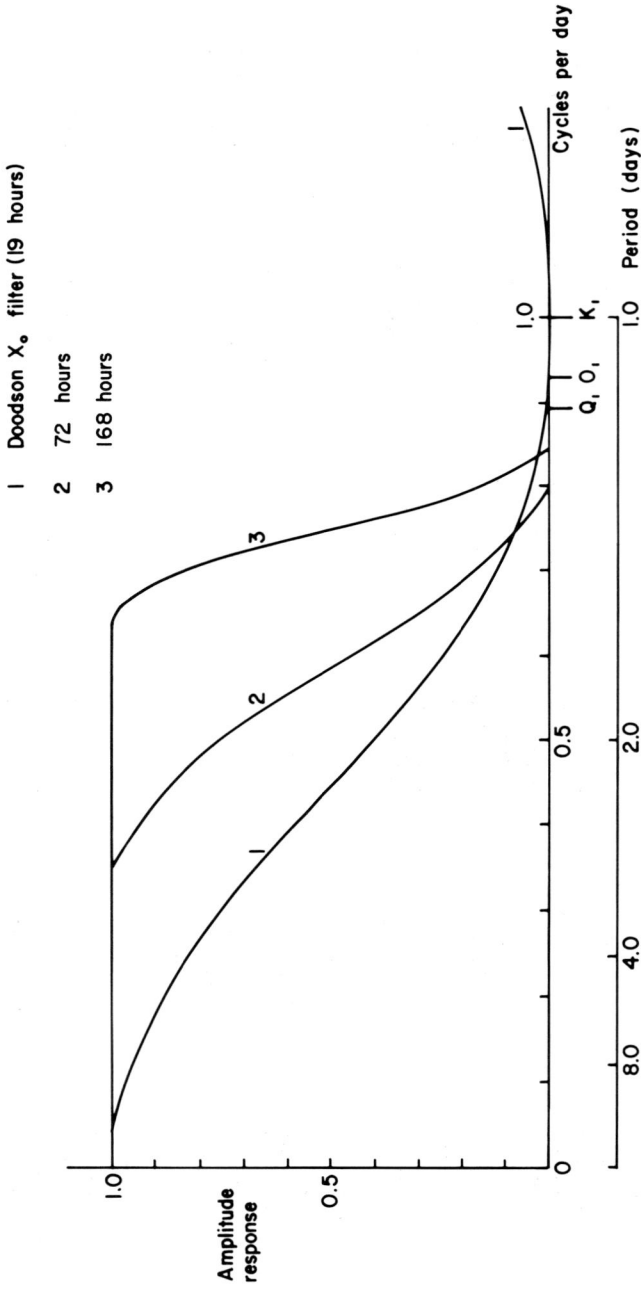

Figure A1:2 Characteristics of three low-pass filters applied to hourly values.

Filters for calculating hourly values

All filters are listed as F_0 in line 0, followed by the other F factors in sets of six and ascending order. For example, for the 5-minute filter $F_{20} = -0.0094346$

5-minute sampling (M = 54)

0	0.0648148					
1	0.0643225	0.0628604	0.0604728	0.0572315	0.0532331	0.0485954
2	0.0434525	0.0379505	0.0322412	0.0264773	0.0208063	0.0153661
3	0.0102800	0.0056529	0.0015685	−0.0019127	−0.0047544	−0.0069445
4	−0.0084938	−0.0094346	−0.0098173	−0.0097074	−0.0091818	−0.0083247
5	−0.0072233	−0.0059642	−0.0046296	−0.0032942	−0.0020225	−0.0008672
6	0.0001321	0.0009493	0.0015716	0.0019984	0.0022398	0.0023148
7	0.0022492	0.0020729	0.0018178	0.0015155	0.0011954	0.0008830
8	0.0005986	0.0003568	0.0001662	0.0000294	−0.0000560	−0.0000970
9	−0.0001032	−0.0000862	−0.0000578	−0.0000288	−0.0000077	0.

10-minute sampling (M = 27)

0	0.1296296					
1	0.1257208	0.1144630	0.0971907	0.0759010	0.0529545	0.0307322
2	0.0113057	−0.0038254	−0.0138889	−0.0188692	−0.0194147	−0.0166494
3	−0.0119285	−0.0065884	−0.0017344	0.0018986	0.0039967	0.0046296
4	0.0041458	0.0030310	0.0017660	0.0007136	0.0000589	−0.0001939
5	−0.0001725	−0.0000576	0.			

15-minute sampling (M = 18)

0	0.1944445					
1	0.1814184	0.1457861	0.0967236	0.0460983	0.0047054	−0.0208333
2	−0.0294517	−0.0249741	−0.0138889	−0.0026017	0.0047148	0.0069445
3	0.0054533	0.0026489	0.0004986	−0.0002909	−0.0001735	0.

30-minute sampling (M = 12)

0	0.3750000	0.1062099	−0.0355647	−0.0625000	−0.0164860	0.0208333
1	0.2898969	0.0000000	−0.0061020	−0.0020433	0.0002740	0.
2	0.0179817					

Low-pass filters applied to hourly values

Doodson X₀ 19-hour filter

0	0.0	1.0	2.0	0.0	1.0
1	2.0	2.0	0.0	1.0	1.0
2	1.0	0.0	0.0	1.0	0.0
3	0.0				
4	1.0				

Divisor = 30.0

72-hour filter

0	0.0486111	0.0477829	0.0467609	0.0453546	0.0435877	0.0414900
1	0.0484032	0.0364465	0.0335833	0.0305528	0.0274025	0.0241809
2	0.0390965	0.0177159	0.0145646	0.0115246	0.0086342	0.0059275
3	0.0209364	0.0011764	−0.0008260	−0.0025611	−0.0040222	−0.0052083
4	0.0034336	−0.0067781	−0.0071852	−0.0073629	−0.0073327	−0.0071186
5	−0.0061239	−0.0062435	−0.0056376	−0.0049562	−0.0042260	−0.0034722
6	−0.0067464	−0.0019851	−0.0012907	−0.0006504	−0.0000761	0.0004234
7	−0.0027183	0.0011787	0.0014322	0.0016059	0.0017049	0.0017361
8	0.0008425	0.0016297	0.0015114	0.0013633	0.0011954	0.0010170
9	0.0017079	0.0006622	0.0004996	0.0003538	0.0002281	0.0001247
10	0.0008368	0.0000145	−0.0000526	−0.0000727	−0.0000782	−0.0000727
11	0.0000440	−0.0000434	−0.0000267	−0.0000126	−0.0000033	0.
12	−0.0000598					

168-hour filter

0	0.0625000	0.0608882	0.0589092	0.0562042	0.0528366	0.0488849
1	0.0620947	0.0396035	0.0344838	0.0291942	0.0238489	0.0185604
2	0.0444400	0.0085771	0.0040724	0.0000000	−0.0035764	−0.0066086
3	0.0134364	−0.0109286	−0.0122017	−0.0129004	−0.0130559	−0.0127126
4	−0.0090645	−0.0107611	−0.0092892	−0.0075864	−0.0057306	−0.0037988
5	−0.0119261	0.0000000	0.0017357	0.0032878	0.0046125	0.0056768
6	−0.0018656	0.0069477	0.0071442	0.0070590	0.0067124	0.0061329
7	0.0064587	0.0044211	0.0033729	0.0022566	0.0011177	0.0000000
8	0.0053557	−0.0020140	−0.0028436	−0.0035204	−0.0040274	−0.0043543
9	−0.0010560	−0.0044643	−0.0042621	−0.0039086	−0.0034248	−0.0028359
10	−0.0044986	−0.0014553	−0.0007225	0.0000000	0.0006853	0.0013091
11	−0.0021697	0.0022943	0.0026273	0.0028429	0.0029389	0.0029178
12	0.0018510	0.0025555	0.0022390	0.0018536	0.0014176	0.0009503
13	0.0027864	0.0000000	−0.0004463	−0.0008515	−0.0012024	−0.0014881
14	0.0004714	−0.0018376	−0.0018959	−0.0018783	−0.0017897	−0.0016374
15	−0.0017013	−0.0011813	−0.0009008	−0.0006021	−0.0002977	0.0000000
16	−0.0014309	0.0005320	0.0007483	0.0009223	0.0010499	0.0011290
17	0.0002799	0.0011431	0.0010836	0.0009862	0.0008571	0.0007035
18	0.0011594	0.0003542	0.0001740	0.0000000	−0.0001614	−0.0003045
19	0.0005333	−0.0005197	−0.0005868	−0.0006256	−0.0006367	−0.0006219
20	−0.0004250	−0.0005259	−0.0004523	−0.0003672	−0.0002751	−0.0001805
21	−0.0005838	0.0000000	0.0000790	0.0001470	0.0002021	0.0002432
22	−0.0000876	0.0002828	0.0002825	0.0002706	0.0002488	0.0002193
23	0.0002700	0.0001460	0.0001065	0.0000680	0.0000320	−0.0000000
24	0.0001843	−0.0000485	−0.0000642	−0.0000740	−0.0000785	−0.0000783
25	−0.0000271	−0.0000669	−0.0000576	−0.0000473	−0.0000366	−0.0000266
26	−0.0000741	−0.0000010	−0.0000041	0.0000000	0.0000025	0.0000036
27	−0.0000175	0.0000029	0.0000020	0.0000010	0.0000003	0.
28	0.0000036					

APPENDIX 2

Response Analysis Inputs and Theory

Physically the locally observed tides are generated by the gravitational forces acting over a very wide area (Chapter 5) and so it is best to take a broadly representative value for the gravitational potential input. Munk and Cartwright (1966) used an expansion of the potential at the surface of the earth in complex spherical harmonics:

$$\Omega(\theta,\lambda,t) = \sum_{n=0}^{\infty} \sum_{m=0}^{\infty} g[a_n^m(t)U_n^m(\theta,\lambda) + b_n^m V_n^m(\theta,\lambda)]$$

where θ, λ and t are north colatitude ($90°$ - latitude), east longitude and time variables respectively. a_n^m and b_n^m are the real and imaginary parts of a complex time-varying coefficient $C_n^m(t)$, computed directly from the ephemerides of lunar and solar motion. $(U_n^m + iV_n^m)$ represents the variation in the potential over the earth's surface. In practice very few values of n and m are necessary. Terms of degree $n = 2$ are much larger than those of degree $n = 3$; $n = 2$ relates to the second term of the Legendre polynomial expansion in equation (3:7) and $n = 3$ relates to the third term. The order m requires values 0, 1, 2 and sometimes 3: these *orders* correspond with the long-period, diurnal and semi-diurnal and third-diurnal *species* of harmonic analysis.

The tidal variations are represented as the weighted sum of past values of each input spherical harmonic of the potential:

$$T(t) = \int_0^{\infty} w(\tau) \, C(t - \tau) \, d\tau$$

where $w(\tau)$ are the weights to be attached to the Ω values at time $(t - \tau)$. For physical systems it is reasonable to expect w to become small as τ becomes large. In the discrete form used in real applications it has been found that two increments of τ of 48 hours give satisfactory results. Thus:

$$T(t) = w(0) \, C(0) + w(-\tau_1) \, C(t - \tau_1) + w(-\tau_2) \, C(t - \tau_2)$$

where τ_1, and τ_2 are 48 hours and 96 hours respectively. The response weights have a physical meaning—they represent the remaining effects at $t = 0$ of the system response to a unit impulse at $t = -\tau$. Although physically difficult to

422

justify, there are mathematical advantages in using both past and future values of the potential, so values of $C(t + \tau_1)$ and $C(t + \tau_2)$ and their weights $w(\tau_1)$ and $w(\tau_2)$ are also used.

The complex admittance of the ocean to these inputs is given as a function of frequency by the Fourier transform of the weights:

$$Z(\omega) = \int_{-\infty}^{\infty} w(\tau) \exp(-i\omega\tau) \, d\tau$$

or in discrete terms:

$$
\begin{aligned}
Z(\omega) = \; & w(-\tau_2) \exp(-i\omega\tau_2) + w(-\tau_1) \exp(-i\omega\tau_1) \\
& + w(0) \\
& + w(\tau_1) \exp(i\omega\tau_1) + w(\tau_2) \exp(i\omega\tau_2)
\end{aligned}
$$

The more weights and lags that are included, the more detailed the frequency variations of the admittance (the 'wigglyness') becomes. In the absence of lags, $w(0)$ would give a uniform admittance across the band.

The amplitude and phase responses are then:

$$|Z(\omega)| \quad \text{and} \quad \arg (Z(\omega))$$

which are easily calculated from the real and imaginary parts.

An alternative but inferior way of deriving the admittance at a site is to evaluate the frequency spectrum of the input and output signals by Fourier transformation:

$$\Omega(\omega) = \int_{-\infty}^{\infty} \Omega(t) \exp(-i\omega t) \, dt$$

$$O(\omega) = \int_{-\infty}^{\infty} O(t) \exp(-i\omega t) \, dt$$

and to determine $Z(\omega)$ as:

$$Z(\omega) = O(\omega)/\Omega(\omega)$$

One advantage of response analysis is the facility to input other forcing functions in addition to the gravitational potential. Munk and Cartwright introduced the idea of a radiation potential, which varies with the radiant solar energy incident on a unit surface in unit time. A harmonic expansion of this radiational potential has been developed to allow calculations of radiational harmonic constituents from the radiational response functions. The radiational function is defined as the quantity of heat per unit area received by the earth's surface (see Figure A2:1):

$$\mathcal{R} = s\left(\frac{PS}{R_s}\right) \cos \varphi \qquad -\frac{\pi}{2} < \varphi < \frac{\pi}{2} \; \text{(day-time)}$$

Earth   Sun

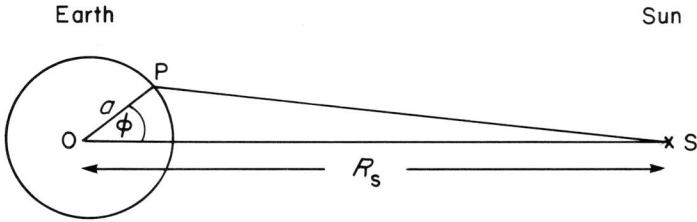

Figure A2:1   Symbols used for defining the earth–sun system.

where s is the 'solar constant', $1380 \, \text{W m}^{-2}$, and

$$\mathscr{R} = 0 \qquad \frac{\pi}{2} < \varphi < \frac{3\pi}{2} \text{ (night-time)}$$

The earth is transparent to gravitational forces but not to solar radiation. As previously (Section 3:2:1):

$$PS^2 = a^2 + R_s^2 - 2aR_s \cos \varphi$$

Expanding in Legendre polynomials, and neglecting terms in $(a/R_s)^2$ the total day/night function of $\mathscr{R}$ is:

$$\mathscr{R} = s \left(\frac{R_s}{R_s} \right) [\tfrac{1}{4} + \tfrac{1}{2} \mathscr{P}_1(\cos \varphi) + \tfrac{5}{16} \mathscr{P}_2(\cos \varphi) + \cdots]$$

The first term gives the net radiation on the whole earth surface $(4\pi a^2 \, (s/4) = \pi a^2 s$, i.e. the cross-section multiplied by the mean radiation). This is balanced by the infra-red back radiation and is part of the overall heat balance of the planet. This term compares with the first term for the gravitational potential, and has no tide-producing effect. The $\tfrac{1}{2}\mathscr{P}_1(\cos \varphi)$ term represents the non-transparency of the earth to solar radiation, and is absent as a gravitational tide. It gives large annual and daily radiational tides. The $\tfrac{5}{16} P_2(\cos \varphi)$ term is similar to the gravitational term. In practice only the $P_1(\cos \varphi)$ and $P_2(\cos \varphi)$ terms are necessary. The coefficients of the expansion are given by Cartwright and Tayler (1971).

APPENDIX 3

Analysis of Currents

A3:1 Current ellipse parameters from component amplitudes and phases

Consider a harmonic constituent whose east-going and north-going components are given by:

$$U \cos (\omega t - g_u)$$
$$V \cos (\omega t - g_v)$$
(speeds and phase angles are in radians)

Then in the usual Cartesian convention (see Figure A3:1) the direction of flow and the current speed are:

$$\theta = \arctan \frac{V \cos (\omega t - g_v)}{U \cos (\omega t - g_u)} \tag{A3:1}$$

$$q = \{U^2 \cos^2 (\omega t - g_u) + V^2 \cos^2 (\omega t - g_v)\}^{\frac{1}{2}} \tag{A3:2}$$

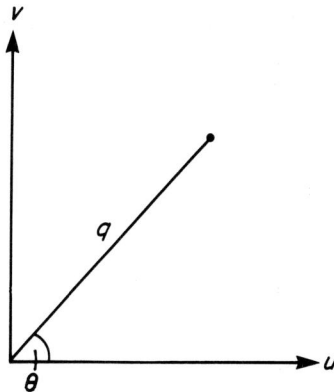

Figure A3:1   Current component axes in the Cartesian system.

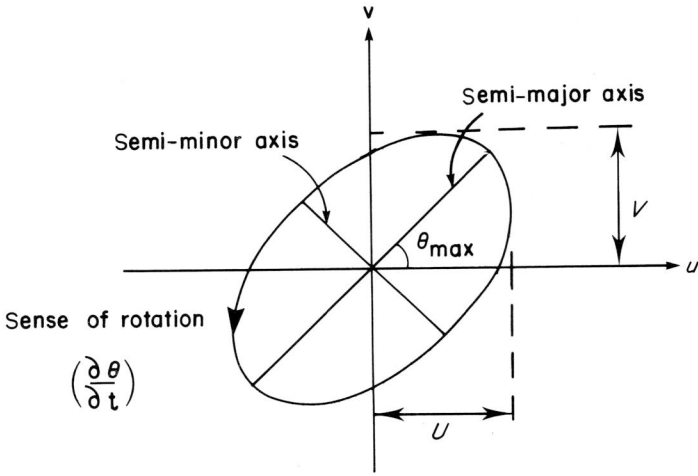

Figure A3:2   Basic parameters of a current ellipse.

Semi-major and semi-minor axes of ellipse

From (A3:2), using the identity:

$$2 \cos^2 \beta = (1 + \cos 2\beta)$$

$$q^2 = U^2 \left[\tfrac{1}{2} + \tfrac{1}{2} \cos 2(\omega t - g_u)\right] + V^2 \left[\tfrac{1}{2} + \tfrac{1}{2} \cos 2(\omega t - g_u + (g_u - g_v))\right]$$

$$= \tfrac{1}{2}(U^2 + V^2)$$
$$+ \tfrac{1}{2}[U^2 + V^2 \cos 2(g_u - g_v)] \cos 2(\omega t - g_u)$$
$$- \tfrac{1}{2}[V^2 \sin 2(g_u - g_v)] \sin 2(\omega t - g_u)$$

which may be written in the form:

$$q^2 = \tfrac{1}{2}(U^2 + V^2) + \tfrac{1}{2}\alpha^2 \cos 2(\omega t - g_u + \delta) \tag{A3:3a}$$

or

$$q^2 = \tfrac{1}{2}(U^2 + V^2 - \alpha^2) + \alpha^2 \cos^2 (\omega t - g_u + \delta) \tag{A3:3b}$$

where:

$$2\delta = \arctan \frac{V^2 \sin 2(g_u - g_v)}{U^2 + V^2 \cos 2(g_u - g_v)} \tag{A3:4a}$$

and

$$\alpha^2 = [(U^2 + V^2 \cos 2(g_u - g_v))^2 + (V^2 \sin 2(g_u - g_v))^2]^{\frac{1}{2}}$$
$$= [U^4 + V^4 + 2U^2 V^2 \cos 2(g_u - g_v)]^{\frac{1}{2}} \tag{A3:4b}$$

The maximum value of current speed is determined from equation (A3:3b) as:

$$q_{max} = \left(\frac{U^2 + V^2 + \alpha^2}{2}\right)^{\frac{1}{2}} = \text{semi-major axis} \tag{A3:5a}$$

and the minimum value:

$$q_{min} = \left(\frac{U^2 + V^2 - \alpha^2}{2}\right)^{\frac{1}{2}} = \text{semi-minor axis} \tag{A3:5b}$$

The eccentricity of the ellipse may be calculated, as in equation (3:13):

$$\frac{q_{max}}{q_{min}} = \frac{1 + e}{1 - e} \quad \text{or} \quad e = \frac{q_{max} - q_{min}}{q_{max} + q_{min}}$$

There is a constant current speed, so that the vector tip follows a circle ($\alpha = 0$) if:

$$g_u - g_v = \frac{\pi}{2}, -\frac{\pi}{2} \quad \text{and} \quad U = V$$

To prove this substitute into (A3:4b):

$$\alpha^2 = \left[U^4 + V^4 + 2U^2 V^2 \cos\left(\begin{array}{c}\pi\\-\pi\end{array}\right)\right]^{\frac{1}{2}}$$

$$= [U^2 - V^2] = 0$$

In this case:

$$q_{max} = q_{min} = \left[\frac{U^2 + V^2}{2}\right]^{\frac{1}{2}}$$

Time of maximum current speeds

From equation (A3:3b), the maximum values of q are obtained when:

$$\omega t - (g_u - \delta) = 0, 2\pi, \cdots, 2m\pi$$

or

$$\omega t - (g_u - \delta) = \pi, 3\pi, \cdots, (2m + 1)\pi \tag{A3:6}$$

More generally if the Equilibrium phase at a specified zero hour is $V_0 + u$:

$$V_0 + u + \omega t = m\pi + (g_u - \delta)$$

and hence the times of maximum currents are:

$$t_{max} = \frac{1}{\omega}\{m\pi - (V_0 + u) + (g_u - \delta)\} \tag{A3:7}$$

There will be two occasions during each cycle when the value of q_{max} is reached; these are in the positive and in the negative directions along the direction of the major axis of the current ellipse.

Direction of maximum current speed

Substituting from (A3:6) into (A3:1) we have for the direction of the semi-major axis:

$$\theta_{max} = \arctan \frac{V \cos((g_u - g_v) - \delta)}{U \cos(\delta)} \qquad (A3:8)$$

Sense of the ellipse vector rotation

Rewriting equation (A3:1) as:

$$\theta = \arctan K$$

and differentiating wrt time:

$$\frac{\partial \theta}{\partial t} = \frac{1}{(1 + K^2)} \frac{\omega UV}{U^2 \cos^2(\omega t - g_u)} \sin(g_v - g_u) \qquad (A3:9)$$

$$= \frac{\omega UV}{q^2} \sin(g_v - g_u)$$

All terms on this expression are always positive except for the $\sin(g_v - g_u)$ factor. Hence:

$$0 < g_v - g_u < \pi, \quad \frac{\partial \theta}{\partial t} \text{ positive; anticlockwise rotation.}$$

$$\pi < g_v - g_u < 2\pi, \quad \frac{\partial \theta}{\partial t} \text{ negative; clockwise rotation.}$$

$$g_v - g_u = 0, \pi, \quad \frac{\partial \theta}{\partial t} \text{ zero; } \quad \text{rectilinear flow.}$$

A3.2   Rotary current components

The ellipse characteristics of a tidal current constituent can be calculated more readily from a slightly different parameterization in terms of two polar vectors, each of constant amplitude, which rotate at the angular speed of the constituent, but in opposite directions. When these vectors are aligned the currents have their maximum speed. A quarter of a revolution later they oppose each other and the current speeds have their minimum value. After a further quarter of a cycle they are again aligned, this time in the opposite direction along the major axis, and the maximum current speeds are again obtained.

If the amplitudes and phases of the clockwise and anticlockwise rotating vectors are (Q_C, g_C) and (Q_{AC}, g_{AC}) their values may be calculated from the (U, g_u) and (V, g_v) component parameters.

$$Q_{AC} = \tfrac{1}{2}[U^2 + V^2 - 2UV \sin(g_v - g_u)]^{\frac{1}{2}}$$

$$Q_{AC} = \tfrac{1}{2}[U^2 + V^2 + 2UV \sin(g_v - g_u)]^{\frac{1}{2}} \qquad (A3:10)$$

$$g_C = \arctan \left\{ \frac{U \sin g_u + V \cos g_v}{U \cos g_u - V \sin g_v} \right\}$$

$$g_{AC} = \arctan \left\{ \frac{-U \sin g_u + V \cos g_v}{U \cos g_u - V \sin g_v} \right\}$$

These parameters may be obtained directly from the observed currents by a process of complex demodulation. In terms of these rotation vector parameters:

$$\text{semi-major axis} = Q_C + Q_{AC}$$
$$\text{semi-minor axis} = |Q_C - Q_{AC}|$$
$$\text{phase of semi-major axis} = -\tfrac{1}{2}(g_{AC} - g_C)$$
$$\text{direction of the semi-major axis} = \tfrac{1}{2}(g_{AC} + g_C)$$

(positive anticlockwise from east).

$$= \frac{\pi}{2} - \tfrac{1}{2}(g_{AC} + g_C) \quad \text{(clockwise from north)}.$$

The total current vector rotates anticlockwise if $Q_{AC} > Q_C$
clockwise if $Q_C > Q_{AC}$

From equation (A3:10), $Q_{AC} > Q_C$ if $\sin(g_v - g_u) > 0$, i.e. $0 < (g_v - g_u) < \pi$, as above.

APPENDIX 4

Theoretical Tidal Dynamics

A4:1 Long progressive waves, no rotation

In Section 5:2 it was shown that a wave of small amplitude in shallow water travels at a speed:

$$c = (gD)^{\frac{1}{2}} \qquad (A4:1)$$

where D is the undisturbed water depth. The associated currents are:

$$u = \zeta(g/D)^{\frac{1}{2}} \qquad (A4:2)$$

so that the maximum currents in the direction of propagation occur at the same time as local high water. Currents at low water are directed in the opposite direction to that of wave propagation.

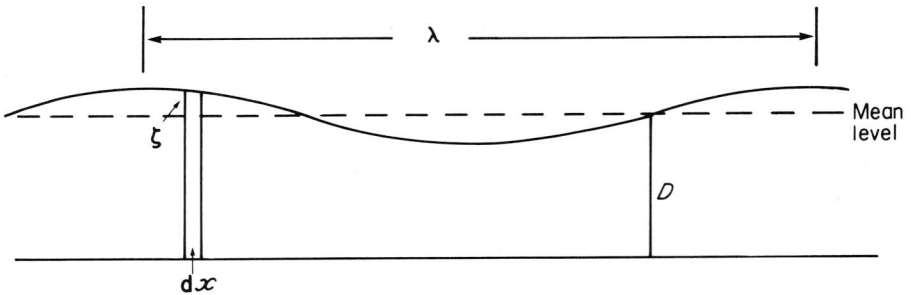

Figure A4:1 Symbols used to define the characteristics of a long wave.

Energetics

Consider the energy associated with the wave (which need not be a simple harmonic) shown in Figure A4:1. The potential energy of a vertical column with unit cross-sectional length along the wavefront, and an incremental width dx is:

$$\rho g(\zeta + D) . \tfrac{1}{2}(\zeta + D) \, dx = \tfrac{1}{2}\rho g(\zeta + D)^2 \, dx$$

430

Over a complete wavelength the potential energy is:

$$\tfrac{1}{2}\rho g \int_0^\lambda (\zeta + D)^2 \, \mathrm{d}x$$

If there is no wave present, the potential energy of the water of uniform depth D is:

$$\tfrac{1}{2}\rho g D^2 \lambda$$

The difference is the potential energy due to the wave:

$$\tfrac{1}{2}\rho g \int_0^\lambda \zeta^2 \, \mathrm{d}x + \rho g D \int_0^\lambda \zeta \, \mathrm{d}x$$

and since the second term is zero when integrated over a complete wavelength, this potential energy becomes:

$$\tfrac{1}{2}\rho g \int_0^\lambda \zeta^2 \, \mathrm{d}x \qquad (A4:3)$$

Similarly we may calculate the kinetic energy per unit cross-sectional length on the long-wave assumption that vertical velocities are small compared with horizontal velocities and $\zeta \ll D$. For the vertical column this is:

$$\tfrac{1}{2}\rho D u^2 \, \mathrm{d}x$$

$$= \tfrac{1}{2}\rho D \zeta^2 \frac{g}{D} \, \mathrm{d}x$$

The kinetic energy integrated over a wavelength is:

$$\tfrac{1}{2}\rho g \int_0^\lambda \zeta^2 \, \mathrm{d}x \qquad (A4:4)$$

which is the same as the potential energy.

The total energy per wavelength is the sum of the potential and kinetic energies:

$$\rho g \int_0^\lambda \zeta^2 \, \mathrm{d}x$$

If the surface disturbance takes the form of a harmonic wave $H_0 \cos kx$ (where k is the wavenumber $2\pi/\lambda$), this total energy is:

$$\tfrac{1}{2}\rho g H_0^2 \lambda$$

and the total energy per unit area of surface is:

$$\tfrac{1}{2}\rho g H_0^2 \qquad (A4:5a)$$

Energy is transmitted by the wave at the group velocity, which for long waves is the speed c, so that the energy flux per unit width of wavefront is:

$$\tfrac{1}{2}\rho g H_0^2 \, (gD)^{\frac{1}{2}} \qquad (A4:5b)$$

If the energy flux is to remain constant as the water depth decreases, the wave amplitude must increase so that the factor $H_0^2 D^{\frac{1}{2}}$ remains constant.

These energy properties have been calculated above for a purely progressive wave. However, the energy fluxes through a section may be considered in a more general way (Taylor, 1919). The energy flux across a section of channel consists of the work done by the water which crosses the section, plus the potential and kinetic energies which it transfers. Using the hydrostatic approximation (equation (3:21)), the average water head pressure in water of depth D is $\frac{1}{2}\rho g(D + \zeta)$. The entering water does work against the pressure at a rate $\frac{1}{2}\rho g(D + \zeta)^2 u$ per unit width of wavefront. The entering water also brings its own potential and kinetic energy. If we now measure the potential energy of the water column relative to a zero defined as the local mean sea-level, the total potential energy per unit area is:

$$\tfrac{1}{2}\rho g \,(\zeta^2 - D^2)$$

so the flow of potential energy per unit width of wavefront is:

$$\tfrac{1}{2}\rho g \,(\zeta^2 - D^2)\, u$$

The kinetic energy flux per unit width is:

$$\tfrac{1}{2}\rho \,(D + \zeta)\, u^2 \cdot u$$

The total influx of energy per unit width of channel, due to all three sources is:

$$\tfrac{1}{2}\rho g\{(D + \zeta)^2 + \zeta^2 - D^2\}\, u + \tfrac{1}{2}\rho(D + \zeta)\, u^3$$
$$= \rho g \,\{D\zeta + \zeta^2\}\, u + \tfrac{1}{2}\rho(D + \zeta)\, u^3 \qquad (A4:6)$$

Normally $\zeta \ll D$, and since $u^2 = \zeta^2(g/D)$, eliminating all terms in ζ^2 gives an energy flux of:

$$\rho g D \zeta u \qquad (A4:7)$$

If the propagating disturbance consists of a tidal harmonic $H_0 \cos(\omega t - g_{\zeta_0})$ at the flux boundary, with currents which also vary harmonically but lag the elevations $U_0 \cos(\omega t - g_{u_0})$, the average energy flux per unit width over a tidal cycle is:

$$\rho g D H_0 U_0 \int_0^{2\pi/\omega} \cos(\omega t - g_{\zeta_0}) \cos(\omega t - g_{u_0})\, dt$$

$$= \tfrac{1}{2}\rho g D H_0 U_0 \cos(g_{\zeta_0} - g_{u_0}) \qquad (A4:8)$$

In the case of a progressive wave, where the currents are a maximum in the positive direction at the same time as high water, there is no phase lag and so the energy flow has a maximum value. This formula for energy flux may be compared with the formula for power transmission in an electrical alternating current circuit.

Equation (A4:7) gives the average energy flux through each unit width of channel for a single harmonic tidal constituent. By writing the elevations and currents in equation (A4:7) as the sum of two or more harmonics of different frequencies, it is possible to show that averaged over a sufficiently long period, the total energy flux due to all the harmonics is equal to the sum of the fluxes in the individual harmonics computed using equation (A4:8). This is analogous to the summation properties for the spectral decomposition of variance, and is obviously a useful way of computing total fluxes from harmonic analyses of elevations and currents.

Reflection and transmission of waves at a depth discontinuity

Consider a plane wave of amplitude H_1 propagating along a channel in water depth D_1 which encounters a step discontinuity in depth (Figure A4:2). Beyond the step the depth is D_2. The wavelength in the shallower water is reduced but the frequency remains the same. H_1' is the amplitude of the reflected wave and H_2' that of the transmitted wave.

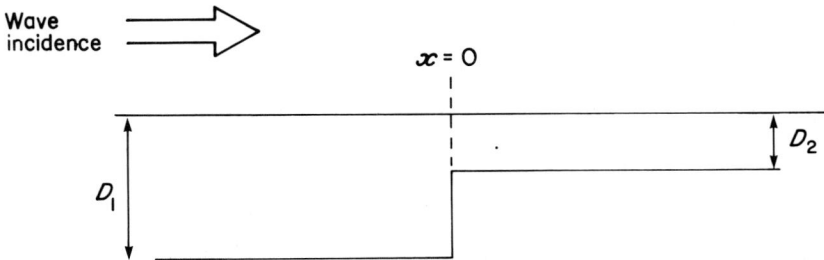

Figure A4:2 Symbols representing a step change of water depth.

Matching elevations at $x = 0$ gives:

$$H_1 + H_1' = H_2' \qquad (A4:9)$$

Similarly for the water fluxes the (current $\times$ depth) must be constant:

$$D_1(g/D_1)^{\frac{1}{2}} (H_1 - H_1') = D_2(g/D_2)^{\frac{1}{2}} H_2'$$

or

$$c_1 (H_1 - H_1') = c_2 H_2' \qquad (A4:10)$$

where c_1 and c_2 are the wave velocities from (A4:1).
From (A4:9) and (A4:10):

$$\frac{H_1'}{H_1} = \frac{\text{reflected wave amplitude}}{\text{incident wave amplitude}} = \frac{c_1 - c_2}{c_1 + c_2}$$

$$\qquad (A4:11)$$

$$\frac{H_2'}{H_1} = \frac{\text{transmitted wave amplitude}}{\text{incident wave amplitude}} = \frac{2c_1}{c_1 + c_2}$$

Note that these satisfy the condition that if there is no discontinuity ($D_1 = D_2$ and so $c_1 = c_2$) there is no reflected wave and the transmitted wave passes unaltered. It is easy to show by applying equation (A4:5b) that the total energy of the incident wave is conserved in the reflected and transmitted waves. For typical ocean and shelf depths of 4000 m and 200 m respectively the reflected and transmitted waves have amplitudes of 0.64 and 1.64 times the amplitude of the incident wave. These values are only indicative of the behaviour of long tidal waves incident on a continental slope because the detailed bathymetry of the slope can only be approximately represented by a step change of depth.

A4:2 Standing waves

From the progressive wave equation (5:4) we may write the following expression for two progressive waves travelling in opposite directions as illustrated in Figure 5:3:

$$H_0 \cos (kx - \omega t) + H_0 \cos (-kx - \omega t) = 2H_0 \cos \omega t \cos kx \quad \text{(A4:12)}$$

This represents a wave motion called a *standing wave*, whose amplitude varies harmonically as a function of the distance x, with wavelength $2\pi/k = \lambda$.

In each wavelength there are two nodes where $\cos kx = 0$, at which the oscillations have zero amplitude. These are located at $x = \frac{1}{4} (n + 1) \lambda$ ($n = 0$, $2 \ldots$) from the reflecting barrier. There are intermediate antinodes, where the oscillations have maximum amplitude at $x = \frac{1}{2}n\lambda$. The currents are given by:

$$H_0(g/D)^{\frac{1}{2}} \cos (kx - \omega t) - H_0(g/D)^{\frac{1}{2}} \cos (-kx - \omega t)$$
$$= 2H_0(g/D)^{\frac{1}{2}} \sin \omega t \sin kx \quad \text{(A4:13)}$$

Maximum currents occur where $\sin kx = 1$, i.e. $x = \frac{1}{4}(n + 1)\lambda$ and zero currents occur where $\sin kx = 0$, i.e. $x = \frac{1}{2}n\lambda$, so there can be vertical walls in these positions. Maximum currents coincide with zero amplitude whereas zero currents occur where the elevation changes are a maximum. This behaviour of standing waves is as shown for oscillation in a box (!) in Figure 5:4.

When a tidal wave is reflected at the head of a basin, the reflected wave will be weaker than the incident wave. The sum of these two can be written:

$$H_I \cos(kx - \omega t) + H_R \cos (kx + \omega t)$$
$$= H_R \cos(kx - \omega t) + H_R \cos (kx + \omega t) + (H_I - H_R) \cos (kx - \omega t)$$
$$= 2H_R \cos \omega t \cos kx + (H_I - H_R) \cos (kx - \omega t) \quad \text{(A4:14)}$$

The result is equivalent to a standing wave of amplitude $2H_R$ and a progressive wave of amplitude $(H_I - H_R)$ which propagates energy towards the head of the basin where it is dissipated by imperfect reflection.

A4:3 Long waves on a rotating earth

The momentum equations in the absence of advective accelerations and friction become (equations (3:23), (3:24))

$$\frac{\partial u}{\partial t} - fv = -g\,\frac{\partial \zeta}{\partial x}$$

$$\frac{\partial v}{\partial t} + fv = -g\,\frac{\partial \zeta}{\partial y}$$

for depth independent currents, using the hydrostatic equation. The continuity equation is

$$\frac{\partial \zeta}{\partial t} + D\left(\frac{\partial u}{\partial x} + \frac{\partial v}{\partial y}\right) = 0$$

Consider a solution which has the special property of no flow along the y-axis ($v = 0$). This would apply for wave propagation along a straight boundary at $y = 0$, such as would be presented for a wave travelling along a coastline. In this case the three equations reduce to:

$$\frac{\partial u}{\partial t} = -g\,\frac{\partial \zeta}{\partial x}$$

$$fu = -g\,\frac{\partial \zeta}{\partial y}$$

$$\frac{\partial \zeta}{\partial t} + D\,\frac{\partial u}{\partial x} = 0$$

We seek a solution of the form:

$$\zeta = \zeta(y)\,\cos(kx - \omega t)$$

$$u = u(y)\,\cos(kx - \omega t)$$

where the amplitudes of the harmonic variations are functions only of the distance for the boundary at $y = 0$. Substituting these into the three simplified hydrodynamic equations yields:

$$\omega u(y)\,\sin\,(kx - \omega t) = gk\zeta(y)\,\sin\,(kx - \omega t)$$

$$fu(y)\,\cos\,(kx - \omega t) = -g\,\frac{\partial \zeta(y)}{\partial y}\,\cos\,(kx - \omega t)$$

$$\omega \zeta(y)\,\sin\,(kx - \omega t) - Dku(y)\,\sin\,(kx - \omega t) = 0$$

from which:

$$\omega u(y) = gk\zeta(y)$$

$$fu(y) = -g \frac{\partial \zeta(y)}{\partial y}$$

$$Dku(y) = \omega \zeta(y)$$

Dividing the first and third of these:

$$\frac{\omega}{Dk} = g \frac{k}{\omega}$$

or

$$\omega^2 = gD$$

But ω/k is the speed of propagation of the wave, c, from which we deduce that $c = (gD)^{\frac{1}{2}}$ in this solution, which is the same as for a wave in a non-rotating system. So too, from the first equation, is the relationship between currents and elevations (equation (A4:2) and equation (A4:1)). The second equation can be rearranged:

$$\frac{\partial \zeta(y)}{\partial y} = -\frac{f}{g} u(y) = -\frac{f}{c} \zeta(y)$$

therefore

$$\zeta(y) = H_0 \exp(-fy/c); \quad u(y) = (g/D)^{\frac{1}{2}} \zeta(y) \tag{A4:15}$$

where H_0 is the amplitude at $y = 0$

These solutions to the hydrodynamic equations on a rotating earth in the presence of a boundary are called Kelvin waves, and they provide a good description of many observed tidal characteristics. The effect of the rotation appears only in the factor $\exp(-fy/c)$, which gives a decay of wave amplitude away from the boundary with a length scale $c/f = [(gD)^{\frac{1}{2}}/f]$, which depends on the latitude and the water depth. This scale is called the Rossby radius of deformation. At a distance $y = c/f$ from the boundary the amplitude has fallen to $0.37H_0$. At $45°$ N in water of 4000 m depth the Rossby radius is 1900 km, but in water 50 m deep this is reduced to 215 km. Kelvin waves are not the only solution to the hydrodynamic equations on a rotating earth. A special solution when $\omega = f$ gives inertial currents (Section 6:4:6). Another more general form, called Poincaré waves, allows finite values of the v current in the presence of a coast. The amplitude varies sinusoidally rather than exponentially in the direction transverse to the direction of wave propagation. The transverse velocity component also varies sinusoidally, which allows Poincaré waves to be realistic solutions provided that coastlines are located along one of the lines of zero transverse velocity.

Away from the coast another form of Poincaré waves, called Sverdrup waves allows both u and v to have finite values, with wave amplitudes independent of y (see Platzman, 1971; Gill, 1982). Poincaré and Sverdrup waves have phase speeds which exceed those of Kelvin waves, but energy, which is transmitted at the group velocity, has a maximum speed of transmission of $(gD)^{\frac{1}{2}}$.

Other solutions, which are of great interest in mathematical studies of tidal behaviour, are beyond the scope of this account.

In the case of a wave reflected at the head of a channel in a non-rotating system, there is a series of nodes, each with zero surface level changes, at distances $x = \frac{1}{4}(n + 1)\lambda$ from the reflecting boundary at $x = 0$. The node extends as a line across the channel. In a rotating system the nodal line is reduced to a nodal point, called the amphidromic point, at the same distance from the reflecting boundary. The position of this nodal point depends on the relative amplitudes of the ingoing and reflected Kelvin waves. By adding two Kelvin waves travelling in opposite directions it is possible to show that the y displacement of the amphidrome from the centre of the channel ($y = 0$) is:

$$y = -\frac{(gD)^{\frac{1}{2}} \ln \alpha}{2f} \tag{A4:16}$$

Because the reflected wave is smaller than the incident wave, the ratio between reflected and ingoing wave amplitudes will be less than one, and so $\ln \alpha$ is negative. Hence the amphidrome displacement y is to the left of the direction of the incoming wave in the northern hemisphere. If the incident and reflected wave have equal amplitude, the nodal point or amphidrome is in the middle of the channel. Figure A4:3 shows how the location of the cancelling point moves as the amplitude of the reflected wave is reduced. If the reflected wave is sufficiently weakened, the point at which the cancelling takes place does not have a real existence, but is apparently located outside the channel in which the waves propagate. In this case the amphidrome is said to be degenerate.

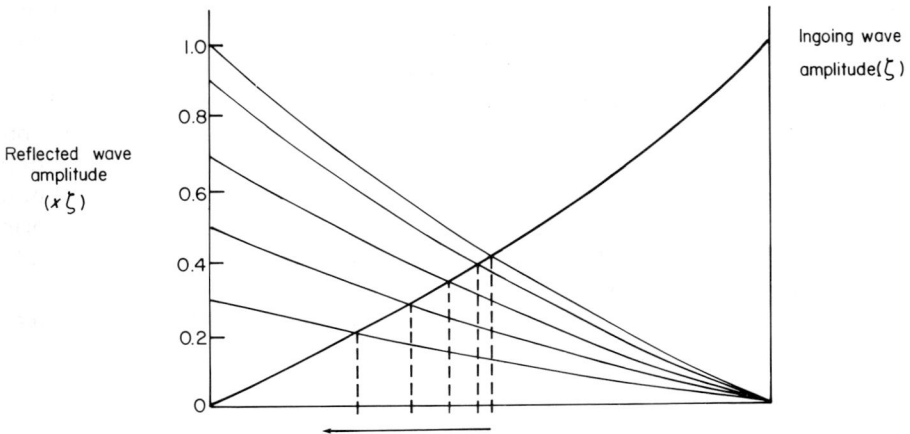

Figure A4:3   Showing how an amphidrome cancelling point moves to the left as the amplitude of the reflected tidal wave decreases. In the northern hemisphere the ingoing wave is moving into the paper.

A4:4 Cotidal and coamplitude lines

The preparation of cotidal and coamplitude charts from observations of sea levels and bottom pressures alone is quite straightforward, but the accuracy with which the lines can be traced offshore is usually limited by the scarcity of available data. This data can be supplemented by knowledge of the harmonic constituents of the depth-averaged currents, which are dynamically related to changes in elevations by the momentum and continuity equations. Solving the equations for an individual harmonic constituent at an offshore station:

$$H_\zeta \cos (\omega t - g_\zeta)$$

gives H_ζ and g_ζ. The amplitudes and phases of two components of currents give four extra parameters, the directions of the cotidal and coamplitude lines and the gradients on the cotidal and coamplitude surfaces (for full details see Proudman and Doodson, 1924; Defant, 1961).

To avoid laborious algebra the direction of the X-axis is taken along the direction of the major axis of the current ellipse. In the usual Cartesian convention of angles being measured positive anticlockwise from this direction, the directions of the cotidal and coamplitude lines are given by:

$$\tan \psi = - \frac{\omega U + fV}{\omega V + fU} \cot (g_\zeta - g_u)$$
(A4:17)

$$\tan \psi' = \frac{\omega U + fV}{\omega V + fU} \tan (g_\zeta - g_u)$$
(A4:18)

where ψ is the direction of the cotidal line, ψ' is the direction of the coamplitude line, ω is the constituent speed, U, V, are the semimajor and semiminor axes of the current ellipse and g_u is the current phase relative to the Equilibrium Tide. The full analysis includes the harmonic constituents of the frictional term, which can be evaluated by harmonic analysis of a time series of friction generated by equation (7:3) from the full time series of tidal currents. The nonlinear form of this relationship between friction and currents means that much of the friction variance appears at higher harmonics. For example $\mathbf{M_2}$ produces large frictional terms at $\mathbf{M_4}$. At the Inner Dowsing (Figures 1:2 and 4:5, 4:6) with strong currents in 20 m depth, including friction changes the values of ψ and ψ' by 9° and 24° respectively.

In the absence of friction, equations (A4:17) and (A4:18) show that the cotidal and coamplitude lines are always at right angles to each other:

$$\psi' = \psi + \pi/2$$

For a progressive wave, high water and flow along the major axis are simultaneous. From equation (A4:17) $\tan \psi = -\infty$, $\psi = \pi/2$ and from equation (A4:18) $\tan \psi' = \tan 0$, and so $\psi = 0$. The cotidal lines are parallel to the crest of the wave, and the coamplitude lines are perpendicular to this direction. This is the relationship which applies for a Kelvin wave. For a standing wave flow along the major axis is zero at high water, and $(g_\zeta - g_u) = \pi/2$. Hence from equation (A4:17) $\tan \psi = 0$, $\psi = 0$ and from equation (A4:18) $\tan \psi' = \infty$, $\psi' = \pi/2$. The cotidal lines are parallel to the major axis and the coamplitude lines are perpendicular.

In the case of a tidal wave near a coastline the flow will be parallel to the coast and so the semiminor axis will be very small by comparison with the semimajor axis. The angles become:

$$\tan \psi = -(\omega/f) \cot (g_\zeta - g_u)$$
$$\tan \psi' = (\omega/f) \tan (g_\zeta - g_u)$$

The lines have constant angles with the direction of the coast. Over an area sufficiently small for $(g_\zeta - g_u)$ to be constant the effect of this is to concentrate cotidal lines onto headlands or capes and to make them more diffuse in bays, as shown in Figure A4:4. This effect also causes the range of the tide to change more rapidly near headlands.

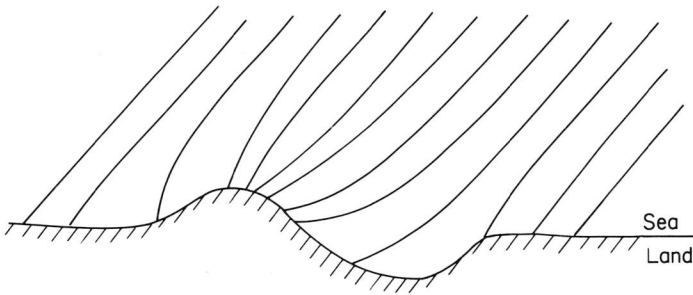

Figure A4:4 Cotidal lines converging on a cape or headland, and diverging in a bay.

Fuller accounts of tidal dynamics are found in Lamb (1932), Proudman (1953), Defant (1961), Cartwright (1977), Gill (1982) and Marchuk and Kagan (1984).

APPENDIX 5

Legal Definitions in the Coastal Zone

The area of coast covered and uncovered as the tide rises and falls is variously called the intertidal zone, the littoral zone, the tidelands or the foreshore. Shalowitz (1962; 1964) has thoroughly investigated the development of legal thinking and its application to the definition of this intermediate territory: the upper and lower limits may be defined in many ways in terms of the local tidal regime. Legal rights within this zone and within the subtidal zone have been interpreted in different ways at different times in different countries. Inevitably, a knowledge of tidal principles has an important role to play in the development and interpretation of legal rules, but the technical aspects of defining tidal boundaries have sometimes been underemphasized (Graber, 1980).

The English legal expert Lord Hale (1609–1676) wrote a treatise *De Jure Maris* which established as settled law in England the principle that the Crown has ownership of the foreshore, which extends landwards as far as it is covered by 'the ordinary flux of the sea'. The common-law term 'ordinary high-water mark' has been generally used to describe the boundary between the Crown's foreshore and the adjoining privately owned lands, but this term has also been subject to several interpretations.

Modern interpretation in the United Kingdom

Foreshore boundaries are shown on large-scale maps published by the Ordnance Survey, which summarizes the official basis on which these boundaries are determined as follows.

In 1854 the Lord Chancellor in giving judgement on the limits of foreshore boundaries around England and Wales defined these boundaries as following the High and Low Water Marks of a medium or average tide. In Scotland there has been no legal definition of foreshore boundaries but ancient custom has decreed that the extent of the foreshore shall be limited by mean spring tides. The Ordnance Survey therefore decided to survey the High and Low Water Marks of a mean or average tide in England and Wales and of an average spring tide in Scotland.

440

Before August 1935 the lines so surveyed in England and Wales were called High and Low Water Marks of Ordinary Tides, and on the 1:2500 maps and 1:10,560 (now 1:10,000) these descriptions, or their abbreviations, were always shown against them. In August 1935 the term "Medium" was adopted in place of "Ordinary" as a better description. In Scotland the lines were called High and Low Water Mark Ordinary Spring Tides.

Since March 1965 the term "Mean" has been adopted in place of "Medium" and the words "Mark" and "Tide" omitted. Thus the present descriptions in England and Wales are Mean High (or Low) Water abbreviated to MH (or L) W; and in Scotland Mean High (or Low) Water Springs abbreviated to MH (or L) WS.

Tidal lines are surveyed by either air or ground survey methods depending upon the type of foreshore. When conditions favour ground survey direct measurements from foreshore detail are used; but where ground conditions are difficult, air photographs are taken on infra-red film. These show the water's edge distinctly and are used for making a survey.

The ground survey or the aerial photography is carried out at a time when the Hydrographer advises that High (or Low) Water will reach a level equal to that of Mean High (or Low) Water or Mean High (or Low) Water Springs, as appropriate. The survey is repeated if the tide fails to run as predicted.

As high tide generally leaves a clear mark until the next high water there is not much difficulty in surveying this line. Low water, however, often presents considerably greater difficulty and its definition cannot be guaranteed to the same degree of accuracy.

The Ordnance Survey does not redefine its tidal boundaries at set intervals of time. Changes in tidal boundaries due to changes in coastal geomorphology and in mean sea-level are incorporated as new surveys are undertaken.

Parts of the foreshore have been granted by the Crown to other bodies or leased, for example, to the Nature Conservancy Council or local authorities (Great Britain, 1979). The public are entitled at common law to exercise certain rights over the seashore, including navigation and fishery. However, there is no right to go on the foreshore to take seaweed without appropriate permission, nor does the public have the right to dig for bait on the foreshore. The legal position is very complicated and subject to increasing scrutiny as the demands for the exploitation of the coastal zone become more extensive.

The United States

The law relating to ownership of the tidelands in the United States varies from state to state. It stems from two separate systems, Roman law which is followed on the continent of Europe, and common law which evolved in England and has been generally adopted by the 13 original states and most of the later-admitted states. After the Revolution, the former English colonies succeeded to the rights of the English Crown in respect of their tidelands. Absolute title to all tidelands passed to the original states, in trust, except where they had previously been granted into private ownership. Each state is free to adopt its own rules of real property, and questions of ownership are determined under state constitution, statutes and case law. Graber (1980 and subsequently) summarizes the complicated current situation. Most American coastal states have adhered to the basic English common law rule that the ordinary high-water mark (now interpreted as the line of mean high water) constitutes the legal boundary between privately owned uplands and state-owned tidelands. As a generalization, 16 coastal states deem the mean high-water line to be the private/state boundary. However, six Atlantic Coast states, Delaware, Massachusetts, Maine, New Hampshire, Pennsylvania and Virginia use the mean low-water line as the boundary. In Louisiana the boundary is the line of the highest winter tide. In Texas, if the source of upland title is a Spanish or Mexican grant predating Texas' independence, the line of mean higher high water is the legal boundary. Hawaii adheres to its aboriginal, customary concept that the private/public boundary is marked by the upper reaches of the wash of the waves. In general, 'gradual, imperceptible' physical changes in the location on the ground of the boundary, natural or artificial, result in a shift of the legal boundary.

Tidal datums are defined by the tidal authority, the National Ocean Service of the National Oceanic and Atmospheric Administration (Hicks, 1985; Swanson and Thurlow, 1979). As the official charts of the United States, NOS nautical charts automatically become legal documents in regard to the portrayal of maritime boundaries and tidal datums. Four datums are commonly calculated from observations for each National Tidal Datum Epoch; the Epoch is a 19-year period which is reviewed annually for possible revision, and must be actively considered for revision every 25 years. The present Epoch is 1960 to 1978. Adjustment to a new Tidal Epoch makes systematic allowances for the changes in mean sea-level or tidal characteristics. The four tidal datums and their uses have been summarized by Hicks (1985).

Mean Higher High Water (MHHW)

The average of the higher high-water heights of each tidal day. The average is the arithmetic mean. To be counted, each higher high water must be 0.10 ft or more above, and must occur 2.0 hours from, the adjacent low waters. The corresponding line is the property boundary between privately owned uplands and state owned tidelands for the State of Texas when the upland parcel's original source of title is a pre-1840 grant by Spain, Mexico, or the Republic of Texas. Prior to 28 November 1980 the mean higher high water line was delineated on nautical charts as the shoreline in areas of predominantly diurnal tides. The values of 0.10 ft and 2.0 hours are consistently applied for computations of all tidal datum levels and are arbitrary cutoffs, based on many years of experience.

Mean High Water (MHW)

The average of all the high water heights. This is used to define the line of the property boundary between privately owned uplands and state owned tidelands for most maritime States. Since 1980 the mean high water line is delineated on nautical charts as the shoreline.

Mean Low Water (MLW)

The average of all the low-water heights. Mean low water is the level used to define the boundary between private and state property in six east coast states. Along the east coast of the United States, the mean low-water line (as delineated on the official large-scale charts of the National Ocean Service) is the boundary between inland waters and the territorial sea. Mean low water has been used as the elevation for chart datum for NOS nautical charts along the east coast of the United States, but will eventually be replaced by MLLW.

Mean Lower Low Water (MLLW)

The average of the lower low water in each tidal day. Along the west and Gulf coasts the mean lower low water line is the boundary between inland waters and territorial seas. It is used as chart datum for hydrographic surveying along the west and Gulf coasts of the United States. With the adoption of the National Tidal Datum Convention of 1980, it is also authorized for use as chart datum for the east coast, and will eventually replace MLW to give a uniform system for defining chart datum for the United States.

For stations with shorter series than the Tidal Datum Epoch, simultaneous observation comparisons are made with a control tidal station to make the appropriate adjustments.

References

Abbott, M. B. (1956). A theory of the propagation of bores in channels and rivers. *Proceedings of the Cambridge Philosophical Society*, **52**, 344–62.

Ackers, P. and Ruxton, T. D. (1975). Extreme levels arising from meteorological surges, pp. 69–86 in vol. 1, *Proceedings of the International Conference on Coastal Engineering, 1974*. Copenhagen: American Society of Civil Engineers.

Alcock, G. A. (1984) Parameterizing extreme still water levels and waves in design level studies. *Institute of Oceanographic Sciences Report No. 183*, 95 pp.

Alcock, G. A. and Carter, D. J. T. (1985). *Extreme events in wave and water levels, in Breakwaters '85*. London: Thomas Telford, pp. 33–48.

Allen, G. P., Salomon, J. C., Bassoullet, P., du Penhoat, Y. and de Grandpre, C. (1980). Effects of tides on mixing and suspended sediment transport in macrotidal estuaries. *Sedimentary Geology*, **26**, 69–90.

Allen, J. R. L. (1981). Paleotidal speeds and ranges estimated from cross-bedding sets with mud drapes. *Nature*, **293**, 394–6.

Allen, J. R. L. (1985a). *Principles of Physical Sedimentology*. London: George Allen & Unwin, 272 pp.

Allen, J. R. L. (1985b). Intertidal drainage and mass-movement processes in the Severn Estuary: rills and creaks (pills). *Journal of the Geological Society of London*, **142**, 849–61.

Amin, M. (1979). A note on extreme tidal levels. *International Hydrographic Review*, **56**, 133–41.

Amin, M. (1982). On analysis and prediction of tides on the west coast of Great Britain. *Geophysical Journal of the Royal Astronomical Society*, **68**, 57–78.

Anderson, D. M., White, D. W. and Boden, D. G. (1985) (eds). *Toxic Dinoflagellates*. New York: Elsevier, 561 pp.

Arur, M. G. and Basir, F. (1981). Yearly mean sea level trends along the Indian coast, pp. 54–61, in *Papers and Proceedings of the Seminar on Hydrography in Exclusive Economic Zones, Demarcation and Survey of its Wealth Potential*. Calcutta: Hugli River Survey Service, 305 pp.

Aubrey, D. G. and Emery, K. O. (1983). Eigenanalysis of recent United States sea levels. *Continental Shelf Research*, **2**, 21–33.

Baker, T. F. (1984). Tidal deformations of the earth. *Science Progress*, **69**, 197–233.

Balazs, E. I. and Douglas, B. C. (1979). Geodetic leveling and the sea level slope along the California coast. *Journal of Geophysical Research*, **84**, 6195–6206.

Balch, W. M. (1986). Are red tides correlated to spring–neap tidal mixing?: use of a historical record to test mechanisms responsible for dinoflagellate blooms, pp. 193–223, in *Tidal Mixing and Plankton Dynamics*, *(ed. M. J. Bowman, C. M. Yentsch and W. T. Peterson)*. New York: Springer-Verlag, 502 pp.

Banal, M. (1982). L'énergie marémotrice en 1982. *La Houille Blanche*, **5**, 433–9.

Barnes, R. S. K. and Hughes, R. N. (1982), *An Introduction to Marine Ecology*. St. Louis, Missouri: Mosby, 234 pp.

Barnett, T. P. (1983a). Recent changes in sea level and their possible causes. *Climatic Change*, **5**, 15–38.

Barnett, T. P. (1983b). Long-term changes in dynamic height. *Journal of Geophysical Research*, **88**, 9547–52.

Barnett T. P. (1984). The estimation of 'global' sea level change: a problem of uniqueness. *Journal of Geophysical Research*, **89**, 7980–88.

Barrick, D. E., Evans, M. W. and Weber, B. L. (1977). Ocean surface currents mapped by radar. *Science*, **198**, 138–44.

Barth M. C. and Titus, J. G. (eds.) (1984). *Greenhouse Effects and Sea Level Rise: A challenge for this generation*. New York: Van Nostrand Reinhold.

444

445

Belderson, R. H., Pingree, R. D. and Griffiths, D. (1986). Low sea level tidal origin of Celtic Sea sand banks—evidence from numerical modelling of M_2 tidal streams. *Marine Geology*, **73**, 99–108.

Berkeley Thorn, R. and Roberts, A. G. (1981). *Sea Defence and Coastal Protection Works: a Guide to Design*. London: Thomas Telford Ltd., 216 pp.

Bird, E. C. F. (1968). *Coasts*. Canberra: Australian National University Press.

Bokuniewicz, H. J. and Gordon, R. B. (1980). Storm and tidal energy in Long Island Sound. *Advances in Geophysics*, **22**, 41–67.

Bowden, K. F. (1983). *The Physical Oceanography of Coastal Waters*. Chichester: Ellis Horwood, 302 pp.

Bowman, M. J., Yentsch, C. M. and Peterson, W. T. (1986). *Tidal Mixing and Plankton Dynamics*. New York: Springer-Verlag, 502 pp.

Brady, J. (ed.) (1982). *Biological Time Keeping*. Cambridge University Press, 197 pp.

Brosche, P. and Sundermann, J. (eds). (1982). *Tidal Friction and the Earth's rotation II*. New York: Springer-Verlag, 344 pp.

Bruun, P. (1978). *Stability of Tidal Inlets. Theory and Engineering*. Amsterdam: Elsevier, 506 pp.

Bruun, P. (1983). Review of conditions for uses of the Bruun rule of erosion. *Dock and Harbour Authority*, **64**, 79–82.

Brusca, R. C. (1980). *Common intertidal invertebrates of the Gulf of California*, second edn. University of Arizona Press, 513 pp.

Burrows, E. M., Conway, E., Lodge, S. M. and Powell, H. T. (1954). The raising of intertidal algal zones on Fair Isle. *Journal of Ecology*, **42**, 283–8.

Bye, J. (1965). Wind-driven circulation in unstratified lakes. *Limnology and Oceanography*, **10**, 451–8.

Bye, J. A. T. and Heath, R. A. (1975). The New Zealand semi-diurnal tide. *Journal of Marine Research*, **33**. 423–42.

Carr, A. P. and Graff, J. (1982). The tidal immersion factor and shore platform development: discussion. *Transaction of the Institute of British Geographers*, N.S. **7**, 240–5.

Carter, W. E., Robinson, D. S., Pyle, T. E. and Diamante, J. (1986). The application of geodetic radio interferometric surveying to the monitoring of sea-level. *Geophysical Journal of the Royal Astronomical Society*, **87**, 3–13.

Cartwright, D. E. (1968). A unified analysis of tides and surges round north and east Britain. *Philosophical Transactions of the Royal Society of London*, **A263**, 1–55.

Cartwright, D. E. (1971). Tides and waves in the vicinity of Saint Helena. *Philosophical Transactions of the Royal Society of London*, **A270**, 603–49.

Cartwright, D. E. (1972). Secular changes in the oceanic tides at Brest, 1711–1936. *Geophysical Journal of the Royal Astronomical Society*, **30**, 433–49.

Cartwright, D. E. (1974). Years of peak astronomical tides. *Nature*, **248**, 656–7.

Cartwright, D. E. (1975). A subharmonic lunar tide in the seas off Western Europe. *Nature*, **257**, 277–80.

Cartwright, D. E. (1977). Ocean tides. *Reports on Progress in Physics*, **40**, 665–708.

Cartwright, D. E. (1978). Oceanic tides. *International Hydrographic Review*, **55**, 35–84. (The postscript updates Cartwright, 1977.)

Cartwright, D. E. (1985). Tidal prediction and modern time scales. *International Hydrographic Review*, **62**, 127–38.

Cartwright, D. E. and Alcock, G. A. (1981). On the precision of sea surface elevations and slopes from SEASAT altimetry of the northeast Atlantic Ocean, pp. 885–95, in *Oceanography from Space*, (ed. J. F. R. Gower). New York: Plenum Press. 978 pp.

Cartwright D. E. and Crease, J. (1963). A comparison of the geodetic reference levels of England and France by means of the sea surface. *Proceedings of the Royal Society of London*, **A273**, 558–80.

Cartwright D. E. and Edden, A. C. (1973). Corrected tables of tidal harmonics, *Geophysical Journal of the Royal Astronomical Society*, **33**, 253–64.

Cartwright D. E. and Tayler, R. J. (1971). New computations of the tide-generating potential. *Geophysical Journal of the Royal Astronomical Society*, **23**, 45–74.

Cartwright D. E., Edden, A. C., Spencer, R. and Vassie, J. M. (1980). The tides of the northeast Atlantic Ocean. *Philosophical Transactions of the Royal Society of London*, **A298**, 87–139.

Cartwright, D. E. and Zetler, B. D. (eds) (1985). Pelagic tidal constants (2). (Compiled by the IAPSO Advisory Committee on Tides and Mean Sea Level). *Publication Scientifique*, International Association for the Physical Sciences of the Ocean, IUGG, *No. 33*, 59 pp.

Chapman, S. and Lindzen, R. S. (1970). *Atmospheric Tides, Thermal and Gravitational.* Dordrecht: Reidel, 200 pp.

Charlier, R. H. (1982). *Tidal Energy.* New York: Van Nostrand, 351 pp.

Clark, J. A. (1980). The reconstruction of the Laurentide ice sheet of North America from sea level data: method and preliminary results. *Journal of Geophysical Research,* **85**, 4307–23.

Close, C. (1922). *The Second Geodetic Levelling of England and Wales 1912-1921.* London: HMSO, 62 pp. and plates.

Colman, J. (1933). The nature of the intertidal zonation of plants and animals. *Journal of the Marine Biological Association of the U.K.,* **18**, 435–76.

Cook A. H. (1973) *Physics of the Earth and Planets.* London: Macmillan. 316 pp.

Csanady, G. T. (1982). *Circulation in the Coastal Ocean.* Dordrecht: Reidel, 279 pp.

Currie, R. G. (1981). Amplitude and phase of the 11-year term in sea-level: Europe. *Geophysical Journal of the Royal Astronomical Society,* **67**, 547–56.

Darwin, G. H. (1911). *The Tides and Kindred Phenomena in the Solar System,* 3rd edn. London: John Murray. 437 pp.

Davies, A. M. and Furnes, G. K. (1980). Observed and computed M_2 tidal currents in the North Sea. *Journal of Physical Oceanography,* **10**, 237–57.

Davies, A. M. and Flather, R. A. (1987). On computing extreme meteorologically induced currents, with application to the north-west European continental shelf. *Continental Shelf Research,* **7**(7).

Davies, J. L. (1964). A morphogenic approach to world shorelines. *Zeitschrift für Geomorphologie,* **8**, 127–42.

Deacon, M. (1971). *Scientists and the Sea, 1650–1900.* London: Academic Press, 445 pp.

Defant, A. (1958). *Ebb and Flow.* Ontario: Ambassador Books, 121 pp.

Defant, A. (1961). *Physical Oceanography. Volume II.* Oxford: Pergamon Press, 598 pp.

Delta Committee, (1962). *Final Report.* State Printing and Publishing Office, The Hague, 100 pp.

Demerliac, A. (1974). Le niveau de la mer. Calcul du niveau moyen journalier. *Annals Hydrographiques,* 5 ser, **2**, 49–57.

Deutsches Hydrographisches Institut (1963). *Handbuch für das Rote Meer und der Golf Von Aden.* Nr. 2034.

Dijkzeul, J. C. M. (1984). Tide filters. *Journal of Hydraulic Engineering,* **110**, 981–7.

Disney, L. P. (1955). Tidal heights along the coasts of the United States. *Proceedings of the American Society of Civil Engineers,* **81**, 1–9.

Dixon, J. (1979). Apparent sea level slopes—Ireland. *Chartered Land Surveyor, Chartered Minerals Surveyor,* **1**, 46–50.

Doake, C. S. M. (1978). Dissipation of tidal energy by Antarctic ice shelves. *Nature,* **275**, 304–5.

Dobson, F., Hasse, L. and Davis, R. (1980). *Air-sea Interaction.* New York: Plenum, 801 pp.

Donovan, D. T. and Jones, E. J. W. (1979). Causes of world-wide changes in sea level. *Journal of the Geological Society of London,* **136**, 187–92.

Doodson, A. T. (1921). Harmonic development of the tide-generating potential. *Proceedings of the Royal Society of London,* **A100**, 305–29.

Doodson, A. T. and Warburg, H. D. (1941). *Admiralty Manual of Tides.* London: HMSO, 270 pp.

Doty, M. S. (1946). Critical tidal factors that are correlated with the vertical distribution of marine algae and other organisms along the Pacific Coast. *Ecology,* **27**, 315–28.

Dronkers, J. J. (1964). *Tidal Computations in Rivers and Coastal Waters.* Amsterdam: North-Holland, 518 pp.

Druehl, L. D. and Green, J. M. (1982). Vertical distribution of intertidal seaweeds as related to patterns of submersion and emersion. *Marine Ecology Progress Series,* **9**, 163–70.

Dyer, K. R. (1973). *Estuaries, a Physical Introduction.* London: John Wiley, 140 pp.

Dyer, K. R. (1986). *Coastal and Estuarine Sediment Dynamics.* Chichester: John Wiley, 342 pp.

Eccles, J. (1901). Account of the operation of the Great Trigonometrical Survey of India, XVI: Details of tidal observations, Dehra Dun: *Survey of India.* 152 pp.

Ellers, F. S. (1982). Advanced offshore oil platforms. *Scientific American,* **246**(4), 30–41.

Eltringham, S. K. (1971). *Life in Mud and Sand.* London: English Universities Press, 218 pp.

Emery, K. O. and Aubrey, D. G. (1985). Glacial rebound and relative sea levels in Europe from tide-gauge records. *Tectonophysics,* **120**, 239–55.

Enfield, D. B. and Allen, J. S. (1980). On the structure and dynamics of monthly sea level anomalies along the Pacific coast of North and South America. *Journal of Physical Oceanography,* **10**, 557–78.

Enfield, D. (1986). An overview of subinertial sea-level variability along the eastern Pacific boundary. *Geophysical Journal of the Royal Astronomical Society,* **87**(1), 119–20.

Etkins, R. and Epstein, E. S. (1982). The rise of global mean sea level as an indication of climate change. *Science*, **215**, 287–9.

Evans, G. (1965). Intertidal flat sediments and their environments of deposition in the Wash. *Quarterly Journal of the Geological Society of London*, **121**, 209–40.

Evans, J. J. and Pugh, D. T. (1982). Analysing clipped sea-level records for harmonic tidal constituents. *International Hydrographic Review*, **59**(2), 115–22.

Evans, P. R. (1979). Adaptations shown by foraging shorebirds by cyclical variations in the activity and availability of their intertidal invertebrate prey, pp. 357–66, in *Cyclic Phenomena in Marine Plants and Animals (ed. E. Naylor and R. G. Hartnoll)*. Oxford: Pergamon Press.

Feare, C. J. and Summers, R. W. (1985). Birds as predators on rocky shores, pp. 249–64, in *The Ecology of Rocky Coasts* (ed. P. G. Moore and R. Seed). London: Hodder & Stoughton, 467 pp.

Flather, R. A. (1976). A tidal model of the North-West European continental shelf. *Mémoires de la Société royal des Sciences de Liège*, **10**, 141–64.

Flather, R. A. (1979). Recent results from a storm surge prediction scheme for the North Sea, pp. 385–409, in *Marine forecasting* (ed. J. C. J. Nihoul). Amsterdam: Elsevier, 493 pp.

Flather, R. A. (1981). Practical surge prediction using numerical models, pp. 21–43 in *Floods Due to High Winds and Tides* (ed. D. H. Peregrine). London: Academic Press, 109 pp.

Flather, R. A. (1987) Estimates of extreme conditions of tide and surge using a numerical model of the north-west European continental shelf. *Estuarine, Coastal and Shelf Science*, **24**, 69–93.

Flemming, N. C. (1978). Holocene eustatic changes and coastal tectonics in the northeast Mediterranean: implications for models of crustal consumption. *Philosophical Transactions of the Royal Society of London*, **A289**, 405–458.

Fogg, G. E., Egan, B., Floodgate, G. D., Jones, D. A., Kassab, Y. K., Lochte, K., Rees, E. I. S., Scrope-Howe, S. and Turley, C. M. (1985). Biological studies in the vicinity of a shallow-sea tidal mixing front VII. The frontal ecosystems, *Philosophical Transactions of the Royal Society of London*, **B310**, 555–71.

Forbes, J. M. and Garrett, H. B. (1979). Theoretical studies of atmospheric tides. *Reviews of Geophysics and Space Physics*, **17**, 1951–81.

Foreman, M. G. G. (1977). Manual for tidal heights analysis and prediction. *Canadian Pacific Marine Science Report No. 77-10*, 10 pp.

Forrester, W. D. (1983). *Canadian Tidal Manual*. Ottawa, Canada: Department of Fisheries and Oceans, 138 pp.

Franco, A. dos S. (1981). *Tides: Fundamentals, Analysis and Prediction*. Instituto de Pesquisas Tecnologicas do Estado de São Paulo, Publication No. 1182, 232 pp.

Gallagher, B. S. and Munk, W. H. (1971). Tides in shallow water: spectroscopy. *Tellus*, **23**, 346–63.

Garland, G. D. (1965). *The Earth's Shape and Gravity*. Oxford: Pergamon Press, 183 pp.

Garrett, C. J. (1984). Tides and tidal power in the Bay of Fundy. *Endeavour*, **8**, 58–64.

Garrett, C. J. and Toulany, B. (1982). Sea level variability due to meteorological forcing in the north east Gulf of St. Lawrence. *Journal of Geophysical Research*, **87**, 1968–78.

Gelfenbaum, G. (1983). Suspended-sediment response to semidiurnal and fortnightly tidal variations in a mesotidal estuary: Columbia River, USA. *Marine Geology*, **52**, 39–57.

Gilbert, S. and Horner, R. (1984). *The Thames Barrier*. London: Thomas Telford, 182 pp.

Gill, A. E. (1979). A simple model for showing the effects of geometry on the ocean tides. *Proceedings of the Royal Society of London*, **A367**, 549–71.

Gill, A. E. (1982). *Atmosphere-ocean Dynamics*. London: Academic Press, 662 pp.

Godin, G. (1972). *The Analysis of Tides*. Liverpool University Press, 264 pp.

Godin, G. (1980). *Cotidal Charts for Canada*. Canada, Marine Sciences and Information Directorate, Manuscript Report Series, No. 55, 93 pp.

Goldreich, P. (1972). Tides and the earth–moon system. *Scientific American*, **226**(4), 42–52.

Gordon, C. M. (1975). Sediment entrainment and suspension in a turbulent tidal flow. *Marine Geology*, **18**, M57–64.

Gornitz, V., Lebedeff, S. and Hansen, J. (1982). Global sea level trends in the past century. *Science*, **215**, 1611–14.

Graber, P. H. F. (1980). The law of the coast in a clamshell, Part I: Overview of an interdisciplinary approach. *Shore and Beach*, **48**(4), 14–20.

Graff, J. (1981). An investigation of the frequency distributions of annual sea level maxima at ports around Great Britain. *Estuarine, Coastal and Shelf Science*, **12**, 389–449.

Gray, T. J. and Gashus, O. K. (eds) (1972). *Tidal Power*. New York: Plenum Press, 630 pp.

Great Britain, Department of Energy (1984). *Offshore Installations; Guidance on Design and Construction*, 3rd edn. London: Her Majesty's Stationery Office, 188 pp.

448

Great Britain, Hydrographic Department. *Annual Tide Tables*. 3 vols.
Great Britain, Hydrographic Department (1940). *Atlas of Tides and Tidal Streams, British Islands and adjacent waters* (2nd edn.) London: Hydrographer of the Navy.
Great Britain, Ministry of Agriculture, Fisheries and Food, Committee on Tide Gauges (1979). *Operating Instructions for Tide Gauges on the National Network*, 22 pp.
Great Britain, Hydrographic Department (1986). The Admiralty Method of Harmonic Tidal Analysis. *N. P. 122(1) for Long Period Observations*, 137 pp. *N. P. 122(3) for Short Period Observations*, 45 pp. London: Hydrographer of the Navy.
Greensmith, J. T. and Tooley, M. J. (eds) (1982). I.G.C.P. Project 61. Sea-level movements during the last deglacial hemicycle (about 15 000 years). Final Report of the United Kingdom Working Group. *Proceedings of the Geologists' Association*, **93**(1), 1–125.
Greer Walker, M., Harden Jones, F. R. and Arnold, G. P. (1978). The movements of plaice (*Pleuronectes platessa* L.) tracked in the open sea, *Journal du Conseil International pour l'Exploration de la Mer*, **38**, 58–86.
Gutenberg, B. (1941). Changes in sea level, postglacial uplift, and mobility of the earth's interior. *Bulletin of the Geological Society of America*, **52**, 721–72.
Harris, R. A. (1897–1907). *Manual of Tides. Appendices to Reports of the U.S. Coast and Geodetic Survey*. Washington: Government Printing Office.
Harris, D. L. (1981). *Tides and Tidal Datums in the United States*. United States Army, Corps of Engineers, Coastal Engineering Research Center, Special Report No. 7, 382 pp.
Hayden, B. P. and Dolan, R. (1979). Barrier islands, lagoons and marshes, *Journal of Sedimentary Petrology*, **49**, 1061–72.
Hayes, M. O. (1980). General morphology and sediment patterns in tidal inlets. *Sedimentary Geology*, **26**, 139–56.
Heaps, N. S. (1967). Storm surges. *Oceanography and Marine Biology—an Annual Review*, **5**, 11–47.
Heaps, N. S. (1983). Storm surges, 1967-1982. *Geophysical Journal of the Royal Astronomical Society*, **74**, 331–76.
Heathershaw, A. D. (1981). Comparison of measured and predicted sediment transport rates in tidal currents. *Marine Geology*, **42**, 75–104.
Heathershaw, A. D., Carr, A. P., Blackley, M. W. L. and Wooldridge, C. F. (1981). Tidal variations in the compaction of beach sediments. *Marine Geology*, **41**, 223–38.
Heathershaw, A. D., New, A. L. and Edwards, P. D. (1987). Internal tides and sediment transport at the shelf-break in the Celtic Sea. *Continental Shelf Research*, **7**, 485–517.
Henderson, F. M. (1964). *Open Channel Flow*. New York: Macmillan, 552 pp.
Hendry, R. M. (1977). Observations of the semidiurnal internal tide in the Western North Atlantic Ocean. *Philosophical Transactions of the Royal Society of London*, **A286**, 1–24.
Hicks, S. D. (1978). An average geopotential sea level series for the United States. *Journal of Geophysical Research*, **83**, 1377–9.
Hicks, S. D. (1984). *Tide and Current Glossary*. United States Dept. of Commerce, National Oceanic and Atmospheric Administration, National Ocean Service, 28 pp.
Hicks, S. D. (1985). Tidal datums and their uses—a summary. *Shore and Beach*, **53**(1), 27–32.
Hide, R. (1984). Rotation of the atmospheres of the Earth and planets. *Philosophical Transactions of the Royal Society of London*, **A313**, 107–21.
Holligan, P. M. (1981). Biological implications of fronts on the northwest European continental shelf. *Philosophical Transactions of the Royal Society of London*, **A302**, 547–62.
Holligan, P. M., Pingree, R. D. and Mardell, G. T. (1985). Oceanic solitons, nutrient pulses and phytoplankton growth. *Nature*, **314**, 348–50.
Holman, R. A. and Bowen, A. J. (1982). Bars, bumps and holes: models for the generation of complex beach topography. *Journal of Geophysical Research*, **87**, 457–68.
Holme, N. A. and Wilson, J. B. (1985). Faunas associated with longitudinal furrows and sand ribbons in a tide-swept area in the English Channel, *Journal of the Marine Biological Association of the U.K.*, **65**, 1051–72.
Horner, R. W. (1985). The Thames Barrier, *Proceedings of the Institution of Civil Engineers*, Part 1, **78**, 15–25.
Howarth, M. J. (1982). Tidal currents of the continental shelf, pp. 10–26, in *Offshore Tidal Sands* (ed. A. H. Stride). London: Chapman & Hall, 222 pp.
Howarth, M. J. and Pugh, D. T. (1983). Observations of tides over the continental shelf of northwest Europe, pp. 135–85, in *Physical Oceanography of Coastal and Shelf Seas* (ed. B. Johns). Amsterdam: Elsevier, 470 pp.
Howarth, M. J. and Huthnance, J. M. (1984). Tidal and residual currents around a Norfolk Sandbank. *Estuarine, Coastal and Shelf Science*, **19**, 105–17.

Howse, H. D. (1985). Some early tidal diagrams. *Revista da Universidade de Coimbra*, **33**, 365–85.

Huntley, D. A. (1980). Tides on the north-west European continental shelf. In *The North West European Shelf Seas: the Sea Bed and the Sea in Motion. II. Physical and Chemical Oceanography, and Physical Resources* (ed. F. T. Banner, M. B. Collins and K. S. Massie). Amsterdam: Elsevier, pp. 301–51.

Huthnance, J. M. (1982). On the formation of sand banks of finite extent. *Estuarine, Coastal and Shelf Science*, **15**, 277–99.

IAPSO (1985). Changes in relative mean sea level. Working Party Report of the International Association for the Physical Sciences of the Ocean (Chairman: D. E. Cartwright). *EOS, Transactions of the American Geophysical Union*, **66**, 754–6.

International Hydrographic Bureau (1930). *Tides, Harmonic Constants*. International Hydrographic Bureau, Special Publication No. 26 and addenda.

International Hydrographic Organization (1974). *Hydrographic Dictionary*, Part 1. (3rd edn.), No. 32, 370 pp. and appendix (English and French).

Intergovernmental Oceanographic Commission, UNESCO. (1985). *Manual on Sea Level Measurement and Interpretation*, UNESCO, IOC Manuals and Guides, No. 14, 83 pp. (available in English, French, Russian and Spanish).

James, I. D. (1983). The effects of wind waves on sea level at the coast. *Institute of Oceanographic Sciences Report, No. 155*, 66 pp.

Jarvinen, B. R. and Lawrence, M. B. (1985). An evaluation of the SLOSH storm-surge model. *Bulletin of the American Meteorological Society*, **66**, 1408–11.

Jeffreys, H. (1976). *The Earth, its Origin, History and Physical Constitution*, 6th edn. Cambridge University Press, 574 pp.

Jelesnianski, C. P. (1972). SPLASH (Special Program to List Amplitudes of Surges from Hurricanes); I. Landfall Storms, *NOAA Technical Memorandum. NWS TDL-46*, 52 pp.

Jelesnianski, C. P. (1978). Storm surges. *Geophysical Predictions*. Washington D.C.: National Academy of Sciences, pp. 185–92.

Jelesnianski, C. P., Chen, J., Shaffer, W. A. and Gilad, A. J. (1984) SLOSH—a hurricane storm surge forecasting model, pp. 314–17 in *Ocean 84*. New York: Institute of Electrical and Electronic Engineers, 1050 pp.

Jones C. W. (1943). *Bedae. Opera de Temporibus*. Massachusetts: The Mediaeval Academy of America, 416 pp.

Jones, P. D., Wigley, T. M. L. and Wright, P. B. (1986). Global temperature variations between 1861 and 1984. *Nature*, **322**, 430–4.

Karunaratne, D. A. (1980). An improved method for interpolating and smoothing hourly sea level data. *International Hydrographic Review*, **57**, 135–48.

Kasahara, K. (1981). *Earthquake Mechanics*. Cambridge University Press, 248 pp.

Kaula, W. M. (1968). *Introduction to Planetary Physics: the Terrestrial Planets*. London: John Wiley, 490 pp.

Kinsman, B. (1965). *Wind Waves: their Generation and Propagation on the Ocean Surface*, New Jersey: Prentice-Hall, 676 pp.

Kirby, R. and Parker, W. R. (1982). A suspended sediment front in the Severn Estuary. *Nature*, **295**, 396–9.

Komar, P. D. (1976). *Beach Processes and Sedimentation*. New Jersey: Prentice-Hall, 464 pp.

Komar, P. D. (ed.) (1983). *CRC Handbook of Coastal Processes and Erosion*. Florida: CRC Press, 305 pp.

Kowalik, Z. (1981). A study of the M_2 tide in the ice-covered Arctic Ocean. *Modelling, Identification and Control*, **2**, 201–23.

Kreeke van de, J. (1985). Stability of tidal inlets—Pass Cavallo, Texas. *Estuarine, Coastal and Shelf Science*, **21**, 33–43.

Ku, L. F., Greenberg, D. A., Garrett, C. J. R. and Dobson, F. W. (1985). Nodal modulation of the lunar semidiurnal tide in the Bay of Fundy and Gulf of Maine. *Science*, **230**, 69–71.

Lamb, H. (1932). *Hydrodynamics*. 6th edn. Cambridge University Press, 738 pp.

Lamb, H. H. (1980). Climate fluctuations in historical times and their connection with transgressions of the sea, storm floods and other coastal changes, pp. 251–90. In *Transgressies en occupatiegeschiedenis in de Kustgebieden van Nederland en Belgie* (ed. A. Verhulst and M. K. E. Gottschalk). Belgisches Centrum voor Landelijke Geschiedenis, Pub. No. 66.

Lambeck, K. (1980). *The Earth's Variable Rotation: Geophysical Causes and Consequences*. Cambridge University Press, 449 pp.

Larsen, L. H., Cannon, G. A. and Choi, B. H. (1985). East China Sea tidal currents. *Continental Shelf Research*, **4**, 77–103.

Lawson, G. W. (1966). The littoral ecology of west Africa, pp. 405–48, in *Oceanography and Marine Biology, Annual Review, Vol. 4* (ed. H. Barnes). London: Allen & Unwin, 505 pp.

LeBlond, P. H. (1978). On tidal propagation in shallow rivers. *Journal of Geophysical Research*, **83**, 4717–21.

LeBlond, P. H. (1979). Forced fortnightly tides in shallow rivers. *Atmosphere-Ocean*, **17**, 253–264.

Leentvaar, J. and Nijboer, S. M. (1986). Ecological impacts of the construction of dams in an estuary. *Water Science and Technology*, **18**, 181–91.

Lees, B. J. (1983). The relationship of sediment transport rates and paths to sandbanks in a tidally dominated area off the coast of East Anglia, U.K. *Sedimentology*, **30**, 461–83.

Lennon, G. W. (1963a). The identification of weather conditions associated with the generation of major storm surges along the west coast of the British Isles. *Quarterly Journal of the Royal Meteorological Society*, **89**, 381–94.

Lennon, G. W. (1963b). A frequency investigation of abnormally high tidal levels at certain west coast ports. *Proceedings of the Institution of Civil Engineers*, **25**, 451–84.

Lennon, G. W. (1971). Sea level instrumentation, its limitations and the optimisation of the performance of conventional gauges in Great Britain, *International Hydrographic Review*, **48**(2) 129–47.

Lerch, F. J., Klosko, S. M., Laubscher, R. E. and Wagner, C. A. (1979). Gravity model improvement using GEOS-3 (Gem 9 and 10). *Journal of Geophysical Research*, **84**, 3897–3916.

Lewis, J. R. (1964). *The Ecology of Rocky Shores*. London: English Universities Press, 323 pp.

Lipa, B. J. and Barrick, D. E. (1986). Tidal and storm surge measurements with single-site CODAR. *IEEE Journal of Oceanic Engineering*, OE-11, 241–5.

Lisitzin, E. (1974). *Sea Level Changes*. Amsterdam: Elsevier, 286 pp. (Volume 8 of 'Elsevier Oceanography Series'.)

Lobban, G. S., Harrison, P. J. and Duncan, M. J. (1985). *The Physiological Ecology of Sea Weeds*. Cambridge University Press, 242 pp.

LoCicero, V. R. (ed.) (1975). *Proceedings of the First International Conference on Toxic Dinoflagellate Blooms*. Massachusetts: The Massachusetts Science and Technology Foundation.

Loder, J. W. and Garrett, C. (1978). The 18.6-year cycle of sea-surface temperature in shallow seas due to tidal mixing. *Journal of Geophysical Research*, **83**, 1967–70.

Loomis, H. G. (1978). Tsunami, pp. 155–65, in *Geophysical Predictions*, Washington D.C.: National Academy of Sciences, 215 pp.

Longuet-Higgins, M. S. and Stewart, R. W. (1964). Radiation stresses in water waves; a physical discussion, with applications. *Deep-Sea Research*, **11**, 529–62.

Lynch, D. K. (1982). Tidal bores. *Scientific American*, **247**(4), 134–43.

McCave, I. N. (1978). Grain-size trends and transport along beaches: example from eastern England. *Marine Geology*, **28**, M43–51.

McDowell, D. M. and O'Connor, B. A. (1977). *Hydraulic Behaviour of Estuaries*. London: Macmillan, 292 pp.

McKinley, I. G., Baxter, M. S., Ellett, D. J. and Jack, W. (1981). Tracer applications of radiocaesium in the Sea of the Hebrides. *Estuarine, Coastal and Shelf Science*, **13**, 69–82.

McLean, S. R. (1981). The role of non-uniform roughness in the formation of sand ribbons. *Marine Geology*, **42**, 269–89.

Macmillan, D. H. (1966). *Tides*. London: CR Books, 240 pp.

Marchuk, G. I. and Kagan, B. A. 1984. *Ocean Tides* (*Mathematical models and numerical experiments*). Oxford: Pergamon Press, 292 pp.

Mardell, G. T. and Pingree, R. D. (1981). Half-wave rectification of tidal vorticity near headlands as determined from current meter measurements. *Oceanologica Acta*, **4**, 63–8.

Marmer, H. A. (1926). *The Tide*. New York: D. Appleton and Company, 282 pp.

Marmer, H. A. (1928). On cotidal maps. *Hydrographic Review*, **5**(2), 195–205.

Marsh J. G., Brenner, A. C., Beckley, B. D. and Martin, T. V. (1986). Global mean sea surface based upon the SEASAT altimeter data. *Journal of Geophysical Research*, **91**, 3501–6.

Maul, G. A. (1986). Linear correlations between Florida Current volume transport and surface speed with Miami sea-level and weather during 1964–70. *Geophysical Journal of the Royal Astronomical Society*, **87**, 55–66.

Maul, G. A., Chew, F., Bushnell, M. and Mayer, D. A. (1985). Sea level variation as an indicator of Florida current volume transport: comparisons with direct measurements. *Science*, **227**, 304–7.

Meier, M. F. (1984). Contribution of small glaciers to global sea level, *Science*, **226**, 1418–21.

Middleton, J. F. and Thompson, K. R. (1986). Return periods of extreme sea levels from short records. *Journal of Geophysical Research*, **91**, 11707–16.

Miller, G. R. (1966). The flux of tidal energy out of the deep oceans. *Journal of Geophysical Research*, **71**, 2485–9.

Miller, S. P. and Wunsch, C. (1973). The pole tide. *Nature Physical Science*, **246**, 98–102.

Mooney, M. J. (1980). Maelstrom, the legend and the reality. *Compass*, **2**, 31–5.

Moore, P. G. and Seed, R. (1985). *The Ecology of Rocky Coasts*. London: Hodder & Stoughton, 467 pp.

Morner, N. A. (1973). Eustatic changes during the last 300 years. *Palaeogeography, Palaeoclimatology, Palaeoecology*, **13**, 1–14.

Morner, N. A. (ed.) (1980). *Earth Rheology, Isostasy and Eustasy*. Chichester: John Wiley, 599 pp.

Muir Wood, A. M. and Flemming, C. A. (1981). *Coastal Hydraulics*. London: Macmillan, 280 pp.

Muller, P. M. and Stephenson, F. R. (1975). The accelerations of the Earth and Moon from early astronomical observations, pp. 459–534, in *Growth Rhythms and History of the Earth's Rotation* (ed. G. D. Rosenberg and S. K. Runcorn). New York: Wiley, 559 pp.

Munk, W. H. and Cartwright, D. E. (1966). Tidal spectroscopy and prediction. *Philosophical Transactions of the Royal Society of London*, **A259**, 533–81.

Munk, W., Snodgrass, F. and Wimbush, M. (1970). Tides off-shore: transition from California coast to deep-sea waters. *Geophysical Fluid Dynamics*, **1**, 161–235.

Murty, T. S. (1977). *Seismic Sea Waves and Tsunamis*. Ottawa, Canada: Department of Fisheries and the Environment, Fisheries and Marine Service. 337 pp. (Bulletin No. 198).

Murty, T. S. (1984). *Storm Surges—Meteorological Ocean Tides*. Ottawa, Canada: Department of Fisheries and Oceans. 897 pp. (Canadian Bulletin of Fisheries and Aquatic Sciences No. 212).

Murty, T. S., Flather, R. A. and Henry, R. F. (1986). The storm surge problem in the Bay of Bengal. *Progress in Oceanography*, **16**, 195–233.

Mysak, L. A., Hsieh, W. W. and Parsons, T. R. (1982). On the relationships between interannual baroclinic waves and fish populations in the northeast pacific. *Biological Oceanography*, **2**, 63–103.

Natural Environment Research Council (1975). *Flood Studies Report, Volume 1: Hydrological Studies*. London: Natural Environment Research Council, 550 pp.

Nature Conservancy Council and Natural Environment Research Council (1979). *Nature Conservation in the Marine Environment*, 65 pp.

Naylor, E. (1982). Tidal and lunar rhythms in animals and plants, pp. 33–48, in *Biological Timekeeping* (ed. J. Brady). Cambridge University Press.

Naylor, E. (1985). Tidally rhythmic behaviour of marine animals. *Symposia of the Society of Experimental Biology*, **39**, 63–93.

Naylor, E. and Hartnoll, R. G. (ed.) (1979). *Cyclic Phenomena in Marine Plants and Animals*. Oxford: Pergamon Press, 477 pp.

Neumann, D. (1981). Tidal and lunar rhythms. pp. 351–80 in *Handbook of behavioural neurobiology. IV. Biological rhythms* (ed. J. Aschoff). New York: Plenum Press, 582 pp.

Neumann, G. (1968). *Ocean Currents*. Amsterdam: Elsevier, 352 pp.

Det Norske Veritas (1977). *Rules for the design, construction and inspection of Offshore Structures*, 67 pp. and appendices.

Norton, T. A. (1985). The zonation of seaweed on rocky shores, pp. 7–21 in *The Ecology of Rocky Coasts* (ed. P. G. Moore and R. Seed). London: Hodder & Stoughton, 467 pp.

Noye, B. J. (1974). Tide-well systems: 3. Improved interpretation of tide-well records. *Journal of Marine Research*, **32**, 183–94.

Nybakken, J. W. (1982). *Marine Biology—an Ecological approach*. Harper & Row, 446 pp.

Oberkommando der Kreigsmarine (1942). Karten der Harmonischen Gezeitkonstanten für das Gebeit der Nordsee. *Marineobservatorium Wilhelmshafen*, Ausgabe A, Nr. 2752.

O'Brien, M. P. (1931). Estuary tidal prisms related to entrance areas. *Civil Engineering*, **1**, 738–9.

Owen, M. W. (1983). The hydraulic design of sea-wall profiles, pp. 185–92, in *Shoreline Protection*. London: Thomas Telford, 248 pp.

Palmer, H. R. (1831). Description of graphical register of tides and winds. *Philosophical Transactions of the Royal Society of London*, **121**, 209–13.

Palmer, J. D. (1974). *Biological Clocks in Marine Organisms: the Control of Physiological and Behavioural Tidal Rhythms*. Wiley: Interscience, 173 pp.

Panikkar, N. K. and Srinivasan, T. M. (1971). The concept of tides in ancient India. *Indian Journal of the History of Science*, **6**, 36–50.

Parke, M. E. (1982). O_1, P_1, N_2 models of the global ocean tide on an elastic earth plus surface potential and spherical harmonic decompositions for M_2, S_2, and K_1. *Marine Geodesy*, **6**, 35–81.

Parke, M. E. and Rao, D. B. (eds) (1983). *Report of NASA Workshop on Tidal Research, 1982.* California Institute of Technology, Jet Propulsion Laboratory, JPL Publication 83-71, 51 pp.

Pattiaratchi, C. B. and Collins, M. B. (1985). Sand transport under the combined influence of waves and tidal currents: an assessment of available formulae. *Marine Geology*, **67**, 83–100.

Pattiaratchi, C. B., Hammond, T. M. and Collins, M. B. (1986). Mapping of tidal currents in the vicinity of an offshore sandbank, using remotely sensed imagery. *International Journal of Remote Sensing*, **7**, 1015–29.

Pattullo, J. G. (1963). Seasonal changes in sea-level, pp. 485–96 in vol. 2, *The Sea* (ed. M. N. Hill). New York: Wiley–Interscience. 554 pp.

Pattullo, J. G., Munk, W. H., Revelle, R. and Strong, E. (1955). The seasonal oscillation of sea level. *Journal of Marine Research*, **14**, 88–155.

Peltier, W. R., Drummond, R. A. and Tushingham, A. M. (1986). Post-glacial rebound and transient lower mantle rheology. *Geophysical Journal of the Royal Astronomical Society*, **87**, 79–116.

Peltier, W. R., Farrell, W. E. and Clark, J. A. (1978). Glacial isostasy and relative sea level: a global finite element model. *Tectonophysics*, **50**, 81–110.

Peregrine, D. H. (1976). Interaction of water waves and currents, pp. 9–117 in *Advances in Applied Mechanics*, 16. New York: Academic Press.

Pillsbury, G. B. (1956). *Tidal Hydraulics*. United States Army, Corps of Engineers, 247 pp.

Pingree, R. D. (1978). The formation of the Shambles and other banks by the tidal stirring of the seas. *Journal of the Marine Biological Association of the UK*, **58**, 211–26.

Pingree, R. D. and Griffiths, D. K. (1979). Sand transport paths around the British Isles resulting from M_2 and M_4 tidal interactions. *Journal of the Marine Biological Association of the UK*, **59**, 497–513.

Pingree, R. D. and Griffiths, D. K. (1981). S_2 tidal simulations on the north-west European shelf. *Journal of the Marine Biological Association of the UK*, **61**, 609–16.

Pingree, R. D., Griffiths, D. K. and Mardell, G. T. (1983). The structure of the internal tide at the Celtic Sea shelf break. *Journal of the Marine Biological Association of the UK*, **64**, 99–113.

Pingree, R. D. and Maddock L. (1978). The M_4 tide in the English Channel derived from a non-linear numerical model of the M_2 tide. *Deep-Sea Research*, **25**, 53–63.

Pingree, R. D. and Maddock, L. (1979). The tidal physics of headland flows and offshore tidal bank formation. *Marine Geology*, **32**, 269–89.

Pingree, R. D. and Maddock L. (1985). Stokes, Euler and Lagrange aspects of residual tidal transports in the English Channel and the Southern Bight of the North Sea. *Journal of the Marine Biological Association of the UK*, **65**, 969–82.

Pingree, R. D., Mardell, G. T. and Maddock, L. (1985). Tidal mixing in the Channel Isles region derived from the results of remote sensing and measurements at sea. *Estuarine, Coastal and Shelf Science*, **20**, 1–18.

Pingree, R. D., Mardell, G. T. and New, A. L. (1986). Propagation of internal tides from the upper slopes of the Bay of Biscay. *Nature*, **321**, 154–8.

Pingree, R. D., Pugh, P. R., Holligan, P. M. and Forster, G. R. (1975). Summer phytoplankton blooms and red tides along tidal fronts in the approaches to the English Channel. *Nature*, **258**, 672–7.

Platzman, G. W. (1971). Ocean tides and related waves, pp. 239–91, in *Mathematical Problems in the Geophysical Sciences* (ed. W. H. Reid), Vol. 2. Providence, Rhode Island: American Mathematical Society, 370 pp.

Platzman, G. W. (1983). World ocean tides synthesised from normal modes. *Science*, **220**, 602–4.

Platzman, G. W. (1984). Normal modes of the World Ocean. Part IV: synthesis of diurnal and semidiurnal tides. *Journal of Physical Oceanography*, **14**,1532–50.

Pond, S. and Pickard, G. L. (1978). *Introductory Dynamic Oceanography*. Oxford: Pergamon Press, 241 pp.

Prandle, D. (1982). The vertical structure of tidal currents and other oscillatory flows. *Geophysical and Astrophysical Fluid Dynamics*, **22**, 24–49.

Prandle, D. (1987) The fine-structure of near-shore tidal and residual circulations revealed by HF radar surface current measurements. *Journal of Physical Oceanography*, **17**, 231–245.

Prandle, D. and Wolf, J. (1978). The interaction of surge and tide in the North Sea and River Thames. *Geophysical Journal of the Royal Astronomical Society*, **55**, 203–16.

Price, J. and Karim, I. (1978). Mitiruku, a Fijian madness: an initial assessment. *British Journal of Psychiatry*, **133**, 228–30.

Price, W. A. and Kendrick, M. P. (1963). Field and model investigations into the reasons for siltation in the Mersey Estuary. *Proceedings of the Institute of Civil Engineers*, **24**, 473–518.

Price, W. A., Motyka, J. M. and Jaffrey, L. J. (1978). The effect of off-shore dredging on coastlines. *Proceedings of 16th Coastal Engineering Conference*, 1978, Vol. 2. 1347–58. New York: American Society of Civil Engineers, 3060 pp.

Proudman, J. (1925). A theorem in tidal dynamics. *Philosophical Magazine*, **49**, 570–79.

Proudman, J. (1942). On Laplace's differential equations for the tides. *Proceedings of the Royal Society of London*, **A179**. 261–88.

Proudman, J. (1953). *Dynamical Oceanography*. London: Methuen and Co., 409 pp.

Proudman, J. and Doodson, A. T. (1924). The principal constituents of the tides of the North Sea. *Philosophical Transactions of the Royal Society of London*, **A224**, 185–219.

Pugh, D. T. (1972). The physics of pneumatic tide gauges. *International Hydrographic Review*, **49**(2), 71–97.

Pugh, D. T. (1979). Sea levels at Aldabra atoll, Mombasa and Mahé, Western equatorial Indian Ocean, related to tides, meteorology and ocean circulation, *Deep-Sea Research*, **26**, 237–58.

Pugh, D. T. (1981). Tidal amphidrome movement and energy dissipation in the Irish Sea. *Geophysical Journal of the Royal Astronomical Society*, **67**, 515–27.

Pugh, D. T. (1982a). Estimating extreme currents by combining tidal and surge probabilities. *Ocean Engineering*, **9**, 361–72.

Pugh, D. T. (1982b). A comparison of recent and historical tides and mean sea-levels off Ireland. *Geophysical Journal of the Royal Astronomical Society*, **71**, 809–15.

Pugh, D. T. and Faull, H. E. (1983). Tides, surges and mean sea level trends, pp. 59–69, in *Shoreline Protection*. London: Thomas Telford, 248 pp.

Pugh D. T. and Rayner, R. F. (1981). The tidal regimes of three Indian Ocean atolls and some ecological implications. *Estuarine, Coastal and Shelf Science*, **13**, 389–407.

Pugh, D. T. and Thompson, K. R. (1986). The subtidal behaviour of the Celtic Sea—1. Sea level and bottom pressures. *Continental Shelf Research*, **5**, 293–319.

Pugh, D. T. and Vassie, J. M. (1976). Tide and surge propagation off-shore in the Dowsing region of the North Sea. *Deutsche Hydrographische Zeitschrift*, **29**, 163–213.

Pugh, D. T. and Vassie, J. M. (1980). Applications of the joint probability method for extreme sea level computations. *Proceedings of the Institution of Civil Engineers*, **9**, 361–72.

Rebert, J. P., Donguy, J. R., Eldin, G. and Wyrtki, K. (1985). Relations between sea level, thermocline depth, heat content and dynamic height in the tropical Pacific Ocean. *Journal of Geophysical Research*, **90**, 11719–25.

Redfield, A. C. (1958). The influence of the continental shelf on the tides of the Atlantic coast of the United States. *Journal of Marine Research*, **17**, 432–58.

Redfield, A. C. (1980). *The Tides of the Waters of New England and New York*. Woods Hole: Woods Hole Oceanographic Institution, 108 pp.

Reid, J. L. (1961). On the temperature, salinity and density difference between the Atlantic and Pacific oceans in the upper kilometre. *Deep-Sea Research*, **7**, 265–75.

Revelle, R. R. (1983). Probable future changes of sea level resulting from increased atmospheric carbon dioxide, Chapter 8 (pp. 433–48), in *Changing Climate*. Washington D.C. National Academy Press.

Richardson, C. A., Crisp, D. J. and Runham, N. W. (1980a). Factors influencing the shell growth in *Cerastoderma edule*. *Proceedings of the Royal Society of London*, **B210**, 513–31.

Richardson, C. A., Crisp, D. J., Runham, N. W. and Gruffydd, Ll. D. (1980b). The use of tidal growth bands in the shell of *Cerastoderma edule* to measure seasoned growth rates under cool temperate and subarctic conditions. *Journal of the Marine Biological Association of the UK*, **60**, 977–89.

Ricketts, E. F. and Calvin, J. (1968). *Between Pacific Tides*, 4th rev. edn. Stanford University Press, 614 pp.

Robin, G. de Q. (1986). Changing the sea level, pp. 323–359 in *The Greenhouse Effect, Climatic Change and Ecosystems* (eds B. Bolin, B. R. Doos, J. Jager and R. A. Warrick). Scope series No. 29. Chichester: John Wiley.

Robinson, I. S. (1979). The tidal dynamics of the Irish and Celtic Seas. *Geophysical Journal of the Royal Astronomical Society*, **56**, 159–97.

Robinson, I. S. (1983). Tidally induced residual flows, pp. 321–56, in *Physical Oceanography of Coastal and Shelf Seas* (ed. B. Johns). Amsterdam: Elsevier, 470 pp.

Robinson, I. S. (1985). *Satellite Oceanography*. Chichester: Ellis Horwood, 455 pp.

Ross, J. C. (1847). *A Voyage of Discovery and Research in the Southern and Antarctic Regions, during the years 1839–43*. London: John Murray, 447 pp.

Ross, J. C. (1854). On the effect of the pressure of the atmosphere on the mean level of the ocean. *Philosophical Transactions of the Royal Society of London*, **144**, 285–96.

Rossiter, J. R. (1958). Note on methods of determining monthly and annual values of mean water level. *International Hydrographic Review*, **35**, 105–15.

Rossiter, J. R. (1961). A routine method of obtaining monthly and annual means of sea level from tidal records. *International Hydrographic Review*, **38**, 42–5.

Rossiter, J. R. (1962). Long-term variations in sea level, pp. 590–610, in vol. 1, *The Sea* (ed. M. N. Hill). New York: Wiley–Interscience, 864 pp.

Rossiter, J. R. (1967). An analysis of annual sea level variations in European waters. *Geophysical Journal of the Royal Astronomical Society*, **12**, 259–99.

Rowbotham, F. W. (1983). *The Severn Bore*. Devon: David & Charles, 104 pp.

Roy, A. E. (1978). *Orbital Motion*. Bristol: Adam Hilger, 489 pp.

Sager, G. and Sammler, R. (1975). *Atlas der Gezeitestrome für die Nordsee, den Kanal und die Irische See*. Deutschen Demokratischen Republic. Seehydrographischer Dienst, Nr. 8736. 3rd edn, ix, 58 pp.

Sandstrom, H. (1980). On the wind-induced sea level changes on the Scotian Shelf. *Journal of Geophysical Research*, **85**, 461–8.

Sarpakaya, T. and Isaacson, M. (1981). *Mechanics of Wave Forces on Off Shore Structures*. New York: Van Nostrand Reinhold, 651 pp.

Schonbeck, M. and Norton, T. A. (1978). Factors controlling the upper limits of fucoid algae on the shore. *Journal of Experimental Marine Biology and Ecology*, **31**, 303–13.

Schureman, P. (1976). *Manual of Harmonic Analysis and Prediction of Tides*. United States Government Printing Office, Washington, 317 pp. (1st edn, 1924; reprinted 1940, 1958, 1976).

Schwiderski, E. W. (1979). *Global Ocean Tides, Part 2: the Semidiurnal Principal Lunar Tide* (M_2), *Atlas of Tidal Charts and Maps* (revised version). United States Naval Surface Weapons Center, Technical Report Center, NSWC-TR 79-414, 87 pp.

Schwiderski, E. W. (1983). Atlas of ocean tidal charts and maps. Part 1: the semidiurnal principal lunar tide M_2. *Marine Geodesy*, **6**(3-4), 219–65.

Seelig, W. N. (1977). Stilling well design for accurate water level measurement. United States Army, Coastal Engineering Research Center, Technical Paper No. 77-2, 21 pp.

Shaffer, W. A., Jelesnianski, C. P. and Chen, J. (1986). Hurricane storm surge forecasting. *Reprints Oceans* 86. Washington D.C., Marine Technology Society and IEEE/Oceanic Engineering Society.

Shalowitz, A. L. (1962). *Shore and Sea Boundaries: Vol. 1*, 420 pp; (1964), *Vol. 2*, 749 pp. US Dept of Commerce, Coast and Geodetic Survey.

Shearman, E. D. R. (1986). A review of methods of remote sensing of sea-surface conditions by HF radar and design considerations for narrow-beam systems. *IEEE Journal of Oceanic Engineering*, OE-11, 150–7.

Silvester, R. (1974). *Coastal Engineering*. Amsterdam: Elsevier, 2 Vols.

Simpson, J. H. (1981). The shelf-sea fronts: implications of their existence and behaviour. *Philosophical Transactions of the Royal Society of London*, **A302**, 531–46.

Simpson, J. H. and Bowers, D. (1979). Shelf sea fronts' adjustments revealed by satellite IR imagery. *Nature*, **280**, 648–51.

Simpson, J. H., Edelsten, D. J., Edwards, A., Morris, N. C. G. and Tett, P. B. (1979). The Islay front: physical structure and phytoplankton distribution. *Estuarine and Coastal Marine Science*, **9**, 713–26.

Simpson, J. H. and Hunter, J. R. (1974). Fronts in the Irish Sea. *Nature*, **250**, 404–6.

Simpson J. H. and James, I. D. (1986). Coastal and estuarine fronts, pp. 63–93, in *Baroclinic Processes on Continental Shelves* (ed. C. N. K. Mooers). Washington D.C.: American Geophysical Union, 130 pp.

Simpson, J. H., Tett, P. B., Argote-Espinoza, M. L., Edwards, A., Jones, K. J. and Savidge, G. (1982). Mixing and phytoplankton growth around an island in a stratified sea. *Continental Shelf Research*, **1**, 15–31.

Smart, W. M. (1940). *Spherical Astronomy*. Cambridge University Press, 430 pp.

Smart, W. M. (1953). *Celestial Mechanics*. London: Longmans, 381 pp.

Smith, R. A. (1980). Golden Gate tidal measurements: 1854-1978. *Journal of the Waterway, Port, Coastal and Ocean Division, Proceedings of the American Society of Civil Engineers*, **106**, 407–10.

Smith, R. L. (1984). Threshold methods for sampling extremes, pp. 621–38, in *Statistical Extremes and Applications* (ed. Tiago de Oliveira). Dordrecht: D. Reidel, 692 pp.

Smith, S. D. and Banke, E. G. (1975). Variation of the sea surface drag coefficient with wind speed. *Quarterly Journal of the Royal Meteorological Society*, **101**, 665–73.

Wyss, M. (1975a). Mean sea level before and after some great strike-slip earthquakes. *Pure and Applied Geophysics*, **113**, 107–18.

Wyss, M. (1975b). A search for precursors to the Sitka, 1972 earthquake: sea level, magnetic field and P-residuals. *Pure and Applied Geophysics*, **113**, 297–309.

Wyss, M. (1976). Local changes of sea level before large earthquakes in South America. *Bulletin of the Seismological Society of America*, **66**, 903–914.

Yalin, M. S. (1977). *Mechanics of Sediment Transport*. Oxford: Pergamon Press, 298 pp.

Yonge, C. M. 1949. *The Sea Shore*. London: Collins, 311 pp.

Young, T. (1807). *A Course of Lectures on Natural Philosophy and the Mechanical Arts*. Volume 1. London: Johnson, 796 pp.

Zetler, B. D. (1971). Radiational ocean tides along the coasts of the United Staes. *Journal of Physical Oceanography*, **1**, 34–8.

Zetler, B. D. (1986). "Arctic Tides" by Rollin A. Harris (1911) revisited. *EOS, Transactions of the American Geophysical Union*, **67**, pp. 73 and 76.

Zilkoski, D. B. (ed.) (1985). NAVD Symposium '85. *Proceedings of the Third International Symposium on the North American Vertical Datum*. Rockville: NOAA, National Ocean Service. 480 pp.

Zimmerman, J. T. F. (1981). Dynamics, diffusion and geomorphological significance of tidal residual eddies. *Nature*, **290**, 549–55.

Glossary

Exact legal definitions, where applied, vary from country to country. More extensive lists are found in the following publications:

Hicks S. D. (1984), *Tide and Current Glossary*, US National Oceanic and Atmospheric Administration, 28 pp.
International Hydrographic Organisation (1974), *Hydrographic Dictionary*, Part 1 (3rd edn), IHO, 32, 370pp.

Age of the tide: old term for the lag between the time of new or full moon and the time of maximum spring tidal range.

Amphidrome: a point in the sea where there is zero tidal amplitude due to cancelling of tidal waves. Cotidal lines radiate from an amphidromic point and corange lines encircle it.

Bore: a tidal wave which propagates as a solitary wave with a steep leading edge up certain rivers. Formation is favoured in wedge-shaped shoaling estuaries at times of spring tides. Other local names include eagre (England, River Trent), pororoca (Brazil), mascaret (France).

Chart Datum: the datum to which levels on a nautical chart and tidal predictions are referred; usually defined in terms of a low-water tidal level, which means that chart datum is not a horizontal surface but it may be considered so over a limited local area.

Coamplitude lines: lines on a tidal chart joining places which have the same tidal range or amplitude; also called corange lines. Usually drawn for a particular tidal constituent or tidal condition (for example, mean spring tides).

Cotidal lines: lines on a tidal chart joining places where the tide has the same phase; for example, where high waters occur at the same time. Usually drawn for a particular tidal constituent or tidal condition.

Critical tidal level: a level on the shore where the emersion/submersion tidal characteristics change sharply. Some biologists have suggested that zonation of plants and animals is controlled by a series of such levels, but detailed analysis of tidal statistics shows that the tidal transitions are seldom as sharply defined as the biological boundaries.

Current profile: the detailed variation of current speed and direction between the sea-bed and the sea surface.

Data reduction: the process of checking, calibration, and preparation necessary to convert raw measurements into a form suitable for analysis and application.

Declination: the angular distance of an astronomical body north or south of the celestial equator, taken as positive when north and negative when south of the equator. The sun moves through a declinational cycle once a year and the moon moves through a cycle in 27.21 mean solar days. The solar declination varies between 23.5 °N and 23.5 °S. The cycles of the lunar declination vary in amplitude over an 18.6-year period from 28.5° to 18.5°.

Degenerate amphidrome: a terrestrial point on a tidal chart from which cotidal lines appear to radiate. 'An imaginary point where nothing happens.'

Ebb current: the movement of a tidal current away from the shore or down a tidal river or estuary.

Endogenous mechanism: an internal biochemical or physiological pacemaker for biological activity.

Entrainment: stimulation of an organism by local environmental factors, to adopt a new phase in its cycle of behaviour. For example, a short cold treatment in the laboratory can initiate new tidal rhythms in crabs which have become inactive.

Equation of time: the difference between real civil time and apparent solar time (see Figure 3:10).

Equilibrium tide: the hypothetical tide which would be produced by the lunar and solar tidal forces in the absence of ocean constraints and dynamics.

Equinoxes: the two points in the celestial sphere where the celestial equator intersects the ecliptic; also the times when the sun crosses the equator at these points. The vernal equinox is the point where the sun crosses the equator from south to north and it occurs about 21 March. Celestial longitude is reckoned eastwards from the vernal equinox, which is also known as the 'first point of Aries'. The autumnal equinox occurs about 23 September.

Eulerian current: flow past a fixed point as measured, for example, by a moored current meter.

First point of Aries: see Equinoxes.

Flood current: the movement of a tidal current towards the shore or up a tidal river or estuary.

Geoid: the equipotential surface that would be assumed by the sea surface in the absence of tides, water density variations, currents and atmospheric effects. It varies above and below the geometrical ellipsoid of revolution by as much as 100 m due to the uneven distribution of mass within the earth. The mean sea-level surface varies about the geoid by typically decimetres, but in some cases by more than a metre.

GLOSS: A worldwide network of sea-level gauges, defined and developed under the auspices of the Intergovernmental Oceanographic Commission. One purpose is to monitor long-term variations in the Global Level of the Sea Surface, by reporting the observations to the Permanent Service for Mean Sea Level.

GPS: a satellite-based Global Positioning System, capable of accurately locating points in a three-dimensional geometric framework. See also VLBI.

Gravitation, Newton's law of: states that all particles in the universe are attracted to other particles with a force which is proportional to the product of their masses and inversely proportional to the square of their distance apart.

Greenhouse effect: the effect, analogous to that which operates in a greenhouse, whereby the earth's surface is maintained at a much higher temperature than that appropriate to balance the incident solar radiation. Solar radiation penetrates the atmosphere, but some of the longer-wavelength return radiation is absorbed by the carbon dioxide, ozone, water vapour, trace gases and aerosols in the atmosphere. Observed increases in the concentrations of atmospheric carbon dioxide due to the burning of fossil fuels, and of other components, could lead to a steady increase of global temperatures: the resulting thermal expansion of the oceans and the melting of ice-caps would increase sea levels. Concern about the possibility of coastal flooding due to this increase has stimulated recent research into climate dynamics.

Greenwich Mean Time: time expressed with respect to the Greenwich Meridian (0°), often used as a standard for comparisons of global geophysical phenomena.

Harmonic analysis: the representation of tidal variations as the sum of several harmonics, each of different period, amplitude and phase. The periods fall into three *tidal species*, long-period, diurnal and semidiurnal. Each tidal species contains *groups* of harmonics which can be separated by analysis of a month of observations. In turn, each group contains *constituents* which can be separated by analysis of a year of observations. In shallow water, harmonics are also generated in the third-diurnal, fourth-diurnal and higher species.

High water: the maximum water level reached in a tidal cycle.

Highest astronomical tide: the highest level which can be predicted to occur under any combination of astronomical conditions.

Hydrodynamic levelling: the transfer of survey datum levels by comparing mean sea-level at two sites, and adjusting them to allow for gradients on the sea surface due to currents, water density, winds and atmospheric pressures.

Internal tides: tidal waves which propagate at density differences within the ocean. They travel slowly compared with surface gravity waves and have wavelengths of only a few tens of kilometres, but they can have amplitudes of tens of metres. The associated internal currents are termed baroclinic motions.

Inverted barometer effect: adjustment of sea-level to changes in barometric pressure; in the case of full adjustment, an increase in barometric pressure of 1 mb corresponds to a fall in sea-level of 0.01 m. If there is this full adjustment, the observed pressures at the sea-bed are unchanged.

Kelvin wave: a long wave in the oceans whose characteristics are altered by the rotation of the earth. In the northern hemisphere the amplitude of the wave decreases from right to left along the crest, viewed in the direction of wave travel.

Lagrangian current: the movement through space of a particle of water as measured by drogues or drifting buoys.

Long wave: a wave whose wavelength from crest to crest is long compared with the water depth. Tides propagate as long waves. The speed of travel is given by:
(water depth × gravitational acceleration)$^{\frac{1}{2}}$

Low water: the minimum water level reached in a tidal cycle.

Lowest astronomical tide: the lowest level which can be predicted to occur under average meteorological conditions and under any combination of astronomical conditions; often used to define Chart Datum where the tides are semidiurnal.

Maelstrom: a tidal whirlpool found between the islands of Moskenesy and Mosken in the Lofoten Islands of northern Norway. The term is generally applied to other tidal whirlpools.

Mean high water springs: average spring tide high-water level, averaged over a sufficiently long period.

Mean higher high water: a tidal datum; the average of the higher high water height of each tidal day, averaged over the United States National Tidal Datum Epoch.

Mean low water springs: a tidal datum, the average spring low-water level averaged over a sufficiently long period.

Mean lower low water: a tidal datum, the average of the lower low water height of each tidal day observed over the United States National Tidal Datum Epoch.

Mean sea level: a tidal datum; the arithmetic mean of hourly heights observed over some specified period (often 19 years).

Mean sea level trends: changes of mean sea-level at a site over long periods of time, typically decades, also called *secular* changes. Global changes due to the increased volume of ocean water are called *eustatic* changes; vertical land movements of regional extent are called *eperiogenic* changes.

Mean tide level: the arithmetic mean of mean high water and mean low water.

National Tidal Datum Epoch: the specific 19-year period adopted by the National Ocean Service as the official time segment over which sea-level observations are taken and reduced to obtain mean values for datum definition. The present Epoch is 1960 through 1978. It is reviewed annually for revision and must be actively considered for revision every 25 years.

Neap tides: tides of small range which occur twice a month, when the moon is in quadrature.

Nodal factors: small adjustments to the amplitudes and phases of harmonic tidal constituents to allow for modulations over the 18.6-year nodal period.

Numerical modelling: calculations of the hydrodynamic responses of real seas and oceans to physical forcing by representing them as a set of discrete connected elements, and solving the hydrodynamic equations for each element in sequence.

Permanent Service for Mean Sea Level: the organization responsible for collection, analysis, interpretation and publication of mean sea-level data from a global network of gauges. PSMSL is one of the Federation of Astronomical and Geophysical Services, under the auspices of the International Council of Scientific Unions.

Pole tide: small variations in sea-level due to the Chandler Wobble of the axis of rotation of the earth. This has a period close to 436 days. Maximum amplitudes of more than 30 mm are found in the Gulf of Bothnia, but elsewhere amplitudes are only a few millimetres.

Pressure tide gauges: instruments which measure the pressure below the sea surface; this pressure may be converted to sea levels if the air pressure, the gravitational acceleration and the water density are known.

Progressive wave: a wave whose travel can be followed by monitoring the movement of the crest. Energy is transmitted but the water particles perform oscillatory motions.

Radiational tides: tides generated by regular periodic meteorological forcing.

Range: the difference between high and low water in a tidal cycle. Ranges greater than 4 m are sometimes termed macrotidal and those less than 2 m are termed microtidal. Intermediate ranges are termed mesotidal.

Rectilinear current: see Reversing current

Resonance: The phenomenon of the large amplitudes which occur when a physical system is forced at its natural period of oscillation. Tidal resonance occurs when the natural period of an ocean or sea is close to the period of the tidal forcing.

Response analysis: the representation of observed tidal variations in terms of the frequency-dependent amplitude and phase responses to input or forcing functions, usually the gravitational potential due to the moon and sun, and the radiational meteorological forcing.

Return period: the average time between events such as the flooding of a particular level. This information may also be expressed as the level which has a particular return period of flooding, for example, 100 years. The inverse of the return period is the statistical probability of an event occurring in any individual year.

Reversing current: a tidal current which flows alternately in approximately opposite directions with a slack water at each reversal of direction. Also called a *rectilinear current*.

Revised Local Reference: a datum level defined by the Permanent Service for Mean Sea Level for each station, relative to which the mean sea-level is approximately 7 m. Measurements at the location, over different periods and to different tide gauge bench-marks are all related to this datum. The value of 7 m was chosen to avoid confusion with other local datum definitions.

Sediment transport: the total sediment transported by a current is the sum of the *bed-load*, material partly supported by the bed as it rolls or bounces along, and the *suspended load* in the water.

Seiche: the oscillation of a body of water at its natural period. Coastal measurements of sea-level often show seiches with amplitudes of a few centimetres and periods of a few minutes due to oscillations of the local harbour, estuary or bay, superimposed on the normal tidal changes.

Sidereal day: the period of rotation of the earth with respect to the vernal equinox (approximately 0.99 727 of a mean solar day).

Slack water: the state of a tidal current when its speed is near zero, especially the time when a reversing (rectilinear) current changes direction and its speed is zero. For a theoretical standing tidal wave, slack water occurs at the times of high and of low water, while for a theoretical progressive tidal wave, slack water occurs midway between high and low water.

Spring tides: semidiurnal tides of large range which occur twice a month, when the moon is new or full.

Standing wave: wave motion in an enclosed or semi-enclosed sea, where the incident and reflected progressive waves combine to give a node of zero tidal amplitude. Maximum tidal amplitudes are found at the head of the basin where reflection occurs. No energy is transmitted in a standing wave, nor is there any progression of the wave pattern.

Steric level differences: sea-level differences due to differences in water density.

Stilling-well gauges: instrument system for measuring sea levels; tidal changes of levels are detected by the movement of a float in a well, which is connected to the open sea by a restricted hole or narrow pipe. Wind waves are eliminated by the constriction of the connection.

Surge: large change in sea-level generated by an extreme meteorological event.

Survey datum: the datum to which levels on land surveys are related; often defined in terms of a mean sea-level. Survey datum is a horizontal surface to within the accuracy of the survey methods.

Tidal prism: the volume of water exchanged between a lagoon or estuary and the open sea in the course of a complete tidal cycle.

Tide Gauge Bench-mark: a stable bench-mark near a gauge, to which tide gauge datum is referred. It is connected to local auxiliary bench-marks to check local stability and to guard against accidental damage. Tide gauge datum is a horizontal plane defined at a fixed arbitrary level below a tide gauge bench-mark.

Tides: periodic movements which have a coherent amplitude and phase relationship to some periodic geophysical force.

Tsunamis: waves generated by seismic activity, also called seismic sea waves. Tsunamis are also popularly, but inaccurately, called tidal waves. When they reach shallow coastal regions, amplitudes may increase to several metres. The Pacific Ocean is particularly vulnerable to tsunamis.

Vernal equinox: see Equinoxes.

VLBI: Very Long Baseline Interferometry, a technique for fixing positions of terrestrial points in a three-dimensional geometric framework by comparing the relative arrival times of radio signals from distant extra-terrestrial sources. In the long term, fixing several tide gauge bench-marks in such a framework will allow a distinction to be made between vertical movements of land and of sea levels at a site.

Wentworth Scale: used to classify sediments in terms of their particle sizes.

Zonation: the pattern of colonization of the sea-shore, whereby individual species flourish in bands associated with particular tidal levels.

Index